Novel Advances in Microsystems Technologies and Their Applications

Devices, Circuits, and Systems

Series Editor
Krzysztof Iniewski
CMOS Emerging Technologies Inc., Vancouver, British Columbia, Canada

FORTHCOMING TITLES:

Nanomaterials: A Guide to Fabrication and Applications
Gordon Harling and Krzysztof Iniewski

Nanopatterning and Nanoscale Devices for Biological Applications
Krzysztof Iniewski and Seila Selimovic

Nanoplasmonics: Advanced Device Applications
James W. M. Chon and Krzysztof Iniewski

Nanoscale Semiconductor Memories: Technology and Applications
Santosh K. Kurinec and Krzysztof Iniewski

Radio Frequency Integrated Circuit Design
Sebastian Magierowski

Semiconductor Device Technology: Silicon and Materials
Tomasz Brozek and Krzysztof Iniewski

Smart Grids: Design, Strategies, and Processes
David Bakken and Krzysztof Iniewski

Soft Errors: From Particles to Circuits
Jean-Luc Autran and Daniela Munteanu

Technologies for Smart Sensors and Sensor Fusion
Kevin Yallup and Krzysztof Iniewski

VLSI: Circuits for Emerging Applications
Tomasz Wojcicki and Krzysztof Iniewski

Novel Advances in Microsystems Technologies and Their Applications

Edited by
Laurent A. Francis
Krzysztof Iniewski

CRC Press
Taylor & Francis Group
Boca Raton   London   New York

CRC Press is an imprint of the
Taylor & Francis Group, an **informa** business

MATLAB® is a trademark of The MathWorks, Inc. and is used with permission. The MathWorks does not warrant the accuracy of the text or exercises in this book. This book's use or discussion of MATLAB® software or related products does not constitute endorsement or sponsorship by The MathWorks of a particular pedagogical approach or particular use of the MATLAB® software.

CRC Press
Taylor & Francis Group
6000 Broken Sound Parkway NW, Suite 300
Boca Raton, FL 33487-2742

First issued in paperback 2017

© 2014 by Taylor & Francis Group, LLC
CRC Press is an imprint of Taylor & Francis Group, an Informa business

No claim to original U.S. Government works
Version Date: 20130531

ISBN 13: 978-1-138-07279-4 (pbk)
ISBN 13: 978-1-4665-6066-6 (hbk)

Contents

PART I   Organic and Flexible Electronics

PART II   Imaging and Display Technologies

PART III   Sensors and Microdevices

PART IV   Point-of-Care and Biosensors

PART V   Ultra-Low-Power Biomedical Systems

Preface

Microsystems technologies have gained in importance over the last decades and have found their way in an impressive variety of applications. This field of research is intrinsically multidisciplinary; it extends the current capabilities of standard integrated circuits in terms of materials and designs and complements them by creating innovative components and smaller systems that require lower power consumption and display better performance.

This book is divided into three parts and covers the state of the art and the applications of microsystems and microelectronics-related technologies. Part I deals with organic and flexible electronics from polymer solar cell to flexible interconnects for the co-integration of MEMS with CMOS, and with imaging and display technologies. It provides in-depth coverage on the MEMS technology at the heart of reflective displays, followed by the fabrication of thin-film transistors on glass substrates. It also demonstrates several new techniques to display high-quality images or to transmit them quickly with full resolution.

Part II is devoted to physical and chemical sensing, with microsystems being especially relevant when it comes to sensing electrical currents, acoustic pressure or acceleration. In particular, inertial devices have reached a sufficient maturity to be networked and to log their data for structural health monitoring aside of others sensors, while the confidence in robust MEMS structures integrated with the appropriate choice of piezoelectric materials allow for power MEMS as energy harvesters needed to electrically feed other microdevices. From another perspective, thin-film transistors prove to be excellent photosensors and temperature sensors, leading eventually to the development of an artificial retina. Finally, two chapters in Part II look into microwave techniques that have an increasing importance for the non-destructive monitoring of critical industrial processes.

Part III provides an overview of biomedical microsystems, which includes biosensors, point-of-care devices, neural stimulation and recording and ultra-low-power biomedical systems.

This book is targeted towards researchers in academia and engineers and graduate students in the fields of electrical and biomedical engineering. We are grateful to all the contributors for sharing with us their passion for new technologies, and we expect to offer here a good balance between academic and industrial researches through the different chapters.

We sincerely hope that this book will be a source of inspiration for new applications and stimulate further the development of microsystems technologies.

Laurent A. Francis
Louvain-la-Neuve, Belgium

Krzysztof (Kris) Iniewski
Coquitlam, British Columbia, Canada

xi

MATLAB® is a registered trademark of The MathWorks, Inc. For product information, please contact:

The MathWorks, Inc.
3 Apple Hill Drive
Natick, MA 01760-2098 USA
Tel: 508-647-7000
Fax: 508-647-7001
E-mail: info@mathworks.com
Web: www.mathworks.com

Editors

Laurent A. Francis, PhD, is an associate professor in the Electrical Engineering Department at the Université Catholique de Louvain, Belgium. His main focus is on co-integrated, ultra-low-power CMOS MEMS sensors for biomedical applications and harsh environments. He was previously researcher at IMEC in Leuven, Belgium, in the field of acoustic and optical biosensors and piezoelectric RF-MEMS. In 2011, he was a visiting professor at the Université de Sherbrooke, Québec, Canada. He has published more than 60 research papers in scientific journals, international conferences and book chapters and holds 1 patent. He is a board member of the Belgian National Committee on Biomedical Engineering and an editorial board member of the *Journal of Sensors* (Hindawi Publishing Corporation). He is also a regular member of the IEEE and can be reached at laurent.francis@uclouvain.be.

Krzysztof (Kris) Iniewski, PhD, is managing R&D at Redlen Technologies Inc., a start-up company in Vancouver, British Columbia, Canada. Redlen's revolutionary production process for advanced semiconductor materials enables a new generation of more accurate, all-digital, radiation-based imaging solutions. Kris is also a president of CMOS Emerging Technologies Research Inc. (www.cmosetr.com), an organization of high-tech events covering communications, microsystems, optoelectronics and sensors. During the course of his career, Dr. Iniewski held numerous faculty and management positions at University of Toronto, University of Alberta, SFU and PMC-Sierra Inc. He has published over 100 research papers in international journals and conferences. He holds 18 international patents granted in the United States, Canada, France, Germany and Japan. He is a frequent invited speaker and has consulted for multiple organizations internationally. He has written and edited several books for IEEE Press, Wiley, CRC Press, McGraw Hill, Artech House and Springer. His personal goal is to contribute to healthy living and sustainability through innovative engineering solutions. In his leisurely time Kris can be found hiking, sailing, skiing or biking in beautiful British Columbia. He can be reached at kris.iniewski@gmail.com.

Contributors

Themistokles Afentakis
Materials & Device Applications
    Laboratory
Sharp Laboratories of America
Camas, Washington DC

Lissan Afilal
Centre de Recherche en Science et
    Technologie de l'Information et de la
    Communication
Université de Reims
    Champagne-Ardenne
Reims, France

Su-Shin Ang
Toumaz Healthcare, Ltd.
Abingdon, United Kingdom

Muhannad Bakir
School of Electrical and Computer
    Engineering
Georgia Institute of Technology
Atlanta, Georgia

Yves Blaquière
Département d'informatique
Université du Québec à Montréal
Montréal, Québec, Canada

Mohammed Bougataya
Département d'informatique
    et d'ingénierie
Université du Québec en Outaouais
Montréal, Québec, Canada

Alper Bozkurt
Department of Electrical and Computer
    Engineering
North Carolina State University
Raleigh, North Carolina

Gil Bub
Department of Physiology Anatomy
    and Genetics
University of Oxford
Oxford, United Kingdom

Alison Burdett
Toumaz Microsystems, Ltd.
Abingdon, United Kingdom

Giuseppe Cannazza
Department of Engineering
    for Innovation
University of Salento
Lecce, Italy

Antoni J. Canós
Microwave Division (Dimas)
Institute for the Applications of
    Advanced Information and
    Communication Technologies
    Research Institute
Universidad Politécnica de Valencia
Valencia, Spain

Jose M. Catalá-Civera
Microwave Division (Dimas)
Institute for the Applications of
    Advanced Information and
    Communication Technologies
    Research Institute
Universidad Politécnica de Valencia
Valencia, Spain

Andrea Cataldo
Department of Engineering
    for Innovation
University of Salento
Lecce, Italy

Shunfeng Cheng
Intel Corporation
Hillsboro, Oregon

Yang-Kyu Choi
Department of Electrical Engineering
Korea Advanced Institute of Science
    and Technology
Daejeon, Republic of Korea

Jordi Colomer-Farrarons
Department of Electronics,
    Bioelectronics and
    Nanobioengineering Research
    Group (SIC-BIO)
University of Barcelona
Barcelona, Spain

A. Conesa-Roca
Department of Electronic Engineering
Universitat Politècnica de Catalunya
    BarcelonaTech (UPC)
Barcelona, Spain

Anis Daami
Printed Electrical Devices Laboratory
Department of Nano Materials
    Technologies
CEA Grenoble, France

Egidio De Benedetto
Department of Engineering
    for Innovation
University of Salento
Lecce, Italy

A. Dompierre
Microengineering Laboratory for MEMS
Department of Mechanical Engineering
Université de Sherbrooke
Sherbrooke, Québec, Canada

Hassen Fourati
Department of Automatic Control
Grenoble University
Grenoble, France

L.G. Fréchette
Microengineering Laboratory for MEMS
Department of Mechanical Engineering
Université de Sherbrooke
Sherbrooke, Québec, Canada

B. García-Baños
Microwave Division (Dimas)
Institute for the Applications of
    Advanced Information and
    Communication Technologies
    Research Institute
Universidad Politécnica de Valencia
Valencia, Spain

Jennifer Gille
Qualcomm MEMS Technologies
San Jose, California

Benoit Gosselin
Department of Electrical and Computer
    Engineering
Université Laval
Québec City, Québec, Canada

Mikaël Guillemot
Groupe de Recherche en
    Microélectronique et Microsystèmes
École Polytechnique de Montréal
Montréal, Québec, Canada

Yves Handrich
Département Ecologie, Physiologie
    et Ethologie
Institut Pluridisciplinaire Hubert
    CURIEN
Université de Strasbourg
Strasbourg, France

Miguel Hernandez-Silveira
Toumaz Healthcare, Ltd.
Abingdon, United Kingdom

Reynald Hoskinson
Department of Electrical and Computer
    Engineering
University of British Columbia
Vancouver, British Columbia, Canada

Farhad Sheikh Hosseini
Department of Electrical and Computer
    Engineering
Université Laval
Québec City, Québec, Canada

Kao-Cheng Huang
Dharma Academy
Taipei, Taiwan, Republic of China

Maesoon Im
Department of Neurosurgery
Massachusetts General Hospital
Harvard Medical School
Cambridge, Massachusetts

Esteve Juanola-Feliu
Department of Electronics,
    Bioelectronics and
    Nanobioengineering Research
    Group (SIC-BIO)
University of Barcelona
Barcelona, Spain

Nassim Khonsari
Qualcomm MEMS Technologies
San Jose, California

Mutsumi Kimura
Department of Electronics and
    Informatics
Ryukoku University
Otsu, Shiga, Japan

Ahmed Lakhssassi
Département d'informatique
    et d'ingénierie
Université du Québec en Outaouais
Montréal, Québec, Canada

Roger Light
School of Electrical Engineering
University of Nottingham
Nottingham, United Kingdom

Noureddine Manamanni
Centre de Recherche en Science et
    Technologie de l'Information et de la
    Communication
Université de Reims
    Champagne-Ardenne
Reims, France

George C. McConnell
Department of Biomedical Engineering
Duke University
Durham, North Carolina

Pere Miribel-Catala
Department of Electronics,
    Bioelectronics and
    Nanobioengineering Research
    Group (SIC-BIO)
University of Barcelona
Barcelona, Spain

Nathan Nebeker
Cordin Scientific Imaging
Salt Lake City, Utah

Hai Nguyen
Groupe de Recherche en
    Microélectronique et Microsystèmes
École Polytechnique de Montréal
Montréal, Québec, Canada

Michael Pecht
Center of Advanced Life Cycle
    Engineering (CALCE)
University of Maryland
College Park, Maryland

Felipe L. Peñaranda-Foix
Microwave Division (Dimas)
Institute for the Applications of
    Advanced Information and
    Communication Technologies
    Research Institute
Universidad Politécnica de Valencia
Valencia, Spain

Qiquan Qiao
Department of Electrical Engineering
South Dakota State University
Brookings, South Dakota

Rashmi Rao
Qualcomm MEMS Technologies
San Jose, California

M. Román-Lumbreras
Department of Electronic Engineering
Universitat Politècnica de Catalunya
    BarcelonaTech (UPC)
Barcelona, Spain

Sébastien Roy
Department of Electrical and Computer
    Engineering
Université Laval
Québec City, Québec, Canada

Seung-Tak Ryu
Department of Electrical Engineering
Korea Advanced Institue of Science and
    Technology
Daejeon, Republic of Korea

Josep Samitier
Department of Electronics,
    Bioelectronics and
    Nanobioengineering Research
    Group (SIC-BIO)
University of Barcelona
and
Nanobioengineering Group
Institute for Bioengineering
    of Catalonia
Barcelona, Spain

and

Biomedical Research Networking
    Center in Bioengineering,
    Biomaterials and Nanomedicine
Zaragoza, Spain

Yvon Savaria
Groupe de Recherche en
    Microélectronique et Microsystèmes
École Polytechnique de Montréal
Montréal, Québec, Canada

Mary Shields
Groupe de Recherche en
    Microélectronique et Microsystèmes
École Polytechnique de Montréal
Montréal, Québec, Canada

Boris Stoeber
Department of Mechanical Engineering
and
Department of Electrical and Computer
    Engineering
University of British Columbia
Vancouver, British Columbia, Canada

Paragkumar Thadesar
School of Electrical and Computer
    Engineering
Georgia Institute of Technology
Atlanta, Georgia

Fares Tounsi
Electrical Engineering Department
National Engineering School of Sfax
Sfax, Tunisia

G. Velasco-Quesada
Department of Electronic Engineering
Universitat Politècnica de Catalunya
    BarcelonaTech (UPC)
Barcelona, Spain

S. Vengallatore
Department of Mechanical Engineering
McGill Institute for Advanced Materials
McGill University
Montréal, Québec, Canada

Swaminathan Venkatesan
Department of Electrical Engineering
South Dakota State University
Brookings, South Dakota

Zhaocheng Wang
Department of Electronic Engineering
Tsinghua University
Haidian, Beijing, People's Republic
    of China

Hyung Suk Yang
School of Electrical and Computer
    Engineering
Georgia Institute of Technology
Atlanta, Georgia

Chaoqi Zhang
School of Electrical and Computer
    Engineering
Georgia Institute of Technology
Atlanta, Georgia

Part I

Organic and Flexible Electronics

.

1 Device Modelling for SPICE Simulations in Organic Technologies

Anis Daami

CONTENTS

Outline

Organic semiconductor technologies have opened a wide range of new applications to industries that are interested in large-area low-cost applications with a certain not negligible lifetime. Many efforts have been made to optimize and stabilize processes and especially render devices air-stable. More or less complex designs are being demonstrated since, with different technologies.

Nevertheless, one essential step between circuit design and process fabrication is still not mature: organic devices Simulation Program with Integrated Circuit Emphasis (SPICE) modelling. Indeed, compared to the silicon industry where public models are well defined and commonly used to provide designers with a relative good description of any process, organic devices are still looking for their complete device models that can fully describe their opto-electrical characteristics. Many studies exist in the literature to understand the physics of these devices in order to mathematically describe their behaviours. We will try to synthesize in this chapter the most accomplished models that are being used nowadays in the organic technology community to describe their respective processes whatever is the final application using these discrete devices. A special focus will be done on the SPICE modelling of organic thin-film transistors (OTFTs) that represent the most important milestone of the chapter subject.

1.1 INTRODUCTION

One essential key in between semiconductor devices processing and complex circuit simulations has always been the accurate description of these later devices, basic electrical and/or optical characteristics by modelling. Indeed, any process modification or optimization has to be taken into account by designers in their simulations, to better secure their circuit designs. And in the opposite way, a given process technology can be optimized from results coming from circuit simulations. This has been the case for decades in silicon technology.

Nevertheless, organic semiconductor technologies that essentially emerged in the last 20 years are still lacking this very important link, even though much effort has been carried out to put into place device modelling for organic devices. Now that organic applications are emerging with a very big interest from industrials, device modelling has to be considered with a great accuracy in order to take the organic semiconductor world a step forwards.

Depending on the organic device to be modelled, more or less device models have been developed and studied in the last decade. OTFTs are widely the components that organic research groups are focusing on today. Beside OTFTs, other devices such as organic photodiodes (OPDs), organic light-emitting diodes (OLEDs), organic photovoltaic cells (OPVs) and other passive components are also being studied, and their electrical and optical functionalities put down into models. In this chapter, we will essentially focus on SPICE models dedicated to OTFTs as they are the most widespread in literature.

At first, silicon-based public SPICE models often used for organic device modelling will be overviewed. We will then focus on physics-based analytical models developed by several groups to better describe carrier transport and mobility bias dependence in OTFTs. Some other empirical approaches of OTFT modelling will finally be considered. The presented models will principally be discussed in terms of current and mobility equations. We will briefly overview the contact resistances, dynamic/capacitance behaviour and extra equations that can optimize the OTFT modelling in general. Finally, other organic device models will also be slightly overviewed.

1.2 SPICE MODELS

SPICE has been developed in the early 1970s in the University of California, Berkeley, inside the Electronics Research Laboratory [1]. Since then, it has become the most popular commercial and industrial circuit integrated simulator in semiconductor technologies. Through time, a variant range of SPICE models have been developed and standardized by the means of an industry working group called the Compact Modelling Council (CMC) [2]. This has permitted their stabilization and their common use in the silicon industry. In 2011, SPICE has finally been named as an IEEE milestone [3].

1.2.1 ORGANIC THIN-FILM TRANSISTORS

1.2.1.1 Silicon-Based Models

1.2.1.1.1 SPICE Level 1 MOSFET Model

In the late 1960s, Shichman and Hodges developed an equivalent circuit model for insulated-gate field-effect transistors [4]. Since then, it has been integrated as a standardized public model into SPICE and referred to as level 1 MOSFET model. The model was built at first to describe the strong inversion operating mode of transistors. It has been shown that level 1 MOS model current–voltage equations are applicable to accumulation operating mode of TFTs [5]. A closer look to the equations of level 1 MOS model shows that only a single parameter (λ) has been added to the well-known ideal current–voltage MOS equations [6] developed for strong inversion. The purpose of this additional parameter was to take into account the channel length modulation and the drain–source coupling.

Level 1 MOS does not model the leakage/cut-off regime of the transistor's current–voltage characteristic. Indeed, it is supposed to be nil under the threshold voltage V_T.

The equations of SPICE level 1 MOSFET model are presented as follows in (1.1) for cut-off, linear and saturation regimes, respectively:

$$I_{DS} = \begin{cases} 0 & \text{if } V_{GS} < V_T \\ \mu C_{ins} \dfrac{W}{L} \left(V_{GS} - V_T - \dfrac{V_{DS}}{2} \right) V_{DS}(1 + \lambda V_{DS}) & \text{if } 0 < V_{DS} \le V_{GS} - V_T \\ \dfrac{1}{2} \mu C_{ins} \dfrac{W}{L} (V_{GS} - V_T)^2 (1 + \lambda V_{DS}) & \text{if } 0 < V_{GS} - V_T \le V_{DS} \end{cases} \tag{1.1}$$

The adjustable parameters in (1.1) are μ as the free carrier mobility, C_{ins} as the gate dielectric capacitance per unit area, V_T as the threshold voltage and λ corresponding to the channel length modulation as stated earlier. W and L correspond to the transistor geometry, while V_{GS} and V_{DS} are the gate to source and drain to source biases.

This simple model has often been used in literature to describe OTFT current–voltage characteristics as its model parameter extraction is very straightforward. Different uses of level 1 model can be found in literature. In 1997, Bao et al. [7] used the saturation regime equation to describe the output of a poly3-hexylthiophene (P3HT)-based TFT. More recently, transient and noise-margin analyses on organic inverters have been carried out, and authors of both studies used level 1 to model single devices [8,9]. Marien et al. [10] explained that the use of this model to design of a fully integrated $\Delta\Sigma$ analogue–digital converter (ADC) gave limitations to their simulation precision. Despite this inaccuracy, level 1 model helped them in the choice of the optimum circuit topology.

1.2.1.1.2 Hydrogenated Amorphous Silicon (a-Si:H) and Polysilicon (Psi) TFT MOS Models

To simulate and match electrical characteristics of applications using hydrogenated amorphous silicon and/or polysilicon TFTs, Shur et al. [11] proposed the use of new models dedicated to these TFTs. The first one called a-Si:H TFT model has been validated on n-channel amorphous TFTs, and the second named Psi TFT model was developed and validated on both n- and p-type polysilicon transistors.

In both models, three different current regimes are distinguished: leakage (I_{leak}), subthreshold (I_{sub}) and above threshold (I_a). Furthermore, these models account for a gate-voltage dependence of carriers' mobility above threshold for each model in its way. Both mobility variations for a-Si:H and Psi TFT models are presented, respectively, in Equations 1.2 and 1.3:

$$\mu_{a-Si:H} = \mu_0 \left(\frac{V_{GS} - V_T}{V_{AA}} \right)^{\gamma} \tag{1.2}$$

The mobility gate-bias dependence of a-Si:H TFT is explained by the fact that the Fermi level moves slowly towards the conduction band from the deep to tail localized states [12]. In the case of Psi TFT mobility gate-bias dependence, it is stated that this power dependence is solved at a silicon grain boundary taking into account trap states [13]:

$$\frac{1}{\mu_{Psi-TFT}} = \frac{1}{\mu_1 \left[\frac{2q(V_{GS} - V_T)}{\eta kT} \right]^m} + \frac{1}{\mu_0} \tag{1.3}$$

For Equations 1.2 and 1.3, μ_0, V_{AA} and γ are mobility parameters to be extracted in the case of a-Si:H TFT model, and μ_0, μ_1, m and η those to be adjusted for the Psi TFT model, respectively. V_T is the threshold voltage in both cases.

Regardless of the equations that describe the three current regimes in the same reference [11] for both models, the total drain current is modelled as follows:

$$I_{DS} = I_{leak} + \frac{1}{(1/I_{sub}) + (1/I_a)} \tag{1.4}$$

Here, I_{leak}, I_{sub} and I_a are the three current regions described earlier. The reader may refer to the reference paper to explore each current regime equation in both models.

As for level 1 MOS model, both a-Si:H TFT and Psi TFT models have been integrated successfully in commercial SPICE circuit simulators. Nowadays, both models are much more known as the Rensselaer Polytechnic Institute (RPI) models. The use of both models to describe OTFTs has been since widely spread in the organic semiconductor community. Bartzch et al. [14] used the a-Si:H TFT model to simulate their thin-film poly(9,9-dioctylfluorene-co-bithiophene) (F8T2) organic transistors, inverters and ring oscillators. Different works on pentacene-based devices have also reported the use of a-Si:H TFT model to simulate the output characteristics. In [15], it is reported that the model fits well the above-threshold and leakage regimes, but the authors suggest the use of Psi TFT model to better adjust the transition regime. Results from [16] show that the same model can also describe with a good accuracy inkjet-printed transistors and simulate the output characteristic of an enhanced-mode inverter. A huge work, done by the research group of Iñiguez [17,18], showed that both models can be easily adapted to OTFT modelling. More recently, zinc tin oxide (ZTO)–based TFT modelling for logic circuits simulations has been reported using the a-Si:H TFT model [19].

1.2.1.2 Physics-Based Analytical Models

Even though silicon-based SPICE models permitted designers a fast access to OTFT characteristics for their simulations, the complete device modelling of these devices still lacked of accuracy. Indeed, charge transport in organic semiconductors differs largely from the inorganic one. Furthermore, organic semiconductors can exhibit a pseudo-crystalline structure or can, at the opposite, show a non-ordered structural configuration as observed in conjugated polymer semiconductors. For all these reasons, drastic efforts have been made to explain the charge transport in these materials. Over the last decades, many papers with different transport and mobility models have been published describing their good agreement with OTFT electrical outputs. We will try as follows to give a picture of the most important transport model developed for this issue.

1.2.1.2.1 Multiple Trapping and Release Mobility Models

Initially the multiple trapping and release (MTR) model has been developed by Le Comber et al. [20] on a-Si:H samples by carrying out conductivity measurements versus temperature. They suggested that, during their transit in the extended states near the conduction band, the carriers interact with localized states through trapping and thermal release. The drift mobility coming out from their analysis is then described as follows:

$$\mu_D = \mu_0 \alpha \exp\left(\frac{-E_T}{kT}\right) \tag{1.5}$$

where

μ_0 is the drift mobility at the conduction band edge

α is the ratio between the density of states (DOS) in the transport band edge and the density of traps

E_T is the thermal activation energy of the trap state

The MTR model has been used by Horowitz et al. [21] in 1995 to explain the transport behaviour in unsubstituted sexithiophene (6T) and end-substituted dihexyl-sexithiophene (DH6T) OTFTs by carrying out conductivity versus temperature measurements. They concluded that the localized trap states in 6T and DH6T semiconductors would probably originate from the grain boundaries. In a similar study, Chen et al. [22] explained the conduction mechanism in pentacene OTFTs by using an exponential trap states distribution in the band tail using the MTR model.

1.2.1.2.2 Polaron-Based Mobility Model

Basically a polaron is a virtual particle resulting from the combination of carrier and the lattice deformation due to the charge of that carrier. This quasiparticle definition has been firstly introduced in [23,24] for inorganic crystals transport. It has been rapidly used to describe the transport in molecular crystals [25]. Two decades later, Fesser et al. [26] and Marcus et al. [27,28] introduced the polaron model to explain the carrier's mobility in conjugated polymers. The carrier mobility proposed by Marcus et al. is expressed as follows:

$$\mu_D = \mu_0 \exp\left(-\frac{E_R}{4kT} - \frac{(aF)^2}{4E_R kT}\right) \frac{\sinh(aF/2kT)}{aF/2kT} \tag{1.6}$$

where
    μ_0 is the mobility at high temperature
    E_R is the intramolecular reorganization energy
    F is the applied electrical field over the molecular distance a

In a similar study, Seki et al. [29] argued that this mobility model has weak temperature activation energy and suggested the use of the disordered model of Bässler et al. [30] to have higher temperature activation.

Considering a perfect polymer chain, Su et al. [31] developed a model describing the alteration of the conjugated polymer bond structure by the presence of charge. This mechanism has been named as soliton formation. They concluded that the soliton formation plays an important role in the charge transfer along the conjugated polymer due to its low creation energy compared to the band excitation needed.

1.2.1.2.3 Variable Range Hopping Mobility Models

Developed earlier by Miller et al. [32] to describe the conduction in low-doped n-type semiconductors, by phonon-induced electron hopping from a donor site to another, the model has been further developed and named variable range hopping (VRH) model by assuming that trapping states are spread all over the energy bandgap. The very first use of the VRH model in OTFT mobility description was carried out by Vissenberg et al. [33] where the transport of charges is supposed to be determined by the tail of the band states at low carrier densities and low temperatures. These tail states have been approximated by the authors as an exponential DOS expressed as follows:

$$g(\varepsilon) = \begin{cases} 0 & \text{if } \varepsilon > 0 \\ \dfrac{N_T}{kT_0} \exp\left(\dfrac{\varepsilon}{kT_0}\right) & \text{if } -\infty < \varepsilon \leq 0 \end{cases} \tag{1.7}$$

where
    N_T is the DOS
    k is the Boltzmann constant
    T_0 is a parameter indicating the width of the exponential distribution

Using the percolation theory to derive the conductivity and taking into account an occupation fraction δ in the tail band, a gate-voltage-dependent mobility model is obtained by Vissenberg et al. as follows:

$$\mu_{FE} = \frac{\sigma_0}{q} \left(\frac{\pi(T_0/T)^3}{(2\alpha)^3 B_c \Gamma(1-T_0/T)\Gamma(1+T_0/T)} \right)^{T_0/T} \left(\frac{(C_i V_{GS})^2}{2kT_0\varepsilon_S} \right)^{T_0/T} \quad (1.8)$$

where

 σ_0 being a conductivity prefactor
 α is the effective overlap of the electronic wave between two sites i and j in the
  percolation theory
 B_c is the percolation onset criterion depending essentially on parameters α and T_0
 C_i is the dielectric capacitance per unit area
 ε_S is the semiconductor's relative permittivity

$$\Gamma(z) \equiv \int_0^{+\infty} \exp(-x)x^{z-1}dx.$$

By comparing their model to experimental results on pentacene and polythienyl-ene vinylene (PTV) TFTs, authors concluded to a good agreement on temperature and gate-bias dependence of the measured and simulated field-effect mobility.

In the same way, Meijer et al. [34], introducing an onset voltage V_{S0} representing the beginning of accumulation in the channel, instead of the commonly used threshold voltage V_T, employed the percolation theory within the gradual channel approximation to attain a drain current description that showed good concordance with experimental results on solution-based pentacene, PTV and P3HT TFTs. The drain current is expressed as

$$I_{DS} = \frac{WV_{DS}\sigma_0\sqrt{2kT_0\varepsilon_S\varepsilon_0}}{qL} \left(\frac{T}{2T_0-T} \right) \left[\frac{(T_0/T)^4 \sin(\pi T/T_0)}{(2\alpha)^3 B_c} \right]^{T_0/T} \left[\frac{C_i(V_{GS}-V_{S0})}{\sqrt{2kT_0\varepsilon_S\varepsilon_0}} \right]^{\frac{2T_0}{T}-1}$$

$$(1.9)$$

One can notice easily that only four parameters are necessary (α, σ_0, V_{S0}, T_0) to model the output characteristics of the transistor with Equation 1.9.

In another approach and keeping in mind SPICE-like modelling, Calvetti et al. [35] used the VRH mobility theory to develop their own OTFT model in linear and saturation operating conditions through simple formulations. The simplified drain current of OTFTs developed by Calvetti et al. is finally defined as

$$I_{DS} = \begin{cases} \beta \dfrac{W}{L} \dfrac{T}{2T_0} \left[(V_{GS}-V_{FB})^{2T_0/T} - (V_{GS}-V_{DS}-V_{FB})^{2T_0/T} \right] & \text{if } (V_{GS}-V_{FB} > V_{DS}) \\[3mm] \beta \dfrac{W}{L} \dfrac{T}{2T_0} (V_{GS}-V_{FB})^{2T_0/T} & \text{if } (V_{GS}-V_{FB} \leq V_{DS}) \end{cases}$$

$$(1.10)$$

where V_{FB} is an equivalent flat-band voltage, accounting for a charge neutrality in the metal–dielectric–semiconductor structure

The authors also precise that their model does not require the threshold voltage definition as an input parameter justifying that the V_T definition is still not well defined in OTFTs. In a complementary study, an identical approach has been adopted to express the subthreshold current by the same team [36].

In a similar way, the OTFT device model developed by Fadlallah et al. is SPICE simulator compatible [37,38]. This model is based on the empirical equation of the conductivity in semiconductor films observed by Brown et al. [39] expressed as

$$\sigma = K.N_A^{\gamma} \tag{1.11}$$

K and γ being constants to be extracted and N_A the carrier density in the semiconductor film. The final drain current has been developed for all operating regimes and is expressed as

$$I_{DS} = \begin{cases} I_{00}.W \\ \dfrac{W}{L}\dfrac{K}{(2m+1)(2m+2)}\dfrac{C_i^{2m+1}}{(2kT\varepsilon_0\varepsilon_S)^m}\left[V_{GST}^{2m+2}-(V_{GST}-V_{DS})^{2m+2}\right] \\ \dfrac{W}{L}\dfrac{K}{(2m+1)(2m+2)}\dfrac{C_i^{2m+1}}{(2kT\varepsilon_0\varepsilon_S)^m}V_{GST}^{2m+2}\left[1+\lambda(V_{DS}-V_{GST})\right] \end{cases}$$

if $V_{GS} < V_T$

if $0 < V_{DS} \leq V_{GS} - V_T$

if $0 < V_{GS} - V_T \leq V_{DS}$ $\tag{1.12}$

Here, I_{00} is the cut-off drain current per channel width assumed to be independent of the gate voltage; K, m and λ are parameters to be extracted; and $V_{GST} = V_{GS} - V_T$. All other parameters have their usual meaning. The authors insisted that their model allows an easy extraction as it reposes on only 19 parameters to adjust. They have stated that its use in simulating inverters and ring oscillators showed good simulation timings and convergence. Finally, they used their model to design an RFID sequencer. An original approach of Schliewe et al. [40] based on the observation that source–drain (S/D) resistance does not saturate at high gate voltages suggests taking this phenomenon into account by a simultaneous consideration of a gate-voltage-dependent mobility as derived from the VRH model and adding a parallel parasitic depletion-mode transistor that would increase the channel resistance when the OTFT channel pinch-off occurs leading to a depletion region at the drain side. Considering that the threshold voltage definition is still ambiguous for OTFTs, Torricelli et al. [41] developed their own model based on the VRH theory and without any simplifying assumptions. One main advantage of this model is the correspondence between the parameters to be extracted and the physical parameters of the material to be studied as announced by the authors. The validation of the model has been carried out by a comparison with exact numerical solutions of the integral charge equations and also with experimental data based on both pentacene- and PTV-fabricated TFTs. The reader may refer to the complete article to better understand the unified current equation developed in [41]. Instead of considering an exponential DOS as stated in [33], Li et al. [42] suggested the use of a sum of two exponential densities of states as expressed in Equation 1.13, arguing that when V_{GS} sweeps from subthreshold to strong accumulation, the Fermi level energy can vary by an order of 1 eV making the approximation of the Gaussian DOS by an exponential one obsolete (figure here):

$$g(\varepsilon) = \begin{cases} 0 & \text{if } \varepsilon > 0 \\ \dfrac{N_T}{kT_0}\exp\left(\dfrac{\varepsilon}{kT_0}\right) + \dfrac{N_D}{kT_1}\exp\left(\dfrac{\varepsilon+\varepsilon_D}{kT_1}\right) & \text{if } -\infty < \varepsilon \leq 0 \end{cases} \tag{1.13}$$

The first term describes the deep tail states as in [33] with the same parameter meanings. The second term corresponds to states at higher energy levels with N_D, T_1 and ε_D being, respectively, the state concentration, the distribution function width and the energy depth. By using the percolation theory for the conductivity of the transistor, and assuming that one term is predominant on the other in the double exponential DOS distribution whether being above or below threshold voltage, Li et al. derived a unified drain current equation expressed as

$$\begin{cases} I = I_{lin} + \dfrac{I_{sub} - I_{lin}}{1 + \exp(V_T - V_{GS})} & \text{where} \\ I_{sub} \equiv \dfrac{W}{L}\left[(V_{GS} - V_{FB})^{2T_0/T} - (V_{GS} - V_{FB} - V_{DS})^{2T_0/T}\right] \\ I_{lin} \equiv \dfrac{W}{L}V_{GT}^{2(T_1/T-1)}\left[V_{GT}^2 - (V_{GT} - V_{DS})^2\right](1+\lambda V_{DS}) \end{cases} \tag{1.14}$$

All parameters earlier have their usual standard meanings. Furthermore, the authors have defined the threshold voltage as

$$V_T = V_{FB} + \dfrac{Q_T}{C_i} \tag{1.15}$$

Here, $Q_T = N_T/t_i$ is the total deep-state charges and t_i the accumulation channel thickness and C_i the well-known insulator capacitance per unit area. The model validation on pentacene-based transistors and inverters showed good agreement between measurements and simulation. Aiming to meet compact modelling requirements for SPICE simulators, the work of Marinov et al. [43] pointed out that through all different previous works, the mobility derived was usually expressed as $\mu \propto (V_G - V_T)\gamma$.

Based on this observation and deriving the established concept of charge drift in TFTs, the authors managed to develop a simple equation for the drain current as follows:

$$I_D = \frac{W}{L}\mu_0 C_i \frac{(V_G - V_T - V_S)^{(\gamma+2)} - (V_G - V_T - V_D)^{(\gamma+2)}}{\gamma + 2} \tag{1.16}$$

It is also noted that this model allows easy modifications to take into account sub- and above-threshold regimes easily by introducing an effective voltage overdrive interpolation function between both regimes. This generic model allows further modifications such as taking into account the channel length modulation or the incorporation of voltage drops accounting for contact resistances at the OTFT terminals. Marinov et al. insist on the fact that their generic model is a solid bridge between the physics-based models and the compact modelling of OTFTs. In an interesting application study, Zaki et al. [44] used the model developed in [43] to study the impact of complementary low-voltage OTFT characteristics on a 3-bit unary current-steering digital-to-analog converter (DAC). Finally, one of the most accomplished compact SPICE model for OTFTs that has been integrated into the public simulator SmartSpice [45] has been developed by Mijalkovic et al. [46]. Based on the VRH mobility description, the model has been ameliorated and since then called universal OTFT model (UOTFT) [47]. It essentially combines the universal charge control model (UCCM) with the specific charge DOS in the OTFTs described by exponential distributions. The UOTFT model has been validated on different OTFT structures and finally put into the Silvaco's model parameter extraction tool UTMOST IV that uses the SmartSpice simulator for optimization procedure (Figure 1.1).

1.2.1.3   Other Models

Different modelling approaches have been addressed earlier to describe the current behaviour of OTFTs; they are more or less based on physical equations. Nevertheless, many other modelling solutions in literature have been published by limiting the impact of, or completely disregarding, the physics of OTFTs.

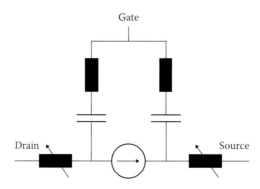

FIGURE 1.1   Equivalent circuit of the UOTFT model. (Redrawn from http://www.silvaco. com/products/analog/spicemodels/models/uotft/uotft.html.)

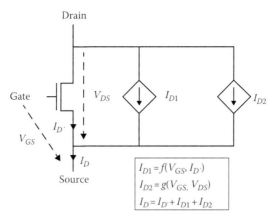

FIGURE 1.2 Equivalent circuit diagram of the OTFT with PSPICE building blocks for current sources. (Redrawn from Meixner, M. et al., *IEEE Trans. Electron. Dev.*, 55(7), 1776, 2008.)

Following a top to down approach, Meixner et al. [48] developed a physical-based PSPICE model. Based on the standard IGFET BSIM public model, two simple voltage-controlled current sources have been added to take into account both the voltage dependence of the mobility and the bulk conductivity in their P3HT TFTs. They stated that five additional parameters easily extractable are sufficient to correctly contribute to a good description the electrical characteristics of the OTFTs (Figure 1.2).

Disregarding any physical background, Gay et al. [49] managed to adjust the electrical behaviour of pentacene-based TFTs. The model in itself uses power series to approximate the shape of the drain current as follows:

$$I_D = K_G.f\left[\frac{V_{DS}}{\tau(V_{GS})}\right].h(V_{GS}) \tag{1.17}$$

where K_G is a geometry-related factor. The functions f, h and τ can be expressed in a generic way as

$$\begin{cases} P(V_i) = k.10^{X_P} \\ X_P = \sum_j a_j^P.\exp\left(-b_j^P.V_i\right) \end{cases} \text{ where } \begin{cases} k = -1, & i = DS \quad \text{for } P = f \\ k = 1, & i = GS \quad \text{for } P = h, \tau \end{cases} \tag{1.18}$$

Using this non-physical model, Gay et al. managed to illustrate with a good agreement output characteristics of three different simple circuits: a cascode amplifier, a differential amplifier and a differential-to-single-ended converter.

In a similar approach, the behavioural model developed by Infiniscale [50] totally disregards the physics of the TFT. Their model is based on more or less complex function classes that are used to handle non-linear phenomena. The model in itself is a black box extracted by minimizing the well-known root mean square error denoted as

$$RMSE = \sqrt{\frac{1}{n}\sum_{i=1}^{n}(f(X_i)-Y_i)^2} \qquad (1.19)$$

In Equation 1.19, X_i and Y_i are, respectively, input and output data that are being linked by the behavioural model function f. In comparison with the a-Si TFT model, Jacob et al. [51] showed that this behavioural model can be a fast and low-cost solution to describe the electrical behaviour of devices. Indeed, a comparable description with a-Si TFT model is obtained on simple logic circuits such as inverters or ring oscillators. Nevertheless, when dealing with very accurate description levels, this type of modelling presents a lack of accuracy. This is the case when looking to first derivatives of the drain current or focusing on the subthreshold leakage of OTFTs. The authors show the impact of this inaccuracy on some simple analogue circuits such as current mirrors or differential pairs.

1.3   OTFT SPICE MODEL OPTIMIZATION

1.3.1   CONTACT RESISTANCES AND CURRENT INJECTION

As detailed in the aforementioned paragraphs, different approaches towards the modelling of OTFTs exist and are more or less accurate. One important point that has been thoroughly studied in literature to optimize the description of the characteristics of OTFTs is the electrical injection through contacts. Indeed, it is at first sight supposed to be ideal, which in reality is not the case. Many authors focused on injection behaviour of current in the channel through contacts and tried to interpret the contact effects through some modifications of the available SPICE models. In a very simple approach, Benor et al. [52] replaced the term V_{DS} by $(V_{DS} - I_{DS}.R_C)$ in the linear region, to take into account the potential drop in drain and source contacts. Considering some mathematical changes, the authors managed to describe the effective mobility of their devices as dependent of the contact resistance and the device geometry in both linear and saturation regimes:

$$\begin{cases} \mu_{eff-lin} = \mu_0 \dfrac{L}{L + W\mu_0 C_G R_C (V_{GS} - V_T)} \\[4mm] \mu_{eff-sat} = \mu_0 \left[1 - \left(\dfrac{W\mu_0 C_G R_C (V_{GS} - V_T)}{L + W\mu_0 C_G R_C (V_{GS} - V_T)} \right)^2 \right] \end{cases} \qquad (1.20)$$

We can also notice the gate dependence of the mobility in both regimes that has been observed in many other works. In a complementary work, Vinciguerra et al. [53] showed and modelled the gate-bias dependence of contact resistance in polycrystalline OTFTs by introducing a Schottky barrier lowering as well as a Poole–Frenkel mechanism at the source and drain contacts, respectively. Similarly, on fully printed p-type OTFTs, Valletta et al. [54] observed a drastic increase of the contact resistance of short-channel transistors, with increasing drain–source bias. Based on the gradual channel approximation and assuming that long-channel OTFTs are not affected by contact effects, the authors show that the contact effect can be modelled as a reverse-biased Schottky diode including a barrier lowering induced by this latter diode. They also demonstrate a gate modulation of this reverse current. In a more detailed study, Rapisarda et al. [55] show the presence of three distinct regimes. At first, with low V_{DS}, the channel and contact resistances behave like gate-modulated resistors; then for intermediate V_{DS}, the barrier lowering due to the reverse-biased Schottky diode at the source side affects the device electrical characteristics; finally, at high V_{DS} values, a pinch-off of the channel appears at the drain side and controls the channel current. This complete developed model has been coded in Verilog-A language and implemented in computer-aided design system without any difficulties and is being used for circuit simulations.

1.3.2 Capacitances and Dynamic Behaviour

In order to have an accurate description of the electrical characteristics of OTFTs, one has also to consider with a big interest their dynamic behaviour. Indeed, any alternating current (AC) or transient simulation of TFTs would require a precise portrayal of the capacitances of the device. Some interesting works can be found in literature concerning the dynamic behaviour of thin-film organic transistors. In a simple derivation, assuming that OTFTs work in accumulation mode, Fadlallah et al. [37,38] described the total charge Q_G on the gate electrode as follows:

$$Q_G = -WC_i \int_0^L (V_G - V_T - V(x))dx \tag{1.21}$$

In the aforementioned equation, the authors assumed that the threshold voltage is the voltage at which accumulation occurs and thus being the flat-band voltage V_{FB}. By integrating (1.21) and deriving the gate to source and gate to drain capacitances as $C_{GS-D} = \partial Q_G / \partial V_{GS-D}$, they finally manage to obtain an analytical description of both capacitances. In a very similar approach, Li et al. [42] managed to derive the gate charge Q_G. Nevertheless, based on a charge partitioning scheme from drain to source in the channel, they determined the drain side charge as

$$Q_D = \frac{W}{L} C_i \int_0^L (V_G - V_T - V(x))dx \tag{1.22}$$

Using then the neutrality principle, they determined the source side charge by

$$Q_S = -Q_D - Q_G \tag{1.23}$$

And finally, all capacitances are easily calculated through the partial derivatives as follows:

$$\begin{cases} C_{ii} = +\dfrac{\partial Q_i}{\partial V_i} \\[4mm] C_{ij} = -\dfrac{\partial Q_i}{\partial V_j} \quad i \neq j \end{cases} \tag{1.24}$$

For a better and total description of the dynamic behaviour of OTFTs, one may refer to the paper of Torricelli et al. [41] where the authors derived a very complete AC model based on the charge-oriented approach.

1.3.3 PARASITICS, NOISE, BIAS STRESS, ETC.

Some disparate modelling studies exist in literature on different effects shown to be present in OTFTs. The contributors tried to model as possible these existing phenomena in order to add them in an easy way in available OTFT SPICE models. An interesting study on rubrene single-crystal back-gate OTFTs, done by Marinov et al. [56], showed that the fringing capacitances between source and drain electrodes and the back of the organic semiconductor can have an impact on the transistor current. The result is an enhancement of the channel current by two components ΔI_D and ΔI_L that authors compare to a channel modulation effect in the saturation effect at high V_{DS} and a space-charge-limited conduction in the leakage current at low V_{GS}. It is demonstrated also that both effects are easily incorporated into a SPICE model by using an additional current source and a diode connected FET in parallel to the OTFT (Figure 1.3).

In a different subject, the same author presented a work [57] on flicker noise in OTFTs. The paper states that using the VRH model we can explain the origin of the *1/f* noise in OTFTs. They state that the origin of this low-frequency noise is a conduction fluctuation in the channel equivalent to a $\Delta\sigma - \Delta N$ noise. This noise

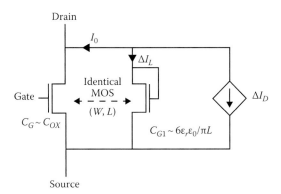

FIGURE 1.3 Equivalent circuit for OTFT modelling with impact of fringing capacitances extracted. (From Marinov, O. et al., *Org. Electron.*, 12, 936, 2011.)

modelling can easily be integrated into existing SPICE models such as a-Si TFT or P-Si TFT as they already have some built-in parameters for taking into account noise in TFTs. It can also be added in other models by means of subcircuit modelling.

Another important issue that appears when characterizing OTFTs is their instability towards a prolonged bias. This, so-called, bias-stress effect has been widely studied and reported in literature [58,59] in order to better understand its origin and mechanisms. In a recent study, Ryu et al. [60] developed an empirical model to describe the shift ΔV that occurs in the OTFT transfer characteristic when bias-stressed. The proposed model for the shift is based on a stretched-exponential equation as follows:

$$\Delta V = \Delta V_{final}\left\{1 - \exp\left(-\left(\frac{t}{\tau}\right)^{\beta}\right)\right\} \tag{1.25}$$

Here, ΔV_{final}, τ and β are parameters that depend on gate to source V_{GS} and drain to source V_{DS} voltages. We can easily imagine that this equation can be incorporated into a SPICE model without any difficulties. Furthermore, for a matter of simplicity, the parameters can be turned to be only fitting parameters.

1.4   OTHER ORGANIC DEVICES

Besides OTFTs, different devices are also processed and their respective characterizations carried out. These devices have also to be modelled in order to be incorporated into simulations. Works on OLEDs can easily be found in literature where SPICE models for these devices are usually presented as subcircuits of different existing public models. Li et al. [61] suggested a simple subcircuit model consisting of a DC voltage source and a non-linear resistance in series with an ideal diode to represent the characteristics of their OLED. They compare their model to a subcircuit using a simple linear resistance in series with an ideal diode explaining that the modifications used in their model are necessary to take into account the built-in voltage in the organic layers and the nonideal ohmic contacts in the device. In a more complete work, on top-emitting positively doped, intrinsic, negatively doped OLEDs (PIN-OLEDS), Cummins et al. [62] investigated the DC and AC behaviours of their device. Their suggested SPICE model takes into account both behaviours. The DC model consists of two separate parallel voltage-controlled current non-linear sources, describing, respectively, different conduction regions of the OLED, in series with a constant voltage source behaving as the built-in device voltage, the whole in parallel with an ideal resistance and a capacitance representing, respectively, the leakage and the geometrical capacitance of the PIN-OLED. Their AC SPICE subcircuit model consisted of 2 RC elements modelling the electron and hole transport layers, respectively, in series with a resistance for the contact electrodes (Figure 1.4).

OPDs and OPVs have also been modelled in literature and different models suggested. As for OLEDs, SPICE subcircuit models are commonly used to describe these OPDs and OPVs. In general, the model consists of an ideal diode in parallel

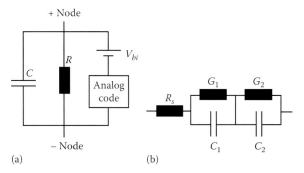

FIGURE 1.4 Equivalent circuit for OLED for DC behaviour (a) and AC behaviour (b) (Redrawn from Cummins, G. et al., *J. Soc. Inf. Display*, 19(4), 360, 2011.)

with a current source that is used to emulate the photocurrent due to the incident light on the device. Some of these suggested models are more accurate than others, when taking into account contact electrodes and additional leakage by adding series and parallel resistances, respectively, for these effects [63,64]. Some other approaches on the modelling of these later devices can also be gathered in literature such as physical behaviour modelling of OLEDS [65] or an opto-electrical analytical model of OPDs and solar cells [66]. Finally, few papers can be found [67,68] on the SPICE modelling of passive devices like resistances, inductances or capacitances in the organic world. Indeed, these devices have been lately focused on, and the available silicon public models are being still sufficient to permit their electrical behaviour and their integration in simulations.

1.5   CONCLUSION

Organic technologies have come to a certain level of maturity that allows the process of first-level electronic circuits. The key between the basic device characteristics and their use in applications has always been an accurate circuit simulation before any fabrication. This is only possible through a precise description of these organic devices in terms of SPICE models. We have presented in this chapter the most important SPICE models found in literature concerning organic-based devices. An important part has been dedicated to OTFTs that are widely studied and characterized. To summarize this synthesis, a huge work has started on organic device models for SPICE simulations and very good results are achieved. Nevertheless, the high dispersion in these models has to be reduced and stabilized in order to let the organic technology take the same departure as it was for silicon some decades ago.

REFERENCES

1. L. W. Nagel et al., SPICE (Simulation Program with Integrated Circuit Emphasis), Memorandum No. ERL-M382, University of California, Berkeley, CA, April 1973.
2. CMC, Compact Modelling Council, http://www.geia.org/
3. http://www.ieeeghn.org/wiki/index.php/Milestones:SPICE_Circuit_Simulation_ Program

4. H. Shichman et al., Modeling and simulation of insulated-gate field-effect transistor switching circuits, *IEEE J. Solid-State Circuits*, 3(3), 285–289, September 1968.

5. G. Horowitz et al., An analytical model for organic-based thin-film transistors, *J. Appl. Phys.*, 70(1), 469–475, July 1991.

6. S. M. Sze, *Physics of Semiconductor Devices*, 2nd edn., John Wiley & Sons, New York, 1981.

7. Z. Bao et al., Soluble and processable regioregular poly(3-hexylthiophene) for thin film field-effect transistor applications with high mobility, *Appl. Phys. Lett.*, 69(26), 4108–4110, December 1996.

8. N. Satyala et al., Simulation and transient analysis of organic/inorganic CMOS inverter circuit, *Proceedings of 41st Southeastern Symposium on System Theory*, University of Tennessee Space Institute, Tullahoma, TN, March 15–17, 2009, T2A.5-324–T2A.5-329.

9. D. Bode et al., Noise margin analysis for organic thin-film complementary technology, *IEEE Trans. Electron. Dev.*, 7(1), 201–208, January 2010.

10. H. Marien et al., A full integrated DS ADC in organic thin-film transistor technology on flexible plastic foil, *IEEE J. Solid-State Circuits*, 46(1), 276–284, January 2011.

11. M. Shur et al., SPICE models for amorphous silicon and polysilicon thin film transistors, *J. Electrochem. Soc.*, 144(8), 2833–2839, August 1997.

12. M. Shur et al., Physics of amorphous silicon based alloy field-effect transistors, *J. Appl. Phys.*, 55(10), 3831–3842, May 1984.

13. J. Y. W. Seto et al., The electrical properties of polycrystalline silicon films, *J. Appl. Phys.*, 46(12), 5247–5254, December 1975.

14. M. Bartzch et al., Device and circuit simulation of printed polymer electronics, *Org. Electron.*, 8, 431–438, February 2007.

15. O. Yaghmazadeh et al., A SPICE-like DC model for organic thin-film transistors, *J. Korean Phys. Soc.*, 54(1), 523–526, January 2009.

16. S. Chung et al., All-inkjet-printed organic thin-film transistor inverter on flexible plastic substrate, *IEEE Electron Dev. Lett.*, 32(8), 1134–1136, August 2001.

17. B. Iniguez et al., Universal compact model for long- and short-channel thin-film transistors, *Solid-State Electron.*, 52, 400–405, December 2007.

18. M. Estrada et al., Accurate modelling and parameter extraction method for organic TFTs, *Solid-State Electron.*, 49, 1009–1016, March 2005.

19. T. D. Joshi, Device modeling and circuit design for ZTO based amorphous metal oxide TFTs, Master Thesis Dissertation, University of Texas, Austin, TX, May 2011.

20. P. G. Le Comber et al., Electronic transport in amorphous silicon film, *Phys. Rev. Lett.*, 25(8), 509–511, August 1970.

21. G. Horowitz et al., Temperature dependence of the field-effect mobility of sexithiophene. Determination of the density of traps, *J. Phys. III France*, 5(4), 355–371, April 1995.

22. H. K. Chen et al., Variable temperature measurement on operating pentacene-based OTFT, *MRS Proceedings*, Vol. 1091, Symposium AA, Cambridge University Press, Cambridge, U.K., pp. 91–98, 2008.

23. J. Yamashita et al., On electronic current in NiO, *J. Chem. Phys. Solids*, 5(1–2), 34–43, 1958.

24. J. Yamashita et al., Heitler-London approach to electrical conductivity and application to d-electron conductions, *J. Phys. Soc. Japan*, 15(5), 802–821, May 1960.

25. T. Holstein, Studies of polaron motion, Part I & II, *Ann. Phys.*, 8(3), 325–389, November 1959.

26. K. Fesser et al., Optical absorption from polarons in a model of polyacetylene, *Phys. Rev. B*, 27(8), 4804–4825, April 1983.

27. R. A. Marcus, Nonadiabatic processes involving quantum-like and classical-like coordinates with applications to nonadiabatic electron transfers, *J. Chem. Phys.*, 81(10), 4494–4500, July 1984.

28. R. A. Marcus et al., Electron transfers in chemistry and biology, *Biochim. Biophys. Acta*, 811(3), 265–322, August 1985.

29. K. Seki et al., Electric field dependence of charge mobility in energetically disordered materials: Polaron aspects, *Phys. Rev. B*, 65(1), 14305–14317, December 2001.

30. H. Bässler, Charge transport in disordered organic photoconductors, a Monte Carlo simulation study, *Phys. Stat. Sol. B*, 175(1), 15–56, January 1993.

31. W. P. Su et al., Solitons in polyacetylene, *Phys. Rev. Lett.*, 42(25), 1698–1701, June 1979.

32. A. Miller et al., Impurity conduction at low concentrations, *Phys. Rev.*, 120(3), 745–755, November 1960.

33. M. C. J. M. Vissenberg et al., Theory of the field-effect mobility in amorphous organic transistors, *Phys. Rev. B*, 57(20), 12964–12967, May 1998.

34. E. J. Meijer et al., Switch-on voltage in disordered organic field-effect transistors, *Appl. Phys. Lett.*, 80(20), 3838–3840, May 2002.

35. E. Calvetti et al., Organic thin film transistors: A DC/dynamic analytical model, *Solid State Electron.*, 49(4), 567–577, January 2005.

36. E. Calvetti et al., Analytical model for organic thin-film transistors operating in the sub-threshold region, *Appl. Phys. Lett.*, 87(22), 223506–223509, November 2005.

37. M. Fadlallah et al., Modeling and characterization of organic thin film transistors, *J. Appl. Phys.*, 99(10), 104504–104510, May 2006.

38. M. Fadlallah et al., DC/AC unified OTFT compact modelling and circuit design for RFID applications, *Solid State Electron.*, 51(7), 1047–1051, May 2007.

39. A. R. Brown et al., Field-effect transistors made from solution-processed organic semi-conductors, *Synth. Met.*, 88(1), 37–55, January 1997.

40. R. R. Schliwe et al., Static model for organic field-effect transistors including both gate-voltage-dependent mobility and depletion effect, *Appl. Phys. Lett.*, 88(23), 233514–233516.

41. F. Torricelli et al., A charge-based OTFT model for circuit simulation, *IEEE Trans. Electron. Dev.*, 56(1), 20–30, January 2009.

42. L. Li et al., Compact model for organic thin-film transistor, *IEEE Electron. Dev. Lett.*, 31(3), 210–212, March 2010.

43. O. Marinov et al., Organic thin-film transistors: Part I – Compact DC modeling, *IEEE Trans. Electron. Dev.*, 56(12), 2952–2961, December 2009.

44. T. Zaki et al., Circuit impact of device and interconnect parasitics in a complementary low-voltage organic thin-film technology, *Proceedings of Semiconductor Conference Dresden (SCD)*, Dresden, Germany, pp. 1–4, September 27–28, 2011.

45. http://www.silvaco.com/products/analog/spicemodels/models/uotft/uotft.html

46. S. Mijalkovic et al., Modelling of organic field-effect transistors for technology and circuit design, *26th International Conference on Microelectronics (MIEL)*, Nis, Serbia, pp. 469–476, May 11–14, 2008.

47. S. Mijalkovic et al., UOTFT: Universal organic TFT model for circuit design, *Digest 6th International Conference Organic Electronics (ICOE)*, Liverpool, U.K., June 2009.

48. R. M. Meixner et al., A physical-based PSPICE compact model for poly(3-heylthio-phene) organic field-effect transistors, *IEEE Trans. Electron. Dev.*, 55(7), 1776–1781, July 2008.

49. N. Gay et al., Analog signal processing with organic FETS, *Solid-State Circuits Conference (ISSCC 2006), Digest Technical Papers, IEEE International*, San Francisco, pp. 1070–1079, February 6–9, 2006.

50. http://www.infiniscale.com

51. S. Jacob et al., Fast behavioral modelling of organic CMOS devices for digital- and analog-circuit applications, *Proceedings of SPIE*, San Diego, CA, Vol. 8117, pp. 81170Q-1–81170Q-7, August 1–5, 2011.

52. A. Benor et al., Contact effects in organic thin film transistors with printed electrodes, *Org. Electron.*, 9, 209–219, November 2007.
53. V. Vinciguerra et al., Modelling the gate bias dependence of contact resistance in staggered polycrystalline organic thin film transistors, *Org. Electron.*, 10, 1074–1081, May 2009.
54. A. Valletta et al., Contact effects in high performance fully printed p-channel organic thin film transistors, *Appl. Phys. Lett.*, 4(12), 233309-1–233309-4.
55. M. Rapisarda et al., Analysis of contact effects in fully printed p-channel organic thin film transistors, *Org. Electron.*, 13, 2017, 2012.
56. O. Marinov et al., Impact of the fringing capacitance at the back of thin-film transistors, *Org. Electron.*, 12, 936–949, March 2011.
57. O. Marinov et al., Flicker noise due to variable range hopping in organic thin-film transistors, *21st International Conference on Noise and Fluctuations (ICNF)*, Toronto, Ontario, Canada, pp. 287–290, June 12–16, 2011.
58. S. J. Zilker et al., Bias stress in organic thin film transistors and logic gates, *Appl. Phys. Lett.*, 79(8), 1124–1126, 2001.
59. H. Zan et al., Stable encapsulated organic TFT with a spin-coated poly(4-vinylphenol-co-methyl methacrylate) dielectric, *IEEE Trans. Electron. Dev.*, 32(8), 1131–1133, August 2011.
60. K. K. Ryu et al., Bias-stress effect in pentacene organic thin-film transistors, *IEEE Trans. Electron. Dev.*, 57(5), 1003–1008, May 2010.
61. Y. Li et al., A novel SPICE compatible current model for OLED circuit simulation, *NSTI Nanotechnology Conference and Trade Show*, Anaheim, CA, Vol. 3, pp. 103–106, May 8–12, 2005.
62. G. Cummins et al., Electrical characterization and modelling of top-emitting PIN-OLEDs, *J. Soc. Inf. Display*, 19(4), 360–367, April 2011.
63. M. Sams et al., Mixed design of integrated circuits and systems (MIXDES), *Proceedings of the 15th International Conference*, Poznan, Poland, pp. 425–429, June 2008.
64. A. Daami et al., A complete SPICE subcircuit-based model library for organic photodiodes, *Solid State. Electron.*, 75, 81–85, 2012.
65. S. Lee et al., Physics-based OLED analog behavior modelling, *J. Information Display*, 10(3), 101–106, November 2010.
66. S. Altazin et al., Analytical modeling of organic solar cells and photodiodes, *Appl. Phys. Lett.*, 99(14), 143301-1–143301-3, October 2011.
67. E. Benevent et al., Microwave characterization and modeling of fully-printed MIM capacitors on flexible plastic foils, *IEEE Trans. Microw. Theory Tech.*, to be published.
68. E. Benevent et al., Full-printed inductances on flexible plastic foils for electromagnetic energy harvesting, fabrication, characterization, modelling, *European Microwave Conference*, Amsterdam, the Netherlands, October 29–November 1, 2012.

2 Nanoscale Phase Separation and Device Engineering in Polymer Solar Cells

Swaminathan Venkatesan and Qiquan Qiao

CONTENTS

2.1 INTRODUCTION

Sunlight is an abundant and renewable energy resource, and converting sunlight into electricity has been regarded as one of the most promising approaches to provide clean alternative energy compared to traditional fossil fuels. Conventional silicon solar cells generally require high purity materials and sophisticated vacuum-based processing making them expensive to fabricate, and hence, the cost per watt is much higher than fossil fuels. On the other hand, organic solar cells have attracted considerable attention in the last decade due to their low-cost fabrication, easy processing and compatibility with flexible substrates. Since most of these cells can be processed and fabricated from solutions, they can be easily manufactured by a variety of low-cost techniques including roll-to-roll printing (Krebs et al. 2009a; Søndergaard et al. 2012), inkjet printing (Schilinsky et al. 2006; Hoth et al. 2007; Aernouts et al. 2008), spray coating (Vak et al. 2007; Green et al. 2008; Kang et al. 2012), dip coating (Li et al. 2011) and screen printing (Krebs et al. 2009b). Polymer–fullerene solar cells have recently achieved efficiency exceeding 10% (Chemicals 2012). In spite of such potential advantages, they still lack commercial realization due to their lower efficiency and operation lifetime compared to single-crystalline silicon solar cells (Manceau et al. 2011; Jørgensen et al. 2012). To solve these issues, researchers are working vigorously to understand photophysics of semiconducting polymers and formulating new designs and improvements to overcome such shortcomings encountered currently by organic solar cells.

2.2   SEMICONDUCTING POLYMERS

Semiconducting properties of polymers arise from conjugation in carbon-based main chains, which is attributed to alternating single and double bonds between carbon atoms. Each carbon atom consists of six electrons, of which four are valence electrons. The six electrons ideally should be $1s^2 2s^2 2p^2$ orbital configuration, but to lower total energy, the 2s and 2p orbitals can hybridize in either sp^3 or $sp^2 pz$ configuration. Traditional insulating polymers (e.g. polyethylene) are in sp^3-hybridized configuration. However, in $sp^2 pz$ configuration (e.g. polyacetylene), three electrons form σ bonds as denoted by sp^2 and one electron, namely, pz electron (Figure 2.1), from each carbon forms π bonds, which can delocalize electrons along polymer chains.

The delocalized π-electron wave functions lead to formation of π-band thereby giving rise to an energy bandgap between the bonding (π-band) and anti-bonding (π^*-band) orbitals. Such polymers are called conjugated polymers and their electronic band structure is shown in Figure 2.1.

In order to fabricate organic solar cells from solution, conjugated polymers are required to have side chains to help dissolve in solvents such as chloroform, toluene, chlorobenzene and dichlorobenzene. In addition, molecular structures of conjugated polymers need to be modified to exhibit low bandgaps, broadband optical absorption, efficient exciton diffusion, high carrier mobility and ease of processability. Early commonly used conjugated polymers include derivatives of polythiophenes (PT) and poly(phenylenevinylenes) (PPV) such as poly[2-methoxy-5-(2′-ethylhexyloxy)-p-phenylenevinylene](MEH-PPV),   poly(2-methoxy-5-(3′-7′-dimethyloctyloxy)-1,4-phenylenevinylene) (MDMO-PPV) and poly(3-hexylthiophene (P3HT). P3HT has

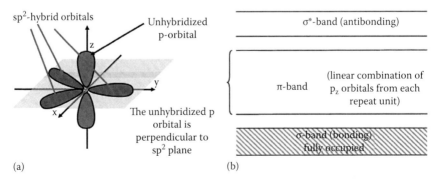

FIGURE 2.1   (a) $sp^2 p_z$-hybridized orbital and (b) schematic band diagram of π-conjugated polymers. (From Heeger, A. J., *Chemical Society Reviews*, 39, 2354, 2010. Reprinted with permission from The Royal Society of Chemistry.)

been to date the most widely studied photovoltaic polymer due to its high solubility, mobility and tendency to crystallize in ordered chains. However, due to its narrow absorption band and shallow highest occupied molecular orbital (HOMO) level, there is need for new polymers having broader optical absorption band that better match solar spectrum and deeper HOMO energy levels. Hence, the search for new polymers with desired properties (e.g. broader band absorption and deeper HOMO) has started and led to synthesis of a variety of copolymers having alternating electron-rich and electron-deficient groups for lower bandgaps and deeper HOMOs (Peet et al. 2009). On the other hand, acceptor components have been shifted from [6,6]-phenyl-C61-butyric acid methyl ester (PC60BM) to [6,6]-phenyl-C-71-butyric acid methyl ester (PC70BM) in recent reports (Wienk et al. 2003; He and Li 2011; Zhang et al. 2012). Also, the fullerene molecular structures were modified with a different functional group (Cheng et al. 2010; He et al. 2010; Zhao et al. 2010; Mikroyannidis et al. 2011) to manipulate their lowest unoccupied molecular orbital (LUMO) and HOMO levels.

2.3 WORKING PRINCIPLE OF POLYMER SOLAR CELLS

Working principle of polymer solar cells is quite different from their counterpart inorganic (e.g. Si) solar cells. Polymer solar cells employ two components, namely, donor and acceptor. The donor is generally a conjugated polymer and the acceptor is a fullerene derivative. Polymer solar cell mechanism involves six steps as shown in Figure 2.2. When donor polymer absorbs incoming photons (step 1), an exciton is generated (step 2). The low dielectric constant in polymers causes considerable attractive coulombic force between the negatively charged electron and positively charged hole in excitons, leading to tightly bound electron–hole pair. When a polymer

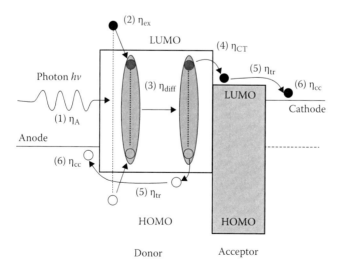

FIGURE 2.2 Polymer solar cell mechanism elucidated in six steps. Here η is the efficiency of each step. (From Siddiki, M. K., Li, J., Galipeau, D., and Qiao, Q., A review of polymer multijunction solar cells, *Energy Environ. Sci.*, 3(7), 867–883, 2010. Reprinted with permission from The Royal Society of Chemistry.)

is mixed with an acceptor, excitons generate and diffuse (step 3) in the donor to donor–acceptor interface for dissociation into electrons and holes (step 4) due to their LUMO level offset. The electrons then flow through the acceptor and holes through the donor (step 5) and are finally collected at the corresponding electrodes (step 6).

2.4 CONTROL OF NANOSCALE PHASE SEPARATION IN POLYMER SOLAR CELLS

Because exciton diffusion length is in the order of a couple of tens of nanometers, it is very important that phase separation between the donor and acceptor components should be of the same length scale. Ideal morphology would be a bicontinuous inter-penetrating network of donor and acceptor phases separated homogenously within the blend for efficient exciton diffusion and dissociation. Device processing conditions are extremely crucial as they control both the lateral and vertical phase separation between the donor and acceptor components. Several reports have focused on various device processing factors that affect the nanomorphology of polymer–fullerene blends (Kim et al. 2005; Reyes-Reyes et al. 2005; Yang et al. 2005; Campoy-Quiles et al. 2008; Zeng et al. 2010).

2.4.1 Effect of Solvent and Thermal Annealing

Several groups have studied the effects of solvent and post-spin-coating conditions on the morphology and device performance. Li et al. (2005, 2007) showed that slow growth of P3HT films leads to better chain packing. This effect was clearly evidenced in the absorption spectra (Figure 2.3) where the films that had been spin coated for shorter time were wet and when covered with petri dish after coating the film dried very slowly. Such films showed broader and higher absorption peaks. Their absorption spectra also consisted of characteristic vibronic peaks at 550 and 610 nm.

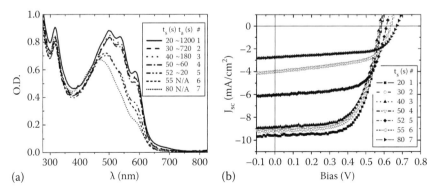

FIGURE 2.3 UV-vis absorption spectra (a) and current density versus voltage curves (b) for P3HT:PCBM solar cells spin casted at different spin-coating time (t_s from 20 s for # 1 to 80 s for # 7) and different solvent annealing time (t_a). (From Li, G., Yao, Y., Yang, H. et al.: Solvent annealing effect in polymer solar cells based on poly(3-hexylthiophene) and methanofuller-enes. *Adv. Funct. Mater.* 2007. 17(10). 1636–1644. Copyright Wiley-VCH Verlag GmbH & Co. KGaA. Reprinted with permission.)

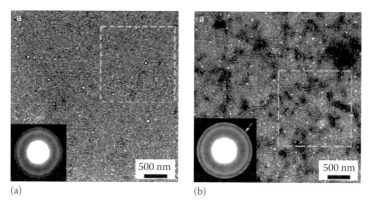

(a)   (b)

FIGURE 2.4   TEM image of (a) as deposited and (b) thermally annealed P3HT:PCBM thin films. (Reprinted with permission from Yang, X., Loos, J., Veenstra, S.C. et al., Nanoscale morphology of high-performance polymer solar cells, *Nano Lett.*, 5(4), 579–583, 2005. Copyright 2005, American Chemical Society.)

These peaks were absent or weaker for films that were spin coated for longer time and films that were kept open in air, which led to fast drying. When such a phenomena was examined using x-ray diffraction (XRD) technique (Fan et al. 2012), it was shown that films that dried slowly showed exhibited self-ordering and higher crystalline fraction. These crystallites give rise to the vibronic peaks observed in the absorption spectra.

Thermal annealing treatment plays two important roles: (1) better self-organization of polymer chains that leads to stronger $\pi–\pi^*$ stacking for more efficient charge transport between chains and (2) induced diffusion of fullerene molecules resulting in homogenous phase separation within the blend. Several groups (Yang et al. 2005; Campoy-Quiles et al. 2008) have studied the effects of thermal annealing on P3HT:PCBM blends and showed that heating of the blend films after spin coating improves crystalline order of P3HT for higher hole mobility and increased interfacial area for exciton dissociation. For P3HT:PCBM system, temperature ranging from 110°C to 160°C (Kim et al. 2005; Reyes-Reyes et al. 2005; Kim et al. 2006a; Zeng et al. 2010) has been reported for annealing with varying annealing times (Figure 2.4).

2.4.2   EFFECT OF SOLVENTS AND ADDITIVES

Solvents used to fabricate polymer solar cells play a significant role in deciding the final morphology of blend films. Factors such as solubility, boiling point, temperature of stirring distinguish the resulting morphology. Hoppe et al. (2004) performed a study on the effects of different solvents on resulting MDMO:PCBM blend film morphology and found that toluene formed large PCBM clusters, while chlorobenzene formed a very fine nanoscale-phase-separated morphology (Figure 2.5). The authors attributed this to higher solubility of PCBM in chlorobenzene than in toluene.

It has also been shown that dichlorobenzene (Yao et al. 2006) serves as a better solvent than chlorobenzene and chloroform because of its higher boiling point that leads to slow drying of film and hence better conformation of polymer chains.

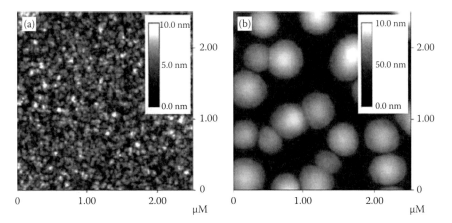

FIGURE 2.5 Topographic AFM images of MDMO:PPV blended in 1:4 ratio casted from (a) chlorobenzene and (b) toluene solvent. (Hoppe, H., Niggemann, M., Winder, C. et al.: Nanoscale morphology of conjugated polymer/fullerene-based bulk-heterojunction solar cells. *Adv. Funct. Mater.* 2004. 14. 1005–1011. Copyright Wiley-VCH Verlag GmbH & Co. KGaA. Reprinted with permission.)

The most commonly used solvents for various polymer solar cells are chloroform, chlorobenzene and dichlorobenzene, a mixture of them, or the three of them mixed with small amount of additives.

In 2006, Peet et al. (2006) showed that addition of small volume percent of alkyl thiols to P3HT:PCBM blend solution resulted in films with much higher mobility and fewer deep-level trap states in the polymer. A similar approach (Peet et al. 2007) was proved to be successful for low-bandgap polymer poly[2,1,3-benzothiadiazole-4,7-diyl[4,4-bis(2-ethylhexyl)-4H-cyclopenta[2,1-b:3,4-b′]dithiophene-2,6-diyl](PCPDTBT) blended with PC70BM. As shown in Figure 2.6, the addition of alkanedithiol additive to parent solvent shows better performance than pristine parent solvent itself. In addition, cell performance also depended on the alkyl side-chain length of alkanedithiol, and octanedithiol showed the highest enhancement in efficiency from 2.8% to 5.5%. The enhancement in short-circuit current density (J_{sc}) and fill factor (FF) is rather noticeable; however, the open-circuit voltage (V_{oc}) decreases a by a small value. The overall increase in performance is attributed to higher degree of phase separation as seen in the tapping-mode AFM image from Figure 2.6b. The general rule of thumb for choosing the additives is to (1) specifically dissolve one of the two components, i.e. either donor polymer or acceptor fullerene, and (2) have higher boiling point than the parent solvents.

Hoven et al. (2010) studied the effects of additives on nanoscale morphology and local photocurrent using conductive atomic force microscopy. Figure 2.7 showed that using 2 volume percent of 1-chloronapthalene (CN) led to a rougher surface, smaller polymer aggregates and higher short-circuit photocurrent measured by a gold-coated tip.

Several other groups used different additives such as 1,8-diiodoctane (DIO) (Son et al. 2011), N-methyl pyrrolidone(NMP) (Sun et al. 2011), nitrobenzene (Chang et al. 2011), dimethyl sulphoxide (DMSO) (Chu et al. 2011) and dimethylformamide (DMF) (Chu et al. 2011). These additives also showed similar enhancement in efficiencies.

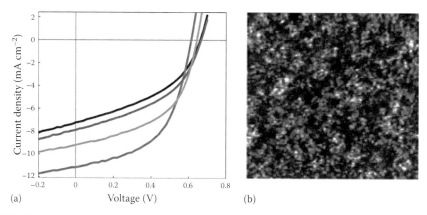

(a)   Voltage (V)   (b)

FIGURE 2.6 (See colour insert.) (a) Device performance of PCPDTBT:C71-PCBM solar cells casted from pristine chlorobenzene (black line) and chlorobenzene containing 24 mg mL^{-1} of butanedithiol (green line), hexanedithiol (orange line) or octanedithiol (red line) additives. (b) AFM image of PCPDTBT:PC$_{71}$BM thin film casted from chlorobenzene containing 24 mg mL^{-1} of octanedithiol. (Reprinted with permission from Peet, J., Kim, J.Y., Coates, N.E. et al., Efficiency enhancement in low-bandgap polymer solar cells by processing with alkane dithiols, *Nat. Mater.*, 6, 497–500, 2007, Copyright 2007, American Institute Physics.)

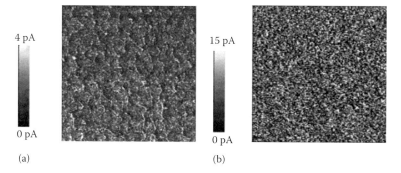

(a)   (b)

FIGURE 2.7 (See colour insert.) Local photocurrent mapping of poly[(4,4-didodecyldithieno[3,2-b:2′,3′-d]silole)-2,6-diyl-alt-(2,1,3-benzoxadiazole)-4,7-diyl] and PCBM bulk heterojunction films casted without any additive (a) and with 2 volume %1-chloronapthalene (b) performed at short-circuit condition. (From Hoven, C.V., Dang, X.-D., Coffin R.C. et al.: Improved performance of polymer bulk heterojunction solar cells through the reduction of phase separation via solvent additives. *Adv. Mater.* 2010. 22(8). E63–E66. Copyright Wiley-VCH Verlag GmbH & Co. KGaA. Reproduced with permission.)

2.5 DEVICE STRUCTURAL ENGINEERING FOR EFFICIENCY AND LIFETIME ENHANCEMENT

2.5.1 INVERTED SOLAR CELLS STRUCTURE

Conventional solar cells have ITO/PEDOT:PSS as anode and Ca or LiF/Al as cathode. The work function of ITO can be tuned from 4.5 to 5.1 eV and therefore can act as either anode or cathode. ITO can serve as anode when deposited with an electron-blocking layer such as PEDOT:PSS and cathode if coated with an n-type layer such as TiO_2 or ZnO. If ITO serves as a cathode instead of anode, device configuration is called inverted solar cell, which has potentially two main advantages: (1) eliminate use of acidic PEDOT:PSS that etches ITO and causes interface instability by diffusion of indium into active layer and (2) eliminate use of low work function metals such as Ca or Ba that are easily oxidized and hence increase series resistance. Both these reasons contribute to cell degradation with lower lifetime. In inverted device architecture, high work function material such as Au or Ag is used as anode. Generally an electron-blocking layer such as PEDOT:PSS, MoO_3 or V_2O_5 thin film is deposited prior to anode deposition.

A variety of n-type metal oxides such as TiOx (Cheng et al. 2011; Liu et al. 2011; Orawan et al. 2011), Al_2O_3 (Zhou et al. 2010), Nb_2O_5 (Orawan et al. 2011), IZO (Kyaw et al. 2011) and ZnO (Yang et al. 2010; Liu et al. 2011; Sun et al. 2011; Small et al. 2012) have become attractive choices for use as hole-blocking layer over ITO surface in inverted solar cells due to their wide bandgap and charge-selective nature (Heo et al. 2004; White et al. 2006; Sun et al. 2011). ZnO has been used most extensively due to its high mobility and transparency, simple sol-gel-based process and smooth surface for active layer deposition. Sol-gel ZnO is synthesized from a zinc salt (generally zinc acetate) using an alcohol and amine as catalyst. Liang et al. (2012) studied the effect of ZnO sol-gel precursor's concentration and ZnO buffer layer morphology on device performance of inverted solar cells (Figure 2.8) and found that device performance was independent on layer thickness, but morphology of buffer layer played an important role. Smoother and compact morphology was desired to avoid decrease of shunt resistance and to lower contact resistance between active layer and ZnO. Heeger and coworkers (Sun et al. 2011) proposed a low-temperature-based approach and showed that a thin layer of amorphous ZnO was sufficient for achieving high efficiency in inverted solar cells.

Small et al. (2012) recently reported using ZnO-PVP (poly(vinyl pyrrolidone)) electron-selective layer for high-efficiency inverted solar cells. The ZnO-PVP nanocomposite consisted of ZnO nanoparticles or clusters grown in PVP matrix. This method of ZnO is more versatile as claimed by the authors as it gave more tenability in cluster size by controlling the Zn2+/PVP ratio and also more uniformity in cluster

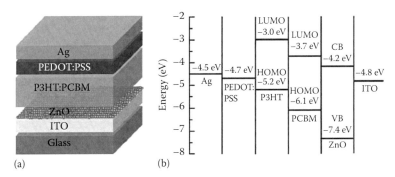

(a)   (b)

FIGURE 2.8   (a) Schematic showing the device structure using sol-gel zinc oxide as buffer layer and (b) corresponding band energy diagram of solar cell. (From Liang, Z., Zhang, Q., Wiranwetchayan, O. et al.: Effects of the morphology of a ZnO buffer layer on the photovoltaic performance of inverted polymer solar cells. *Adv. Funct. Mater.* 2012. 22(10). 2194–2201. Copyright Wiley-VCH Verlag GmbH & Co. KGaA. Reprinted with permission.)

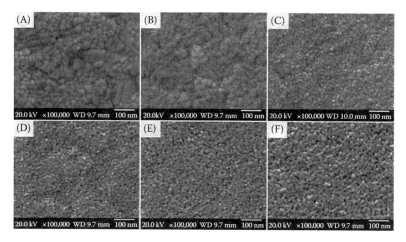

FIGURE 2.8 (continued)   (A), (B), (C), (D), (E) and (F) are the SEM images of zinc oxide with 0.02 M, 0.05 M, 0.1 M, 0.3 M, 0.6 M and 1 M concentration, respectively. (From Liang, Z., Zhang, Q., Wiranwetchayan, O. et al.: Effects of the morphology of a ZnO buffer layer on the photovoltaic performance of inverted polymer solar cells. *Adv. Funct. Mater.* 2012. 22(10). 2194–2201. Copyright Wiley-VCH Verlag GmbH & Co. KGaA. Reprinted with permission.)

size due to the absence of agglomeration as their growth is mediated by PVP matrix. However, the PVP component leads to a contact barrier between PC70BM and ZnO. This could be avoided by using UV-ozone treatment. Effect of UV-ozone treatment time was studied on solar cell efficiency and found that 10 min ozone treatment was the optimum as shown in Figure 2.9.

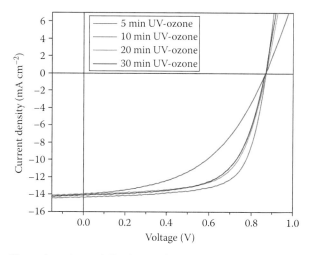

FIGURE 2.9   **(See colour insert.)** Device performance of inverted PDTG-TPD:PC71BM solar cells with UV-ozone-treated ZnO-PVP nanocomposite films as buffer layer for various treatment times (5, 10, 20 and 30 min). (Reprinted with permission from Macmillan Publishers Ltd., *Nat. Photon.*, Small, C.E., Chen, S., Subbiah, J. et al., High-efficiency inverted dithieno-germole-thienopyrrolodione-based polymer solar cells 6(2), 115–120, 2012, Copyright 2012.)

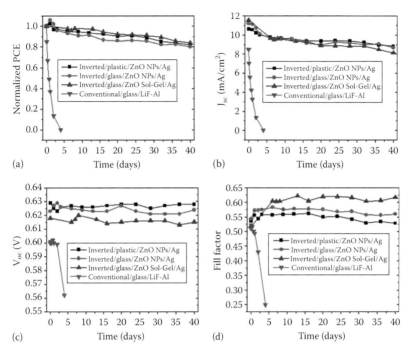

FIGURE 2.10   (See colour insert.) Variation in (a) power conversion efficiency (PCE), (b) short-circuit current density (J_{sc}), (c) open-circuit voltage (V_{oc}) and (d) FF with days in air for conventional and inverted solar cells with different zinc oxide–processed buffer layers on glass and plastic substrates. (Reprinted with permission from Hau, S.K., Yip, H.-L., Baek, N.S. et al., Air-stable inverted flexible polymer solar cells using zinc oxide nanoparticles as an electron selective layer, *Appl. Phys. Lett.*, 92(25): 253301, 2008a. Copyright 2008, American Institute of Physics.)

Stability of polymer solar cells in air with ZnO buffer layer was much higher than conventional device structures. Hau et al. (2008a) showed that unencapsulated inverted solar cells with ZnO nanoparticles on either plastic or glass substrates retained more than 80% of their initial efficiency after 40 days of fabrication (Figure 2.10a). Such results show promising application in low-temperature roll-to-roll printed flexible solar cells.

2.5.2   INTERLAYERS

Interfaces in polymer solar cells are crucial as they affect charge extraction and recombination. Interfaces could be between active layer and the cathode or anode. In conventional structure, PEDOT:PSS acts as electron-blocking layer between active layer and ITO, while a low work function metal is used as hole-blocking layer between PCBM and Al interface and secondly increases the internal electric field and hence the V_{oc} (Mihailetchi et al. 2003). Reese et al. (2008) studied the effects of various low work function metals (Table 2.1) on device performance. Ba/Al and

TABLE 2.1

Variation in Device Performance Parameters of P3HT:PCBM Solar Cells with Different Cathodes along with Their Work Function

Electrode	Work Function (eV)	V_{oc} (meV)	J_{sc} (mA/cm²)	FF (%)	Normalized Efficiency
Ag	4.26	403 ± 27	9.04 ± 1.16	36.6 ± 3.7	0.36 ± 0.08
Al	4.28	420 ± 15	10.02 ± 0.33	38.2 ± 2.6	0.41 ± 0.03
Mg:Ag/Ag	3.66	565 ± 6	10.30 ± 0.39	50.5 ± 2.8	0.75 ± 0.06
LiF/Al	2.9	580 ± 2	10.42 ± 0.22	57.3 ± 1.2	0.89 ± 0.03
Ca/Al	2.87	601 ± 2	10.42 ± 0.18	60.6 ± 2.1	0.96 ± 0.04
Ba/Al	2.7	600 ± 2	10.47 ± 0.36	62.2 ± 1.5	1.00 ± 0.04

Source: Reese, M.O. et al., *Appl. Phys. Lett.*, 92, 053307, 2008.

Note: The work function values listed here are used from literature reports.

Ca/Al were proved to be the best cathode. Although Ca-/Al-based solar cells show slightly lower efficiency than Ba/Al, their stability tested over a 6-week period was higher than Ba/Al electrode. Barium is more reactive than calcium and needs to be stored in oil to prevent any oxidation or reaction. Hence, most polymer solar cells use Ca/Al as cathode.

Another approach to improve charge extraction in polymer solar cells is to use titanium suboxide (TiOx) as an interfacial layer between active layer and Al. The TiOx layer significantly increased incident photon-to-electron efficiency (IPCE) due to two reasons: (1) suppress back recombination from Al to PCBM because it is an n-type oxide having deep valence band (~8.1 eV [Kim et al. 2006b]) and (2) act as an optical spacer to increase absorption because it spatially redistributes light intensity within the device (Figure 2.11).

Several n-type metal oxides such as ZnO (White et al. 2006; Gilot et al. 2007; Yang et al. 2010; Sun et al. 2011), Nb_2O_5 (Orawan et al. 2011) and $Cs:TiO_2$ (Sista et al. 2010) and p-type metal oxides such as MoO_3 (Liu et al. 2010; Girotto et al. 2011; Sun et al. 2011), V_2O_5 (Shrotriya et al. 2006) and WO_3 (Han et al. 2009) were used as hole- and electron-blocking layers, respectively, in both conventional and inverted polymer solar cells.

Hau et al. (2008) used a self-assembled monolayer (SAM) of C60 on sol-gel zinc oxide layer, which led to enhanced electronic coupling between zinc oxide and polymer with higher current density and FF in P3HT:PCBM inverted solar cells. Later on they studied the effect of chemical functionalization of fullerenes and used modified fullerenes in similar SAM structure on ZnO thin layer for inverted P3HT:PCBM devices (Hau et al. 2010). The functionalized C60-SAMs resulted in improved interface dipole moment in which carboxylic-acid-based C60-SAMs showed the highest enhancement. Another way is to modify ITO chemically to tune its work function and improve interfacial dipole moment. Armstrong et al. performed chemical modification of ITO and studied its surface composition when activated by plasma cleaning, solvent cleaning and etching with strong haloacids (Brumbach et al. 2007;

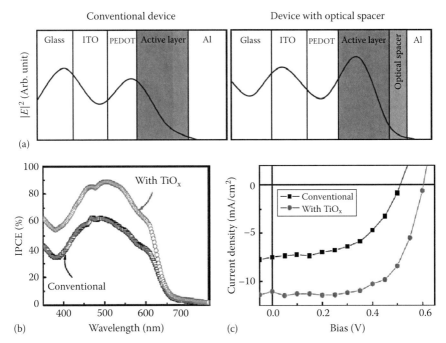

FIGURE 2.11 (a) Optical electrical field within a polymer solar cell device with and without optical spacer, (b) IPCE spectra of devices with and without the TiO$_x$ layer and (c) current–voltage curve of cells with and without TiO$_x$ layer. (From Kim, J.Y., Kim, S.H., Lee, H.H. et al.: New architecture for high-efficiency polymer photovoltaic cells using solution-based titanium oxide as an optical spacer. *Adv. Mater.* 2006. 18(5). 572–576. Copyright Wiley-VCH Verlag GmbH & Co. KGaA. Reprinted with permission.)

Armstrong et al. 2009). ITO etched with hydrochloric and hydroiodic acid showed the highest nanoscale conductivity and thus served as an ideal electrode for electrochemical deposition of P3HT. Irwin et al. (2009) studied the effects of HCl treatment and UV-ozone treatment on P3HT:PCBM solar cell performance. An optimal time of 20 min for HCl treatment and 10 min for UV-ozone treatment exhibited the highest device efficiency. Sun et al. (2011) reported sodium-based compounds as another surface modification technique to tune ITO work function. ITO work function was reduced by ~1 eV using NaOH due to the interaction between anions of sodium compounds and cations (e.g. Sn4+ and In3+) of ITO.

Recently, many groups (Luo et al. 2009; He et al. 2010, 2011; Seo et al. 2011) have reported use of conjugated polyelectrolytes (CPEs) as an interfacial layer in polymer photovoltaic devices due to ease of room temperature solution processing from common solvents, introduction of interfacial dipole layer and elimination of use of low work function metals. Seo et al. (Seo et al. 2011) reported that simple solution-processed deposition of PT-based CPE from methanol solution led to significant increase in current density and FF as well as small improvement in open-circuit voltage for PCDTBT:PC71BM solar cells. He et al. (He et al. 2011b) used alcohol- and water-soluble CPE, poly [(9,9-bis(3'-(N,N-dimethylamino) propyl)-2,7-fluorene)-alt-2,7-(9,9-dioctylfluorene)] (PFN), as an interlayer that led to substantial

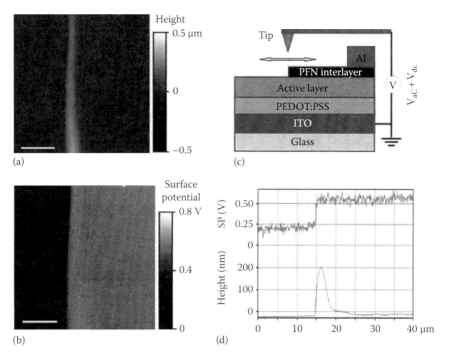

FIGURE 2.12   **(See colour insert.)** (a) Tapping mode AFM image and (b) KPFM image of active layer partially covered with PFN interlayer as shown in the (c) schematic of experimental setup and (d) cross-sectional profile of topography and surface potential measured using (a) and (b). (He, Z., Zhong, C., Huang, X. et al.: Simultaneous enhancement of open-circuit voltage, short-circuit current density, and fill factor in polymer solar cells. *Adv. Mater.* 2011. 23. 4636–4643. Copyright Wiley-VCH Verlag GmbH & Co. KGaA. Reprinted with permission.)

improvement in device performance in poly[[4,8-bis[(2-ethylhexyl)oxy]benzo[1,2-b:4,5-b′]dithiophene-2,6-diyl][3-fluoro-2-[(2-ethylhexyl)carbonyl]thieno[3,4-b]thio-phenediyl]](PTB7) and PC71BM bulk heterojunction solar cells leading to certified 8.37% efficiency. This improvement was attributed to enhancement in built-in potential across the device due to interface dipole formation and suppressed back recombination of charges from the electrode to active layer. The interfacial dipole moment was measured using Kelvin probe force microscopy (KPFM) across the active and PFN interlayer interface as seen in Figure 2.12 as an increase of 300 meV in contact potential difference between the tip and active layer and PFN interface.

2.5.3 Tandem Solar Cell Structure

Unlike inorganic semiconductors that absorb any photon having energy higher than bandgap, organic semiconductors only have a band of absorption that results in some solar spectrum not being absorbed. Also due to their low mobility, polymer thin films are kept fairly thin, which results in lower absorption efficiency. Thus, the absorption loss is significantly higher in polymer single-junction solar cells.

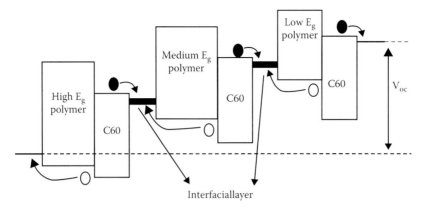

FIGURE 2.13   Energy-level diagram of a triple-junction polymer solar cell in a series connection under the open-circuit voltage condition showing how the electrons in one subcell and holes in its adjacent subcell will diffuse to the middle contact (interfacial) layer to recombine. (Reprinted from Siddiki, M.K. et al., *Energy Environ. Sci.*, 3(7), 867, 2010. With permission.)

The photons higher than the bandgap of polymers lead to hot carrier and give rise to another loss mechanism called thermalization loss. One of the obvious solutions to this major loss mechanism is to use two or more donor materials having bandgaps with complementary absorption to fabricate series double or multijunction solar cells. In a double or multijunction cell, each subcell is separated by an interfacial layer where electrons and holes from the top and bottom subcells tunnel or recombine (Peumans et al. 2003) (Figure 2.13). The overall J_{sc} is limited by the subcell having the lowest J_{sc}, but the overall V_{oc} is the sum of that of each subcell. Siddiki et al. (2010) did a comprehensive review on double or multijunction solar cells and showed the dependency of cell performance (V_{oc}, J_{sc} and η) on the absorption of back subcell for a double, triple and quadruple junction, respectively (Figure 2.14). V_{oc} decreased and J_{sc} increased as the optical back cell's light absorption spectra were broadened to longer wavelength regions. One needs to carefully select materials for each subcell to strike balance between J_{sc} and V_{oc} to achieve the highest possible efficiency.

Selection and identification of recombination layers are important in double and multijunction solar cells. The recombination layers need to be transparent (or semi-transparent) and charge-selective and have good adhesion and wettability with the films below and above. In 2007, Kim et al. (2007) reported the first successful high-efficiency double-junction polymer tandem cells using TiOx and PEDOT:PSS as recombination layers. The TiOx thin layer, deposited by sol-gel process (Kim et al. 2006b), served as electron transport layer and protected the hydrophobic front cell from the hole transport layer (PEDOT:PSS) that was casted from the water solvent. P3HT:PC70BM and PCPDTBT:PCBM (Figure 2.15) were used as top and bottom subcells. Several other materials such as TiO2/Al/PEDOT (Sista et al. 2010), Au/Sm/PEDOT:PSS (Hadipour et al. 2007), Nb2O5/PEDOT:PSS (Siddiki et al. 2012) and ZnO/PEDOT:PSS (Gilot et al. 2010; Moet et al. 2010; Gevaerts et al. 2012; Kouijzer et al. 2012) have also been tried by other researchers.

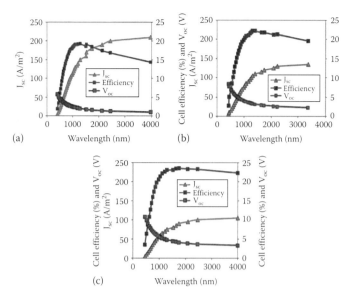

FIGURE 2.14   Theoretical dependence of cell efficiency, open-circuit voltage (V_{oc}) and short-circuit current density (J_{sc}) plotted based on the bandgap of back cell in a (a) double-junction, (b) triple-junction and (c) quadruple-junction solar cell. (Reprinted from Siddiki, M.K. et al., *Energy Environ. Sci.*, 3(7), 867, 2010. With permission.)

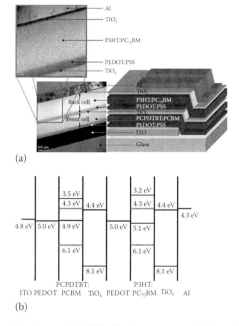

FIGURE 2.15   **(See colour insert.)** (a) The device structure (right) and TEM cross-sectional image (left) of the polymer tandem solar cell. (b) Energy-level diagram showing the HOMO and LUMO energies of each of the component materials. (Reprinted from Kim, J.Y. et al., *Science*, 317(5835), 222–225, 2007. With permission from Science.)

Inverted tandem cells consisting of P3HT:PCBM and PSBTBT:PC70BM as front and back cells, respectively, were investigated by Yang et al. (Chou et al. 2011). The cells separated by MoO_3 (10 nm)/Al (1 nm)/ZnO (30 nm) recombination layers achieved 5.1% efficiency with V_{oc} of 1.2 V. In this structure, ITO served as cathode and MoO_3/Al as cathode. As mentioned earlier in Section 5.1, inverted structure employing transparent metal oxides not only provides improved stability against oxygen and moisture but also prevents bottom cells from dissolution while coating top cells. Then in 2012 they published another report on double-junction polymer solar cells with efficiency of 8.62% (Dou et al. 2012). P3HT:ICBA was utilized as bottom cell and a low-bandgap polymer PBDTT-DPP (Eg = 1.3 eV) was used as top cell (Figure 2.16). ZnO and modified PEDOT:PSS were used as recombination layers. Modified PEDOT:PSS was made by mixing a stock solution of polystyrene sulphonate and commercial PEDOT:PSS in 16:1 volume ratio leading to a 90 nm thick interlayer. A cell efficiency of 8.62% certified by the National Renewable

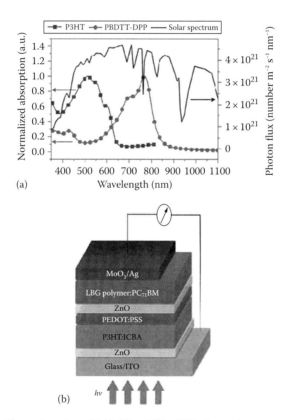

FIGURE 2.16 **(See colour insert.)** (a) UV–visible–NIR absorption spectra of PBDTT-DPP and P3HT thin films along with the AM 1.5 spectra; (b) schematic of the inverted tandem device structure. (Reprinted with permission from Macmillan Publishers Ltd., *Nat. Photon.*, Dou, L., You, J., Yang, J. et al., Tandem polymer solar cells featuring a spectrally matched low-bandgap polymer, 6(3), 180–185, Copyright 2012.)

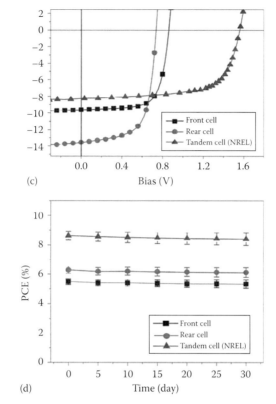

(c)

(d)

FIGURE 2.16 (continued) **(See colour insert.)** (c) current density–voltage characteristics of front cell, rear cell and inverted tandem solar cell, tested at NREL; and (d) variation of power conversion efficiency of cells with days of storage. (Reprinted with permission from Macmillan Publishers Ltd., *Nat. Photon.*, Dou, L., You, J., Yang, J. et al., Tandem polymer solar cells featuring a spectrally matched low-bandgap polymer, 6(3), 180–185, Copyright 2012.)

Energy Laboratory (NREL) was achieved. The cells retained more than 95% of their original efficiency after 30 days of storage in nitrogen environment. Yang group then worked with Sumitomo chemicals on an undisclosed polymer in tandem cell and obtained an efficiency of 10.6% certified by the NREL (Chemicals 2012).

2.6 CONCLUSIONS

Overall there has been a significant progress in polymer solar cells since the advent of semiconducting polymers for photovoltaic application. In this chapter, a variety of parameters that affect active layer properties and device performance were reviewed. The role and importance of post-deposition processing was shown with emphasis on the structural changes that lead to higher photon-to-electron conversion efficiency. Also the role of charge-selective buffer layers to attain higher power conversion efficiency and lifetime was discussed. Despite their merits of low-cost and simple fabrication methods, single-junction polymer solar cells still suffer from low efficiency and stability. To overcome this issue, tandem solar cells were reviewed and were shown to reduce thermalization and absorption loss. It was shown that using careful selection of materials, interfacial layers and optimum device processing efficiencies, higher efficiency could be achieved. Such theoretical predictions weave the path for researchers to better understand the physics and chemistry of organic photovoltaic materials and devices for future low-cost optoelectronic devices.

REFERENCES

Aernouts, T., T. Aleksandrov, C. Girotto, J. Genoe, and J. Poortmans (2008). Polymer based organic solar cells using ink-jet printed active layers. *Applied Physics Letters* 92(3): 033306–033303.

Armstrong, N. R., P. A. Veneman, E. Ratcliff, D. Placencia, and M. Brumbach (2009). Oxide contacts in organic photovoltaics: Characterization and control of near-surface composition in Indium–Tin oxide (ITO) electrodes. *Accounts of Chemical Research* 42(11): 1748–1757.

Brumbach, M., P. A. Veneman, F. S. Marrikar et al. (2007). Surface composition and electrical and electrochemical properties of freshly deposited and acid-etched Indium Tin oxide electrodes. *Langmuir* 23(22): 11089–11099.

Campoy-Quiles, M., T. Ferenczi, T. Agostinelli et al. (2008). Morphology evolution via self-organization and lateral and vertical diffusion in polymer:fullerene solar cell blends. *Nature Materials* 7(2): 158–164.

Chang, L., H. W. A. Lademann, J.-B. Bonekamp, K. Meerholz, and A. J. Moulé (2011). Effect of trace solvent on the morphology of P3HT:PCBM bulk heterojunction solar cells. *Advanced Functional Materials* 21(10): 1779–1787.

Chemicals, S. (2012). Material from sumitomo chemical employed in UCLA's tandem polymer photovoltaic cell achieving power conversion efficiency of 10.6%, 2012, from http://www.sumitomo-chem.co.jp/english/newsreleases/docs/20120214e.pdf. (accessed on February 14, 2012.)

Cheng, Y.-J., F.-Y. Cao, W.-C. Lin, C.-H. Chen, and C.-H. Hsieh (2011). Self-assembled and cross-linked fullerene interlayer on titanium oxide for highly efficient inverted polymer solar cells. *Chemistry of Materials* 23(6): 1512–1518.

Cheng, Y.-J., C.-H. Hsieh, Y. He, C.-S. Hsu, and Y. Li (2010). Combination of indene-C60 bis-adduct and cross-linked fullerene interlayer leading to highly efficient inverted polymer solar cells. *Journal of the American Chemical Society* 132(49): 17381–17383.

Chou, C.-H., W. L. Kwan, Z. Hong, L.-M. Chen, and Y. Yang (2011). A metal-oxide interconnection layer for polymer tandem solar cells with an inverted architecture. *Advanced Materials* 23(10): 1282–1286.

Chu, T.-Y., S. Alem, S.-W. Tsang et al. (2011). Morphology control in polycarbazole based bulk heterojunction solar cells and its impact on device performance. *Applied Physics Letters* 98(25): 253301.

Dou, L., J. You, J. Yang et al. (2012). Tandem polymer solar cells featuring a spectrally matched low-bandgap polymer. *Nature Photonics* 6(3): 180–185.

Fan, X., C. Cui, G. Fang et al. (2012). Efficient polymer solar cells based on poly(3-hexylthiophene): Indene-C70 bisadduct with a MoO3 buffer layer. *Advanced Functional Materials* 22(3): 585–590.

Gevaerts, V. S., A. Furlan, M. M. Wienk, M. Turbiez, and R. A. J. Janssen (2012). Solution processed polymer tandem solar cell using efficient small and wide bandgap polymer:fullerene blends. *Advanced Materials* 24(16): 2130–2134.

Gilot, J., I. Barbu, M. M. Wienk, and R. A. J. Janssen (2007). The use of ZnO as optical spacer in polymer solar cells: Theoretical and experimental study. *Applied Physics Letters* 91(11): 113520.

Gilot, J., M. M. Wienk, and R. A. J. Janssen (2010). Optimizing polymer tandem solar cells. *Advanced Materials* 22(8): E67–E71.

Girotto, C., E. Voroshazi, D. Cheyns, P. Heremans, and B. P. Rand (2011). Solution-processed MoO₃ thin films as a hole-injection layer for organic solar cells. *ACS Applied Materials & Interfaces* 3(9): 3244–3247.

Green, R., A. Morfa, A. J. Ferguson et al. (2008). Performance of bulk heterojunction photovoltaic devices prepared by airbrush spray deposition. *Applied Physics Letters* 92(3): 033301.

Hadipour, A., B. de Boer, and P. W. M. Blom (2007). Solution-processed organic tandem solar cells with embedded optical spacers. *Journal of Applied Physics* 102(7): 074506.

Han, S., W. S. Shin, M. Seo et al. (2009). Improving performance of organic solar cells using amorphous tungsten oxides as an interfacial buffer layer on transparent anodes. *Organic Electronics* 10(5): 791–797.

Hau, S. K., Y.-J. Cheng, H.-L. Yip et al. (2010). Effect of chemical modification of fullerene-based self-assembled monolayers on the performance of inverted polymer solar cells. *ACS Applied Materials & Interfaces* 2(7): 1892–1902.

Hau, S. K., H.-L. Yip, N. S. Baek et al. (2008a). Air-stable inverted flexible polymer solar cells using zinc oxide nanoparticles as an electron selective layer. *Applied Physics Letters* 92(25): 253301.

Hau, S. K., H.-L. Yip, H. Ma, and A. K.-Y. Jen (2008b). High performance ambient processed inverted polymer solar cells through interfacial modification with a fullerene self-assembled monolayer. *Applied Physics Letters* 93(23): 233304.

He, Y., H.-Y. Chen, J. Hou, and Y. Li (2010). Indene–C60 bisadduct: A new acceptor for high-performance polymer solar cells. *Journal of the American Chemical Society* 132(4): 1377–1382.

He, Y. and Y. Li (2011). Fullerene derivative acceptors for high performance polymer solar cells. *Physical Chemistry Chemical Physics* 13(6): 1970–1983.

He, Z., C. Zhong, X. Huang et al. (2011). Simultaneous enhancement of open-circuit voltage, short-circuit current density, and fill factor in polymer solar cells. *Advanced Materials* 23(40): 4636–4643.

He, C., C. Zhong, H. Wu et al. (2010). Origin of the enhanced open-circuit voltage in polymer solar cells via interfacial modification using conjugated polyelectrolytes. *Journal of Materials Chemistry* 20(13): 2617–2622.

Heo, Y. W., D. P. Norton, L. C. Tien et al. (2004). ZnO nanowire growth and devices. *Materials Science and Engineering: R: Reports* 47(1–2): 1–47.

Hoppe, H., M. Niggemann, C. Winder et al. (2004). Nanoscale morphology of conjugated polymer/fullerene-based bulk-heterojunction solar cells. *Advanced Functional Materials* 14(10): 1005–1011.

Hoth, C. N., S. A. Choulis, P. Schilinsky, and C. J. Brabec (2007). High photovoltaic performance of inkjet printed polymer:fullerene blends. *Advanced Materials* 19(22): 3973–3978.

Hoven, C. V., X.-D. Dang, R. C. Coffin et al. (2010). Improved performance of polymer bulk heterojunction solar cells through the reduction of phase separation via solvent additives. *Advanced Materials* 22(8): E63–E66.

Irwin, M. D., J. Liu, B. J. Leever et al. (2009). Consequences of anode interfacial layer deletion. HCl-treated ITO in P3HT:PCBM-based bulk-heterojunction organic photovoltaic devices. *Langmuir* 26(4): 2584–2591.

Jørgensen, M., K. Norrman, S. A. Gevorgyan et al. (2012). Stability of polymer solar cells. *Advanced Materials* 24(5): 580–612.

Kang, J.-W., Y.-J. Kang, S. Jung et al. (2012). Fully spray-coated inverted organic solar cells. *Solar Energy Materials and Solar Cells* 103(0): 76–79.

Kim, Y., S. A. Choulis, J. Nelson et al. (2005). Device annealing effect in organic solar cells with blends of regioregular poly(3-hexylthiophene) and soluble fullerene. *Applied Physics Letters* 86(6): 063502.

Kim, Y., S. Cook, S. M. Tuladhar et al. (2006a). A strong regioregularity effect in self-organizing conjugated polymer films and high-efficiency polythiophene:fullerene solar cells. *Nature Materials* 5(3): 197–203.

Kim, J. Y., S. H. Kim, H. H. Lee et al. (2006b). New architecture for high-efficiency polymer photovoltaic cells using solution-based titanium oxide as an optical spacer. *Advanced Materials* 18(5): 572–576.

Kim, J. Y., K. Lee, N. E. Coates et al. (2007). Efficient tandem polymer solar cells fabricated by all-solution processing. *Science* 317(5835): 222–225.

Kouijzer, S., S. Esiner, C. H. Frijters et al. (2012). Efficient inverted tandem polymer solar cells with a solution-processed recombination layer. *Advanced Energy Materials*, 2, 945–949.

Krebs, F. C., S. A. Gevorgyan, and J. Alstrup (2009a). A roll-to-roll process to flexible polymer solar cells: Model studies, manufacture and operational stability studies. *Journal of Materials Chemistry* 19(30): 5442–5451.

Krebs, F. C., M. Jørgensen, K. Norrman et al. (2009b). A complete process for production of flexible large area polymer solar cells entirely using screen printing – First public demonstration. *Solar Energy Materials and Solar Cells* 93(4): 422–441.

Kyaw, A. K. K., Y. Wang, D. W. Zhao et al. (2011). The properties of sol–gel processed indium-doped zinc oxide semiconductor film and its application in organic solar cells. *Physica Status Solidi (A)* 208(11): 2635–2642.

Li, G., V. Shrotriya, J. Huang et al. (2005). High-efficiency solution processable polymer photovoltaic cells by self-organization of polymer blends. *Nature Materials* 4(11): 864–868.

Li, Y., M. Wang, H. Huang et al. (2011). Influence on open-circuit voltage by optical heterogeneity in three-dimensional organic photovoltaics. *Physical Review B* 84(8): 085206.

Li, G., Y. Yao, H. Yang et al. (2007). Solvent annealing effect in polymer solar cells based on poly(3-hexylthiophene) and methanofullerenes. *Advanced Functional Materials* 17(10): 1636–1644.

Liang, Z., Q. Zhang, O. Wiranwetchayan et al. (2012). Effects of the morphology of a ZnO buffer layer on the photovoltaic performance of inverted polymer solar cells. *Advanced Functional Materials* 22(10): 2194–2201.

Liu, J.-P., K.-L. Choy, and X.-H. Hou (2011). Charge transport in flexible solar cells based on conjugated polymer and ZnO nanoparticulate thin films. *Journal of Materials Chemistry* 21(6): 1966–1969.

Liu, F., S. Shao, X. Guo, Y. Zhao, and Z. Xie (2010). Efficient polymer photovoltaic cells using solution-processed MoO3 as anode buffer layer. *Solar Energy Materials and Solar Cells* 94(5): 842–845.

Luo, J., H. Wu, C. He et al. (2009). Enhanced open-circuit voltage in polymer solar cells. *Applied Physics Letters* 95(4): 043301–043303.

Manceau, M., E. Bundgaard, J. E. Carle et al. (2011). Photochemical stability of [small pi]-conjugated polymers for polymer solar cells: A rule of thumb. *Journal of Materials Chemistry* 21(12): 4132–4141.

Mihailetchi, V. D., P. W. M. Blom, J. C. Hummelen, and M. T. Rispens (2003). Cathode dependence of the open-circuit voltage of polymer:fullerene bulk heterojunction solar cells. *Journal of Applied Physics* 94(10): 6849–6854.

Mikroyannidis, J. A., A. N. Kabanakis, S. S. Sharma, and G. D. Sharma (2011). A simple and effective modification of PCBM for use as an electron acceptor in efficient bulk heterojunction solar cells. *Advanced Functional Materials* 21(4): 746–755.

Moet, D. J. D., P. de Bruyn, and P. W. M. Blom (2010). High work function transparent middle electrode for organic tandem solar cells. *Applied Physics Letters* 96(15): 153504.

Peet, J., A. J. Heeger, and G. C. Bazan (2009). "Plastic" solar cells: Self-assembly of bulk heterojunction nanomaterials by spontaneous phase separation. *Accounts of Chemical Research* 42(11): 1700–1708.

Peet, J., J. Y. Kim, N. E. Coates et al. (2007). Efficiency enhancement in low-bandgap polymer solar cells by processing with alkane dithiols. *Nature Materials* 6(7): 497–500.

Peet, J., C. Soci, R. C. Coffin et al. (2006). Method for increasing the photoconductive response in conjugated polymer/fullerene composites. *Applied Physics Letters* 89(25): 252105.

Peumans, P., A. Yakimov, and S. R. Forrest (2003). Small molecular weight organic thin-film photodetectors and solar cells. *Journal of Applied Physics* 93(7): 3693–3723.

Reese, M. O., M. S. White, G. Rumbles, D. S. Ginley, and S. E. Shaheen (2008). Optimal negative electrodes for poly(3-hexylthiophene): [6,6]-phenyl C61-butyric acid methyl ester bulk heterojunction photovoltaic devices. *Applied Physics Letters* 92(5): 053307.

Reyes-Reyes, M., K. Kim, and D. L. Carroll (2005). High-efficiency photovoltaic devices based on annealed poly(3-hexylthiophene) and 1-(3-methoxycarbonyl)-propyl-1-phenyl-(6,6)C[sub 61] blends. *Applied Physics Letters* 87(8): 083506.

Schilinsky, P., C. Waldauf, and C. J. Brabec (2006). Performance analysis of printed bulk heterojunction solar cells. *Advanced Functional Materials* 16(13): 1669–1672.

Seo, J. H., A. Gutacker, Y. Sun et al. (2011). Improved high-efficiency organic solar cells via incorporation of a conjugated polyelectrolyte interlayer. *Journal of the American Chemical Society* 133(22): 8416–8419.

Shrotriya, V., G. Li, Y. Yao, C.-W. Chu, and Y. Yang (2006). Transition metal oxides as the buffer layer for polymer photovoltaic cells. *Applied Physics Letters* 88(7): 073508.

Siddiki, M. K., J. Li, D. Galipeau, and Q. Qiao (2010). A review of polymer multijunction solar cells. *Energy & Environmental Science* 3(7): 867–883.

Siddiki, M. K., S. Venkatesan, and Q. Qiao (2012). Nb2O5 as a new electron transport layer for double junction polymer solar cells. *Physical Chemistry Chemical Physics* 14(14): 4682–4686.

Sista, S., M.-H. Park, Z. Hong et al. (2010). Highly efficient tandem polymer photovoltaic cells. *Advanced Materials* 22(3): 380–383.

Small, C. E., S. Chen, J. Subbiah et al. (2012). High-efficiency inverted dithienogermole-thienopyrrolodione-based polymer solar cells. *Nature Photonics* 6(2): 115–120.

Son, H. J., W. Wang, T. Xu et al. (2011). Synthesis of fluorinated polythienothiophene-co-benzodithiophenes and effect of fluorination on the photovoltaic properties. *Journal of the American Chemical Society* 133(6): 1885–1894.

Søndergaard, R., M. Hösel, D. Angmo, T. T. Larsen-Olsen, and F. C. Krebs (2012). Roll-to-roll fabrication of polymer solar cells. *Materials Today* 15(1–2): 36–49.

Sun, Y., C. Cui, H. Wang, and Y. Li (2011). Efficiency enhancement of polymer solar cells based on poly(3-hexylthiophene)/indene-C70 bisadduct via methylthiophene additive. *Advanced Energy Materials* 1(6): 1058–1061.

Sun, Y., J. H. Seo, C. J. Takacs, J. Seifter, and A. J. Heeger (2011). Inverted polymer solar cells integrated with a low-temperature-annealed sol-gel-derived ZnO film as an electron transport layer. *Advanced Materials* 23(14): 1679–1683.

Sun, Y., C. J. Takacs, S. R. Cowan et al. (2011). Efficient, air-stable bulk heterojunction polymer solar cells using MoOx as the anode interfacial layer. *Advanced Materials* 23(19): 2226–2230.

Sun, K., H. Zhang, and J. Ouyang (2011). Indium tin oxide modified with sodium compounds as cathode of inverted polymer solar cells. *Journal of Materials Chemistry* 21(45): 18339–18346.

Vak, D., S.-S. Kim, J. Jo et al. (2007). Fabrication of organic bulk heterojunction solar cells by a spray deposition method for low-cost power generation. *Applied Physics Letters* 91(8): 081102.

White, M. S., D. C. Olson, S. E. Shaheen, N. Kopidakis, and D. S. Ginley (2006). Inverted bulk-heterojunction organic photovoltaic device using a solution-derived ZnO underlayer. *Applied Physics Letters* 89(14): 143517.

Wienk, M. M., J. M. Kroon, W. J. H. Verhees et al. (2003). Efficient methano[70]fullerene/MDMO-PPV bulk heterojunction photovoltaic cells. *Angewandte Chemie* 115(29): 3493–3497.

Wiranwetchayan, O., Z. Liang, Q. Zhang, G. Cao, and P. Singjai (2011). The role of oxide thin layer in inverted structure polymer solar cells. *Materials Sciences and Applications* 2(12): 5.

Yang, T., W. Cai, D. Qin et al. (2010). Solution-processed zinc oxide thin film as a buffer layer for polymer solar cells with an inverted device structure. *The Journal of Physical Chemistry C* 114(14): 6849–6853.

Yang, X., J. Loos, S. C. Veenstra et al. (2005). Nanoscale morphology of high-performance polymer solar cells. *Nano Letters* 5(4): 579–583.

Yao, Y., C. Shi, G. Li et al. (2006). Effects of C[sub 70] derivative in low band gap polymer photovoltaic devices: Spectral complementation and morphology optimization. *Applied Physics Letters* 89(15): 153507.

Zeng, L., C. W. Tang, and S. H. Chen (2010). Effects of active layer thickness and thermal annealing on polythiophene: Fullerene bulk heterojunction photovoltaic devices. *Applied Physics Letters* 97(5): 053305.

Zhang, F., Z. Zhuo, J. Zhang et al. (2012). Influence of PC60BM or PC70BM as electron acceptor on the performance of polymer solar cells. *Solar Energy Materials and Solar Cells* 97(0): 71–77.

Zhao, G., Y. He, and Y. Li (2010). 6.5% Efficiency of polymer solar cells based on poly(3-hexylthiophene) and indene-C60 bisadduct by device optimization. *Advanced Materials* 22(39): 4355–4358.

Zhou, Y., H. Cheun, J. W. J. Potscavage et al. (2010). Inverted organic solar cells with ITO electrodes modified with an ultrathin Al2O3 buffer layer deposited by atomic layer deposition. *Journal of Materials Chemistry* 20(29): 6189–6194.

3 Mechanically Flexible Interconnects and TSVs
Applications in CMOS/ MEMS Integration

Hyung Suk Yang, Paragkumar Thadesar, Chaoqi Zhang and Muhannad Bakir

CONTENTS

3.1 INTRODUCTION

The steady growth of the microelectromechanical systems (MEMS) industry over the last two decades has been nothing short of incredible. Currently, it is an $8B industry (as of 2010), and by the year 2015 it is projected to more than double and become a $17B industry (Eloy, 2010). Furthermore, what was a technology with a very limited number of commercial applications available for decades, namely, inkjet heads, digital light projectors (DLP) and automobile sensors systems, is now becoming ubiquitous (Marek and Gómez, 2012).

However, despite the rapid growth of the industry, it is also hard to ignore the fact that the market is dominated by a very limited number of device types and also by very few MEMS powerhouses such as STMicroelectronics and Texas Instruments. Considering the vast interest and huge potential of MEMS technology demonstrated by the universities and research labs around the world, it is disappointing to see that the commercial world has not yet encompassed the full potential of MEMS device technology; emerging MEMS technology represents only 10% of MEMS market, and even those 10% is related to the reused existing devices in new applications or packaging or integration of the existing devices in a new way (*MEMS Market Overview*, 2010).

What is preventing new MEMS device technologies from being commercialized? One possible answer is the *cost*. MEMS devices, especially ones that require the use of novel materials or unconventional processes, are extremely costly to turn into a commercial product due to what Yole Development call *MEMS law* – 'One product, one process, one package.' (*MEMS Market Overview*, 2010) This *MEMS law* refers to the observed trend that fabrication processes and packages needed by MEMS devices are so unique to those devices that both the fabrication processes and packages cannot be standardized, and therefore, both need to be custom designed for each unique product. Compared to the microelectronics industry where many small successful fabless companies exist, taking advantage of dedicated foundry like TSMC to handle fabrication and packaging needs, many of the MEMS companies require a significant initial investment.

This makes it difficult for a completely new MEMS technology or new companies to enter the market. Naturally more effort is being spent in finding new applications for existing devices and packaging existing devices in more efficient and cost-effective manner as noted previously.

However, there are emerging integration and packaging technologies for MEMS that aim to address this issue in the market. These technologies aim to create a generic integration scheme and packaging platform that can be used by a wide range of MEMS (and sensors) devices without a significant modification or engineering.

Specifically, by leveraging new advances in flexible I/O technologies and through-silicon vertical interconnect access (via) (TSV) technologies, one can create a generic integration platform for state-of-the-art complementary metal oxide semiconductor (CMOS) and arbitrary MEMS devices.

3.2   NEED FOR INTEGRATION OF MEMS AND CIRCUITRY

In order for a system incorporating MEMS devices to operate, the devices must be connected to a read-out (sensors) or a driving circuitry (actuators). This is because signals from the MEMS transducer devices produce very small signals and require signal conditioning, amplification and in many cases conversion to digital signals (Baltes et al., 2005).

For example, a modern capacitive accelerometer device designed by Jiangfeng et al. (2004) has a device output sensitivity of 0.6 fF/g and a linear range of +/−6 g. This means that the circuit must be capable of resolving capacitance changes in the ~10 aF range. Of course, the ability to resolve such a small signal is often a function of both circuit design and device design; fundamentally however, performance of the interconnect technology connecting the device and the circuit also plays a significant role.

Specifically, in capacitive sensing systems, parasitic capacitance of the circuit is correlated to the minimum detectable capacitance (resolution) (Yazdi et al., 2004; Seraji and Yavari, 2011). Despite the differences in the degree of sensitivity to the parasitic capacitance in various types of circuits used, it is possible to discern that an increase in the parasitic capacitance will increase the minimum detectable capacitance change in all cases, thereby resulting in the reduction of the sensitivity and overall resolution of the system (Yazdi et al., 2004). For some circuits where the effective parasitic capacitance is not reduced using techniques such as bootstrapping, it can also attenuate the signal at the input of the amplifier circuit, in turn affecting the sensitivity and resolution of the system even more. For high-frequency systems, other parasitic parameters play an important role; Joseph et al. (2008) shows that inductance of interconnects plays an important role in determining the performance of the RF system.

What is evident is that, for a system that involves MEMS or sensors, the interconnect performance plays a vital role in determining the overall performance of the system; for an integration and packaging platform capable of providing low parasitic interconnects is just as important as the performance of individual devices.

3.3 CONVENTIONAL METHOD OF INTEGRATION

For reasons stated in the previous section, the integration technologies for MEMS is a vital part of the system development that determines what type of interconnects are available for use. In modern systems, one can broadly categorize packaging and integration methods of MEMS into two categories: monolithic integration and hybrid integration. Most MEMS products currently in the market use one of these two methods, each with both disadvantages and advantages.

3.3.1 MONOLITHIC INTEGRATION

Monolithic integration is when the CMOS and MEMS are fabricated in the same silicon chip as shown in Figure 3.1. Due to the use of on-chip interconnects for connecting MEMS devices to required circuits, this method of integration has a very low parasitic capacitance; though the capacitance depends on the length of interconnects, Fedder (1998) reports on-chip interconnect capacitance of 0.017 pF and approximately 0.3 pF/mm (Krishnamoorthy et al., 2011). Also, due to the fine pitch wires available with the CMOS IC's back-end-of-the-line (BEOL) process, even a large array of MEMS/sensors can be individually interconnected. Monolithically integrated chips, which contain both CMOS and MEMS, are fabricated at the wafer level; therefore, the unit cost can also be lowered significantly.

There are three general approaches in monolithic integration of CMOS and MEMS:

1. Pre-CMOS – MEMS before CMOS
2. Intra-CMOS – MEMS between FEOL and BEOL
3. Post-CMOS – MEMS after CMOS

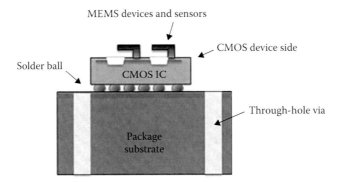

FIGURE 3.1   Monolithic integration of CMOS and MEMS.

The main differences between the three are related to when the MEMS devices are fabricated with respect to CMOS' front-end-of-the-line (FEOL) and BEOL processes. The choice of the approach used will largely depend on the material and processes required for the fabrication of MEMS devices. This is due to the fact that presence of two dissimilar technologies in the same substrate imposes many restrictions on processes and materials one can use for CMOS and MEMS processes. The restriction is often based on technical reasons including

- Different doping requirement of the Si substrate by CMOS and MEMS; both CMOS device and bulk Si etch common in MEMS are sensitive to doping type and concentration and require unconventional starting Si wafer to be used (Muller et al., 2000).
- Thermal budget available during the post-processing; long thermal process and/or high-temperature process can change the doping profile and alter device characteristics (Sedky et al., 2001). Studies like Huang et al. (2008) show an example additional effort required in trying to keep thermal processes in MEMS under certain temperature.
- Permitted material available due to process compatibility concerns; materials that introduce impurity and/or are incompatible with CMOS process, such as metals, can only be used selectively. Materials such as poly-Si are commonly used as conductors.

Sometimes the restriction is based on practical and economic reason as well; for example, a state-of-the-art CMOS foundry will be unlikely to process your wafers with MEMS already present on it due to contamination concerns. This means that without an in-house fabrication capability, post-CMOS MEMS process is the only likely option available for most MEMS designers. As a result, monolithic integration's nonrecurring cost, which includes research and development cost, is quiet high.

In summary, the monolithic integration provides low parasitic interconnections, a high-performance circuitry, low unit cost and ability to integrate a large array of MEMS/sensor devices; however, it is a complex and expensive process to develop that may impose significant restriction on the fabrication of MEMS devices. As a result, monolithic integration is used by MEMS products that require high-performance integration and by products that expect a large sale volume.

3.3.2 HYBRID INTEGRATION

On the opposite end of spectrum to monolithic integration is the hybrid integration, also known as the package-based integration. Hybrid integration refers to configurations where MEMS and CMOS are fabricated on separate chips using completely independent processes and then assembled onto the same package substrate.

This is currently the most commonly used method of integrating CMOS and MEMS (Witvrouw, 2006). This is because, unlike monolithic integration, the integration method allows arbitrary MEMS chips and state-of-the-art CMOS ICs to be integrated with relative ease; CMOS ICs and MEMS chips are fabricated independently, and as a result, both CMOS and MEMS fabrications can be done without being limited to specific materials or processes as it is the case with the monolithic integration.

Unfortunately, unlike monolithic integration where MEMS and CMOS interconnections are provided by short, low parasitic on-chip interconnects, hybrid integration requires the use of either

1. Long on-chip interconnect to redistribute signal to chip's perimeter and wire bond (Figure 3.2) or
2. Flip-chip bonding with long wires on the package substrate (Figure 3.3)

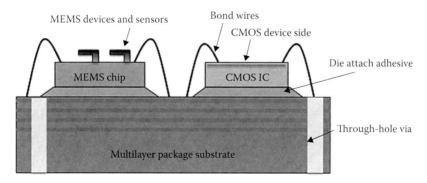

FIGURE 3.2   Hybrid integration of CMOS and MEMS using wirebonds.

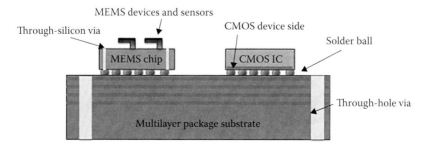

FIGURE 3.3   Hybrid integration of CMOS and MEMS using flip chip.

Both flip-chip and wire bonding suffers from the large parasitic capacitance; the main source of the capacitance is the pad capacitance, which depends on the size of the pad area. Pad size of the bond wire can be as large as 100 μm × 100 μm, and the resulting capacitance ranges from 300 to 500 fF (Krishnamoorthy and Goossen, 1998; Kisiel and Szczepanski, 2005). For a more advanced assembly process, 35 μm × 35 μm pitch pads are possible and result in 200–300 fF in pad capacitance (Agarwal and Karim, n.d.). Also, these parasitics are in addition to the on-chip wire parasitics that may be used in redistributing signals to the edge of the chip or even to the pads slightly away from the devices.

Due to the limited wire pitch possible on the package substrate and also the limited bond wire density, this method of integration makes it difficult to integrate a large array of MEMS/sensors. Also, hybrid integration cannot be done in batch, as it is possible with monolithic integration; each chip must be assembled individually to the package substrate, and if the wire bonding technology is used, individual bond wire must be interconnected one at a time using a serial process. This serial nature of the process increases the unit cost significantly.

3.3.3 Emerging Method of Integration and 3D Integration of CMOS and MEMS

Despite the availability of monolithic and hybrid integration methods in the industry however, there exists a huge segment of market that is yet to be exploited. This stems from the fact that the two integration methods force drastic compromise between performance and cost. Monolithic provides high-performance integration at the cost of material and process flexibility and integration complexity, while hybrid integration provides relatively simple integration with a significant reduction in performance.

Recently, a new method of integrating MEMS has been proposed (Yang et al., 2010; Lee et al., 2011); by leveraging the 3D integration technologies, it may be possible to provide the performance matching the monolithic integration, and at the same time allow separate fabrication of CMOS IC and MEMS, lifting the severe material and process limitations. By fabricating CMOS IC and MEMS chip independently, assembling them on top of each other and making vertical interconnections, MEMS designers are no longer restricted to a narrow process window available with monolithic integration nor the low-performance routing and redistribution wires used in a package-based integration (Figure 3.4).

However, 3D integration of CMOS and MEMS has potential to do more than just address the problems of conventional integration methods – it can provide new features. For example, as discussed in Yang and Bakir (2010) and Yang et al. (2010), separating the MEMS and CMOS IC into physically different chips makes it possible for one to discard and replace one of the chips during both the assembly process and during the lifetime of the system. For biosensor systems, where sensors may be contaminated often, this means that only sensor chip can be replaced while the CMOS IC is reused.

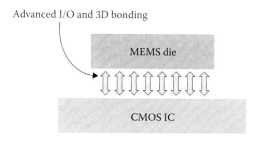

FIGURE 3.4 3D integration of CMOS and MEMS require advanced I/O and 3D bonding technologies.

The 3D integration of MEMS also provides a good opportunity to address common issues in MEMS integration – stress. By interconnecting the MEMS die with CMOS IC using an advanced interconnect technology that allows isolation of stress, it may be possible to reduce the thermo-mechanical stress experienced by the MEMS devices significantly. For example, flexible interconnects have a potential to provide such capability (Figures 3.5 and 3.6).

However, before one can leverage 3D integration for MEMS integration, many interconnect challenges must be resolved. The first is the problem of making sure that the MEMS and sensors are facing away from the package. This is important as some sensors require exposure to the environment or the sealing techniques for MEMS require that the chips face outwards. If the chip faces towards the package, the presence of underfill or die-attach adhesive may have an unwanted effect on the MEMS device. This is a challenge, because stacking two chips require that signals from one side of the chip be routed to the other side the chip – a TSV technology is needed.

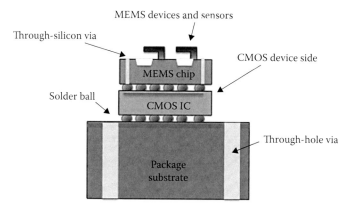

FIGURE 3.5 3D integration of CMOS and MEMS using solder ball array.

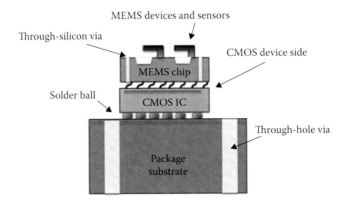

FIGURE 3.6   3D integration of CMOS and MEMS using MFI technology.

3.4   FLEXIBLE I/Os AND MFIS

The concept of flexible structures as interconnects has been around for a while. The motivation for such technology is due to the presence of the package substrate that plays a vital role in protecting MEMS chips from various contaminants and provides mechanical support; unfortunately, one side effect of having a composite structure with various materials, fabricated, assembled and operating at various temperatures, is *stress*.

Beginning with multiple generations of works called sea of leads (SoL) (Bakir et al., 2002; Dang et al., 2006), SoL aimed to mitigate thermo-mechanical stress issues mainly by providing lateral compliance and lateral range of motion, and later work also included ways to provide few micrometres of stand-off height and therefore some vertical compliance, by using a sacrificial layer process. G-helix (Lo and Sitaraman, 2004), FlexConnect (Kacker and Sitaraman, 2008, 2009) and β-helix (Zhu et al., 2003) are other examples of continued works in this area. Other examples of compliant interconnects are discussed in Fjelstad (1998), Novitsky and Pedersen (1999), Fjelstad et al. (2000), Kim et al. (2001) and Bakir and Meindl (2008).

However, these flexible interconnects have been designed to address a very specific problem present in the CMOS IC packaging; many focus on the reliability issues due to thermo-mechanical stress. However, unlike stress in MEMS where even a small amount of stress can drastically change the device characteristics, for CMOS, one was more worried about physical destruction of the IC. The mechanical characteristics of the flexible I/O for MEMS integration required were very different, and a new flexible I/O technology was needed.

Exactly how significant does the stress affect MEMS? Package-induced stress actually may have a devastating effect on the performance of MEMS device, and various methods have been proposed and are being used to reduce the effect of stress on the MEMS device. For example, in Lishchynska et al. (2007), as much as 37% change in the performance was reported as a result of packaging stress.

One proposed solution for addressing the stress issue is by taking into account the effect of the package-induced stress during the design stage of MEMS devices. By utilizing finite element models (FEM), it has been shown to be able to predict the chip warpage caused by both die-attach adhesive and ball grid array flip-chip processes and take into account the change in the geometry of the MEMS devices after the packaging. For example, the effects of package-induced stress were determined using simulations as shown in studies by Walwadkar et al. (2003) where 3.5 mm by 3.5 mm silicon dies were attached to ceramic package substrates using silver glass and polymide adhesives.

Once the simulation shows how and to what degree the package-induced warpage of the chip affects the MEMS devices, one can either modify the device design to reduce the impact of the warpage to the performance of the MEMS chip or sometimes even incorporate the warpage in determining the MEMS geometries.

However, these simulations do not take into account the dynamic nature of the stress; the degree of chip warpage caused by the stress changes not just during the manufacturing process but also throughout the life cycle of the chip due to the viscoelastic nature of the adhesive material used, as well as varying conditions in which the system operates.

For example, in Joo and Choa (2007), the coefficient of thermal expansion (CTE) mismatch between Si chip and package substrate caused, on average, 80 Hz resonance frequency shift in MEMS gyroscopes, where a frequency shift of 30 Hz was considered to substantially degrade the sensor performance such as sensitivity and phase change resulting in yield loss or failure. It was only after several changes in the design and the material used for the device that the authors were able to achieve a frequency shift of 20.7 Hz.

In MEMS integration, flexible interconnects can provide benefits in multiple fronts other than relieving stress.

First, flexible interconnects can be used to compensate for nonplanar surfaces that may exist, and source of nonplanarity could be on the substrate as well as from the inherent limitation of solder deposition method (Basavanhally et al., 2007) or other bonding mechanisms.

Second, flexible interconnects can be used to make temporary interconnections enabling the idea of reusable electronics where the CMOS IC can be reused while the sensor chip is disposed. For areas where sensors are often irreversibly contaminated, or if the cleaning of the sensors does not make an economic sense, this ability to replace the sensor chip only can potentially bring down the cost of operation. An example of such area is the biosensor application, where sensors can be contaminated by blood or other biohazardous materials (Ravindran et al., 2010; Yang et al., 2010). It can also be used for interconnecting a macro-chip with nanophotonics and proximity I/Os (Shubin et al., 2009).

In both instances, one requires flexible interconnect structures to have a higher stand-off height and also a vertical range of motion that utilizes all of the available stand-off height.

CASE STUDY: MFIS

Mechanically flexible interconnect (MFI) is a flexible I/O technology developed at Georgia Tech that aims to provide MFI as much vertical range of motion as possible (50 μm) and also to provide benefits discussed in the previous section. In order to do this, the latest generation of MFIs has incorporated several design features.

A. Tapered Interconnect Structure

In order to minimize the plastic deformation of the flexible interconnect structure during vertical deformation, a tapered interconnect design was used instead of a more common constant width design; by linearly varying the width of the beam, it is possible to distribute the stress more uniformly. This lowers the maximum stress experienced by the beam as shown in the ANSYS simulation. Figure 3.7 shows the reduction in the peak stress experienced by the structure due to the tapering shape.

B. Curved Beam Design

In order to allow the 100% of the stand-off height to be utilized, it was also necessary to diverge from the conventional cantilever design as shown in Figures 3.8 through 3.10. With such design, the range of motion would be restricted to the height of material deposited on the tip of the beam, which in this case was the height of the solder ball. By having a curved beam design, this problem can be avoided and it is the design used for the MFIs.

C. Use of High-Yield Point Material and Oxidation Prevention

Copper is a good material for interconnects as it has a very low resistivity. However, there are two main challenges for using copper for the flexible interconnects. The first is the low-yield point of copper, which can cause reliability issues at large deformation, and the second is the issue of oxidation, as copper readily oxidizes at room temperature. For the new generation of MFIs, an alternative material (nickel/tungsten alloy) was used to improve the yield characteristic, and surfaces of the structures were passivated by electroless gold plating to minimize oxidation (Zhang et al., 2012).

Solder Confinement

Though MFIs can be used to provide temporary interconnects as discussed previously, if it is to be used for permanent interconnection, the ability to fabricate solder ball on it becomes critical. However, due to the entire structure being metal, solder must be properly confined so that it does wet the entire structure during the bonding process; wetting the entire interconnect structure would cause unexpected mechanical behaviour and therefore inconsistent assembly results. A polymer ring was formed on the pad area as shown in Figure 3.11, and the solder was deposited in the middle. Figure 3.12 shows that the solder is confined to the pad area only after reflowing. The polymer rings also allow electroplating of various UBM metals underneath the solder; in this work, nickel is used as a UBM.

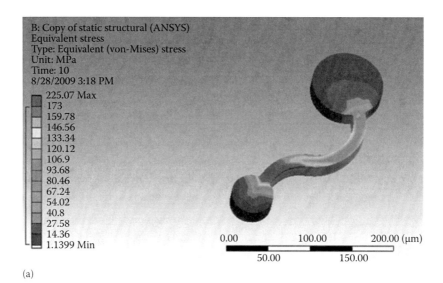

(a)

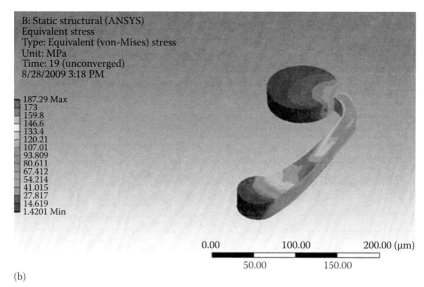

(b)

FIGURE 3.7 (See colour insert.) ANSYS FEM simulation comparing the tapered design (a) with the constant width design (b) of the interconnect structures. The tapered design results in more uniform distribution of stress. (© 2012 IEEE.)

Standard beam

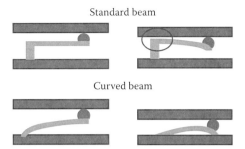

Curved beam

FIGURE 3.8 Comparing curved beam design and conventional cantilever design for use in flexible interconnects. (© 2012 IEEE.)

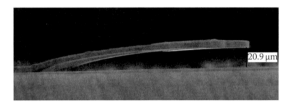

FIGURE 3.9 Scanning electron microscope (SEM) image showing the curved profile of a MFIs. (© 2012 IEEE.)

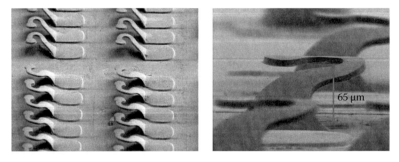

FIGURE 3.10 Microscope and SEM images showing various versions of MFIs. (© 2012 IEEE.)

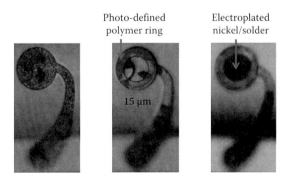

FIGURE 3.11 (See colour insert.) Confining of a solder ball at the tip of MFIs using polymer rings. (© 2012 IEEE.)

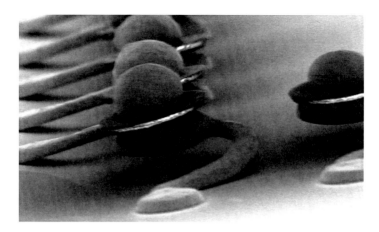

FIGURE 3.12 SEM showing the confined solder ball after the reflow process. (© 2012 IEEE.)

3.4.1 FABRICATION OF MFI

The process can be performed at the wafer level and are processes that can be implemented following the end of the semiconductor BEOL processes. This allows MFIs to be fabricated on CMOS chips. Figure 3.13 shows an overview of the MFI fabrication process.

The first part of the process is the fabrication of the curved polymer surface. This is done by spin coating a photo-definable sacrificial polymer and then reflowing it. Though the shape of the curved polymer surface is created almost instantly,

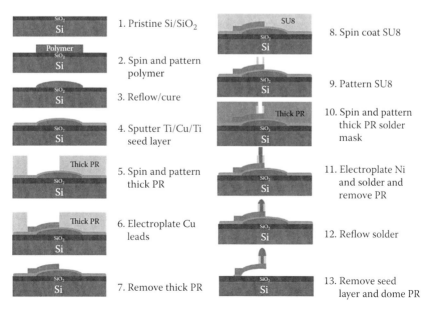

FIGURE 3.13 Process flow for the fabrication of MFIs. (© 2012 IEEE.)

the reflowed polymer then needs to be cured at 150°C in order to increase the glass transition temperature (T_g) and to remove excess solvent.

Increasing the glass transition temperature of the polymer is critical, as initial glass transition temperature is below many of the baking temperatures of the photoresists used in the following processes. The optimal curing time and temperature were experimentally determined. Details of the experiment as well as the fabrication process can be found in Yang and Bakir (2012). The second part of the process is to deposit an electroplating seed layer on top of the reflowed polymer. The third part of the process is to spin coat and pattern an electroplating mould for the electroplating of the interconnect beam structure. Nickel/tungsten alloy (or copper) is then electroplated. After the electroplating process, the electroplating mould is removed. SU-8 polymer ring is then formed and another electroplating mould with an opening inside the polymer ring is formed. Nickel and solder are then electroplated, respectively. Finally, the seed layer is removed followed by the removal of the sacrificial polymer (using acetone), which releases the MFIs. MFIs are then coated with gold using electroless gold plating process.

3.4.2 Mechanical Testing of MFI

Mechanical testing of the MFIs was done with an indenter setup (Figure 3.14) that can measure the vertical displacement as a function of the force applied. The results are shown in the Figure 3.15. Two important results can be seen from the aforementioned results. The first is that the compliance of the MFIs is predictable using simulations (ANSYS FEM) and can be engineered easily by changing the thickness, which can be done quite easily by adjusting the electroplating time required. The second is that the range of the compliance that can be achieved with MFIs is quiet wide.

MFI compliance requirements are not the same for all applications. Variables such as the size of the chip and number of I/Os determine the mechanical requirements of individual MFI properties, and the ability to predict the properties allows MFI to be used for wide range of applications.

In order to verify that the MFIs are not yielding during the operation, one of the MFIs was indented multiple times. The graph in Figure 3.16 shows that the mechanical characteristics of the MFI remain unchanged.

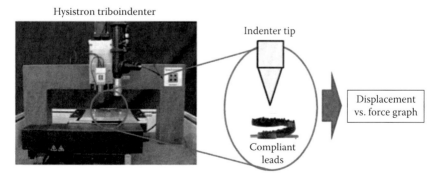

FIGURE 3.14   Test setup for measuring the compliance of a single MFI. (© 2012 IEEE.)

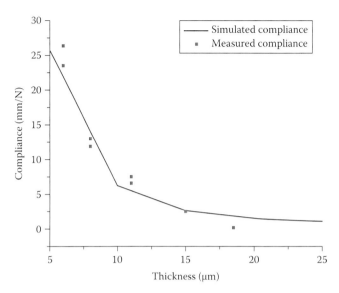

FIGURE 3.15 Graph showing the simulated and measured compliance of MFIs. (© 2012 IEEE.)

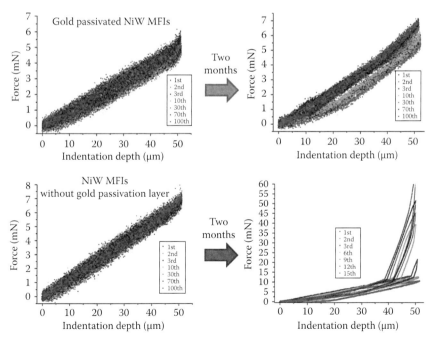

FIGURE 3.16 Graph showing the change in the mechanical compliance of MFIs after 2 months with and without gold coating. (© 2012 IEEE.)

3.5   TSV FOR MEMS

As discussed in the previous section, TSV is a vital technology if MEMS is to be integrated vertically (3D). This is because, there exists many cases where the MEMS devices must be placed on the top chip facing outwards; for example, the MEMS sensor may require an interaction with the material it is trying to sense.

Though many TSV technologies have been explored by many for fabrication in CMOS ICs, TSV in MEMS chip presents vastly different challenges. First, TSV technology for MEMS must be compatible with thicker chips (typically thicker than 300 µm), as MEMS chips are rarely thinned down like a CMOS IC. Second, in order to not restrict the MEMS fabrication in terms of processes and materials, TSV must be able to be fabricated after the MEMS device.

3.5.1   Challenges of Fabricating TSVs in Thick Chip

3.5.1.1   Stress

A TSV generally consists of a metal conductor in vertical direction through silicon and thin dielectric liner between the metal and the silicon. Various materials like copper, tungsten, nickel and aluminium can be used as metal in TSVs. Copper is mostly used for the TSV metal part because of the ease of fabricating high-aspect-ratio TSVs in chips as well as in silicon interposer packages using electroplating, better electro-migration resistance and comparatively lower resistivity. But CTE of copper is almost seven times higher than that of silicon. This induces higher stress in silicon surrounding the copper as well as causes reliability issues for the TSV structure when thermal load is exerted on TSVs. Moreover, the previous section discusses the significant effect stress has on the performance of MEMS devices.

TSV stresses increase as TSV diameter increases and the nature of these stresses is dependent on the relative arrangement of TSVs (Lu et al., 2009; Jung et al., 2011), and because the processes in TSVs are aspect ratio limited, a TSV in thick MEMS chips requires that the diameter be very large compared to TSVs in thin CMOS ICs.

Increased TSV stress can affect five regions in and around TSVs: silicon, copper, silicon dioxide, copper–silicon dioxide interface and BEOL layers near TSVs. Firstly, due to increased stress in silicon, mobility of carriers in silicon changes that can affect the operation of MOSFETS in the regions near TSVs. This creates requirement of keep-out zones around TSVs to ensure the desirable operation of MOSFETS as well as MEMS devices. The keep-out zone increases as TSV stresses increase (Lu et al., 2009; Athikulwongse et al., 2010). Moreover, higher stress in silicon may also lead to crack formation and propagation in silicon (Lu et al., 2009). Secondly, due to increased TSV stresses, cohesive cracks can form and propagate in silicon dioxide as well as in copper. Thirdly, increased stresses at copper–silicon dioxide interface can lead to interfacial crack propagation and interfacial delamination of the copper in TSVs (Andry et al., 2008; Liu et al., 2009).

To reduce the effect of TSV stresses, there are several ways including effective stress-aware placement of TSVs (Liu et al., 2009; Jung et al., 2011), using TSV conducting material (e.g. tungsten) with CTE comparable to that of silicon (Andry et al., 2008; Bauer et al., 2009), introducing and optimizing pre-chemical mechanical polishing (CMP) anneal to reduce copper pumping (Malta et al., 2011; Wolf et al., 2011) or using a thick stress buffer polymer cladding between copper and silicon in TSVs instead of thin silicon dioxide liner to form polymer-clad TSVs (Parekh et al., 2011). In polymer-clad TSVs, when a thick polymer material is selected with lower Young's modulus compared to silicon and copper, the polymer can absorb stress caused by the CTE mismatch between the silicon and the copper. Various modelling results have been shown for polymer-clad TSVs to show reduction in TSV stresses (Chen et al., 2009; Jung et al., 2011; Ryu et al., 2011). In addition to reduction in TSV stresses, a significant reduction in TSV dielectric capacitance can be obtained using polymer material with lower dielectric constant and with appropriate cladding thickness (Civale et al., 2011).

For cladding purpose, various materials have been investigated in literature including SU-8, parylene, BCB, epoxy and polymide. The cladding fabrication process can be by vapour deposition of polymer (e.g. parylene coating) (Majeed et al., 2008), filling of etched areas in silicon followed by selective silicon etching to form vias with cladding (Civale et al., 2011) or photodefinition of polymer filled in etched vias in silicon (Thadesar et al., 2012). In case of parylene deposition, limited thickness may be obtained. In case of filling etched trenches, the filling process would be dependent on aspect ratio of trenches (the higher the aspect ratio, the difficult is the filling) as well as on viscosity of polymer to be filled (difficult to fill polymers with higher viscosity). However, the process of filling via openings with polymer and later photo-defining the polymer can be used to fabricate high-aspect-ratio polymer-clad TSVs for chips as well as silicon interposer packages, using polymers with various range of viscosities (e.g. SU-8).

Thadesar et al., (2012) fabricated SU-8-clad TSVs with 120 μm outer diameter, 80 μm inner diameter (20 μm thick annulus-shaped SU-8 cladding) and 390 μm tall for silicon interposer application (Figures 3.17 and 3.18). The 80 μm inner diameter of TSVs meets ITRS 2010 projection of 80 μm diameter vias for silicon interposers for high-performance computing systems for the year 2017. SU-8 is a widely used photoresist for fabrication of high-aspect-ratio structures (Campo and Greiner, 2007). Young's modulus of SU-8 is very low (4 GPa) as compared to that of silicon (185 GPa)

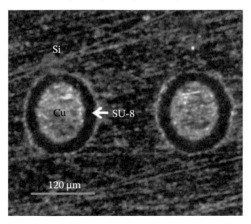

FIGURE 3.17 Microscope image showing the SU-8-clad TSVs (top view). (From Thadesar, P.A. and Bakir, M.S., Silicon interposer featuring novel electrical and optical TSVs, in *ASME International Mechanical Engineering Congress and Exposition (IMECE)*, 2012, Houston, TX.)

FIGURE 3.18   Microscope image showing the SU-8-clad TSVs (cross-sectional view). (From Thadesar, P.A. and Bakir, M.S., Silicon interposer featuring novel electrical and optical TSVs, in *ASME International Mechanical Engineering Congress and Exposition (IMECE)*, 2012, Houston, TX.)

and copper (117 GPa). Consequently, SU-8 can act as a stress buffer layer between silicon and copper relieving TSV stresses. Along with stress reduction, due to 20 µm thick cladding of SU-8 with relative dielectric constant ~3 (silicon dioxide relative dielectric constant is 3.9), considerable reduction in TSV capacitance can be obtained compared to TSVs with same copper diameter, same length and with thin silicon dioxide liner.

3.6   SEED-LAYER FABRICATION

Large-diameter TSVs that result due to the thick chips create a fabrication challenge. Typically a TSV is filled using an electroplating process and requires a seed layer to be fabricated. Many conventional TSV processes have fabricated this seed layer as shown in the Figure 3.19 as follows.

After the via hole etch using a DRIE, the seed layer is formed by first depositing a metal layer around the via area and electroplating until the via holes are closed. The thick bulk metal formed is then polished before via is filled. Both the 'pinch off' time and the polishing time is a very time-consuming process.

Lai et al. (2010) introduced an alternative method called mesh process that eliminates this process. This process starts with the deposition of SiO_2 layer where the TSVs will be formed. SiO_2 is used as the stop layer for the DRIE via hole etch. In the suspended SiO_2 layer, a mesh is patterned and etched as shown in the Figure 3.19. A metal layer is evaporated on top and then electroplated. A short electroplating session is enough to close up the mesh holes, creating a seed layer through the mesh layer in which the TSVs can be electroplated from. The fabrication results show that the presence of mesh does not create voids during the electroplating process.

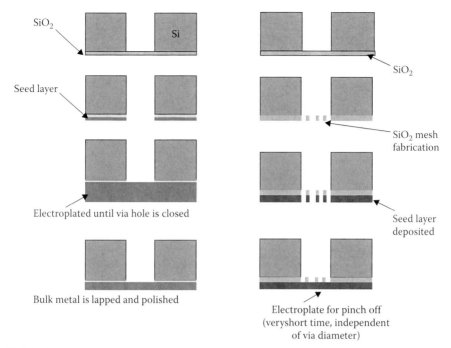

FIGURE 3.19 Conventional process flow for fabricating seed layer for electroplating up TSV in thick wafers. (From Lai, J.-H., Yang, H.S., Chen, H. et al., 2010, A "mesh" seed layer for improved through-silicon-via fabrication, *J. Micromech. Microeng.*, 20, 025016, 2012. Copyright 2012, Institute of Physics.)

3.7 ELIMINATION OF CMP FOR POST-MEMS TSV FABRICATION

As discussed previously, fabricating TSV must be done post-MEMS in order to not restrict MEMS devices in terms of processes and materials that can be utilized. However, most TSV processes require chemical and mechanical planarization (CMP) to be performed on the side where sensitive MEMS device may be present. Conventional TSV process flow requiring CMP process is shown in Figure 3.20. For a post-MEMS fabrication of TSVs, a new planarization method is needed.

In Lai et al. (2010), a chemical planarization method is introduced that can be used safely with a wide range of MEMS devices already present on the substrate. The process involves the use of second type of metal, different from the via material, and exploits the mesh process introduced in the previous section. The process begins after the fabrication of seed layer. Instead of using the seed layer to fill up the via completely with copper, a thin layer of nickel is electroplated first. After the nickel electroplating, the copper is electroplated. Then, the side without SiO_2 is covered and the sample is placed in a chemical copper etchant bath, and because nickel does not etch in the copper etchant, the etch stops when the copper seed layer is removed and nickel layer is exposed in the via. The process flow is shown in the Figure 3.21.

Without the mesh process, there exist two mechanical polish steps on one side (the side with the seed layer) and another one on the opposite side. By using mesh to

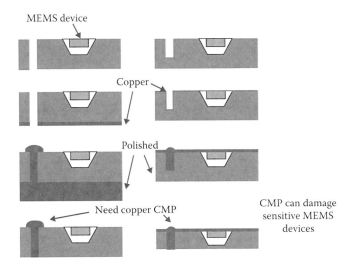

FIGURE 3.20 Conventional process flow for filling via holes for TSV requires CMP that can damage sensitive devices. (From Lai, J.-H., Yang, H.S., Chen, H. et al., 2010, A "mesh" seed layer for improved through-silicon-via fabrication, *J. Micromech. Microeng.*, 20, 025016, 2012. Copyright 2012, Institute of Physics.)

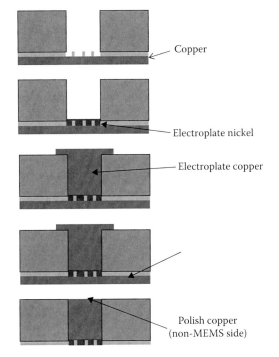

FIGURE 3.21 TSV fabrication using the mesh seed-layer process eliminates the need for the CMP process. (From Lai, J.-H., Yang, H.S., Chen, H. et al., 2010, A "mesh" seed layer for improved through-silicon-via fabrication, *J. Micromech. Microeng.*, 20, 025016, 2012. Copyright 2012, Institute of Physics.)

pinch off, the need for thinning down the bulk metal seed layer is eliminated, and by using two-metal chemical planarization technique, the need for the mechanical planarization is also eliminated. Therefore, if one fabricates MEMS devices on the side where the mesh is present, this process allows TSVs to be fabricated post-MEMS.

3.8 SUMMARY

The 3D integration of CMOS and MEMS provides the performance of monolithic integration and the process simplicity of hybrid integration. The key to exploiting all the benefits of 3D integration for CMOS and MEMS is in leveraging advanced interconnect technologies such as flexible interconnects and TSVs. In this chapter, the motivation and need for such interconnects are discussed as well as overview of challenges involved in design and fabrication of such interconnects.

REFERENCES

Andry, P.S., Tsang, C.K., Webb, B.C. et al., 2008. Fabrication and characterization of robust through-silicon vias for silicon-carrier applications. *IBM Journal of Research and Development* 52, 571–581.

Athikulwongse, K., Chakraborty, A., Yang, J.-S., Pan, D.Z., and Lim, S.K., 2010. Stress-driven 3D-IC placement with TSV keep-out zone and regularity study, in: *Proceedings of the International Conference on Computer-Aided Design, ICCAD '10*, IEEE Press, Piscataway, NJ, pp. 669–674.

Bakir, M.S. and Meindl, J.D., 2008. *Integrated Interconnect Technologies for 3D Nanoelectronic Systems*, 1st edn. Artech House Publishers, Boston, MA.

Bakir, M.S., Thacker, H.D., Zhou, Z. et al., 2002. Sea of leads microwave characterization and process integration with FEOL and BEOL, in: *International Interconnect Technology Conference, 2002*. Burlingame, CA, pp. 116–118.

Baltes, H., Brand, O., Fedder, G.K. et al., 2005. *CMOS-MEMS: Advanced Micro and Nanosystems*. Wiley-VCH, Weinheim, Germany.

Basavanhally, N., Lopez, D., Aksyuk, V. et al., 2007. High-density solder bump interconnect for MEMS hybrid integration. *IEEE Transactions on Advanced Packaging* 30, 622–628.

Bauer, T.M., Shinde, S.L., Massad, J.E. et al., 2009. Front end of line integration of high density, electrically isolated, metallized through silicon vias, in: *59th Electronic Components and Technology Conference (ECTC 2009)*, San Diego, CA, pp. 1165–1169.

del Campo, A. and Greiner, C., 2007. SU-8: A photoresist for high-aspect-ratio and 3D submicron lithography. *Journal of Micromechanics and Microengineering* 17, R81.

Chen, Z., Song, X., and Liu, S., 2009. Thermo-mechanical characterization of copper filled and polymer filled tsvs considering nonlinear material behaviors, in: *59th Electronic Components and Technology Conference (ECTC 2009)*, San Diego, CA, pp. 1374–1380.

Civale, Y., Tezcan, D.S., Philipsen, H.G.G. et al., 2011. 3-D Wafer-level packaging die stacking using spin-on-dielectric polymer liner through-silicon vias. *IEEE Transactions on Components, Packaging and Manufacturing Technology* 1, 833–840.

Dang, B., Bakir, M.S., Patel, C.S. et al., 2006. Sea-of-Leads MEMS I/O interconnects for low-k IC packaging. *Journal Microelectromechanical System* 15, 523–530.

Eloy, Y.D., 2010. Status of the MEMS Industry 2010.

Fedder, G.K., 1998. Integrated MEMS in conventional CMOS, in: *Tribology Issues and Opportunities in MEMS: Proceedings of the NSF/AFOSR/ASME Workshop on Tribology Issues and Opportunities in MEMS*, Columbus, OH, November 9–11, 1997, p. 17.

Fjelstad, J., 1998. WAVETM technology for wafer level packaging of ICs, in: *Proceedings of 2nd Electronics Packaging Technology Conference*, Singapore, 1998. pp. 214–218.

Fjelstad, J., DiStefano, T., and Faraci, A., 2000. Wafer level packaging of compliant, chip size ICs. *Microelectronics International* 17, 23–27.

Huang, W.L., Ren, Z., Lin, Y.W. et al., 2008. Fully monolithic CMOS nickel micromechanical resonator oscillator, in: *IEEE 21st International Conference on Micro Electro Mechanical Systems (MEMS 2008)*, Tucson, AZ, pp. 10–13.

Jiangfeng, W., Fedder, G.K., and Carley, L.R., 2004. A low-noise low-offset capacitive sensing amplifier for a 50-/spl mu/g//spl radic/Hz monolithic CMOS MEMS accelerometer. *IEEE Journal of Solid-State Circuits* 39, 722–730.

Joo, J.-W. and Choa, S.-H., 2007. Deformation behavior of MEMS gyroscope sensor package subjected to temperature change. *IEEE Transactions on Components and Packaging Technologies* 30, 346–354.

Joseph, A.J., Gillis, J.D., Doherty, M. et al., 2008. Through-silicon vias enable next-generation SiGe power amplifiers for wireless communications. *IBM Journal of Research Development* 52, 635–648.

Jung, M., Liu, X., Sitaraman, S.K. et al., 2011. Full-chip through-silicon-via interfacial crack analysis and optimization for 3D IC, in: *Proceedings of the International Conference on Computer-Aided Design, ICCAD 2011*, IEEE Press, Piscataway, NJ, pp. 563–570.

Kacker, K. and Sitaraman, S.K., 2008. Design and fabrication of flex connects: A cost-Effective implementation of compliant chip-to-substrate interconnects. *IEEE Transactions on Components Packaging Technology* 31, 816–823.

Kacker, K. and Sitaraman, S.K., 2009. Electrical/mechanical modeling, reliability assessment, and fabrication of flex connects: A MEMS-based compliant chip-to-substrate interconnect. *Journal of Microelectromechanical System* 18, 322–331.

Karim, N. and Agarwal, A.P. Plastic packages' electrical performance: Reduced bond wire diameter, *Amkor Application Note*, http://www.amkor.com/services/electrical/newabstr.pdf

Kim, Y.G., Mohammed, I., Seol, B.S. et al., 2001. Wide area vertical expansion (WAVETM) package design for high speed application: Reliability and performance, in: *Proceedings of 51st Electronic Components and Technology Conference*, Orlando, FL, pp. 54–62.

Kisiel, R. and Szczepanski, Z., 2005. Trends in assembling of advanced IC packages. *Journal Telecommunication and Information Technology* 1, 63–69.

Krishnamoorthy, A.V. and Goossen, K.W., 1998. Optoelectronic-VLSI: Photonics integrated with VLSI circuits. *IEEE Journal of Selected Topics in Quantum Electronics*, 4, 899–912.

Krishnamoorthy, A.V., Goossen, K.W., Jan, W. et al., 2011. Progress in low-power switched optical interconnects. *IEEE Journal of Selected Topics in Quantum Electronics* 17, 357–376.

Lai, J.-H., Yang, H.S., Chen, H. et al., 2010. A "mesh" seed layer for improved through-silicon-via fabrication. *Journal of Micromechanical Microengineering* 20, 025016.

Lee, K.-W., Noriki, A., Kiyoyama, K. et al., 2011. Three-dimensional hybrid integration technology of CMOS, MEMS, and photonics circuits for optoelectronic heterogeneous integrated systems. *IEEE Transactions on Electron Devices* 58, 748–757.

Lishchynska, M., O'Mahony, C., Slattery, O. et al., 2007. Evaluation of packaging effect on MEMS performance: Simulation and experimental study. *IEEE Transactions on Advanced Packaging* 30, 629–635.

Liu, X., Chen, Q., Dixit, P. et al., 2009. Failure mechanisms and optimum design for electroplated copper through-silicon vias (TSV), in: *Electronic Components and Technology Conference*, Piscataway, NJ, pp. 624–629.

Lo, G. and Sitaraman, S.K., 2004. G-helix: Lithography-based wafer-level compliant chip-to-substrate interconnects, in: *Proceedings of the 54th Electronic Components and Technology Conference 2004*, Las Vegas, NV, pp. 320–325.

Lu, K.H., Zhang, X., Ryu, S.K. et al., 2009. Thermo-mechanical reliability of 3-D ICs containing through silicon vias, in: *IEEE 59th Electronic Components and Technology Conference (ECTC 2009)*, San Diego, CA, pp. 630–634.

Muller, T., Brandl, M., Brand, O. et al., 2000. An industrial CMOS process family adapted for the fabrication of smart silicon sensors. *Sensors and Actuators A: Physical* 84, 126–133.

Majeed, B., Pham, N.P., Tezcan, D.S. et al., 2008. Parylene N as a dielectric material for through silicon vias, in: *IEEE 58th Electronic Components and Technology Conference (ECTC 2008)*, Lake Buena Vista, FL, pp. 1556–1561.

Malta, D., Gregory, C., Lueck, M. et al., 2011. Characterization of thermo-mechanical stress and reliability issues for Cu-filled TSVs, in: *IEEE 61st Electronic Components and Technology Conference (ECTC), 2011*, Lake Buena Vista, FL, pp. 1815–1821.

Marek, J. and Gómez, U.M., 2012. MEMS (micro-electro-mechanical systems) for automotive and consumer electronics. *Chips* 2020, 293–314.

MEMS Market Overview, MicroTech, Anaheim, CA, June 2010. http://www.ardi-rhonealpes.fr/c/document_library/get_file?uuid=0f069af7-9bb4-4c16-8afc-222c38dd4c73&groupId=10136

Novitsky, J. and Pedersen, D., 1999. Form Factor introduces an integrated process for wafer-level packaging, burn-in test, and module level assembly, in: *Proceedings International Symposium on Advanced Packaging Materials: Processes, Properties and Interfaces, 1999*, Braselton, Georgia, pp. 226–231.

Ravindran, R., Sadie, J.A., Scarberry, K.E. et al., 2010. Biochemical sensing with an arrayed silicon nanowire platform, in: *IEEE 60th Electronic Components and Technology Conference (ECTC), 2010* Las Vegas, NV, pp. 1015–1020.

Ryu, S.-K., Lu, K.-H., Zhang, X. et al., 2011. Impact of near-surface thermal stresses on interfacial reliability of through-silicon vias for 3-D interconnects. *IEEE Transactions on Device and Materials Reliability* 11, 35–43.

Sedky, S., Witvrouw, A., Bender, H. et al., 2001. Experimental determination of the maximum post-process annealing temperature for standard CMOS wafers. *IEEE Transactions on Electron Devices* 48, 377–385.

Seraji, N.E. and Yavari, M., 2011. Minimum detectable capacitance in capacitive readout circuits, in: *IEEE 54th International Midwest Symposium on Circuits and Systems (MWSCAS), 2011*, Seoul, Korea, pp. 1–4.

Shubin, I., Chow, E.M., Cunningham, J. et al., 2009. Novel packaging with rematable spring interconnect chips for MCM, in: *59th Electronic Components and Technology Conference. Presented at the 2009 IEEE 59th Electronic Components and Technology Conference (ECTC 2009)*, San Diego, CA, pp. 1053–1058.

Thadesar, P.A. and Bakir, M.S., 2012. Silicon interposer featuring novel electrical and optical TSVs, in: *ASME International Mechanical Engineering Congress and Exposition (IMECE), 2012*, Houston, TX.

Walwadkar, S., Farrell, P., Felton, L.E. et al., 2003. Effect of die-attach adhesives on the stress evolution in MEMS packaging, in: *Proceedings of the 36th International Symposium Microelectronics (IMAPS'03)*, Boston, MA, p. 847.

Witvrouw, A., 2006. CMOS-MEMS integration: Why, how and what? in: *Proceedings of the 2006 IEEE/ACM International Conference on Computer-aided Design*, ACM, San Jose, CA, pp. 826–827.

Wolf, I.D., Croes, K., Pedreira, O.V. et al., 2011. Cu pumping in TSVs: Effect of pre-CMP thermal budget. *Microelectronics Reliability* 51, 1856–1859.

Yang, H.S. and Bakir, M.S., 2010. 3D Integration of CMOS and MEMS using mechanically flexible interconnects (MFI) and through silicon vias (TSV), in *Presented at the 2010 IEEE 60th Electronic Components and Technology Conference (ECTC 2010)*, Las Vegas, NV.

Yang, H.S. and Bakir, M.S., 2012. Design, fabrication, and characterization of freestanding mechanically flexible interconnects using curved sacrificial layer. *IEEE Transactions on Components, Packaging and Manufacturing Technology* 2(4), 561–568.

Yang, H.S., Ravindran, R., Bakir, M.S., and Meindl, J.D., 2010. A 3D interconnect system for large biosensor array and CMOS signal-processing IC integration, in: *2010 IEEE International Interconnect Technology Conference. Presented at the 2010 IEEE International Interconnect Technology Conference – IITC*, Burlingame, CA, pp. 1–3.

Yazdi, N., Kulah, H., and Najafi, K., 2004. Precision readout circuits for capacitive microaccelerometers, in: *Proceedings of IEEE Sensors, 2004*, Vienna, Austria, pp. 28–31.

Zhang, C., Yang, H.S., and Bakir, M., 2012. Gold passivated mechanically flexible interconnects (MFIs) with high elastic deformation, in: *IEEE 62nd Presented at the Electronic Components and Technology Conference (ECTC), 2012*, San Diego, CA.

Zhu, Q., Ma, L., and Sitaraman, S.K., 2003. β-Helix: A lithography-based compliant off-chip interconnect. *IEEE Transactions on Components and Packaging Technology* 26, 582–590.

4 Wafer-Scale Rapid Electronic Systems Prototyping Platform

User Support Tools and Thermo-Mechanical Validation

Mikaël Guillemot, Hai Nguyen,
Mohammed Bougataya,
Yves Blaquière, Ahmed Lakhssassi,
Mary Shields and Yvon Savaria

CONTENTS

4.1   INTRODUCTION

Electronic system design methods are in constant evolution with access to technologies enabling the creation of increasingly complex circuits that not only are miniaturized and consume less energy at the same time but also provide more functionality. The design and prototyping phases of today's systems represent a significant part of the overall cost of many products, particularly when production volumes are moderate or low. At the printed circuit board (PCB) level, it is common to deal with complex integrated circuits (ICs) where the prototyping of such circuits is a long and expensive process, even when abstracted models are used. The DreamWafer™ project proposes a new approach to the rapid prototyping of electronic systems with a wafer-scale prototyping platform called WaferBoard™ [1].

The core of this platform is a 200 mm wafer-scale silicon IC called WaferIC™. The WaferIC™ has a large number of tiny contacts (1,245,184) on its top surface, called NanoPads. User ICs (uICs) are manually placed by the user on this surface, which is covered with an anisotropic contact film. Closing the cover allows to apply mechanical pressure to the uICs, firmly holding them in place and ensuring good electrical contact through the anisotropic film. An interconnection network similar to that in commercial SRAM-based FPGAs is programmably configured to connect any NanoPad to any other. The WaferIC™ structure is made of several layers assembled with miniature electronic components that require power to be supplied in a limited space, making device thermal behaviour prediction a major issue for reliable operation of this thin and fragile wafer-scale IC. An accurate and fast evaluation of heat flow patterns becomes an essential step in the overall design validation.

The main objective of the proposed platform is to prototype electronic systems, equivalent to a PCB, in minutes. It is supported by an essential user Computer-Aided Design (CAD) tool, WaferConnect™, that is developed by our team. That tool allows users to configure and interact with the system via a graphical user interface (GUI). This tool uses algorithms that not only are efficient and provide to the users a large quantity of useful information such as component placement and interconnections but also offer an extended set of tools that are easily inserted into the design flow of electronic systems.

In this chapter, an overview of the WaferBoard™ prototyping platform is first presented. Its user support tool is then outlined with emphasis on the technical and architectural choices used to design and develop a high-performance and evolutionary software adapted to the modern microelectronic field. The last section presents thermo-mechanical validation issues that have been resolved with steady-state thermo-mechanical analysis using a combined mechanical and thermal heat transfer approach for thermal and distortion behaviour analysis.

4.2 WAFERBOARD™ AND WAFERIC™ OVERVIEW

The 200 mm silicon wafer-scale WaferIC™ is made from a grid of identical reticule images interconnected to each other with a reticule stitching technique as depicted in Figure 4.1a. Each group of four reticules is powered by a set of miniature devices, called PowerBlocks, assembled on the back side of the WaferIC™ through 4864 regularly

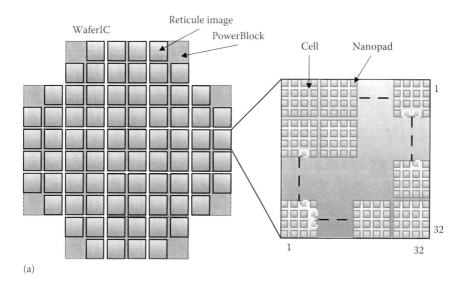

(a)

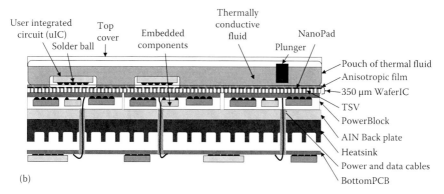

(b)

FIGURE 4.1 WaferIC™ structure (a) and WaferBoard™ layer structure (b) (not to scale).

distributed through-silicon vertical interconnect access (vias) (TSVs). Each PowerBlock also provides input/output (I/O) signals used for configuration of the WaferIC™.

Each reticule image is composed of a tile of 32 × 32 cells and each cell embeds an array of 4 × 4 NanoPads for a total of about 1.25 million NanoPads over the WaferIC™.

The active WaferIC™ includes two different networks, a joint test action group (JTAG) defect-tolerant network and the WaferNet™. The JTAG defect-tolerant network is accessed from the backside of the WaferIC™ and is used to configure the WaferNet™ as well as the configurable NanoPads. Each NanoPad can be in contact with any type of uIC pin and can be therefore configured as a floating pin, a digital I/O, a power supply or a ground. This JTAG network links each cell to its four neighbouring cells by two connections (Figure 4.2a). The second network, WaferNet™, is a defect-tolerant scalable multidimensional mesh network. Once configured according to a netlist provided by the user, specific paths are established, allowing any NanoPad to be linked to any others (Figure 4.2b). Each cell is linked to 6 other cells in each physical direction (north–south–east–west, N-S-E-W), where the length of the link grows according to a geometric series, i.e. to the 1st, 2nd, 4th, 8th, 16th and 32nd neighbouring cells. A dedicated electronic circuit board, called BottomPCB,

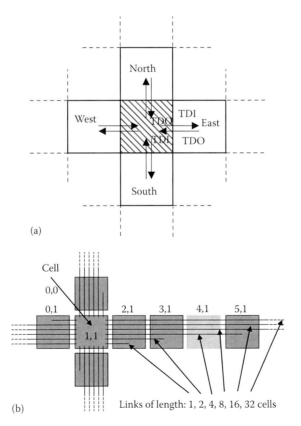

(a)

(b)

FIGURE 4.2   The mesh JTAG network structure (a) and WaferNet structure (b).

provides the communication between the WaferConnect™ software (running on a workstation) and the WaferBoard™. The BottomPCB also manages sensors and regulation of the power supplied via the PowerBlocks to the WaferIC™.

4.3   WAFERCONNECT: USER SUPPORT SOFTWARE TOOLS

User support software aims at providing tools to users for configuring their target electronic systems using the available configurable interconnection and support circuitry. These tools comprise a set of functions allowing users to interact with the prototyping platform through a configuration process. These interactions are realized via a GUI designed to, first, facilitate the execution of a well-defined user workflow and, second, allow data visualization for validation and debugging purposes.

This section is organized as follows: an overview of the software will be first presented, followed by the presentation of the software backbone elements including the core data structure (WaferBoard™ Hardware Representation, *WBHR*), JTAG-based hardware configuration and communication, workflow manager and GUI. Finally, specific functional modules called *Hardware Diagnosis, Pin Detection, Package Recognition* and *Routing*, all grouped in an entity called workflow modules, which implement essential operations for the configuration, are briefly presented.

4.3.1   SOFTWARE OVERVIEW

WaferConnect™ software architecture was developed according to the following constraints and objectives:

- Manage the complex interactions with the hardware (WaferIC, PowerBlock, BottomPCB, etc.)
- Provide acceptable response time
- Support integration of research tool
- Adapt easily to any modifications in WaferIC™ and BottomPCB hardware specifications
- Allow progress in spite of frequent changes of development personnel (tool mostly developed by students)
- Provide a GUI adapted to casual users, as well as to advanced users

Some solutions to these constraints and objectives are

- Interaction with the hardware must be isolated.
- Evolutionary and adaptable software structure.
- Optimization efforts are focused on critical processes and parallel processing is widely used.
- GUI is designed to be flexible and reconfigurable.

The normal sequence of user operations with the platform is as follows (Figure 4.3a): First, the user places the electronic components (uICs) composing the system to be prototyped on the WaferBoard™, closes the cover and then powers up the

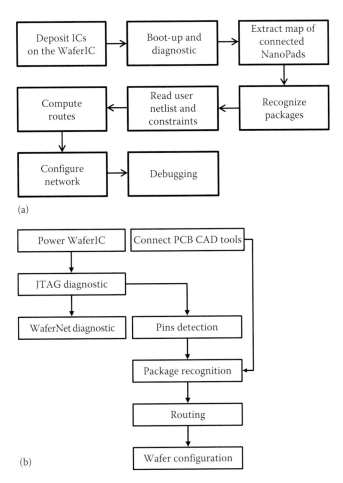

FIGURE 4.3 The proposed WaferConnect workflow (a), dependencies between workflow module (b).

WaferBoard™. Notice that constant pressure applied after closing the cover must be maintained in order to ensure good electrical contacts between the uIC pins and WaferIC™'s NanoPads. The user then imports prototype schematic data such as component footprints and system netlist. The WaferConnect™ software performs a JTAG diagnosis and a WaferNet diagnosis of the WaferIC™, then subsequently applies the pin detection step. The package recognition operation is next executed, taking as input imported prototype footprints and the circuit footprint obtained from the pin detection process, resulting in the identification of electronic components placed on the wafer. The software now runs the routing process, using the imported netlist and netlist constraints, and configures the WaferNet to establish electrical links between component pins that conform to the design netlist. Finally, the circuit is ready to be prototyped and debugged with dedicated tools.

A modular software design approach has been used to reduce the software complexity while allowing collaboration between researchers and developers.

The software offers a guided operational mode for generic users that works according to a well-defined user workflow, or a custom operational mode for advanced users, where actions are defined by the users who provide the needed inputs for each step. Figure 4.4 shows the organization of these software modules, which are presented in more detail in the following sections.

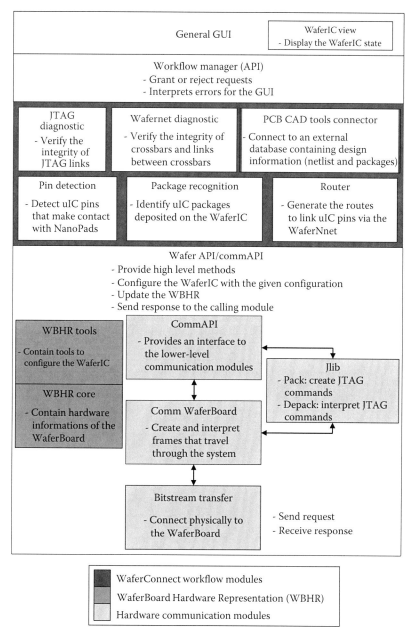

FIGURE 4.4 WaferConnect architecture.

The software comprises functional modules and driver modules. This split allows software developers with limited knowledge in microelectronics to take software development in charge. This is possible with the creation of an application programming interface, namely, WaferIC™ access API (WAPI), which raises the low-level configuration mechanisms to a higher abstraction level. For example, WAPI allows users to issue a request for a route configuration involving only high-level parameters independent from the WaferIC™ register status. This approach allows developers to think of hardware in terms of high-level abstract objects, facilitating the development of sophisticated algorithms involving complex interactions with the hardware.

As explained in Figure 4.4, the application is divided into four layers:

1. Presentation (GUI)
2. Control (workflow manager)
3. Processing (workflow modules)
4. Driver (WAPI)

Functionalities associated with the layers are not exclusive. The GUI periodically accesses data for display purposes, without duplicating it. The driver hides complex functionality for managing (with WBHR) and accessing (through hardware communication modules) hardware-related data. It also supports the creation of data used by the logical layer and that is derived from raw hardware-related data (stored in WBHR) using JLIB.

Compared with typical N-tier architectures, the access constraints between layers are relaxed in the proposed architecture. The relatively open layer access allows isolating the interpretation of hardware response while keeping open the client/server model prospect (in case where several prototypes coexist on the wafer).

It is noteworthy that the WaferConnect™ software is developed using current popular platforms, GNU/Linux and Microsoft Windows, supporting a wide range of potential customers who design and prototype electronic systems. The choice of the C++ programming language meets the need for performance and facilitates the interactions with the hardware. Furthermore, the multiplatform development under Linux/Windows is also facilitated by the use of technologies such as QT, STL and Boost [2–4].

4.3.2   WaferBoard™ Hardware Representation

A central data structure in the WaferConnect™ software is used to represent the WaferBoard™. This data structure is called the *WBHR*. It allows different modules to be synchronized with each other and is used as a bridge for configuring the hardware platform. This important data structure has the following properties:

- It is sharable between several software modules.
- It offers rapid and simple access.
- It is well organized and available for real-time graphical display.
- It is structurally capable of managing a large quantity of information.
- It is simple to integrate into the WaferConnect™ software.

WBHR takes into account the needs for integration, display and simplicity of access. It contains all information necessary to monitor the platform (e.g. temperature, pressure, power supply current values) and to configure the WaferIC™, such as an image of all its registers. The WBHR data structure is accessed by all software modules with dedicated APIs. Its contents can be visually represented with a GUI.

The WBHR data structure is mainly based on the physical hierarchy of the WaferIC™:

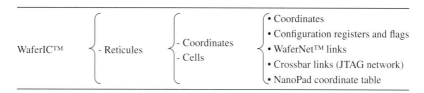

The WBHR data structure easily lends itself to the use of code-generation techniques, despite the criticality of this module in terms of performance. The workload represented by an automatic code generation could be large, but its overhead is quickly absorbed in the context of research and evolving specifications.

4.3.3 WaferIC Access API: Abstraction and Communication

The configuration and communication with the hardware platform is ensured by the module called *WAPI*. It interfaces the high-level software (workflow modules and GUI (Figure 4.4)) with the low-level software (WBHR and hardware communication modules), dedicated to drive the hardware. The *WAPI* module first provides hardware access commands to configure the WaferIC™. Subsequently, these commands are converted into related register states. If the configuration of the WaferIC™ is successful, the *WAPI* module then updates the *WBHR* object, decodes the WaferIC™ responses and converts them to results, which are then interpreted by high-level calling modules. This approach hides hardware technical details allowing high-level modules to focus on their specific tasks.

The four main transactions supported by the *WAPI* are for

1. Configuration and requests for contact detection
2. JTAG link configuration and check from the JTAG Diagnosis module
3. WaferNet configuration and check from the WaferNet Diagnosis module
4. WaferNet configuration coming from the Routing module

From a specific configuration task, a valid JTAG path starting from TDI and going to TDO, containing cells to be configured, is established. A library, called *JLIB*, translates the high-level configuration information contained in the JTAG path into chains of JTAG commands. Each JTAG command chain is organized into macro-chains and micro-chains. Macro-chains specify how cells to be configured can be accessed, while micro-chains contain specific JTAG instructions specifying how to configure these cells. The resulting JTAG command chain is then compressed and sent to the WaferIC™ hardware over the communication link through the BottomPCB.

WaferBoard platform

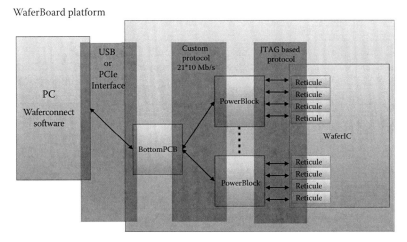

FIGURE 4.5   WaferBoard communication structure.

The software–hardware communication process is managed by the *commAPI* module. This module covers all necessary elements to access hardware registers, such as registers in the WaferIC™ as well as those found in the BottomPCB and PowerBlocks. commAPI reads or write values stored in various registers that relate to:

- Sensors (temperature, pressure, fan speed, etc.)
- Energy consumption (voltage and current sensors)
- WaferIC™ registers
- Fan speed control
- PowerBlock and BottomPCB regulators

As previously stated, each WaferIC™ reticule is configured using a JTAG-based serial protocol. The JTAG bitstream is compressed and encapsulated into frames and forwarded through the BottomPCB, to each PowerBlock's field-programmable gate array (FPGA) from which they are uncompressed and then sent to WaferIC™ (Figure 4.5). The WaferIC™ configuration is carried out, reticule by reticule, and processed in two steps. At first, a JTAG configuration path is constructed, then the WaferIC™ configuration registers are set. Each PowerBlock is independent of the others, which allows for parallel configurations. When a particular JTAG configuration path is diagnosed as defective, a different path can be set up or the reticule can be configured from any adjacent functional reticule. Defect tolerance is effectively a must for wafer-scale circuits. This redundancy feature provides a great flexibility to operate the proposed platform and is automatically handled by the communication mechanism in order to make it transparent to users and developers.

4.3.4   WAFERCONNECT WORKFLOW MANAGER

The software is designed to allow users to effectively control the hardware platform. The algorithms alone, implemented in different modules, do not allow such a control, even with a basic command-line interface. Users can also manage hardware faults and errors where, in most cases, they can be diagnosed and fixed using defect-tolerant mechanisms. The user workflow should also be able to validate the automatic identification of the uIC packages deposited on the WaferIC™ and display the results of the routing process. A proposed user workflow for our electronic system prototyping platform is shown in Figure 4.3a [5].

The first two steps of the workflow verify the integrity of the WaferBoard™ platform. The boot-up phase tests the communication links between the WaferIC™ and the workstation, retrieves the hardware platform status and initiates the power-up sequence. The diagnostic step detects functional elements of the WaferIC™ (reticules, cells, JTAG paths, WaferNet links, etc.). The two following workflow steps consist of recovering information related to uIC components placed on the WaferIC™ surface. Pins are first detected and clustered into entities called components.

These components are identified by comparing their footprints with the ones in a component library. User input in the identification process occurs when the automatic procedure fails. When damaged components are placed on the WaferIC™ or the surface of the WaferIC™ is damaged, uICs can be replaced or moved. The next two workflow steps consist of importing information related to routing (user netlists and constraints) and carrying out the routing itself. When these steps are completed, the rapid prototyping platform is ready to be configured and used to debug the user electronic system.

The dependencies that exist between the various workflow modules are shown in Figure 4.3b. For example, the JTAG network must first be diagnosed, then used to detect NanoPads in contact, which is a precursor step for pin detection. Component packages can be recognized only when uIC footprints are available in the PCB CAD tool library of components and the circuit footprint have been extracted through the WaferConnect™ pin detection process. Finally, the routing process must know the WaferNet status, the pin locations on the WaferIC™ and the user netlist and its constraints. The existing PCB CAD tools connector module allows importing the component footprints and the user netlists in the XML format.

A modular architecture has been adopted to satisfy the functional interdependencies of the modules. Each workflow step is encapsulated and designed to request permission to execute from the workflow manager, which then allocates the necessary resources to the module to execute. This mechanism allows controlling the command flow and ensures the security of the software and the hardware platform. The software can only get necessary resources using this mechanism. Adding a new software command is also secure, because until it is referenced its execution cannot be allowed. This verification step ensures a more coherent process for software development.

Each command invocation is analysed by the workflow manager module and classified according to the following groups:

- Forbidden commands (never to be executed) because they compromise the system security)
- Shell commands (Linux/Windows)
- Local commands (workflow commands)

Forbidden commands such as mkpkg, yes (unix),… are not part of the normal flow, but they are necessary for system control or they were useful through the debug process.

Once a command is identified, its constraints are then verified. If the constraint elements are valid, the execution will be authorized.

Finally, any newly created command will always have a message linked to the GUI and error management modules. Consequently, if a software developer forgets to manage some errors in the code, an exception will be raised, allowing for detection and correction.

4.3.5   WaferConnect GUI

WaferConnect™ software is developed to control the hardware platform, specifically the WaferIC™. User interactions with the software occur through a GUI that is designed with two main objectives:

- Allowing user flow of actions on the hardware platform through intuitive and coherent user interactions with the software
- Offering data visualization capability for a quick and easy data interpretation and assessment to effectively manage hardware faults and errors

4.3.5.1   Structure of the GUI

One of the main objectives in the GUI design is to not only offer users an ergonomic interface that is easy and simple to use but also offers, through a hierarchical simple-to-complex design approach, a more detailed and complete control of the hardware platform for advanced users. The main challenge was to offer an efficient and fast GUI capable of managing the large number of graphical objects found in the wafer-scale IC.

The upper central part of the initial WaferConnect™ view, as shown in Figure 4.6, displays the WaferIC™ object with its main attributes: reticules, cells, uIC pins, uIC footprints, hardware elements with defect, etc. User interactions within the display zone are limited to a number of specific actions: left-clicking the mouse selects the attribute, moving the mouse while keeping the right-click down implements view translations and rolling the middle mouse button is used for zooming. A set of buttons representing the user workflow is placed at the upper left part of the GUI, namely, from top to bottom, *WaferBoard Bootup, JTAG Diagnosis, WaferNET Diagnosis, Netlist, Constraints, Pin Detection, Package Recognition, Routing, WaferBoard Configuration* and *Debugging*. Detailed description of the functions related to each button is presented in Section 4.3.6. The upper right part shows a *Help* window that is used to either guide the user through specific actions at the current phase of the workflow or to offer more general help related to the utilization of the software. Some other visible windows are the *Layer Selection* window that allows choosing which graphical attributes are to be displayed and the *Package Browser* window that allows selecting some uIC package(s) placed on the WaferIC™. A console window is placed at the lower part of the GUI. This console window allows the execution of all user workflow actions through text commands.

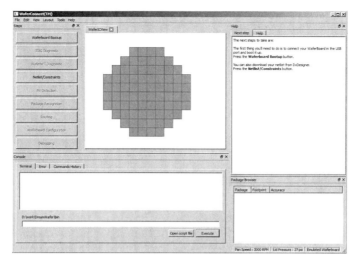

FIGURE 4.6   Default layout of the WaferConnect GUI.

Furthermore, system commands and user custom scripts can also be executed in this console. This minimum set of GUI elements is sufficient for generic users to execute their own specific tasks.

For advanced users, a more sophisticated WaferIC™ display window is offered as shown in Figure 4.7. It allows quick access to all visual tools and displays detailed information about the selected elements of the WaferIC™. Additionally, the view of the WaferIC™ can be completely customized and saved according to the user's

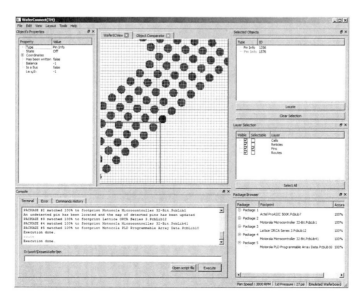

FIGURE 4.7   Advanced layout of the WaferConnect GUI.

preferences. The customization option for the user workspace is common in today's software and meets user expectations in shared working environments in which users can quickly initialize their own work environment and become fully functional within a very short amount of time. Finally, more complex control of the hardware can be implemented through more elaborate user custom scripts that are then executed in the console window.

Any command executed from the console is registered not only in its local environment but also in the local environment related to the GUI and vice versa. Furthermore, all executed commands are registered within the session manager. In this manner, the application remains stable and the information presented to users remains coherent despite the redundancy of the execution environment.

4.3.5.2   WaferIC Visual Structure

Visual elements of the WaferIC™ include reticules, cells, NanoPads, pins, footprints, routes, JTAG links and crossbar links. The display and selection of the WaferIC™ visual elements are organized through a layer system in which selections can be structured and displayed and are controlled by the user requirements. The *Layer Selection* window shows the type of objects to be displayed. Each layer has two properties that can be dynamically modified by the user: displayable and selectable. Choosing option *Visible* for one layer means that objects in this layer will be displayed, while option *Selectable* means that they can be selected by using mouse clicks. Properties of selected objects can be viewed in two dedicated windows: *Object Properties* and *Object Comparator*. With this layer structure, only the objects of interest are displayed and examined at a given time, thus avoiding overcrowding the graphical representation while offering clear object-related information to the user.

4.3.5.3   Error Management

Unlike the constraints of a standard PCB system, the reconfiguration possibilities of the WaferBoard™ can lead to a multi-user working environment. Each user may have to reconfigure the WaferBoard™ for his prototype circuit every time he or she accesses the platform, so reconfiguration has to be completed efficiently. Also, the user is able to react quickly to the platform status and component anomalies.

To cope with these requirements, an error management and error display system is implemented in the software and manipulated through the GUI. The characteristics of such a system are summarized as follows:

- Through a regular and systematic hardware diagnosis procedure, defect zones on the WaferIC™ are identified and can be visually presented to users. Acknowledging these zones allows users to choose the best zones to implement their circuit.
- All failures during operation, either hardware or software, will be reported to users with proposals for solutions. A hierarchical categorization of failures and their associated proposals for actions are implemented in the error management system, allowing appropriate user actions to be taken. At the top level, catastrophic failures will shut down the whole platform or prevent

users from manipulating it for safety, while at the lowest level, the error management system provides information to users in order that they can identify and fix the bugs themselves. For example, short circuits caused by metallic dust can be easily fixed by users if they are informed precisely where the short occurs. Poor pin contact is another example of a user-fixable fault that can easily be rectified with information concerning the specific component and pin identification.

4.3.6 WORKFLOW MODULES

Workflow modules use the WaferConnect™ backbone structure to achieve specific tasks, allowing the prototyping system to fulfil its functions. They include

- JTAG diagnosis
- WaferNet diagnosis
- Pin detection
- Routing
- Package recognition
- Connection with PCB CAD software
- Debug

Each workflow module demands algorithms to request appropriate data, to process them and to generate configuration commands for the platform. Some of these algorithms have been published elsewhere [6] and others will be the subject of future publications.

4.3.6.1 JTAG Diagnosis Module

JTAG diagnosis is an important step for JTAG-based hardware configuration applications such as the WaferConnect™ software. The objective of JTAG diagnosis is to detect functional JTAG links and the functional JTAG controller embedded in each cell. The JTAG and WaferNet networks include defect-tolerant mechanisms to bypass defects, which unavoidably occur in the wafer manufacturing process. The JTAG links have to be configured accordingly. Figure 4.2a shows the eight JTAG links connected to each cell.

Each JTAG link is classified in one of the four categories: functional, non-functional, tested or unknown. A functional JTAG link allows the signal to propagate to the immediate neighbour cells in its specific direction, N, S, W or E, while a non-functional link inhibits this propagation. A link is characterized as tested when it is part of a non-functional path generated during a software diagnosis, while a link with status unknown means that it is not included in any tested path. The manufacturing yield of JTAG links is estimated as very high, so a link under test has very small probability of being defective when it is still an unknown link. This module provides a complete map of defective JTAG links and a set of possible connections between the cells. This information is memorized in a WBHR object and used in all other external workflow modules that need to configure the WaferIC™ such as Routing and Pin Detection.

4.3.6.2 WaferNet Diagnosis Module

As stated in Section 4.2, cells in the WaferIC™ are interconnected through a multi-dimensional mesh network called WaferNet. Any NanoPad configured as an I/O can be linked to any others through the configurable WaferNet, similar to a programmable printed circuit board.

WaferNet is implemented as a 26 × 26 programmable crossbar structure in which each crossbar is associated with a cell. As shown in Figure 4.2a, the crossbar has six incoming links and six outgoing links in each of the four directions: N, S, W and E. The links have lengths that grow geometrically and connect each cell to the 1st, 2nd, 4th, 8th, 16th and 32nd neighbouring cells. Furthermore, each crossbar has also two more inputs and outputs connected to a programmable array of 4 × 4 NanoPads. These connections allow two uIC pins placed against the NanoPad array to be connected to the input or output of the crossbar.

The WaferNet is also subject to contain some defects due to the manufacturing processes involved. These defects have to be precisely localized in order to optimize the route generation process. The WaferNet configuration registers are accessible through the JTAG network. A relatively fast algorithm has been implemented, allowing the detection of defective links by using the 30 control and 30 observation registers on the crossbar inputs and outputs, respectively. The diagnosis algorithm for stuck-at defect model runs in time complexity of $O(n^2)$ where n is the number of cells. Details of the algorithm and results can be found in [6]. Ongoing research is in progress to improve the diagnosis efficiency, using methods such as using concurrent processing.

4.3.6.3 Pin Detection Module

User ICs are placed on the WaferIC™ and their metallic solder balls are pressed against the NanoPad mesh, creating short circuits between NanoPads under the balls. A contact detection procedure provided by the WAPI module is used to detect these NanoPads in contact. It represents NanoPad mesh as a binary image in which NanoPads in contact with a ball are represented by 1 and those that are not in contact with a ball are at 0. Pin detection is a step of the process that leads to uIC pin identification: contact NanoPads are organized into groups, where each group represents a ball or contact pin of the uIC. If sets of locally connected NanoPads are disjointed, a standard connected component labelling algorithm can be used to solve the pin detection problem. However, modern chip packages (FPGAs, processors, RAMs, etc.) may have a very high pin density, while the unit distance between neighbouring NanoPads is fixed. Thus, for a given NanoPad array design, as pin density increases, there is a point where locally connected NanoPads corresponding to one pin may touch those that relate to a neighbouring pin, as illustrated in Figure 4.8a.

Segmenting touching objects in an image is a challenging problem in the image processing field, in which watershed transform emerges as a reference technique due to its stability in a wide range of applications. The watershed transform is based on the topological representation of grey level images (its gradient images) or images resulting from the distance transformation of binary images. It simulates the process of water flooding in a digital elevation model (DEM) for topological

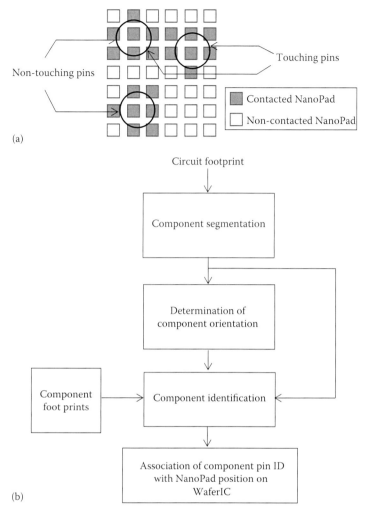

FIGURE 4.8 Touching uIC pins and non-touching uIC pins (a) and steps of the package recognition algorithm (b).

images. In the end, watershed lines (representing image boundaries) delimiting catchment basins (representing image regions) are then obtained, resulting in the segmentation of images.

A variation of the watershed algorithm [7] applied to binary images was implemented to solve the pin touching problem.

4.3.6.4 Package Recognition Module

Package recognition is an operation consisting of identifying user IC (uIC) packages placed on the WaferIC™. The Package Recognition module assumes that a library of package footprint models is available. This module tries to match footprint models on parts of the circuit footprint provided by the Pin Detection module. The matching operation is processed according to the following steps (see [5] for more details):

- Regrouping pins of the circuit footprint into component instances. This step allows a higher level of data abstraction: from uIC pin representation to component representation. This step is carried out by regrouping close-enough uIC pins through a thresholding operation in which the threshold value is empirically determined according to the maximum pitch computed from the available footprint models. This step is based on the assumption that uICs placed on the WaferBoard™ are distanced far enough from each other compared to the component pitches. This assumption is generally true in practice.
- Computing uIC orientations by exploiting the fact that most uIC packages are rectangular and most uIC pins are aligned. The orientation is computed from uIC extreme positions in N, S, W and E directions.
- Finally, recognizing packages by using discriminating attributes such as the number of pins, the pin pitch and the pin size (Figure 4.8b).

The output of the Package Recognition module is a list of associations *pin IDs – NanoPad positions* for each uIC. Combining this list with the netlist provided from the user, a routing netlist can be generated. It contains point-to-point (NanoPad-to-NanoPad) connections on the WaferIC™, conforming to the netlist provided by the user. This netlist will then be inputted to the Routing module, which determines in detail how routes will be created through the WaferNet.

4.3.6.5   Routing Module

The Routing module computes nonconflicting routes through the interconnection network. It takes a user PCB-like netlist as input and its routing algorithms are similar to those used in FPGA routers, with some differences as the network architecture is slightly different. Mainly, links in WaferNet are designed with a large degree of redundancy and multiple paths are possible for pin-to-pin connections. The optimization problem that the routing algorithm solves must satisfy the following constraints:

- *Fault-tolerance*: The computed path must not include any defective links, which were identified by the WaferNet Diagnosis module.
- *Minimization of propagation delay*: WaferNet is implemented with integrated wires and repeaters that introduce delays several times longer than those produced by same length PCB copper traces, exacerbating the necessity for an algorithm that focuses on minimizing propagation delays.
- *Respecting timing constraints of the user netlist*: Some specific sets of connections in the user netlist may have their own timing constraints (e.g. bus structures). The timing constraints of such sets need to be balanced such that these nets must have almost the same propagation delay, within a tolerance.
- *Nonconflicting routes*: A link in the WaferNet cannot belong to more than one route. If it happens, a conflicting situation between routes occurs and must be resolved.

A routing algorithm that aims at satisfying these constraints has been developed. It is processed in three steps:

1. Compute individually the shortest delay path for each route. This is a non-trivial problem because of the non-monotonic propagation delay of the WaferNet. The route conflicting issue is not considered in this step.
2. Solve conflicting routes by permutations using random searches on the same-delay space. The permutation process helps to solve a number of conflicts in highly congested areas, leaving remaining conflicts to be taken in charge by the A* algorithm with a cost function defined only on conflicting links, thus allowing permutations generating conflicting paths with conflicting lengths shorter than the ones that belonged to the path they replace.
3. Paths with some remaining conflicting links, generated in the previous step, are forwarded to an additional routing process in which each conflicting link is replaced by a longer nonconflicting one that starts and ends at the same cell. If some conflicting links is still not solved by this process, associated remaining paths will be computed using a standard maze router.

The proposed routing algorithm was shown to be fast and to generate quality routes (based on resulting time delay) for network size equivalent to the actual WaferIC™ (76 reticule images, 1024 cells per reticule). It is worth mentioning that generated routes in step 3 do not have minimal delay (because they are somewhat longer than the ones they replace), but while they represent as little as 1% of the total wire length, they allow to speed up the overall routing time by more than one hundred times.

4.3.6.6   Connection with PCB CAD Software

The WaferConnect™ software is designed for interacting with selected commercially available PCB CAD software in order to import the user netlist of the prototyped electronic system and the footprint models of electronic components in the circuit. Two CAD software suites, DxDesigner and PCB Expedition, both from Mentor Graphics Inc, a recognized leader in electronic design automation software tool, were selected for ease of netlist import/export. Interfaces for these CAD tools were integrated into the WaferConnect™ software platform.

It is of interest that most electronic systems are not only composed of ICs but typically they need additional passive components connected to the uICs, which the WaferBoard™ does not have to support. Such components may serve as pull-up or impedance matching resistors, decoupling capacitors, resonant circuits and connectors. A user netlist developed for conventional PCB integration could contain many such components. As they are not needed when a system is implemented on the WaferBoard™, these components must be pruned from the netlist. A tool was developed for automatically pruning netlists targeting the WaferBoard™ platform. In addition to producing a new netlist containing uICs only, the tool augments that netlist with commands necessary for configuring the NanoPads, which can be set as floating, pull up, pull down, power supply or I/O with any voltages between 0 and 3.3 V [8].

4.3.7   Summary on WaferConnect: User Support Software Tools

A set of user support tools that compose the WaferConnect™ toolbox has been presented. These tools help users configure their prototype electronic system in a WaferBoard™ Rapid Prototyping for Electronic Systems. The toolbox includes

- Hardware diagnosis for the JTAG network and the WaferNet to detect defective links in these two networks and exclude these regions for prototyping.
- JTAG-based hardware configuration module to configure the WaferIC™.
- A set of modules that must be executed in a well-defined order as specified by a user workflow manager. These modules include Pin Detection, Package Recognition, Routing and Debugging.
- Communication modules interfacing the tools and the WaferBoard™ hardware. These modules are structured according to a multilevel communication architecture that ranges from high-level communication functions interacting with high-level software modules to functions for manipulating the bitstreams driving the hardware level.

The following section addresses a completely different challenge with the WaferBoard™ platform. If not properly managed, the effects of thermo-mechanical stress on the thin and fragile multilayered WaferIC™ structure could induce large values of stress, distortion and warpage. Techniques for accurate evaluation of excessive heat and mechanical stress and their results to circumvent premature failure are presented.

4.4   WAFERIC THERMO-MECHANICAL INVESTIGATIONS

The WaferBoard™ includes a WaferIC™ in a mechanical chamber with a flat pouch of thermally conductive fluid (Figure 4.1b). This chamber accommodates significant variations in the height of uICs with adjustable uniform downward pressure on all uICs, controlled by a plunger. The top of the chamber is a large heat sink that can support considerable pressure (up to 10 atm). The WaferBoard™ platform has a BottomPCB that contains permanently affixed components to power the WaferIC™ and uICs and to provide/get signals to/from WaferIC™ through flexible PCB connections. The WaferIC™ cannot be powered with direct connections to the active surface. The power is supplied from the back of the WaferIC using TSVs. The power supply voltage is regulated by PowerBlocks that embed electronic components such as voltage regulators, capacitors and inductors, surface mounted on an aluminium nitride (AlN) ceramic substrate. To ensure good electrical contact between uIC and NanoPads, a strong force is applied on each uIC with a uniform pressure. Solid mechanical support, uniformly distributed along the bottom of the silicon wafer, is therefore required in order for the fragile mechanical structure to resist such pressure.

The proximity effect of heat sources is important in the multilayer WaferIC structure. In addition to the normal component junction temperature issue, improper control of the temperature can degrade its performance or even induce enough mechanical stress to damage the silicon wafer itself. So the maximum power indicated by the manufacturer of each component that compose the user system must be respected at any moment during the operation of the electronic system prototyped with the platform. Excessive temperature can destroy the crystal lattice of the semiconductor wafer. The heat developed inside the material must therefore be evacuated by means of the heat sink to ensure the maximum junction temperature T_{jmax} is never exceeded.

The multilayered WaferIC™ structure is also made of materials that have different properties, specifically different coefficient of thermal expansions (CTE). Thermal stresses, distortion and warping are therefore a source of concern. Much effort could be spent in performing experiments to measure the stress and temperature generated in this multilevel device. This work is however costly and time consuming. Due to the complexity of these problems, no analytical tool has been developed to adequately solve them. The finite element method (FEM) has been widely used by civil and mechanical engineers to build models in order to estimate thermal stresses and distortion of structures under a variety of loading conditions. In the case of thin multilayered devices such as the WaferIC™, the surface may be assumed to be in a state of plane stress [9]. Hence, an extensive finite element analysis is needed to investigate the WaferIC™ surface, especially at solder ball interfaces. The approach presented in Section 4.1 captures the thermo-mechanical stress singularity at the WaferIC™ level to predict the critical tensile stress. This approach is only valid for steady-state heat conduction and is limited by simplifying assumptions. Hence, measured hot spot temperature and realistic associated stress distribution can be used to guide thermo-mechanical design. The thermo-mechanical stability of the WaferIC structure also depends on PowerBlocks and the thermal peak produced on the WaferIC. The analysis of the PowerBlock thermal behaviour is therefore presented in Section 4.5.

4.4.1 WaferIC Thermal Analysis

4.4.1.1 Thermo-Mechanical Boundary Conditions

One of the most challenging issues in creating compact thermal models is to use an appropriate set of boundary conditions (BCs) for generating 'data' with a detailed finite element model representing the thermal envelope of the application.

In this study, the NISA finite element program (Numerical Integrated elements for System Analysis) [10] has been used to predict thermal behaviour in the multilayered WaferIC™ structure. A wide variety of BCs can be applied using NISA. However, the BC on the vertical sides of the simulation region is somewhat problematic. Placing a fixed temperature as BC on these surfaces produces an incorrect result [11], unless a very large simulation region is used at the expense of very long simulation run times. A more natural BC is a zero flow condition across these tiny surfaces (adiabatic BCs). The uniform heat removal at the bottom and top is modelled by the heat flux exchange coefficient h [W/m² K]. The power dissipated in heat sources (components) placed on the WaferIC™ is modelled by the heat flux produced into the uICs.

The remaining BCs to be defined are on the bottom and top surfaces of the WaferBoard™, representing the heat-sink interfaces. Heat flows happen mainly in the vertical direction because the WaferIC™ is relatively thin and silicon and solder are good thermal conductors, so the BCs in both horizontal directions can be considered adiabatic.

4.4.1.2   Thermal Analysis

The WaferIC™ thermal analysis depends on

- The cooling options (one or two radiators (top and bottom)) applied
- The location/vicinity and power of its heat-dissipating neighbours
- The thermal conductivity of the WaferBoard materials: PCB, heat sink, package, substrate and heat spreader

Final results depend on several parameters and thermo-mechanical components of this platform, such as the package and heat sinks (geometry and materials). The main heat source is the uICs. The bulk of the power consumed by the WaferIC™ is dissipated directly underneath these uICs. Figure 4.9a shows that the vertical heat path dominates and that it is quite independent for each component, so for this prototype, the components can be considered individually. The specific benchmark for which results are presented in this section is a 2×60 W uIC (equivalent to a Pentium as reference) for which the package has a 40×40 mm^2 area (0.0016 m^2) and 2×20 W (equivalent to the size of Xilinx Virtex IV packages). The total power dissipated is 160 W. Both cooling from the top surface of ICs with pouch contact and the bottom heat sink with forced convection at 25°C were considered. Figure 4.9a shows thermal results for a WaferIC™ with two microprocessors and two FPGAs mounted on its top face.

In the case considered here, the first impediment to heat dissipation is the pouch of material over the uICs that maintains a uniform mechanical pressure on the components (Figure 4.1b). It is assumed that the top of the pouch would only offer a small contribution to the thermal impedance, as the thermal fluid acts as a thermal spreader.

The heat generated within the WaferIC™ has an even easier path to escape, with only 15 W (WaferIC™ total power supply heat dissipation at one-fourth solder ball) under a Pentium-sized chip and no polymer pouch to traverse. The uIC solder balls placed on the WaferIC™ are very thin and thus present no significant thermal impedance. A bottom aluminium heat sink would handle only 40 W from the whole WaferIC™ (excluding the uIC) and is sitting in an airflow. Figure 4.9a shows isothermal contours in the WaferBoard™ structure under these assumptions. Figure 4.9b shows a cut view of the material stack under the components, with identification of layer BCs. Hence, as shown in Figure 4.9c, thermal peaks occur at Y1-uIC solder balls–WaferIC™. Finally, Figure 4.9d shows a schematic of the WaferBoard structure.

Figure 4.9c shows the evolution of the temperature along Y1 at uIC solder balls–WaferIC™ junction. Hence, a non-uniform temperature profile and thermal peaks of 42°C exist through the WaferIC™ thickness at the uIC solder balls–WaferIC interface. Under these conditions, the maximum thermal variation ΔT_{max} is 17°C (42–25) in the structure. However, at the uIC solder balls–WaferIC™ interface, the maximum local thermal variation $\Delta T_{max(loc)}$ is 3°C induced locally by WaferIC™ power distribution. Between the uICs, there is spatial relaxation on temperature gradients through the structure leading to an essentially uniform temperature variation (Figure 4.9a). The cooling of the device is essentially controlled by conduction through the WaferIC™ and through the uICs.

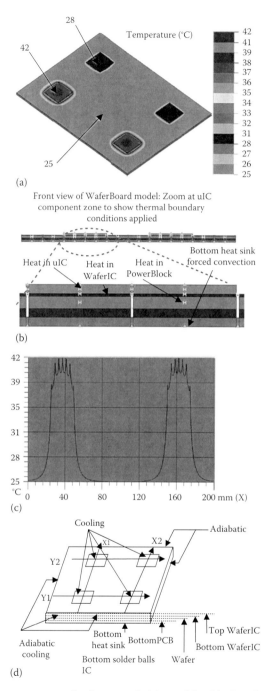

FIGURE 4.9 Temperature results for a sandwich model with forced convection cooling through the bottom heat sink and uICs on its top surface (a), cut view of the material stack under uICs with layers BCs identification (b). sandwich forced convection through backplate surface: temperature profile Y1-uIC solder balls–WaferIC™ (c) and schematic WaferBoard structure (d).

At this stage of the investigation, it is interesting to evaluate the effect of the WaferIC™ thermal energy off the uIC solder balls on the WaferIC™ temperature variation. Figure 4.9c shows the evolution of the temperature at the uIC solder ball interface. Hence, the maximum local thermal variation $\Delta T_{max(loc)}$ is 5°C induced by a 60 W power dissipation. However, larger peaks are observed. They are induced ($\Delta T_{(loc)} = 3°C$) by thermal coupling with the WaferIC™ heat energy. Therefore, the large temperature gradient across the interfaces region can be expected to contribute to the generation of critical thermal stress.

Accurate thermal analysis was necessary to maximize the ability of the platform to support high local power densities. A most likely form of damage that could result due to thermal shear stress is delamination [12] of solder balls in the WaferIC/TSV–PowerBlock interface. The generated stress depends on the induced thermal heat in the WaferIC™ structure. Based on our reported analysis results, the maximum stress occurs at the uIC solder balls. The levels of thermal peaks induced by uIC and WaferIC™ heating have been evaluated. Thermal effects of combined heat sources over the thermal peaks can capture the thermo-mechanical stress singularity at the WaferIC™ level to predict the critical tensile or shear stress in WaferIC™, as shown in the following section.

4.4.2   WaferIC™ Thermo-Mechanical Analysis

This section presents results of thermo-mechanical analysis conducted during the design of the multilayered WaferIC™ structure. The nodal temperature results calculated during the thermal analysis were used to perform a thermal stress analysis of the whole WaferIC™ structure. Some of thermal analysis results were presented in previous papers [13]. However, in this section, the predicted induced thermal stress is combined with the intrinsic one due to the fabrication processes, pressure from the heat transfer fluid pouch, and the stress due to the clamping mechanical structure. For large ICs such as the WaferIC, the shear stress is also a significant element of the total possible thermo-mechanical stress problem.

4.4.2.1   Stresses Convention for WaferIC Thermo-Mechanical Analysis

Figure 4.10 provides the components of stress tensor in the global and/or local XYZ Cartesian coordinate system. Local components may be provided for shell elements or for orthotropic materials. In order to establish the directions of stress components, an infinitesimal cube at the point of interest was considered. The positive stresses act on the positive faces of the cube in the positive direction. The stress tensor components are the direct stresses Sxx, Syy, Szz and the shear stresses Sxy, Syz, Sxz (Syx, Szy, Szx, respectively). The first subscript indicates the face on which the stress is acting and the second subscript is its direction. For example, Sxx is the direct stress acting on the +X face and in the positive X direction. The following stress intensities were computed from the stress components. They provide useful information for interpreting the behaviour of the material according to the various theories that predict failures.

Max shear stress is defined as S_{max} = maximum of [½|S_1−S_2|, ½|S_2−S_3|, ½|S_3−S_1|].
Von Mises stress is defined as S_{eq} = (1/sqrt(2)).[(S_1−S_2)²+ (S_2−S_3)²+ (S_3−S_1)²]$^{1/2}$.

S_1, S_2 and S_3 are the principal stresses in the directions 1, 2 and 3 (not necessarily x, y and z).

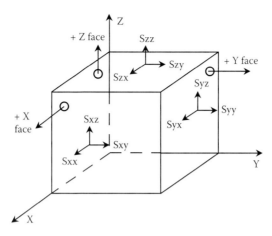

FIGURE 4.10   Components of stress tensor in the global and/or local XYZ Cartesian coordinate system.

TABLE 4.1
Materials Thermo-Mechanical Parameters

Materials	uIC (Die)	Moulding Compound	uIC Solder Ball	AlN
Young's modulus (GPa)	131	16	17	70
Poisson's ratio	0.3	0.25	0.4	0.33
CTE ppm/°C	2.8	15	21	22.4
Thermal conductivity W/m°C	150	65	50	160
Density kg/m³	2330	1660	8460	2700
Specific heat J/kg°C	712	1672	957	960

In general, the compressive stresses result in the development of large stress tensor components, σ_{xx} and σ_{yy}. In the tensor notation, σ_{ij} is the force per unit area in the ith Cartesian direction on a surface, which has a normal in the ith Cartesian direction. The σ_{ij} is the stress on the i face acting in the j direction. All such forces contribute to the stress tensor σ_{ij}, which is a second-order tensor. The first subscript i represents the plane normal over which the stress is acting, and the second subscript j represents the direction of the force. For example, σ_{xx} will be a normal stress since the stress direction and the plane normal are in the same direction, and σ_{xy} will be a shear stress. The component stresses can be either normal (i.e. tension or compression) or shear stresses.

In general, combined stress refers to cases in which two or more types of stress are exerted on a given point at the same time. Table 4.1 shows thermo-mechanical parameters for different materials used for the simulations.

4.4.2.2   Thermo-Mechanical Stress Results

Thermo-mechanical analysis for the sandwich model with forced convection and simply supported BCs on the bottom of the structure was conducted. Then a force was applied emulating the plunger on the top by applying pressure locally on the top surface of uIC. However, the induced thermal stress was combined with the intrinsic one due to the fabrication processes and the stress due to the assembling mechanism. This model allows the exploration of several scenarios to minimize the thermal gradient in the critical areas, especially at the uIC's solder ball level.

Final results depend on several parameters and thermo-mechanical components of this system, such as the uIC packages and heat sinks (geometry and materials), that was designed at a later stage of this research. Considering the fact that the structure has high-thermal-conductivity regions more closely spaced than the size of the highest-power components, the vertical heat path dominates and is relatively independent for each component. For this prototype, the components are thus considered individually. The same uIC benchmarks as in Section 4.11 were used (equivalent to a Pentium 4, 3.4 GHz).

Along Y1 cut (Figure 4.9d), our numerous simulation results showed that maximum shear stress is of the order of +34.6 MPa, acting on solder balls between PowerBlock–WaferIC™. The temperature is maximum in the WaferIC™ substrate, therefore the maximum thermal variation is found at this solder ball interface. Furthermore, a maximum shear stress of +32.6 MPa was found on the top side of WaferIC™. Finally, a maximum shear stress of +68 MPa was found on the uIC bottom face at the solder ball interface.

Considering these observed stresses, cumulative damage is likely to occur at the critical interface regions. As such layers are very thin, any imperfection in structure may lead to cracking and subsequent shear-initiated delamination of the WaferIC™ material. These results reveal that such interface materials must be carefully considered in multilevel device design, especially for intense activities with the platform (repetitive and thermo-mechanical multicycling). The levels of stress and distortion generated by intense power heating of the WaferIC show that solutions must be provided to reduce them. Moreover, during design and final WaferIC™ mechanical configuration selection, safety margins must be applied to ensure safe operating conditions. The level of stress reported in this work and possible stress concentration point in the WaferIC™ structure cannot be attributed to a singularity of the geometrical description.

4.5   THERMAL ANALYSIS OF THE WAFERIC–POWERBLOCK

One of the main and critical sources of thermo-mechanical stress in the WaferIC structure is the PowerBlock. This device embeds voltage regulators and acts as thermal conductor up to the main heat sink (Figure 4.1b). Several factors impact the mechanical design of the PowerBlock device. In addition to the electrical and mechanical concerns, thermal considerations are also critical to the entire WaferIC™ and the power device for thermo-mechanical stability [13]. In this section, a methodology to evaluate and predict a thermal peak of PowerBlock matched to the WaferIC™ is presented. This methodology is applied to investigate important factors contributing to the device's thermal heating.

The method used to perform thermal analysis and investigations on the power device, carried out via a 3D thermal model and thermal simulations are introduced in Section 4.5.2 and investigation results are presented in Section 4.5.3. It is shown that the PowerBlock meets its mechanical and thermal performance requirements.

4.5.1 PowerBlock Design Constraints

Several design constraints were taken into account to improve the PowerBlock performance, such as the choice of proper materials, its dimensions and the selection of electronic components that need to be embedded inside:

- *Proper materials*: The selection criteria for the PowerBlock materials are based on ensuring a good mechanical support to the WaferIC and on maximizing the thermal conductivity of the PowerBlock device. Two materials have been investigated for the PowerBlock assembly: copper and AlN ceramic. Copper is widely used in the electronic industry for its large thermal conductivity and its strength coefficient, which are close to that of steels used by mechanical engineers to construct mechanical parts [14]. The second material considered is the AlN ceramic. It has a high thermal conductivity and a TCE close to that of silicon. It reduces the expansion mismatch between the PowerBlock and the WaferIC [15,16], in addition to offering better thermal conductivity.
- *Dimensions*: The PowerBlock is enclosed between the WaferIC™ and the bottom heat sink (Figure 4.1b). The PowerBlock thermal conductance is inversely proportional to its height. Thus, if the height is reduced, the thermal conductance increases. The height depends upon the height of the electronic components embedded inside the PowerBlock; therefore, the height of the components is an important consideration for improving the PowerBlock's thermal conductance. On the other hand, the PowerBlock area, which is a multiple of the size of a reticule image (2 by 2), is 36×36 mm^2, and it embeds all necessary power electronic components, connectors and heat-sink copper spacers. Finally, these dimensions have a strong influence on the choice of the electronic components.
- *Electronic components*: Based on an analysis of the platform requirements, it was specified that each PowerBlock device must have the ability to provide up to 100 W of electrical power to the region of the WaferIC™ to which it feeds power. The power dissipation in the PowerBlock can generate a significant self-heating that considerably increases its internal temperature and impact its thermal performance. Therefore, the PowerBlock design must take into account the electronic components temperature limit and their size (the smaller, the better, in this case).

The PowerBlock consists of multilayer thick films on an aluminium nitride substrate, four layers, each approximately 0.15 mm thick, carrying signals and current at the appropriate voltages. Two voltages are provided by the power device using switching buck regulators: 3.3 V at 25 A for powering components placed on the wafer and 1.8 V at 5 A for use by the circuitry within the silicon wafer itself. Consequently, the PowerBlock can conduct around 100 W of electrical power and the estimated power loss in the power device is around 10 W. In addition, an FPGA is also mounted on the power device to encode the transmitted data signal coming from the outside to the wafer. Figure 4.11a and b shows, respectively, a simplified schematic for the PowerBlock and the PowerBlock's diagram.

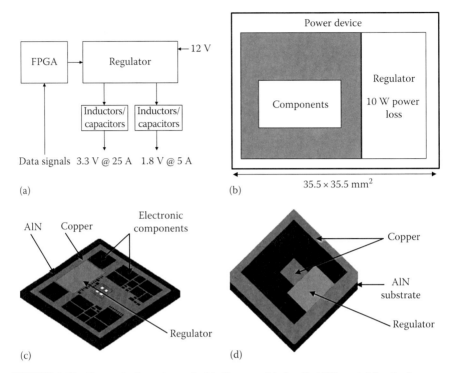

FIGURE 4.11 Power device schematic (a), diagram (b), detailed FE model for the first configuration (c) and simplified for the second configuration (d).

The power reaches the WaferIC™ through TSVs. The back side of the WaferIC™ has a dense array of solder balls that bond the WaferIC to the top of the PowerBlock (Figure 4.1b). This requires the PowerBlock to provide good mechanical support to reduce the bow of the WaferIC™ [13,14]. The use of a material with high thermal conductivity and TCE close to that of silicon, in this case AlN (aluminium nitride), to reduce expansion mismatch is crucial for the integrity of the WaferIC™ [16]. In addition to the mechanical support, the PowerBlock provides power to the WaferIC™; the WaferIC™ needs a downward path for heat dissipation so the PowerBlock device must provide a low thermal resistance path to the heat sink beneath it.

4.5.2   METHODOLOGY FOR POWERBLOCK THERMAL ANALYSIS

The electronic components that are surface mounted in the PowerBlock reduce the total area available to conduct heat. The materials from which the PowerBlock is built must conduct heat to the heat sink, but where components and air gaps are present, heat must flow laterally to reach the areas of good thermal conductivity. This greatly exacerbates the challenges associated with thermal management. Overheating in some areas would cause hot spots that not only reduce circuit life but also induce large thermal stress. A coupled fluid–heat transfer thermal analysis was done. In this case, the thermal behaviour depends on the PowerBlock geometry and materials, junction structure and physical heat sources distribution.

The mixed fluid–heat transfer approach for thermal analysis of a thermal path considers large PowerBlocks, lateral heat flow through the thin structures, vertical heat flow through the limited thermally conductive area and an estimated convection coefficient of the heat sink. Based on the FEM, the approach combines fluid flow and heat transfer analysis to predict working temperature of the wafer-scale active surface. Based on this analysis, the effects of power density, position and heat-sink characteristics during thermal response can be investigated. The adopted mixed approach can be used for accurate rating of semiconductor devices or heat-sink systems when designing large circuits.

As mentioned before, one of the most difficult issues in creating compact thermal models is to use an appropriate set of BC for the finite element model. The accuracy of the calculated thermo-mechanical distributions depends of the BC that must be selected to reflect the thermal envelope of the application of interest. The thermal analysis depends on the following:

- The applied cooling options
- The location/power of heat-dissipating devices
- The thermal conductivity of the various materials and components such as the printed circuit board, heat sink, components package, substrate and heat spreader

As shown in Figure 4.12, the objective of the approach is to extract the equivalent forced convection coefficient to be applied at the device level. This is achieved in

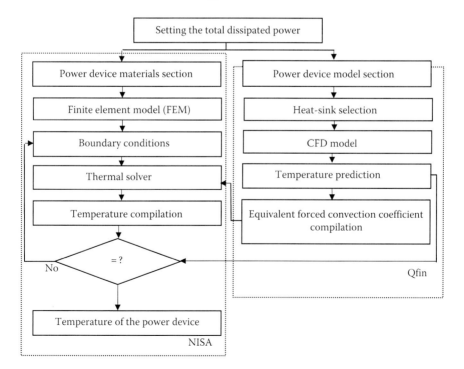

FIGURE 4.12   Flow chart of mixed fluid–heat transfer approach.

two steps. In the first step, the Qfin software (computational fluid dynamics (CFD) thermal analysis) [17] is used to determine the heat-sink performance. The forced convection coefficient and heat-sink configuration are then used to predict detailed PowerBlock temperature distribution.

Qfin is used to compute the thermal equivalent forced convection coefficient, which is then used to solve the assembly configuration. The calculation of the heat transfer coefficient is in turn dependent on the type of convection that the assembly is subjected to, as well as the ambient conditions. Hence, an equivalent forced convection coefficient for the whole heat-sink model is used as an input to complete the thermal BCs. In the second step of the analysis, NISA is again used to obtain the same temperature obtained with the equivalent forced convection coefficient computed in step 1.

4.5.3 EXAMPLES OF POWERBLOCK THERMAL INVESTIGATION RESULTS

Two configurations with different floor plans for the power PowerBlock are presented. The difference between these configurations is that in the first one, a copper rectangle is fixed to one of the four side of the PowerBlock device. The role of this copper rectangle is to add mechanical support to the wafer, as well as to increase the thermal conductivity of the PowerBlock device. For the second configuration, a square copper support was added to the centre of the PowerBlock device. Figure 4.11c and d shows, respectively, a detailed PowerBlock FE model with electronic components and the PowerBlock simplified model for the second configuration with copper at the centre.

For thermal investigations, various BCs and approaches (CFD and FEM) were tried to perform a detailed and accurate thermal analysis of the power device feeding nominal power. The convection BCs represented by h (forced convection coefficient) applied on the bottom face of the PowerBlock device were typically between 10 and 50 W/m^2°C). The worst-case thermal simulation scenario is a free air convection, which is typically between 3 and 12 W/m^2°C, when the power loss in the PowerBlock device is 10 W (when the PowerBlock provides 100 W to the WaferIC, the regulator and the other passive components in the PowerBlock device consume 10 W, the specified power loss) and the power dissipated over the WaferIC™ is 60 W.

The analyses were done assuming forced convection through the bottom PCB. Two configurations are designed to minimize temperature variation and for efficient thermal energy evacuation.

Figure 4.13b and c shows, respectively, the temperature distributions for the first and second PowerBlock device configuration, respectively, at the bottom of the device with a bottom film convection coefficient of 50 W/m^2°C. Tables 4.2 and 4.3 summarize the minimum and maximum temperatures in the PowerBlock device for different bottom film coefficients, ranging from 10 to 50 W/m^2°C.

Tables 4.2 and 4.3 show, respectively, temperature results with different film coefficients for the first configuration and results with for the second power device configuration.

The reported results show that in the first configuration, which is without a central block and with copper on the surface of the PowerBlock device, the maximal

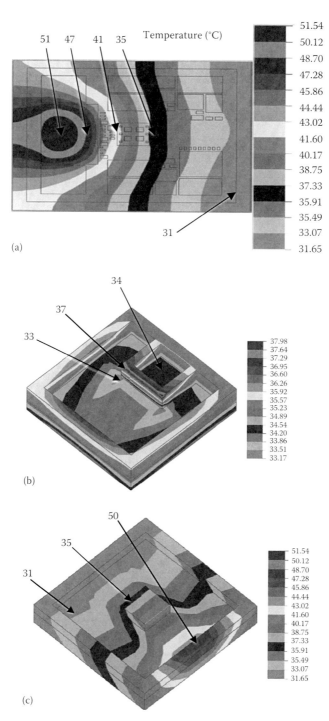

FIGURE 4.13 **(See colour insert.)** Power device thermal profile (a), temperature distribution for the first (b) and second (c) PowerBlock device configuration.

TABLE 4.2

Temperature Results with Different Film Coefficients for the First PowerBlock Configuration (Shown in Figure 4.13a)

Bottom Film Convection Coefficient	T_{min} (°C)	T_{max} (°C)
50 W/m²°C	33.17	37.98
40 W/m²°C	35.34	40.16
30 W/m²°C	38.97	43.80
20 W/m²°C	46.24	51.09
10 W/m²°C	68.09	72.95

TABLE 4.3

Results with Different Film Coefficients for the Second PowerBlock Configuration (Shown in Figure 4.13c)

Bottom Film Convection Coefficient	T_{min} (°C)	T_{max} (°C)
50 W/m²°C	31.65	51.54
40 W/m²°C	34.02	54.53
30 W/m²°C	38.14	59.34
20 W/m²°C	46.68	68.71
10 W/m²°C	73.09	96.09

temperature with a bottom film convection coefficient of 50 W/m²°C is around 38°C, while the minimal temperature is around 33°C. By contrast, for the second configuration, the maximal temperature increases to around 51.5°C, which is caused by the heat being concentrated behind the regulator, located on the right side of the power device. This reduces the beneficial effect of the central block to reduce the temperatures in the power device. These effects are clearly shown in Figure 4.13b and c, respectively, with the heat spreading across the PowerBlock device and the concentration of heat on its right side. A second explanation for these results is the absence of copper on the surface of the PowerBlock in the second configuration, which impedes the desired steady heat spread across the surface of the PowerBlock device.

4.6 CONCLUSION

In this chapter, a complete user tool for WaferBoard™, a rapid reconfigurable electronic prototyping system and a methodology to evaluate and predict a steady-state thermo-mechanical behaviour of reconfigurable wafer-scale IC (WaferIC) have been presented. Users configure their electronic systems by interacting with the WaferIC™ through accessing features offered by the WaferConnect™ software via its GUI. These features include procedures for hardware diagnosis, uIC package identification, routing and WaferIC configuration. All these operations are coordinated by the user workflow manager using security features in order to protect the system from unexpected and unsafe user actions.

Concerning the evaluation of the thermo-mechanical behaviours in the proposed reconfigurable wafer-scale integrated system, a new approach has been developed that combines FEM and mixed fluid–heat transfer approach for analysing thermo-mechanical stress, distortion and warpage. It provides a methodology to evaluate thermal stress and distortion in electronic devices and identifies the most important factors contributing to the WaferIC's thermal failure. Furthermore, a methodology for the evaluation and prediction of a steady-state thermal behaviour of a PowerBlock matched to a silicon wafer was also presented. The important factors contributing to the device's thermal heating were characterized. The temperature distribution of the PowerBlock when supplying power to a high-performance uIC in a standard package was determined. The modelling approach reported in this study can also be applied to predict the peak thermal stress of any miniature printed circuit board matched to a silicon wafer. Finally, the obtained thermal results permitted evaluating and comparing two configurations of the PowerBlock device, which is a necessary step for the design of the power distribution device.

The DreamWafer™ project, after 4 years of research in both hardware and software, should soon lead to a complete functional prototype of the WaferBoard™ technology. The first prototype of the system is planned to be released at the end of this year (2012), including some software infrastructures necessary for making use of that platform. Several innovations were necessary to arrive at such a functional prototype [18,19] and the research so far has answered many questions but opens up many areas for future research.

REFERENCES

1. R. Norman, O. Valorge, Y. Blaquiere, E. Lepercq, Y. Basile-Bellavance, Y. El-Alaoui, R. Prytula, and Y. Savaria, An active reconfigurable circuit board, In *2008 Joint IEEE North-East Workshop on Circuits and Systems and TAISA Conference, NEWCAS-TAISA,* 22–25 June, 2008, Montreal, Québec, Canada, 2008, pp. 351–354.
2. Qt - Cross-platform application and UI framework—Qt–A cross-platform application and UI framework. [Online]. Available: http://qt.nokia.com/. [Accessed: April 24, 2012].
3. Standard Template Library Programmer's Guide. [Online]. Available: http://www.sgi.com/tech/stl/. [Accessed: 24 April, 2012].
4. Boost C++ Libraries. [Online]. Available: http://www.boost.org/. [Accessed: April 24, 2012].
5. E. Lepercq, Y. Blaquiere, R. Norman, and Y. Savaria, Workflow for an electronic configurable prototyping system, In *IEEE International Symposium on Circuits and Systems, 2009. ISCAS 2009,* Taipei, Taiwan, 2009, pp. 2005–2008.
6. Y. Basile-Bellavance, Y. Blaquiere, and Y. Savaria, Faults diagnosis methodology for the WaferNet interconnection network, In *Circuits and Systems and TAISA Conference, 2009. NEWCAS-TAISA '09. Joint IEEE North-East Workshop on,* Toulouse, France, 2009, pp. 1–4.
7. F. Meyer, Topographic distance and watershed lines, *Signal Processing,* 38(1), 113–125, 1994.

8. N. Laflamme-Mayer, M. Sawan, and Y. Blaquiere, A dual-power rail, low-dropout, fast-response linear regulator dedicated to a wafer-scale electronic systems prototyping platform, In *2011 IEEE 9th International New Circuits and Systems Conference, NEWCAS 2011*, 26–29 June, 2011, Bordeaux, France, 2011, pp. 438–441.

9. C. Perret, J. Boussey, C. Schaeffer, and M. Coyaud, Analytic modeling, optimization, and realization of cooling devices in silicon technology, *IEEE Transactions on Components and Packaging Technologies*, 23(4), 665–672, 2000.

10. EMRC, *NISA II User's Manual*, (Engineering Mechanics Research Corporation), Troy, MI, 1998.

11. A. Lakhsasi and A. Skorek, Dynamic finite element approach for analyzing stress and distortion in multilevel devices, *Solid-State Electronics*, 46(6), 925–932, 2002.

12. M. S. Bakir, B. Dang, R. Emery, G. Vandentop, P. A. Kohl, and J. D. Meindl, Sea of leads compliant I/O interconnect process integration for the ultimate enabling of chips with low-k interlayer dielectrics, *IEEE Transactions on Advanced Packaging*, 28(3), 488–494, 2005.

13. M. Bougataya, A. Lakhsasi, R. Norman, R. Prytula, Y. Blaquiere, and Y. Savaria, Steady state thermal analysis of a reconfigurable wafer-scale circuit board, In *IEEE Canadian Conference on Electrical and Computer Engineering, CCECE 2008*, 4–7 May, 2008, Niagara Falls, Ontario, Canada, 2008, pp. 411–415.

14. J. M. Gere and B. J. Goodno, *Mechanics of Materials*, 7th Edn., CL-Engineering, 2008.

15. W. Werdecker, Metallizing of aluminum nitride substrates, In *Fifth European Hybrid Microelectronics Conference Proceedings*, Stresa, Italy, 1985, pp. 472–488.

16. N. Kuramoto and H. Taniguchi, Transparent AIN ceramics, *Journal of Materials Science Letters*, 3(6), 471–474, 1984.

17. *Qfin 4 advanced user manual*. Qfinsoft Technology Inc., 2011.

18. Y. Blaquiere, Y. Savaria, Y. Basile-Bellavance, O. Valorge, A. Lakhsasi, W. André, N. Laflamme-Mayer, M. Bougataya, and M. Sawan, Methods, apparatus and system to support large-scale micro-systems including embedded and distributed power supply, thermal regulation, multi-distributed-sensors and electrical signal propagation, International Patent PCT/CA2011/050537, 16 March, 2012.

19. R. Norman, Reprogrammable circuit board with alignment insensitive support for multiple component contact types, U.S. Patent US 8,124,429 B2, 28 February, 2012.

Part II

Imaging and Display Technologies

.

5 MIRASOL Displays
MEMS-Based Reflective Display Systems

Rashmi Rao, Jennifer Gille and Nassim Khonsari

CONTENTS

5.1 PREVIEW

Qualcomm's mirasol displays are an alternative to current pervasive flat-panel display (FPD) technology; technologies such as liquid crystal display (LCD) and organic light-emitting diode (OLED) are based on interferometric modulator (IMOD) and utilize microelectromechanical systems (MEMS) elements. These elements modulate

105

visible light based on the principles of interference. IMOD technology has unique benefits. Because they are reflective, mirasol displays demonstrate excellent visibility from dim to bright conditions, including direct sunlight. Colour generation is achieved via interference of light, eliminating the need for colour filters or polarizers for either colour generation or light modulation. Additionally, because the switching speed of MEMS pixel elements is intrinsically fast, a variety of refresh rates can be achieved, including video. Early commercialization of mirasol displays has demonstrated the successful transition of these MEMS devices from fabrication at the semiconductor wafer scale to large area manufacturing in what would be considered more conventional FPD fabs.

List of abbreviations: CRT, cathode ray tube; FL, front light; FOS, front of screen; FPD, flat panel display; IMOD, interferometric modulator; LCD, liquid crystal display; MEMS, microelectromechanical systems; mirasol display, brand name of Qualcomm's reflective display devices based on IMOD technology; OLED, organic light-emitting diode; PECVD, plasma-enhanced chemical vapour deposition; TFT, thin-film transistor.

5.2   MIRASOL DISPLAY/IMOD TECHNOLOGY BASIS

Colouration in nature is generally realized in one of two ways: through pigmentation or by iridescence[1,2]. Iridescence (originated from the Greek word 'irides' – 'rainbows') is an optical phenomenon caused by multiple reflections from multiple layers of optical films in which phase shift and interference of the reflections modulate the incident light by amplifying or attenuating certain wavelengths more than others. Some examples of iridescence in nature include colouring in some butterfly wings, feathers of some birds and seashells [3]. There have been ongoing efforts to leverage and mimic these natural colour generation techniques to apply them to man-made devices. Recent developments in micro- and nanofabrication and photonics allow the fabrication of structures that mimic iridescent properties found in nature by manipulating light in a controlled manner via the use of MEMS/nanoelectromechanical systems (NEMS) architectures. These developments make a variety of novel devices and applications possible [4–6], including Qualcomm's reflective mirasol display [7–10].

At the core of the technology, the IMOD pixel is based on two fundamental concepts. First, thin-film interferometric modulation optics are used to produce colour in the reflected spectrum of a (white) ambient light, and second, MEMS switching mechanisms change the pixel state.

The basic MEMS device consists of a suspended conductive membrane serving as a mirror, suspended over a partially reflective optical stack. The gap separating the membrane and optical stack is only hundreds of nanometres wide and is filled with air. Interference between incident light reflected from the mirror and light reflected off the optical stack generates vibrant colour. The image on a mirasol display can switch between colour and black by changing the membrane state on the MEMS device. This is achieved by applying a voltage to the thin-film stack, which is electrically conducting and is protected by an insulating layer. When the voltage is applied, electrostatic forces cause the membrane to collapse. The change in the optical cavity now results in constructive interference at ultraviolet wavelengths, which are not visible to the human eye. Hence, the image on the screen appears black.

In situations where the display image is unchanged, mirasol display's design allows for near-zero power consumption due to its bistable nature. The bistability of mirasol displays comes from the inherent hysteresis derived from the technology's electromechanical properties. More specifically, it derives from an inherent imbalance between the linear restorative forces of the mechanical membrane and the nonlinear forces of the applied electric field. As will be explained in detail in Section 5.5, the resulting electro-opto-mechanical behaviour is hysteretic in nature and provides a built-in 'memory' effect.

The unique functionality and construction of the mirasol display offer significant power advantages because (i) reflected light (energy) is modulated from an ambient lighting source and (ii) low-voltage MEMS pixel switching automatically means low power consumption during display array addressing. Therefore, this IMOD technology offers significant benefits in viewability, ruggedness and power reduction, which make it attractive for a wide variety of consumer electronics applications. While LCDs in particular have also been used to make direct-view reflective displays, they have some inherent limitation especially due to the requirement of polarizer for light modulation because of which less than half of the light gets used to display images.

In this chapter, basic electromechanical physics of the IMOD device and principles of colour generation and modulation are described. Device fabrication and manufacturing utilizing mostly existing materials and tools available from thin-film transistor (TFT) display industry will also be discussed as well as necessary modifications due to the MEMS nature of the technology. A brief discussion on front-of-screen (FOS) components with particular focus on supplemental illumination needed for these reflective displays in darker ambient will also be included.

5.3 PRINCIPLE OF COLOUR GENERATION IN MIRASOL DISPLAYS

An IMOD pixel structure is analogous to a Fabry–Pérot interferometer [11,12] operating in reflection mode (Figures 5.1 and 5.3). It has an airgap between the two reflectors that shape the spectrum of reflected light where some wavelengths constructively interfere and others destructively interfere. In Figure 5.2, the example reflection spectra for IMODs for airgaps (on the order of 40–400 nm) corresponding to red, green, blue and black states, respectively, are shown. The range of colours created by the device as a function of the airgap can be computed using the transfer-matrix method [11] and is shown on a standard CIE 1976 chromaticity diagram [13] for a given thin-film optical stack of several tens of nanometres (see Figure 5.3). A typical optical oxide stack has a thickness of several tens of nanometres.

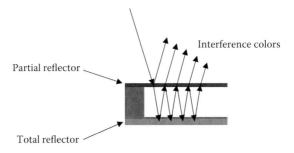

FIGURE 5.1   'Folded' etalon.

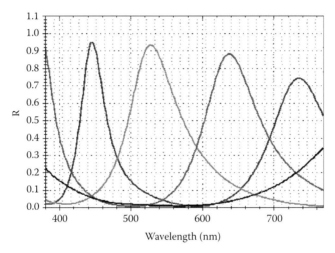

FIGURE 5.2   (See colour insert.) Reflection spectra for four different airgaps correspond-ing to red, green, blue and black colours, respectively.

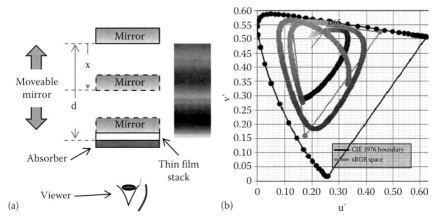

FIGURE 5.3   (See colour insert.) IMOD colours for varied airgaps: (a) IMOD device struc-ture and colour vs. gap and (b) IMOD colour spiral on CIE u′-v′ chart.

5.4 LIGHT MODULATION AND ELECTROMECHANICS

The colour in a mirasol display, as seen by the human observer, is changed by modulating the state of the device's two conductive membranes via the application of voltage. When a voltage is applied, electrostatic forces cause the membrane to collapse. This results in a change in size to the airgap of the IMOD device and hence a change in the reflected wavelength of light. The electrostatic force of attraction between the membranes is given by

$$F_E = \frac{1}{2}QE = \frac{1}{2}A\varepsilon_0\left(\frac{V}{d-x}\right)^2 = F_M = Kx$$

where

F_E is the electrostatic force of attraction between the two plates
Q is the magnitude of charge on the plates
E is the electric field between them
A is the area of the plates
V is the applied voltage
d is the electrical gap in undeflected state
x is the deflection

Under equilibrium, this force is balanced by the mechanical restoring force F_M that follows Hooke's law:

$$F_M = K \cdot x$$

where K is the effective spring constant. This results in a cubic equation in x solving, which results in the three solutions shown in Figure 5.4 (see inset labelled as actuated, released and unstable, respectively). Only two of the three solutions correspond to a stable equilibrium. With reference to Figure 5.4, as the voltage increases in magnitude starting from zero volts, the IMOD device begins in 'released' state ('A'), while the moveable plate gradually deflects towards the fixed plate. Once a threshold, referred to as the pull-in or actuation voltage ('B'), is crossed, the moveable plate collapses towards the fixed plate reducing the airgap thickness to nearly zero nanometres ('C') (the 'actuated' state). The transition to the actuated state is accompanied by a nonlinear increase of the electrostatic attraction, which increases rapidly as the airgap decreases. When the voltage is decreased, the electrostatic force must be reduced significantly before the moveable plate is able to snap back to the released state again. The threshold voltage level where the transition to the released state ('E' → 'F') occurs is referred to as the release voltage. When a voltage between the actuation and release voltages, referred to as the 'bias voltage', is applied, both actuated and released states are stable, akin to a memory effect where each pixel preserves the last state to which it was set.

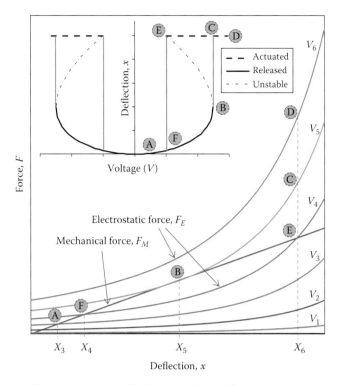

FIGURE 5.4   Balance between electrical and mechanical forces leads to hysteresis.

The dynamical response time of the IMOD device can be modelled by including the terms for acceleration and squeezed-film damping into the force equation [14,15]:

$$m\frac{d^2x}{dt^2} + \frac{\alpha\eta_0 A^2}{\left(1+(9.638)K_n^{1.159}\right)(g-x)^3}\frac{dx}{dt} + Kx = \frac{A\varepsilon_0}{2}\left(\frac{V}{d-x}\right)^2$$

where
    m is the effective mass of the mirror
    K_n is the Knudsen number
    η_0 is the coefficient of the viscosity of air
    α is a constant that is dependent on the geometry of the device [15]

Mechanical response times of a few microseconds are easily achieved (compared to 2–40 ms for LCD); Figure 5.5 demonstrates a good match between model predictions for the response time and device measurements across a range of bias voltages.

Hysteresis behaviour of the device is desirable for low-power operation because, unlike conventional LCDs, the content image written on the panel does not require refresh at high frame rate [16].

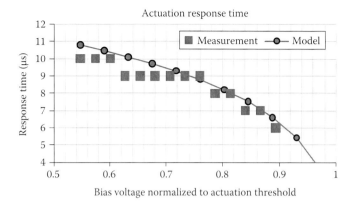

FIGURE 5.5   Response time as a function of bias voltage, model vs. measurements.

To change the state of the IMOD device, a voltage pulse is applied that takes it from the 'hold' or bias voltage to either an actuated or released state. Because smaller magnitude voltages are needed for this change than the bias voltage, low-power image update operation is achieved.

5.5   ADDRESSING METHODS AND CONTROL ELECTRONICS

In its current state of development, each pixel of a mirasol display, similar to other display technologies, typically consists of three sub-pixels for red, green and blue colour channels, respectively. The sub-pixels are independently controlled between the selected primary colour state and a black state. A common process sequence producing three separate airgaps is used to fabricate the individual sub-pixels. The colour points for the sub-pixels are chosen from the IMOD colour 'spiral' shown in Figure 5.3 with consideration to white-balancing constraint and the trade-off between brightness and gamut pertinent to typical reflective display technology.

Given the bistable or 1-bit nature of the device, intermediate colours are generated for each sub-pixel by dividing it into multiple area-weighted mirrors and/or by using spatial dithering techniques [17]. Each colour sub-pixel could be divided into several mirrors to provide some level of grey scale. One example of such a structure is illustrated in Figure 5.6. IMOD pixel sizes used are optimized to avoid

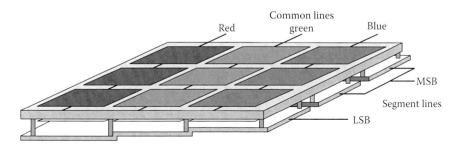

FIGURE 5.6   **(See colour insert.)** Example colour pixel structure (schematic).

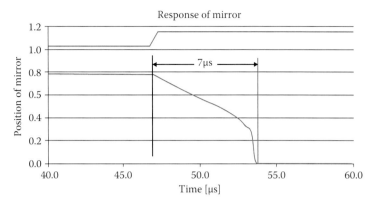

FIGURE 5.7   IMOD pixel response.

dither artefacts while still maintaining a reasonably high fill factor. Device sizes ranging from 30 to 60 µm are typically used for high-resolution displays in the market today.

The temporal response of an IMOD pixel is shown in Figure 5.7. These data show the vertical position of the mirror membrane as it changes with time. When the voltage between the two conductive membranes (indicated by the upper trace) exceeds the actuation voltage, the movable mirror membrane moves rapidly towards the fixed mirror until it contacts the stop layer. The mirror membrane completes its movement in approximately 7 µs. This is sufficient to allow video-rate update of a high-resolution (1000 lines or more) display.

5.5.1   CONVENTIONAL DRIVE METHOD

The conventional drive scheme used in currently available mirasol displays is a variant on passive matrix addressing. It is relatively simple and works well with small-format displays.

5.5.1.1   Image Update

The drive waveforms used to update the mirasol display using a conventional scheme are illustrated in Figure 5.8. The segment voltages are always held at either the positive or negative hold voltage (+VH or −VH, typically around the centre of the positive and negative hysteresis windows, respectively) and switch quickly enough so that the pixels do not release. Unselected common lines are held at 0 V, and addressed lines are biased to +VH or −VH. The addressing sequence is as follows:

- The segment lines are set to either +VH or −VH, depending on the data to be written.
- The first common line addressed is biased to +VH.
- IMOD pixels for which both the common and segment lines are at +VH see a net of 0 V between the mirrors and release. IMOD pixels for which the segment line is at −VH while the common line is at +VH see a net of 2 VH between mirrors and actuate.

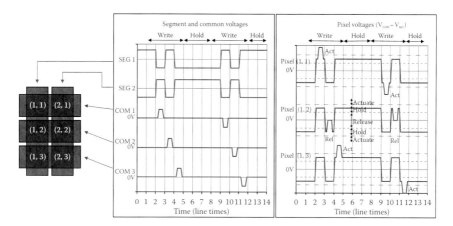

FIGURE 5.8 High-voltage addressing.

The addressing signal polarities are reversed between display updates to minimize charging effects.

5.5.1.2 Hold Mode

In hold mode, the common lines all remain at 0 V. The segments remain at either +VH or –VH, and the image is retained. Periodic rewrites are used to reverse pixel polarity.

5.5.2 LOW-POWER DRIVE METHOD

The low-power drive scheme [21] is designed to allow mirasol displays to be updated at high speed with low-power requirements.

5.5.2.1 Image Update

Drive waveforms are used to update the mirasol display and illustrated in Figure 5.9. Note that the high-frequency data signals applied to the segment lines have a low-voltage swing in order to minimize power requirements. Conversely, the larger bias voltages required to maintain the IMOD pixels within the hysteresis windows are applied via the common lines, which switch at a low rate. The state of a given IMOD pixel is set as follows:

- The common addressed line is switched from one bias polarity to ground. The segment voltage swing stays low enough that all IMOD pixels along the common line release, regardless of the segment voltage.
- The common line is then switched to the opposite bias polarity. The IMOD pixels remain released.
- A small write pulse is applied to the common line. Depending on the state of the corresponding segment voltage, this write pulse will apply a voltage to the pixel that is either slightly above the actuation voltage (actuating the IMOD pixel) or a slightly below actuation (leaving it released) as shown in Figure 5.9. In this way, the state of all IMOD pixels along the addressed common line can be set.

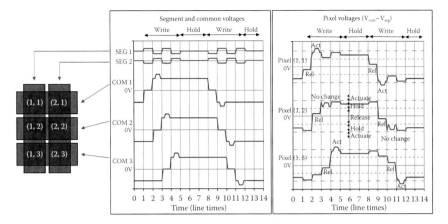

FIGURE 5.9   Low-voltage drive method.

Figure 5.9 also shows how the effective line time of the drive waveforms can be minimized. While one line is being updated, the next is being released in preparation for its own update.

This basic drive scheme reverses the bias polarity every time the image is updated, minimizing charging effects. Partial display updates in which only those lines with pixels that change state are addressed are also possible, again resulting in reduced power requirements.

5.5.2.2   Hold Mode

In hold mode, the voltage across all pixels in the mirasol display is maintained within the hysteresis window and the image is retained without the need for refresh. Periodically, the image is rewritten to reverse the polarity of the voltage applied to each pixel. The time between such rewrites, however, is long enough that there is minimal impact on power requirements.

Much of the drive system used during image update is not required during hold mode and can be powered down or placed in a low-power mode to save energy. In this low-power mode, the most significant power draw from the host battery is the result of common bias voltage generators. The proprietary design of mirasol displays keeps the overall power requirement below 1 mW.

5.5.3   Drive Voltage Adjustment

Like many other display technologies, mirasol displays require some degree of voltage tuning with temperature to achieve optimum operation. This is generally achieved in other technologies by using a look-up table (LUT) or simple algorithmic adjustment of voltage with measured temperature and manually tuned at the point of manufacture. While a similar approach is possible with mirasol displays, a more efficient, automated and low-cost alternative also exists.

The optical (reflectance–voltage) and electrical (capacitance–voltage) character-istics of individual IMOD pixels are very closely correlated. Both the reflectance

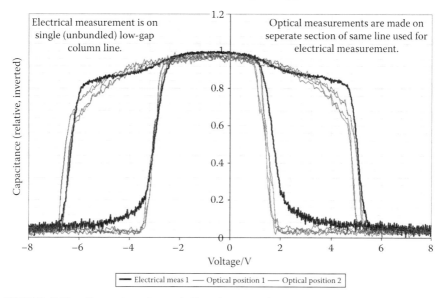

Capacitance measurement on single low-gap column line

FIGURE 5.10 (See colour insert.) Capacitance–voltage (blue) curve and reflectance–voltage (red, green) curve for one common line of mirasol pixels.

and the capacitance of an individual IMOD pixel are determined by the separation of the same two conducting mirrors; thus, as the gap between the two mirrors is changed by the applied voltage, both the reflectance and capacitance change together. Figure 5.10 shows measured optical and electrical characteristics for one row of mirrors in a mirasol display, with the close correlation between the two is clear. In particular, the switching voltages between the actuated and released states are the same in both the reflectance–voltage and capacitance–voltage curves. Once these switching voltages are detected, appropriate drive voltages for the display can be easily calculated.

The use of a charge integration method that minimizes sensitivity to electrical noise delivers accurate measurements for IMOD device capacitance. Figure 5.11 illustrates the circuit used. In this case, one full row (common line) of IMOD pixels is measured with the following sequence:

- A low or high voltage is applied to set all IMOD pixels along the common line to a known state (released or actuated).
- A test voltage, which may either actuate or release IMOD pixels, is applied to all pixels.
- The common line is switched to the input of an integrator, and a small test pulse is applied to the segment lines. The charge sensed by the integrator is proportional to the total capacitance of the pixels along the line. If some of the IMOD pixels have changed state as a result of the applied voltage, this is detected as a change in the line capacitance.

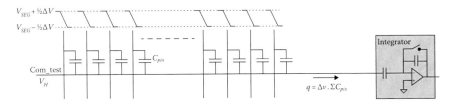

FIGURE 5.11   Simplified capacitance measurement circuit for IMOD pixel.

- The sequence is repeated until the full C–V characteristics (or whatever section of the characteristics is needed for tuning) have been measured.
- The image data are restored to the common line.

The measurement and control circuitry (test pulse generator, integrator, switching matrix and microcontroller) are relatively small and easily integrated into the overall display controller/power management integrated circuit (IC). Measurements are performed periodically to set and maintain the drive voltages with no visible impact on the displayed image and at negligible power cost. This allows continuous optimization of display performance while adapting to a range of changes to the IMOD pixels.

5.5.4   SYSTEM CONFIGURATION

The drive ICs for mirasol displays are shown in Figure 5.12. The configuration is very similar to that used in active-matrix LCDs. There are a larger number of voltages used by the common driver compared to an active-matrix LCD gate driver, but the segment driver is significantly more simple than an active-matrix LCD source driver. This is because it only outputs three different voltages. Moreover, the data are only one bit per mirror, and therefore, there exists a much more narrow data path from panel controller to segment driver than its active-matrix LCD counterpart. The input to the panel controller is the full data width. The controller performs the colour processing, to include spatial halftoning, which reduces the bit depth to a single bit per mirror. The design provides for full greyscale imagery. The power supply IC provides the voltages necessary for the segment and common drivers.

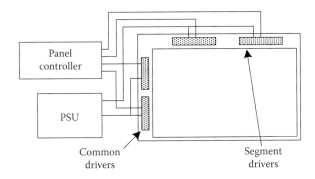

FIGURE 5.12   mirasol display panel architecture.

5.6   OVERVIEW OF MIRASOL DEVICE MATERIALS AND DISPLAY MANUFACTURING

Currently in production on G4.5 (730 × 920 mm) glass substrates, mirasol displays represent a significant development in the MEMS industry that has been traditionally based on smaller-size silicon wafers (Figure 5.13). As a result of this strategy, the production mirasol MEMS device architectures had to be adapted to fit the capabilities of TFT LCD-type fabrication infrastructure leveraging the set of materials and processes available in FPD factories in order to enable a scalable and cost-effective manufacturing platform [19].

In this section, materials and processing aspects involved in fabricating the IMOD MEMS pixel array glass panel are examined. The remaining steps towards producing finished mirasol displays are then described.

5.6.1   PIXEL ARRAY MATERIALS AND PROCESS FLOW

The fabrication of the mirasol display is based on surface micromachining of materials and process steps largely common to TFT array fabrication: large area deposition (plasma-enhanced chemical vapour deposition (PECVD), sputter, etc.), photolithographic patterning and etching (dry, wet) of dielectric and metallic layers [20]. In addition, MEMS-specific processes include forming the airgap below the moveable membrane, which is accomplished by depositing a sacrificial layer that is later etched out through appropriately designed openings in the membrane. Typical thickness of the films in the IMOD pixel layers range from a several nanometres to submicron dimensions. A simplified process flow for IMOD pixel fabrication is shown in Figure 5.14.

While there are significant similarities to standard processes used in TFT array manufacturing, a number of additional controls and optimizations are needed in order to ensure that the resulting MEMS device meets the desired operational and reliability specifications. Given the electro-opto-mechanical nature of the mirasol display device, it is not surprising that additional attention is dedicated to engineering the combined optical, electrical, mechanical and surface properties of the various layers forming the device. Some examples include the residual stress of the moveable membrane, the surface energy and topography of the surfaces on each side of the airgap [21] and the refractive index of each of the layers in the optical path [22].

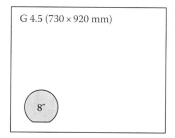

Gen	Glass size (mm)
G5	1100 × 1300
G6	1500 × 1850
G7	1870 × 2200
G8	2160 × 2460

FIGURE 5.13   Substrate size comparison – FPD glass generations and 8″ (200 mm) wafer.

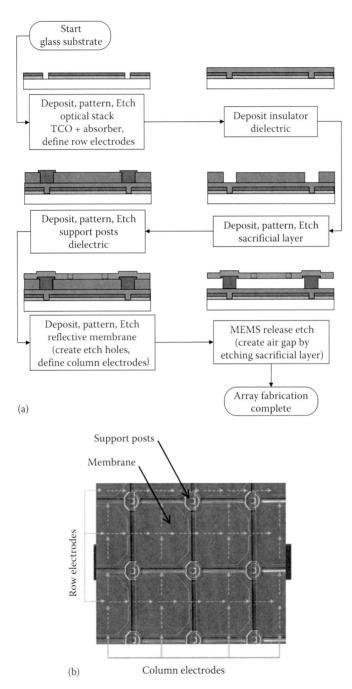

FIGURE 5.14 (See colour insert.) (a) Pixel array glass fabrication process. (b) Scanning electron microscopy (SEM) plan view from the process side of a group of IMOD pixels.

5.6.2 Packaging and Encapsulation Process Flow

The next step in mirasol display module manufacturing is the display panel fabrication. mirasol display panels are made by encapsulating the array glass with a back glass in order to protect the MEMS structures against physical damage or environment contamination (Figure 5.15).

As with the pixel array fabrication, the strategy for packaging of mirasol displays is also based on leveraging the scalable, high-volume solutions that have been developed for the FPD industry. The mirasol display array glass encapsulation is done at G4.5 plate level by using a back glass substrate (item 2 in Figure 5.15), followed by singulation of the glass bilayer to yield the individual display panels. The characteristics of the environment inside the panel package need to be controlled in order to meet the device lifetime requirements for mobile display applications. Typically, the seal is formed by using thin curable polymeric layers, with the package being at near atmospheric pressure and including a desiccant designed to have a moisture capture capacity sufficient for the display lifetime. It should also be noted that vacuum encapsulation is not required as in other traditional MEMS devices, since the pixel response time is already very fast for display applications as discussed in the previous section.

5.7 DISPLAY MODULE ASSEMBLY AND FRONT-OF-SCREEN COMPONENTS

A key enabler for designing and fabricating displays with the performance and functionality required by a target application is the ability to integrate the basic display panel with appropriately selected complementary components. FOS display components are stacked on the viewer side of the display panel and play a major role in optimizing both the display optical performance and user experience. The currently developed mirasol display product, in its fully finished state, is assembled from a number of components that complement the mirasol display panel. The intrinsic attribute of mirasol display technology – interferometric modulation and reflection of ambient light – eliminates the need for a number of traditional display

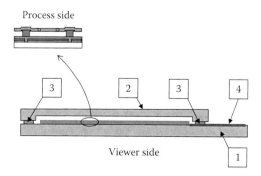

FIGURE 5.15 Typical mirasol display panel structure: (1) IMOD pixel array substrate, (2) glass backplate, (3) seal layer, and (4) routing and pads for display driver attachment.

Mirror

FIGURE 5.16 **(See colour insert.)** Typical mirasol display module stack.

components such as polarizers, backlights and colour filters. As a result, they reduce the complexity of the module architectures, as shown in Figure 5.16.

Processes used for the mirasol display module assembly closely resemble established LCD-type infrastructure. Some examples include driver attach with chip-on-glass or chip-on-flex solutions, film lamination using pressure-sensitive adhesives and touch or cover plate assembly with airgaps (perimeter gasket) or with optically clear resins or adhesive films in a bonded configuration. As described in Table 5.1, at minimum, the display panel needs to be integrated with the driver chip and connected to the display interface.

The most important consideration in the design of a product based on the mirasol reflective display is managing the light path to and from the display. The highest display contrast (and consequently best colour reproduction) is the result of minimizing reflections from the surfaces in front of the display elements. In practice, depending on the particular requirements for the display product, additional components are integrated within the module. These include optical diffusion layers to manage reflective view angle, antireflection (AR) optical films in order to maintain the intrinsic contrast ratio (CR) of the display, touch panels for enabling advanced user interactions, internal light sources (front light (FL)) to enable the use of the display in dark ambient conditions and a cover lens. Some of these components will be discussed in detail in the following.

5.7.1 Antireflection Films/Coatings

Ambient light reflections from the front surface of a display can direct a distracting glare to the user and degrade the overall image quality of the display. The optical performance of reflective displays is directly influenced by the magnitude of the FOS stack reflections [23]. To reap the benefits of the reflective nature of the mirasol display, AR optical films are often added in order to maintain the intrinsic CR and overall benefits of the mirasol display by reducing Fresnel surface reflections. Antireflection coatings (ARCs) are created by depositing film layers thinner than a single wavelength. A thin-film stack, engineered in this way, causes destructive optical interference across the reflected visible spectrum. The typical 4%–5% FSR is reduced to less than 2% for a single-layer ARC or less than 0.5% for multilayer coatings, as shown in Figure 4IB. When selecting between various AR coatings,

TABLE 5.1

Typical mirasol Display Module and FOS Components

Component or Treatments	Description	Function in Display Module
1. Display driver chip, flex attach	The display panel is connected to the display driver chip via necessary flex circuits to drive the content on the display panel.	Display driver
2. Cover lens (glass or plastic)	Top-most component in the display stack. Strengthened glass is a popular choice for smartphones and tablets.	Mechanical robustness and module surface protection
3. Hard coatings	Materials deposited on plastics to improve scratch resistance. Most typical for plastic cover lenses and films	
4. Anticrack coatings	Passivation of glass microcracks. Low-cost solution for improving glass robustness	
5. Diffuser	Volumetric or surface relief diffusing layers to tailor the view cone for ambient conditions with directed, or narrow-angle, illumination sources	Optical performance enhancements
6. ARCs	Thin-film optical stacks or graded refractive index structures. Aids sunlight readability	
7. Antiglare coatings	Surface diffusive microstructures used to eliminate display glare indoors and outdoors	
8. Anti-fingerprint, anti-smudge	Low surface energy coatings, oleophobic and hydrophobic, which reduce display surface contamination (fingerprints, residues, etc.)	
9. Touch sensing	Detect finger or pen touch locations for user input. Currently the preferred input interface for portable devices	Input interface
10. Front illumination	Internal light source for reflective displays. Enable devices with reflective displays to be used in dimly lit ambients	Supplemental illumination
11. Bonding adhesives and processes	Film-to-glass and glass-to-glass lamination to enable integration of FOS components in the display module	FOS assembly

consideration should be given to the magnitude of reflectance as well as the reflection colour tint, the perceived colour variations related to process control or view angle changes and cost. Examples of typical reflectance spectra from representative AR coating designs are shown in Figure 5.17.

5.7.2 Diffuser Films/Coatings

IMOD pixels are intrinsically specular reflectors, and therefore, diffusing layers are added to the display module stack to establish the viewing cone for ambient conditions with directed or narrow-angle illumination sources. The angular profile of the

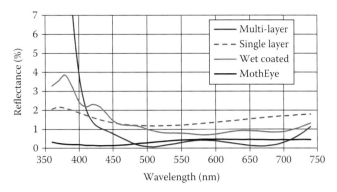

FIGURE 5.17   Spectral reflection curves of various types of AR coatings.

reflection can be managed by controlling the diffusion level and perceived brightness in the desired view cone can be tailored. This presents an advantage for displays in mobile devices where incoming light can be reflected in a smaller view cone, typically +/45°, to increase perceived brightness. The choice of the diffuser level is based on particular usage models (actual ambients are a mixture of diffuse and directional light sources), and their impact on the display appearance can be modelled by taking into account that diffusers interact with light twice – they diffuse incoming light sources and further diffuse the reflection of the mirasol display array, resulting in a desired view cone.

5.7.3   User Interface/Touch Sensing

Touch sensing has taken centre stage as the preferred user input method for mobile devices, with projective capacitive (pro-cap) touch sensing being the dominant technology. However, with reflective displays, the light reaching the user must pass through the touch screen twice. It is critically important to manage the light path through the FOS components, including the touch screen, to ensure optimal image quality [24]. Increasingly superior FOS performance (lower reflections, lower sensor pattern visibility, reduced thickness) is being enabled for pro-cap touch screens through optimizations of the sensor design and architecture, material choices and assembly strategies. Examples include replacing airgaps between the various FOS components with bonding adhesives, elimination of the EMI shield layer between sensor and the display panel, preference for sensors with single ITO layers and the development of transparent conductive materials with low reflection (optically matched ITO [25], alternative transparent conductors [26], etc.). These advances are enabling the adoption of pro-cap touch screens for reflective displays.

In recent years, alternate touch technologies have been proposed that present promising solutions for reflective displays. These newer touch technologies include waveguide optical touch [27] acoustic bending wave-based solutions [28,29] and force-based sensing [30]. These promising solutions are particularly ideal because no material is placed in front of the reflective display module, thus eliminating any chance of image quality degradation.

5.7.4 Supplemental Illumination/Front Lighting

mirasol displays, since they are reflective, utilize ambient light for illumination most of the time. However, under dark conditions, supplemental illumination must be provided by an FL. In display modules, the FL is positioned between the viewer and the display. One of the challenges for FL design is to make it both invisible to the viewer when ambient lighting is sufficient and functional when ambient light for viewing is insufficient.

A list of typical performance limitations introduced by the use of conventional edge-lit FL in a reflective display [31] could include both spatial and view angle non-uniformities as well as a reduction of the reflective and emissive CR of the display module due to airgaps and to light leakage towards the observer.

A novel, high-contrast, LED edge-lit FL technology has been developed in particular for the mirasol display modules [32]. Since the FL has a large and uniform view cone as well as high contrast in both emissive and reflective modes, it is not necessary to require an airgap to ensure the FL lightguide operation. The lightguide is comprised of unidirectionally reflective light-turning microstructures (Figure 5.18), with high reflectivity towards the display side and a low reflectivity towards the viewer side. This lightguide design eliminates the light leakage issue of traditional backlight-derived FL because the turning features are opaque. They use total internal reflection (TIR) for light extraction and are thus key for allowing high emissive CRs in a wide view cone. Simultaneously, the reflective mode

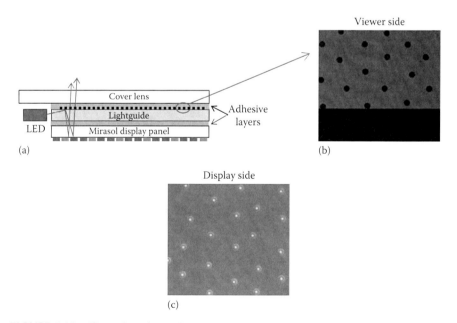

FIGURE 5.18 **(See colour insert.)** (a) Fully laminated, front-lit mirasol module architecture based on a glass lightguide shown in optical micrographs with (b) low reflection towards the viewer and with (c) high reflectivity towards the display side.

contrast of the front-lit display can be now maximized by eliminating all air interfaces inside the display stack since, unlike typical FLs relying on TIR extraction, this novel lightguide can be fully laminated to the display stack and other FOS components without impacting the light-turning directionality and efficiency of the reflective microstructures.

While a variety of substrates are compatible with the fabrication of the unidirectionally reflective microstructures introduced in the previous sections, glass substrates have been the focus for the FL lightguide. Compared to traditional plastics used in display illumination components, glass provides significant FL manufacturability benefits. This is due to superior robustness (environmental, mechanical and optical stability), broad chemical compatibility for cleaning and reworking and superior cosmetic quality. As such, the FL is fully compatible with established liquid or dry film optically clear adhesive bonding processes, with a typical module configuration shown in Figure 5.18.

The design strategies used for this unique FL technology, as described earlier, address the view angle limitation of the conventional FL technologies [31]; a large view angle with a nearly Lambertian emission profile in fully laminated, front-lit mirasol module prototypes was produced while simultaneously maintaining a diffused specular reflection (gain) property in ambient illumination conditions [24,33] as shown in Figure 5.19.

This advancement in novel FL technology for mirasol displays has paved the way for applications compatible with the current optical, mechanical and industrial design requirements of high-performance portable devices, including smartphones.

The rich ecosystem of FOS components is a key resource available to display module designers for meeting the diverse and ever-evolving application demands. Specifically, optimizing display performance (optical, mechanical, environmental) and providing required functionalities (such as touch input or front lighting) are of key importance.

5.7.5   Strengthened Cover Glass

A common requirement of display products is the protection of the display panel during normal device use or in accidental events. Particular attention goes towards specifying the top surface of the module to ensure adequate levels of scratch resistance, chemical compatibility and cleanability and tolerance to mechanical load and/or shock. A popular solution in many of today's portable devices is the use of a strengthened cover glass (CG) substrate placed above the display, replacing traditional solutions based on hard-coated plastics.

Compared to plastic, the adoption of CG has been stimulated by advantages of superior hardness and scratch resistance, coupled with a higher perceived industrial design value. To increase the robustness of glass to required levels, chemical strengthening is the dominant approach. This strengthening is achieved where a compressive surface layer is formed via a high-temperature ion-exchange process where sodium ions are replaced by larger volume ions such as potassium [34].

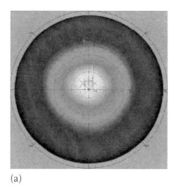

(a)

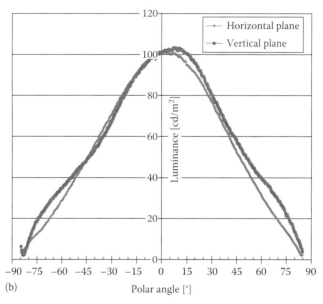

(b)

FIGURE 5.19   (a) Angular luminance profile from a 5.7″ diagonal size prototype front-lit mirasol display and (b) corresponding 2D cross sections in orthogonal planes aligned with the axes of the display.

Significant robustness gains are obtained by optimizing the built-in stress and thickness of this compressive layer (e.g. by ensuring it is thicker than typical surface microcracks that could otherwise become fracture initiation points during a mechanical event).

The mechanical benefits of the CG are significantly enhanced in a 'bonded' configuration characterized by being free of airgaps between the individual layers of the display module. Since there are additional optical performance gains from removing airgaps from the total display module stack, a number of materials, lamination and bonding processes have been developed [35,36] to enable an effective integration of the various FOS layers with the display panel. The table in the following describes these processes.

5.8 SUMMARY

In this chapter, basic electromechanical physics of the mirasol display, principles of interferometric colour generation and modulation and device fabrication and manufacturing were discussed with specific emphasis on demonstrating how materials and tools available from TFT display industry are being applied to the fabrication of mirasol display. A brief discussion was also included on FOS components, with a particular focus on touch input interface and supplemental illumination needed for reflective displays in darker ambient conditions.

Qualcomm's mirasol displays are based on a unique technology called interferometric modulation that mimics the way nature generates bright colour. This technology enables low-power reflective displays for mobile applications with a consistent viewing experience under a wide range of ambient lighting conditions including bright sunlight [7,8]. Interferometric modulation is used in a novel display pixel architecture based on microelectromechanical devices. Qualcomm MEMS Technologies [7] develops and commercializes mirasol displays from the early phases of research to module manufacturing.

REFERENCES

1. P. Vukusic, Manipulating the flow of light with photonic crystals, *Physics Today*, 59(10), 82, October 2006.
2. J. B. Sampsell, Causes of color: Especially interference color, *Proceedings of the Fourteenth IST Color Imaging Conference*, Scottsdale, AZ, 2006, p. 90.
3. P. Vukusic and J. R. Sambles, Photonic structures in biology, *Nature*, 424, 852–855, August 14, 2003.
4. G. T. A. Kovacs, N. I. Maluf, and K. E. Petersen, Bulk micromachining of silicon, *Proceedings of the IEEE*, 86(8), 1536, 1998. doi:10.1109/5.704259.
5. J. M. Bustillo, R. T. Howe, and R. S. Muller, Surface micromachining for microelectromechanical systems, *Proceedings of the IEEE*, 86, 1552–1574, 1998.
6. M. J. Madou, Fundamentals of micro-fabrication and nanotechnology, in *Three-Volume Set: From MEMS to Bio-MEMS and Bio-NEMS: Manufacturing Techniques and Applications*, 3rd edn., CRC Press, Boca Raton, FL, June 13, 2011.
7. J. B. Sampsell, MEMS-based display technology drives next-generation FPDs for mobile applications, *Information Display*, 22(6), 24, June 2006.
8. M. W. Miles, Toward an iMoD ecosystem, Hilton Head 2006 – A solid-state sensors, actuators and microsystems workshop, June 2006.
9. M. W. Miles, E. Larson, C. Chui, M. Kothari, B. Gally, and J. Batey, Digital paper for refractive displays, *SID Symposium Digest*, 33, 115, 2002.
10. D. Felnhofer, K. Khazeni, M. Mignard, Y. J. Tung, J. R. Webster, C. Chui, and E. P. Gusev, Device physics of capacitive MEMS, *Microelectronics Engineering*, 84, 2158, 2007.
11. H. A. Macleod, *Thin-Film Optical Filters*, 4th edn., CRC Press, Boca Raton, FL, 2010.
12. H. K. Pulker, *Coatings on Glass*, 2nd edn., Elsevier Science, Amsterdam, the Netherlands, 1999.
13. J. Schanda, *Colorimetry: Understanding the CIE System*, Wiley-Interscience, Hoboken, NJ, 2007.
14. T. Veijola, Equivalent circuit models of the squeezed film in a silicon accelerometer, *Sensors and Actuators* A48, 239–248, 1995.
15. M. Bao, *Analysis and Design Principles of MEMS Devices*, Elsevier Science, Amsterdam, the Netherlands, 2005.

16. P. M. Alt and P. Pleshko, Scanning limitations in liquid crystal displays, *IEEE Transactions on Electron Devices*, ED-21, 146, 1974.
17. D. L. Lau and G. R. Arce, *Modern Digital Halftoning*, 2nd edn., CRC Press, Boca Raton, FL, 2008.
18. R. Martin, A. Lewis, A. Govil et al., Driving mirasol displays: Addressing methods and control electronics, *SID 2011 Proceedings*, 26, 330–333, 2011.
19. P. D. Floyd, D. Heald, B. Arbuckle et al., IMOD display manufacturing, *SID 2006 Proceedings*, 37, 1980–1983, 2006.
20. A. Londergan, E. P. Gusev, and C. Chui, Advanced processes for MEMS-based displays, *Proceedings of the Asia Display SID*, 1, 107–112, 2007.
21. Y. Du, W. A. de Groot, L. Kogut, Y. J. Tung, and E. P. Gusev, Effects of contacting surfaces on MEMS device reliability, *Journal of Adhesion Science and Technology*, 24, 2397–2413, 2010.
22. B. J. Gally, White-gamut color reflective displays using IMOD interference technology, *Proceedings of SID*, 2004, 103.
23. E. F. Kelley, M. Lindfors, J. Penczek, Display daylight ambient contrast measurement methods and daylight readability, *Journal of the SID*, 14(11), 1019, 2006.
24. W. Cummings, The impact of materials and system design choices on reflective display quality for mobile device applications, SID Digest, Paper 63.1, 2010.
25. S. Kajiya et al., High-transmission optically matched conductive film with sub-wavelength nano-structures, *SID Symposium Digest*, paper 44.1, p. 595, 2012.
26. B. Mackey, Trends and materials in touch sensing, *SID Symposium Digest*, paper 43.1, p. 617, 2011.
27. http://www.rpo.biz/DWT_White_Paper.pdf
28. http://www.elotouch.com/Products/Touchscreens/AcousticPulseRecognition/default.asp
29. http://www.sensitive-object.com/spip.php?lang = en
30. http://www.f-origin.com/
31. H. J. Cornelissen, H. Greiner, and M. J. J. Dona, Frontlights for reflective liquid crystal displays based on lightguides with micro-grooves, *SID Digest*, Paper 42.03, 2000.
32. R. Rao, R. Gruhlke, M. Mienko, and I. Bita, Emerging front-light technologies for reflective displays, *Proceedings of IDW 2010*, Japan, MEET3-2, 2010.
33. R. Rao et al., High contrast edgelit frontlight solution for reflective displays, *SID Symposium Digest (2012)*, paper 66–4, p. 905, 2012.
34. S. Gomez, A. J. Ellison, and M. J. Dejneka, Designing strong glass for mobile devices, *SID Symposium Digest (2009)*, paper 69.2, p. 1045, 2009.
35. R. Smith-Gillespie and W. Bandel, LCD ruggedization in displays with optically bonded AR glass lamination, Americas display engineering and applications conference digest (ADEAC 2006, SID), paper 7.3, p. 78, 2006.
36. Y. Shinya, K. Kamiya, T. Toyoda, Y. Endo, N. Soudakova, and H. Kondo, A new optical elasticity resin for mobile LCD modules, *Journal of the SID*, 17(4), 331, 2009.

6 Thin-Film Transistors Fabricated on Intentionally Agglomerated, Laser-Crystallized Silicon on Insulator

Themistokles Afentakis

CONTENTS

6.1 INTRODUCTION

Laser-induced melting and recrystallization of silicon films, typically referred to as laser crystallization, have been established as the preferred method to obtain high-mobility silicon metal-oxide-semiconductor field-effect transistors (MOSFETs) on glass substrates. The fabrication of such semiconductor devices on glass and other low-temperature-compatible substrates is of primary importance for the flat-panel display industry. Active-matrix (AM) displays, employing one or more thin-film transistors (TFTs) per pixel, are the standard circuit architecture in high-resolution displays, for all mainstream display technologies used to emit or modulate light (such as liquid crystal and light-emitting diode).

Silicon TFTs on glass cover a wide performance range depending on the fabrication process, ranging from low–electron mobility (<1 cm^2/V·s) amorphous silicon devices to very–high mobility ($\sim$500 cm^2/V·s) polycrystalline ones. Mobility is increased primarily by enhancing the microstructure of the active semiconductor film (larger grains, fewer grain boundaries).

Although for an AM liquid-crystal display, high-mobility (>100 cm^2/V·s) transistors are not absolutely necessary, there are several important advantages from the circuit design point of view in incorporating them in a design. Among these, having smaller TFT dimensions for a given drain current (thus smaller pixel dimensions and higher-resolution arrays) and having faster transistors both for pixel and for peripheral driver and signal processing circuitry are of prime importance. Also, medium- to high-mobility polycrystalline silicon TFTs can be made p-type (PMOS) or n-type (NMOS), thus imparting complementary MOS (CMOS) circuit functionality, while low-mobility amorphous Si devices are only n-type. For all these reasons, polycrystalline Si TFTs have been commercially utilized in the manufacturing of high-resolution, portable device displays, and they continue to be an active research topic in high-performance circuitry on low-temperature substrates.

Typically, excimer lasers emitting in the UV are used to crystallize amorphous or polycrystalline silicon precursor films; silicon has a strong absorption of UV radiation and a short diffusion length, resulting in a high-energy transfer to the film. In this process, temperatures at the surface of the film exceed the melting point of silicon, while the substrate (typically glass) is not appreciably heated ($<400°$C). Excimer laser crystallization has resulted in very high field-effect mobility polysilicon TFTs on glass, typically achieving 300 cm^2/V·s [1–4] for n-type devices and often exceeding 600 (n-type) and 400 cm^2/V·s (p-type) on glass [5]; the process has found industrial use in the production of high-resolution small-size AM-LCDs. Laser-crystallized Si TFTs with high performance (electron mobility >100 cm^2/V·s) have also been reported on flexible substrates [6,7].

For static, 'flood' irradiation by an excimer laser source, the resulting crystal structure of the semiconductor film is largely dependent on the laser pulse energy. Three main regions of crystal growth have been identified [8], as shown in Figure 6.1: at low laser energies, the resulting film is composed of a fine grained bottom layer and a larger grain upper layer generated by vertical solidification [9]. At medium energies, the entirety of the film is melted except a few dispersed crystalline clusters; these are acting as seeds for the lateral growth of crystalline grain as the film cools down. Grain boundaries form when the solidification fronts impinge on each other. Large grains and high–electron mobility films are obtained in this region [10]. At high energies, complete melting is induced, resulting in small grains and low carrier mobility [8]. Beyond this region, the film agglomerates from the surface, creating voids and discontinuities [11]. Thus, for the purpose of obtaining high-quality devices, low- and high-energy regions are, by default, discarded [12]. However, the correct processing window for excimer laser annealing (ELA) is very narrow because of the steep dependence of polysilicon grain size on laser energy. Consequently, laser systems with very high illumination uniformity and very low pulse-to-pulse fluctuation are required [13].

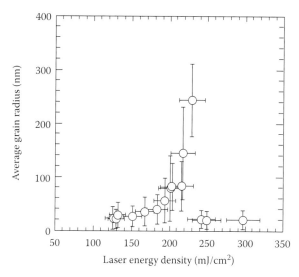

FIGURE 6.1 Variation of average grain radius of crystallized Si films on laser energy density. The radius (r) of a grain is obtained from the image-analysed area of the grain by letting $r = (area/\pi)^{1/2}$. (Reprinted from Im, J.S.H. et al., *Appl. Phys. Lett.*, 63, 1969, 1993. With permission.)

This chapter describes a method of using intentionally agglomerated silicon for the active semiconductor film of TFTs. By *intentional agglomeration*, we mean irradiating the Si film at very high laser energies, causing it to completely melt and de-wet from the substrate it has been deposited onto, forming a regular pattern. Although film agglomeration is very undesirable in ELA, we show that for certain laser irradiation conditions, very uniform ribbon or wirelike structures result. This type of self-forming agglomerated topography, with an approximately truncated cylindrical shape, is found to possess two very desirable properties, not found in typical, conventionally crystallized ELA films.

First, there are no steep angles in the channel cross section, where high-field and gate oxide step-coverage-related problems originate. Second, of the wirelike structures that have so far been analysed, the microstructure consists of relatively larger and fewer grains compared with similarly sized areas of directionally crystallized material; there are few to no grain boundaries present over distances on the order of tens of microns, as measured using electron backscattering diffraction (EBSD). In addition to those advantages, we have noted that films of the same morphology are generated by laser irradiating other semiconductor materials, such as compound semiconductors and germanium films.

This chapter is organized as follows: In the first part of the Experimental section, the results of laser irradiation silicon films of various thicknesses, with the aim of producing the desired agglomerated structure, are shown. This is followed by detailing the fabrication process for CMOS transistors on fused silica substrates. In the Results section, the electrical characteristics of the TFTs are shown.

In the discussion section, the mechanism of formation of this repeatable 3D film structure is discussed, and additional results from the irradiation of other semi-conductors are shown. Finally, the last section discusses our preliminary experimental results aimed at addressing the main issues with the excimer step-and-scan approach, towards achieving a high-throughput, highly uniform controlled-agglomeration process.

6.2 EXPERIMENTAL

6.2.1 EXCIMER LASER AGGLOMERATION OF Si

6.2.1.1 Uncapped Si Films on Quartz

Our excimer laser setup utilizes a reticle that shapes the beam to a rectangular shape of various dimensions. For this experiment, beam shapes 5 mm long and 4, 6, 10 or 25 μm wide were used. As the crystallization process proceeds, the sample-holding stage is stepped in the y-axis a fraction of the beam width, in a step-and-overlap fashion. A step size of 0.10, 0.25, 0.50 and 1.00 μm was used for each tested beam width, resulting in 16 beam-width/step-size combinations.

Initially, the dependence of the optimum laser irradiation conditions (those which result in the desired morphology) on the precursor Si film thickness was investigated. Three Si film thicknesses (20, 30 and 50 nm) were studied in this experiment. The Si films were deposited on the fused silica substrates via plasma-enhanced, chemical vapour deposition (PECVD) at 400°C from a silane source. Prior to laser irradiation, the Si film was dehydrogenated by a furnace anneal at 500°C for 2 h.

For each of the three samples, a 16 × 10 matrix of different laser conditions was run: the 16 beam-width/step-size combinations mentioned earlier and 10 laser fluence levels, ranging from 308 to 614 mJ/cm². Figure 6.2 shows the gradual transformation of the uniform Si to the desired parallel-stripe geometry, and its subsequent degradation as irradiation energy is increased.

A partial set of the results of this experiment is shown in Table 6.1. Because of the large volume of data collected (160 conditions for each Si film thickness), Table 6.1 shows the rows of the experimental matrix in which crystallization in the desired topography was noted. The total energy E (in mJ/cm²) of the laser beam for a particular matrix condition can be calculated according to the following formula (6.1):

$$E = \text{Fluence} \times \left(\frac{\text{Beam width}}{\text{Step size}} \right) \qquad (6.1)$$

The basic morphology of the parallel Si stripes obtained in this experiment is shown in Figure 6.3. Figure 6.3a shows atomic force microscope (AFM) and scanning electron microscope (SEM) pictures of the parallel-stripe structure obtained from a 50 nm thick precursor Si film. Note that due to the step-and-scan process of irradiation, there is surface roughness along the top ridge of each stripe, which is also the direction of current flow in the transistor. It should also be mentioned that, for

FIGURE 6.2 Optical microscope images showing the formation of parallel stripes in 20 nm thick precursor Si film at increasing energy levels, starting from top left (uniform film, $E = 259$ mJ/cm^2) and advancing clockwise to 336, 364 and 394 mJ/cm^2.

the laser irradiation conditions that lead to the formation of the parallel stripes, the stripes run unbroken for several cm of length. Figure 6.3b plots the basic geometric properties of the agglomerated stripes as a function of the precursor silicon film thickness. Stripe thickness varies from 100 to 300 nm, roughly five times the precursor film thickness. We noted that stripe geometry is a function of only the precursor Si film and does not depend on beam width and step size, provided that the latter is suitable for inducing the stripe pattern in the first place. All experiments were run at our constant-wavelength (308 nm) excimer laser tool, so the dependence of stripe geometry on laser wavelength was not investigated.

Since the formation of the parallel stripes is driven by a capillary instability mechanism (as discussed in Section 6.4), it is almost certain that stripe geometry should depend on the surface properties of the substrate material as well. However, all our experiments were run with Si films on PECVD-deposited tetraethyl orthosilicate (TEOS) SiO$_2$ base-coated substrates.

The Si grain microstructure of the resulting stripes was investigated with EBSD. In general, thicker films yielded better microstructure stripes, with fewer grain boundaries. Figure 6.4 shows the EBSD inverse pole figure (IPF) of a typical group of three stripes, formed from an initial 50 nm thick Si film. Stripe crystallinity is very good, although there is random crystal orientation from stripe to stripe.

TABLE 6.1
Partial Set of Laser Irradiation Results for 50, 30 and 20 nm Thick Si Films

Si Film Thickness (nm)	Beam Width (μm)	Step Size (μm)	Shots per Area	Laser Fluence (mJ/cm²)									
				307.8	336.1	364.4	394.0	435.3	468.5	501.2	537.1	579.3	614.3
50	4	0.1	40	U	U	PS	PS	PS	PS	PS	PS	PS	PS
50	6	0.1	60	U	U	U	U	U	U	PS	PS	PS	PS
50	25	0.1	250	U	U	U	PS	PS	PS	X	X	X	X
30	4	0.25	16	PS	X	X	X	X	X	X	X	X	X
30	4	0.25	24	X	D	PS	X	X	X	X	X	X	X
20	4	0.1	40	D	PS	PS	X	X	X	X	X	X	X

Note: The number of shots per area equals the ratio of beam width over the step size. Experiment results are labelled 'PS' (periodic parallel stripes), 'X' (agglomeration resulting in stripes with many discontinuities or ultimately de-wetted Si droplets on the substrate), 'U' (uniform film, no agglomeration observed) and 'D' (localized de-wetting in an otherwise uniform film). Within the scope of this work, only outcome PS is desirable.

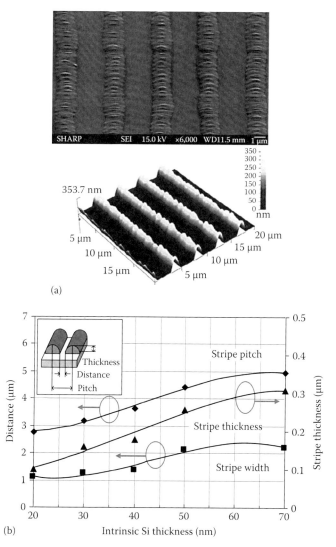

FIGURE 6.3 (a) SEM photo of agglomerated Si film – initially 50 nm thick – after laser irradiation and AFM scan of same area. (b) Stripe geometry characteristics (pitch, thickness from stripe top to substrate and distance between stripe edges) as a function of the precursor silicon film thickness t_{Si}.

6.2.1.2 Si Films on Quartz Capped with SiO₂

As indicated by the data of Table 6.1, in order to generate the desired parallel-stripe geometry, a fairly large shots/area ratio is required (>20). Agglomeration conditions for the 30 nm thick film are an exception, but in this case, the laser fluence window is very small. Based on this observation, another experiment was run, this time comparing uncapped, 50 nm thick Si films on quartz substrates with identical Si films capped by 10 or 20 nm thick SiO_2 films. The thin SiO_2 films were PEVCD deposited

FIGURE 6.4 (**See colour insert.**) EBSD IPF images of three agglomerated stripes in 50 nm thick Si, in the direction normal to the substrate surface. Colours indicate crystal orientation.

at 400°C from a TEOS source. It was also hoped that the capping layer will suppress the roughness that develops along the top of the stripes (as seen in Figure 6.3).

The results of this experiment are shown in Table 6.2. Comparing the samples having the 10 and the 20 nm thick SiO_2 cap layer with the bare Si film, we see a noticeable improvement: in both cases, the desired parallel-stripe structure is induced at steps/area ratios as low as 16 (while for the uncapped film, the minimum steps/area ratio is 40).

Our findings also suggest that the thicker cap film results in an extended range of laser fluence that leads to the formation of the Si stripes. In the case of the 10 nm thick sample, the fluence range for the formation of the parallel stripes is about 100 mJ/cm², while for the 20 nm thick sample, the fluence range is about 200 mJ/cm², for the same scanning speed.

Although substantial benefits in throughput were achieved with the 10 and the 20 nm thick cap oxide, surface roughness measurements of the agglomerated samples (after stripping the cap layer off) indicated no change in the roughness along the stripes' top.

6.2.2 THIN-FILM TRANSISTOR FABRICATION

Two quartz-substrate samples were processed for transistor fabrication. After cleaning the wafers, a 300 nm thick layer of TEOS SiO_2 base coat was deposited via PECVD at 300°C and then furnace annealed at 650°C. The active film was deposited, dehydrogenated and laser annealed as described in the previous section. A square area 42 mm × 42 mm was irradiated in the centre of each wafer. The TEOS SiO_2 gate insulator was then PECVD deposited, 50 nm thick, followed by the Si gate electrode, 200 nm thick.

The Si gate was ion implanted n+ in the first sample and p+ in the second (for negative-channel metal oxide semiconductor [NMOS] and positive-channel metal oxide semiconductor [PMOS] TFTs). After patterning the gate, the drain/source regions were ion implanted (phosphorus, 3×10^{15} ions/cm², 80 keV, or boron, 5×10^{15} ions/cm², 35 keV). The dopants were activated in the furnace at 650°C. Then, the passivation oxide (TEOS) was PECVD deposited at 400°C, 300 nm thick, followed by contact hole patterning and etching and metal deposition. The metal was a standard Ti/TiN/Al

TABLE 6.2

Partial Set of Laser Irradiation Results for 50 nm Thick Si Films Capped with a Thin SiO$_2$ Layer

Cap SiO$_2$ Thickness (nm)	Beam Width (µm)	Step Size (µm)	Shots per Area	Laser Fluence (mJ/cm^2)									
				381.9	435.3	486.4	537.1	579.3	634.1	682.5	735.9	779.9	832.4
0	4	0.1	40	PS	PS	PS	PS	PS	X	X	X	X	X
0	6	0.1	60	PS	PS	PS	PS	PS	X	X	X	X	X
0	25	0.1	250	PS	PS	PS	X	X	X	X	X	X	X
10	4	0.25	16	U	D	PS	PS	PS	X	X	X	X	X
10	4	0.1	40	PS	PS	PS	PS	PS	PS	X	X	X	X
20	4	0.25	16	PS	PS	PS	PS	PS	X	X	X	X	X
20	4	0.1	40	PS	PS	PS	PS	X	X	X	X	X	X

Note: Experiment results are labelled 'PS' (periodic parallel stripes), 'X' (agglomeration resulting in stripes with many discontinuities or ultimately de-wetted Si droplets on the substrate), 'U' (uniform film, no agglomeration observed) and 'D' (localized de-wetting in an otherwise uniform film).

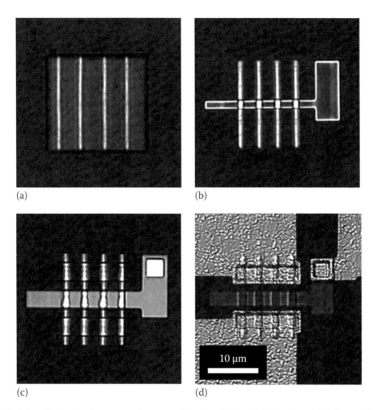

FIGURE 6.5   Optical microscope images showing device fabrication. (a) After active layer patterning (photoresist block present), (b) after gate electrode patterning, (c) after contact hole etching and (d) finished device after top metal patterning.

stack deposited at 380°C. After metal patterning, the wafers were plasma hydrogenated for 10 min. Figure 6.5 shows details of the fabricated TFTs after each processing step.

6.3   RESULTS

TFTs of various geometries were present in this design and measured. For the interest of brevity, this chapter only shows results from devices comprised of two parallel Si stripes; this is the minimum number of agglomerated Si sections per device in this photomask set. Since considerable redistribution of mass takes place during agglomeration, the thickness of each section is considerable, about five times the precursor Si film thickness. In order to avoid inadvertent doping of the channel during drain/source ion implantation, which took place after the gate photoresist was removed, the ion implant conditions resulted in a 'core' section of each stripe left intrinsic.

A total of 60 TFTs of each polarity were measured. Threshold voltage, effective mobility and inverse subthreshold slope statistics are shown in Table 6.3. High-resolution I_D–V_G and I_D–V_D curves of a typical L = 2 μm NMOS are shown in Figure 6.6. The inset in Figure 6.6 shows superimposed all I_D–V_G plots of all measured NMOS and PMOS devices.

TABLE 6.3
Mean Electrical Parameters of Fabricated TFTs

	Threshold Voltage (V)	Inverse Subthreshold Slope (mV/Decade)	Effective Mobility (cm²/V · s)
NMOS	+0.406 (0.24)	196 (79)	257.6 (44.6)
PMOS	−0.469 (0.28)	276 (43)	54.6 (7.4)

Note: Parameters extracted from linear I_D–V_G plots ($|V_D| = 100$ mV). Numbers inside parentheses indicate standard deviation. Minimum subthreshold slopes were 133 mV/decade (NMOS) and 179 mV/decade (PMOS). Maximum mobility was 360.0 cm²/V · s (NMOS) and 72.9 cm²/V · s (PMOS).

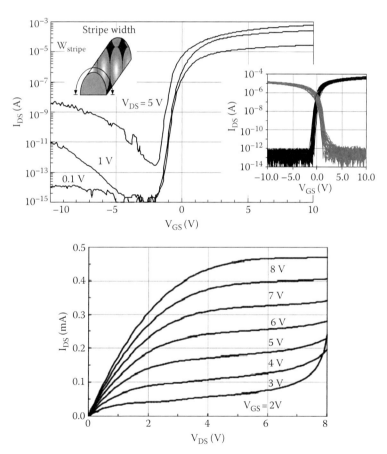

FIGURE 6.6   Top: I_D–V_G plot of typical, two-'stripe' NMOS with L = 2 μm at $V_D = 0.1$, 1 and 5 V. Inset on the top left shows geometric calculation of stripe width W_{stripe} for mobility extraction; inset on the right shows lower-resolution I_D–V_G plots of 60 NMOS (black lines) and 60 PMOS TFTs (grey lines) at $|V_D| = 100$ mV. Bottom: I_D–V_D curves of same NMOS TFT at V_G ranging from 2 to 8 V, in 1 V steps.

The channel width used for mobility calculation is the mean stripe surface width W_{stripe} obtained from AFM measurements, as shown in the inset of Figure 6.6, multiplied by the number of Si stripe sections in the device n. For the TFTs of Figure 6.6, $W_{stripe} = 3.55$ μm and n = 2; thus, a width of 7.1 μm was used for mobility extraction. The high effective mobility is due to the good crystallinity of the semiconductor and the fact that the channel covers the entire curved hemispherical section of the wire(s). The wide range of mobility (about 17%) is due to crystal structure and wire geometry variations, the former being common in laser-crystallized TFTs.

We believe that the negative V_{TH} shift in the I_D–V_G plot, which is observed for drain bias higher than 4 V, is due to a drain-induced barrier lowering effect. Although the channel microstructure is relatively free of grain boundaries, TFT mobility is not on par with single-crystal Si performance. We believe that this is partly due to the surface roughness of the stripes and also to microstructure defects (i.e. grain boundaries) within the TFT channel area.

6.4 DISCUSSION

Surface texturing of bulk or thin-film silicon on insulator (SOI) is a well-known phenomenon, observed after the partial or complete melting of the semiconductor under laser irradiation. Surface textures created by continuous or pulsed laser irradiation [14,15] (so-called laser-induced periodic surface structures (LIPSS)) have been studied extensively using nanosecond, picosecond and femtosecond lasers. The latter tool is well-suited for this task, because of its ultrashort pulse duration and very high peak energy, which allows for precise morphology control, especially with very thin films [16].

A number of different mechanisms have been proposed to account for the formation of this periodic surface morphology; one theory [17] postulates that a surface acoustic wave is generated, caused by the thermoelasticity of Si as it contracts upon irradiation. Another theory [18] proposes that, as the irradiated Si area partially melts, a surface tension gradient is caused between the high- and low-temperature areas of the material; this causes material to flow away from the low-surface tension, irradiated area. As the laser scans the surface, a steady-state condition where material flow away from the beam is balanced by material flow towards the beam due to gravity is established, leading to ripple creation. Other authors [19] have commented that, although capillary forces (i.e. formation and freezing of capillary waves) are present in ripple formation, they might not be the sole cause of ripple formation. For example, capillary melting and solidification models do not explain the strong influence of laser wavelength in the formation of surface ripples. A third theory [20] addresses this point, proposing that ripples are caused by interference between a scattering-caused surface wave and the incident light. Scattering or diffraction of light occurs due to any surface roughness of the irradiated material, defects, etc. Interference between the surface wave and the incident light reflected from the surface causes a standing wave, which modulates the energy absorbed by the material. Thus, the modulated melting of Si where the melting threshold is exceeded gives rise to ripple formation.

Silicon de-wetting to a periodic structure has been reported on the buried oxide layer of SOI wafers [21–23] in the case of very thin (<30 nm) semiconductor

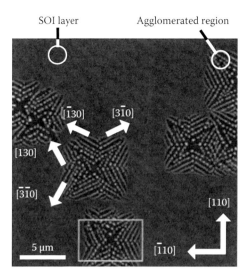

FIGURE 6.7 Typical AFM image taken on a partially agglomerated 7 nm thick (001) SOI layer after UHV anneal. De-wetted Si droplets are single crystal, and they maintain the crystallographic orientation of the initial SOI layer. (Reprinted with permission from Ishikawa, Y. et al., *Appl. Phys. Lett.*, 83, 15, 3162, 2003.)

films annealed in an ultrahigh vacuum (UHV) furnace. Molten silicon de-wets from the SiO_2 surface initially in isolated pits, then into 'stripe' of 'finger' shapes and eventually into droplets. Oriented, periodic stripes (and Si droplets) have been reported for single-crystal Si on SiO_2 [24], where the stripe (and, eventually, droplet) orientation corresponds to regular crystallographic directions, as seen in Figure 6.7.

The long-range periodic structures described in our work bare closer resemblance to those de-wetted Si structures obtained by UHV annealing than to conventional LIPSS formations, for two main reasons: First, because the stripe pitch (see Figure 6.3) is much longer than the laser wavelength (308 nm), as the interference/diffraction model predicts (this discrepancy between the interference model's prediction of stripe pitch and experimental results has also been noted by Yu et al. [25], on excimer laser-irradiated SiO_2-capped Si wafers) and secondly, because we do not observe a stripe pitch dependence on laser fluence, as noted in [25].

However, it should be noted that due to the short time available for Si melting and resolidification, the irradiated film in our process does not reach a surface tension equilibrium as is the case with the UHV-annealed materials.

Danielson et al. have shown [26,27] that UHV-annealed Si films on SiO_2 agglomerate by assuming periodic stripes grown from an initial void formation. A film imperfection or defect can be such a point of origin, and the capillary instability of the receding Si film front is responsible for the stripe formation. Eventually the stripes break down to droplets (the onset of which is shown in the last stage of Figure 6.1), due to a generalized Rayleigh instability that minimizes the surface energy of the stripes. In this model [26], it is shown that the stripe pitch λ_{stripe} is proportional to the Si film thickness:

$$\lambda_{stripe} \propto t_{Si} \qquad (6.2)$$

This model also predicts that λ_{stripe} is independent of temperature, which is consistent with our results.

The mechanism of stripe formation is of course more complex in our case, because of the step-and-scan, melt-and-solidify process of the excimer scanning. As shown in Figure 6.8a, as the Si film resolidifies, two distinct waves appear on the surface: a high-frequency one whose propagation axis is in parallel with the laser scan direction (from top to bottom) and a low-frequency one from left to right. The former is due to the step-and-scan process and eventually causes the roughness along the stripe axis. At a sufficiently high energy, void formation occurs on the Si film and stripes take shape as described by the capillary instability model; the subsequent scan of the laser propagates the stripes only in one direction.

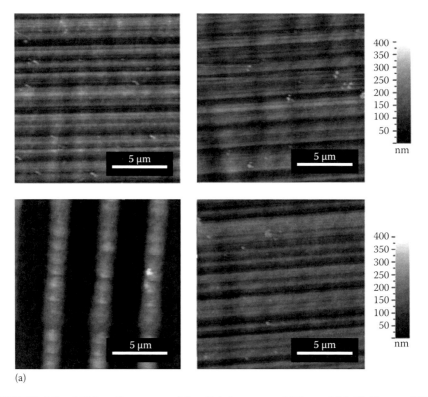

(a)

FIGURE 6.8 AFM surface scans of irradiated, uncapped 50 nm thick Si film on SiO_2. (a) Progression of surface morphology at a constant beam width/step size (4 µm beam size, 0.1 µm step size), with increasing fluence levels: from top left, going clockwise: 258, 308, 336 and 364 mJ/cm^2.

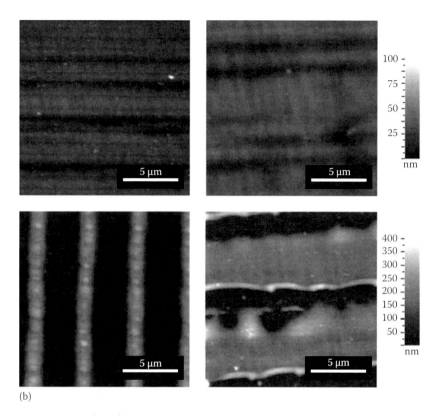

(b)

FIGURE 6.8 (continued) AFM surface scans of irradiated, uncapped 50 nm thick Si film on SiO_2. (b) Progression of surface morphology at a constant fluence (435 mJ/cm^2) and beam width (4 μm) with decreasing step size: from top left, going clockwise: 1.0, 0.5, 0.25 and 0.1 μm (i.e. 4, 8, 16 and 40 shots per area, respectively).

Parallel stripes with irregular edges can also form along the direction of the high-frequency wave (i.e. running perpendicular to the laser scan direction), as shown in Figure 6.8b. In the case of uncapped 50 nm thick Si, this occurred at step sizes >0.2 μm. This is not a desired outcome from the Si microstructure point of view; only when stripes propagate along the scan direction can long Si grains form.

Our experimental results suggest that stripe formation in the preferred orientation (along the scan direction) is primarily a function of step size and fluence. Capping the films with a thin SiO_2 layer can extend the range of step-size values that result in the formation of the desired geometry.

6.5 IMPROVING SI MORPHOLOGY

As it was mentioned earlier, the low throughput of the excimer step-and-scan laser irradiation process (particularly because of the short step size) and the resulting roughness of the Si stripes along the scan axis (which is also the drain-to-source current flow path in a TFT) are the main issues of this process. It was shown earlier that the addition of a SiO_2 capping layer did not improve the roughness issue, although a modest expansion of the step-size range for which the preferred structure occurs was achieved, which increased throughput.

An obvious alternative to the current nanosecond laser system would be to utilize a continuous-wave (CW) laser. Our preliminary experiments with thicker Si films, in the order of 100 nm, have shown that a substantial enhancement in both crystallized stripe, smoothness and process throughput is possible with this approach. Due to the large thickness of the Si layer in those initial experiments, there were no TFTs fabricated on the agglomerated films; establishing the CW irradiation conditions that generate the preferred structure on thinner (<50 nm) precursor Si films is the next step towards the fabrication of improved agglomerated CMOS TFTs.

For this experiment, 100 nm thick films deposited on quartz substrates and capped with a PECVD-deposited, TEOS SiO_2 layer were tested. A Coherent Verdi V-18 (18 W) high-power CW diode-pumped solid-state laser (532 nm wavelength) was used for this purpose. Multiple scan speeds in the range of 25–200 mm/s were tried. For each scan speed, laser power was varied until the desired Si structure was generated. Table 6.4 shows a partial set of the tried laser scan conditions and the Si film morphology they resulted to.

Figure 6.9 shows a typical result of the stripe's morphology and crystallinity obtained with CW laser agglomeration. In addition to the very fast throughput of the CW system, the surface roughness of the stripes is greatly suppressed, as shown in Table 6.5. Clearly, the CW laser approach is a more advantageous approach to generating agglomerated semiconductor stripes on insulator.

TABLE 6.4

Partial Set of CW Laser Irradiation Results for 100 nm Thick Si Films Capped with a SiO_2 Layer

Cap SiO_2 Thickness (nm)	Scan Speed (mm/s)	Laser Energy (W)						
		2.8	3.4	4.0	5.2	5.6	6.2	6.8
50	50	D	D	D	PS	PS	X	X
50	75	U	D	D	PS	PS	PS	X
50	100	U	U	D	D	PS	PS	PS
260	50	D	D	PS	PS	PS	PS	PS
260	75	D	D	D	PS	PS	PS	PS
260	100	D	D	PS	PS	PS	PS	PS

Note: Experiment results are labelled 'PS' (periodic parallel stripes), 'X' (agglomeration resulting in stripes with many discontinuities or ultimately de-wetted Si on the substrate), 'U' (uniform film, no agglomeration observed) and 'D' (localized irregular de-wetting in an otherwise uniform film).

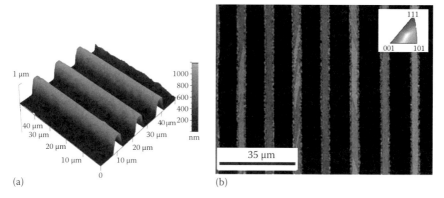

(a)   (b)

FIGURE 6.9   (**See colour insert.**) Agglomerated stripes of 100 nm thick Si film capped with 260 nm of SiO_2 using CW laser. (a) AFM image of stripes after cap oxide removal. (b) EBSD 001 IPF image of stripes; inset shows colour/orientation key for IPFs.

TABLE 6.5

Surface Roughness of Successfully Agglomerated Films Presented in This Chapter

Precursor Si Film Thickness (nm)	Cap SiO_2 Film Thickness (nm)	Crystallisation	Surface Roughness (%)
20	0	Excimer	32.4
50	0	Excimer	18.1
129	50	CW	10.6
129	270	CW	2.9

Note: Surface roughness is measured by calculating the standard deviation of the agglomerated Si thickness along the top ridge of the stripe and expressed as a percentage of the mean agglomerated stripe thickness.

6.6   CONCLUSIONS

This chapter described the first attempts at using an intentionally agglomerated Si film for TFT fabrication. It was shown that, for certain laser-annealing conditions, a very repeatable structure of high crystallinity is created. Stripe formation via this approach is a semi-equilibrium, capillary instability-driven process, which – for a particular Si film thickness – depends primarily on step size and fluence. Capping the Si film with a thin SiO_2 layer was shown to extend the range of laser conditions that result in the preferred structure. Initially, TFTs were fabricated using uncapped Si films.

Based on the electrical characteristics presented here, three areas of improvement were identified: (a) developing a controlled-agglomeration process with a thinner precursor film in order to obtain fully depleted devices, (b) reducing the surface roughness along the ridge of the Si stripes and (c) increasing the laser irradiation process throughput. Since stripe roughness is caused by the step-and-scan laser process, a CW laser should be a better choice for both reduced roughness and increased throughput. Indeed, preliminary experimental results with thicker, SiO_2-capped precursor Si films showed very smooth stripe generation with very good crystallinity at high scan speeds.

REFERENCES

1. G. Giust, T. Sigmon, P. Carey, B. Weiss, and G. Davis, Low-temperature polysilicon thin-film transistors fabricated from laser-processed sputtered-silicon films, *IEEE Electron Device Letters*, 19(9), 343–344, 1998.
2. C.-W. Lin, L.-J. Cheng, Y.-L. Lu, Y.-S. Lee, and H.-C. Cheng, High-performance low-temperature poly-Si TFTs crystallized by excimer laser irradiation with recessed-channel structure, *IEEE Electron Device Letters*, 22(6), 269–271, 2001.
3. S. Higashi, D. Abe, Y. Hiroshima, K. Miyashita, T. Kawamura, S. Inoue, and T. Shimoda, Development of high-performance polycrystalline silicon thin-film transistors (TFTs) using defect control process technologies, *IEEE Electron Device Letters*, 223(7), 407–409, 2002.
4. L. Mariucci, F. Giacometti, A. Pecora, F. Massussi, G. Fortunato, M. Valdinoci, and L. Colalongo, Numerical analysis of electrical characteristics of polysilicon thin film transistors fabricated by excimer laser crystallization, *Electronics Letters*, 34(9), 924–926, 1998.
5. A. Kohno, T. Sameshima, N. Sano, M. Sekiya, and M. Hara, High performance poly-Si TFTs fabricated using pulsed laser annealing and remote plasma CVD with low temperature processing, *IEEE Transactions on Electron Devices*, 42(2), 251–257, 1995.
6. T. Serikawa and F. Omata, High-mobility poly-Si TFTs fabricated on flexible stainless-steel substrates, *IEEE Electron Device Letters*, 20(11), 574–576, 1999.
7. T. Afentakis, M. Hatalis, A. Voutsas, and J. Hartzell, Design and fabrication of high-performance polycrystalline silicon thin-film transistor circuits on flexible steel foils, *IEEE Transactions on Electron Devices*, 53(4), 815–822, 2006.
8. J.S. Im, H.J. Kim, and M. Thompson, Phase transformation mechanisms involved in excimer laser crystallization of amorphous silicon films, *Applied Physics Letters*, 63, 1969–1971, 1993.
9. M. Thompson, G. Galvin, J. Mayer, P. Peercy, J. Poate, D. Jacobson, A. Cullis, and N. Chew, Melting temperature and explosive crystallization of amorphous silicon during pulsed laser irradiation, *Physical Review Letters*, 52, 2360–2363, 1984.
10. H.J. Kim, J.S. Im, and M. Thompson, Excimer laser induced crystallization of thin amorphous Si films on SiO2: Implications of crystallized microstructures for phase transformation mechanisms, *Materials Research Society Symposium Proceedings*, 283, 703–708, 1993.
11. M. He, R. Ishihara, W. Metselaar, and K. Beenaker, Agglomeration of amorphous silicon film with high energy density laser irradiation, *Thin Solid Films*, 515, 2872–2878, 2007.
12. C.R. Kagan and P. Andry (eds.), *Thin Film Transistors*, Marcel Decker Inc., New York, p. 170, 2003.
13. R. Paetzel, L. Herbst, and F. Simon, Laser annealing of LTPS, *Proceedings of SPIE; Photon Processing in Microelectronics and Photonics*, San Jose, CA, Vol. 6106, pp. 61060A1–61060A9, 2006.
14. J.E. Young, J.E. Sipe, and H.M. van Driel, Laser-induced periodic surface structure. III. Fluence regimes, the role of feedback, and details of the induced topography in germanium, *Physical Review B*, 30, 2001–2015, 1984.

15. B. Tull, J. Carey, E. Mazur, J. McDonald, and S. Yalisove, Silicon surface morphologies after femtosecond laser irradiation, *MRS Bulletin*, 31, 626–633, 2006.
16. L. Jiang and H.L. Tsai, Femtosecond laser ablation: Challenges and opportunities, *Proceeding of NSF Workshop on Research Needs in Thermal Aspects of Material Removal*, Stillwater, OK, pp. 163–177, 2003.
17. W. Arnold, B. Betz, and B. Hoffmann, Efficient generation of surface acoustic waves by thermoelasticity, *Applied Physics Letters*, 47(7), 672–674, 1985.
18. T.R. Anthony and H.E. Cline, Surface rippling induced by surface-tension gradients during laser surface melting and alloying, *Journal of Applied Physics*, 48(9), 3888–3894, 1977.
19. K. Kolasinski, Solid structure formation during the liquid/solid state transition, *Current Opinion in Solid State and Materials Science*, 11, 76–85, 2007.
20. A.M. Ozkan, A.P. Mashle, T.A. Railkar, W.D. Brown, M.D. Shirk, and P.A. Molian, Femtosecond laser-induced periodic structure writing on diamond crystals and microclusters, *Applied Physics Letters*, 75(23), 3716–3718, 1999.
21. B. Legrand, V. Agache, J.P. Nys, V. Senez, and D. Stievenard, Formation of silicon islands on a silicon on insulator substrate upon thermal annealing, *Applied Physics Letters*, 76, 3271, 2000.
22. R. Nuryadi, Y. Ishikawa, and M. Tabe, Formation and ordering of self-assembled Si islands by ultrahigh vacuum annealing of ultrathin bonded silicon-on-insulator, *Applied Surface Science*, 159/160, 121–125, 2000.
23. H. Ikeda, Y. Ishikawa, Y. Homma, and M. Tabe, In-situ observation of formation process of self-assembled Si islands on buried SiO2 and their crystallographic structures, *2003 International Microprocesses and Nanotechnology Conference, 2003 Digest of Papers*, 18–19, 2003.
24. Y. Ishikawa, Y. Imai, H. Ikeda, and M. Tabe, Pattern-induced alignment of silicon islands on buried oxide layer of silicon-on-insulator structure, *Applied Physics Letters*, 83, 15, 3162–3164, 2003.
25. J. Yu and Y.F. Lu, Laser engineered rippling interfaces for developing microtextures, adherent coatings and surface coupling, *SPIE Conference in Photonic Systems and Applications*, 3898, 252–262, 1999.
26. D.T. Danielson, J. Michael, and L.C. Kimerling, Thermal agglomeration of ultra-thin SOI and SSOI films: A quantitative stability study and physical model to guide ultra-thin SOI process design, *209th ECS Meeting, ECS Transactions*, 2(2), 375–389, 2006.
27. D.T. Danielson, D.K. Sparacin, J. Michael, and L.C. Kimerling, Surface-energy driven dewetting theory of silicon-on-insulator agglomeration, *Journal of Applied Physics*, 100, 083507–1, 2006.

7 New Approaches for High-Speed, High-Resolution Imaging

Gil Bub, Nathan Nebeker and Roger Light

CONTENTS

7.1   INTRODUCTION

Advances in image capture technologies have changed the way life science research is conducted. It is now relatively routine to capture image sequences from living cells and tissues at high rates, which allows scientists to directly visualize complex dynamic processes, such as embryo development, cell growth and division, and high-speed communication between excitable cells such as neurons and myocytes. However, image acquisition is still a bottleneck in many research projects. Living systems are inherently complex and operate over a wide range of space and timescales. Dynamics at subcellular space scales both influence and are influenced by tissue-level events. Fully capturing the dynamics of living systems is often beyond the capabilities of currently available imaging systems.

An example of the challenges and eventual promise of new technology is the effort to experimentally measure cardiac rhythm disturbances in a lab setting. Cardiac activation is relatively challenging to capture, as electrical transients occur over millisecond timescales and generate complex spatial patterns that are directly relevant to potentially fatal diseases. About 30 years ago, newly developed imaging technology based on low-resolution photodiode array detectors (e.g. the 12 by 12 Centronics MD-144-4PV and later the 16 by 16 Hamamatsu C4675-302) revolutionized experimental cardiac science. For the first time, researchers were able to image complex activation patterns in cardiac tissue at high speed using voltage-sensitive probes; the technique resulted in hundreds of publications and increased the understanding of cardiac disease significantly [1,2]. More recently, higher-resolution devices (in the form of 80×80 pixel CCD detectors and 100×100 pixel CMOS devices running at kHz rates) have largely displaced the photodiode array, although many are still actively used in research environments. As electrical activity in cardiac tissue has a relatively large intrinsic space scale (on the order of 3 mm), it is tempting to assume that current imaging modalities can give us a full understanding of cardiac arrhythmogenesis [3,4]. However, recent findings by several groups suggest the relevant space scales may be several orders of magnitude smaller. For example, subcellular alternans is a disease state where two halves of a single cardiac cell can display different activation patterns. Experiments performed on single cells show that the relevant space scale for observing this type of behaviour is around 2 μm. Recently, subcellular alternans have been experimentally observed in intact hearts [5,6], and there are potentially conditions where the activity can modulate activity at the whole heart level [7]. A camera with $20,000 \times 20,000$ pixels running at kHz rates (ignoring for now issues with resolution of the optical system) for several minutes would be needed to capture the evolution of subcellular alternans on the whole heart, but this is well beyond the current state of the art (see [4] for an in-depth discussion of spatial resolution requirements in a cardiac model system).

High-speed communication in excitable cell networks may represent the fastest events generated by living cells, but there are other biophysical processes that can take advantage of high-speed imaging technologies. Bubble formation in living tissue is an example of a process that is important in several biomedical applications. Bubble-enhanced sonoporation is a process where cell membranes are ruptured in a controlled way to enhance delivery of pharmacological agents. Ultrasound-driven bubbles have been used to enhance contrast during ultrasonic imaging, allowing real-time evaluation of myocardial function. Similarly, shock waves created by bubble formation play a role in kidney stone treatments [8]. Bubble formation and destruction are extremely rapid events, and experimental investigation requires imaging rates approaching 1,000,000 fps.

Another area where high imaging speeds are essential is lifetime fluorescence microscopy (FLIM) [9]. FLIM measures the fluorescence decay of probes to sensitively determine biologically valuable parameters such as pH, calcium and oxygen concentration. In addition, fluorescence decay constants can sensitively discriminate between fluorophores with overlapping spectra, which are very useful for discriminating the desired signal from autofluorescent compounds within a cell. Fluorescence lifetime is also largely independent of probe concentration, which allows quantitative measurements in the presence of uneven dye loading. The insensitivity of FLIM to dye concentration and background fluorescence has been exploited to measure ion concentrations in cells directly with high sensitivity. This has major advantages over traditional fluorescence methods, which require taking two measurements at different wavelengths along with complex calibration procedures. A related technique, called Forster resonance energy transfer (FRET, also known as fluorescence resonance energy transfer) microscopy, is based on measuring energy transfer between donor and acceptor probes. FRET occurs when the donor and acceptor probes are in close proximity to each other and the emission spectrum of the donor probe overlaps with the absorption spectra of the acceptor probe. FRET can be used to quantify the proximity of a fluorescent probe with a nonfluorescent acceptor molecule in tissue: in a combined FLIM-FRET imaging experiment, the presence of the acceptors can be measured by the change in fluorescence lifetime of the donor probe. Since FRET efficiency drops off quickly with distance, it can be used to measure interactions between important biomolecules with nanometre accuracy.

Fluorescence decay typically occurs at very high speeds (most often pico- to microsecond timescales, depending on the probe). Accurate gated intensified cameras and pulsed light sources are required to measure the decay constants by shifting the gate time relative to the light pulse, one point on the curve for each camera frame. This means that several time-offset frames are required for each curve, which both is time consuming and requires multiple potentially damaging light pulses per measurement. Ideally, a single pulse, coupled with a fast camera, could measure a fluorescence lifetime curve in one step. Cameras capable of MHz frame rates exist, but they are specialist devices that often require cumbersome optics and illumination for routine biological measurements. The many potential benefits of using a high-speed camera for FLIM measurements were demonstrated using a 40,000 fps CMOS camera and europium FLIM probes, which have very slow (2 ms) decay times [10].

An emerging area, which can benefit from extremely high frame rates, is microscopy based on time-of-flight spectroscopy measurements. Microscope-mode spatial imaging mass spectrometry captures the distribution and identity of molecules over the imaged area by measuring the mass of particles released from a sample (desorption) by laser bombardment. The distribution of desorbed ions is projected onto microchannel plates (MCPs) and phosphor screen by an electric field that converts the signal to light, which is captured by a 2D array detector. The masses (or, more accurately, the ion mass/charge ratio) can be measured by measuring the time of flight to the detector. The ability to resolve 2D spatial profiles of ions using this technique offers advantages over traditional time-of-flight sensors (either single-point sensors with accurate time to digital converters or streak cameras that translate arrival time differences into spatial deflection on a sensor by application of a strong electric field). Conventional CCD detectors with gated image intensifiers can be used to measure time of flight of desorbed fragments, but here the slow frame rate of the camera necessitates separate experimental runs for each detected mass range. Acquiring a representative set of different ion species on a biological sample therefore requires multiple potentially damaging laser exposures, which is less than ideal. Further, since image intensifier gating times must be set before the experimental run, the researcher needs prior knowledge of the masses expected in the sample or must run additional experiments to determine experimental parameters. In contrast, a detector with a very high frame rate could potentially measure all the desorbed particles after a single laser pulse. The feasibility of using a high-speed detector for this purpose was explored by Nomerotski and colleagues [11,12], who used a low-resolution MHz imaging camera to capture 2D mass profiles from a manufactured test sample.

7.2   CHALLENGES AND APPROACHES

Current imaging techniques are largely based on the following scheme. A 2D (n by m) array of pixels is exposed to light for a period of time (tau). The exposed frame is then digitized and the data are stored in memory. Capturing dynamic events at high speed requires repeating this process as frequently and quickly as possible. If we assume that light levels are not a limiting factor, the main bottlenecks in this process occur during storage and digitization.

First, there are real limits to the amount of data that can be digitized in a given time interval. Modern consumer grade sensors have over a million pixels – indeed, some smartphones have resolutions in excess of 16 million pixels. The time needed to digitize a single frame with a single digitizer is on the order of 100 ms. Conventional cameras typically use pixel skipping (where every nth pixel is digitized) or subwindowing (where a smaller continuous subset of pixels is digitized) to reduce the pixel count in order to record at video (25 fps) rates.

More specialized systems achieve order of magnitude increases in frame rate by parallelization: several digitizers operate on different parts of the imaging array simultaneously. Here, both the imaging chip and supporting hardware are designed to optimize fast digitization speeds. Specialized cameras such as the Fastcam SA5 can achieve 7500 fps capture rates at megapixel resolutions, a level that has been consistently increased with each new generation of camera. The high bandwidth of these systems gives rise to a second technical challenge – once digitized, image data must be stored. The most common strategy in high-speed cameras is to store a sequence of images in RAM on the camera head itself, as transferring data at gigapixel rates is impractical. Fastcam SA5 camera configurations with 32 GB of onboard RAM are available. However, 32 GB are only able to store 3 s worth of data before being filled and must be downloaded to a host machine, which takes several minutes over standard CameraLink interfaces. Even if camera-to-computer data transfer limitation is overcome, data rates will be quickly limited by storage capacity. For example, peripheral component interconnect (PCI) Express can be used as a high-speed external interface. The highest speed currently available is the ×16 interface, which allows 8 GB/s transfers. A 1 million pixel camera operating with this interface used to capacity would be capable of 8000 fps and would fill a 1 TB hard drive in only 125 s.

Finally, all imaging systems must take into account a fundamental limit imposed by high acquisition rates on signal quality. Imaging chips and digitization generate noise (termed read noise) that is proportional to the bandwidth of the signal. Read noise is a catch all term that encompasses several on-chip noise sources related to converting charge to voltage and digitization. The primary sources of noise in this process are thermal and are proportional to bandwidth and temperature. Johnston noise and reset noise have the form $E_t = \sqrt{4k_B TBR}$ where K_B is the Boltzmann's constant, T is the absolute temperature, B is the bandwidth in Hz and R is the source resistance for reset noise or output impedance of the amplifier for Johnson noise.

Simply put, all other factors being equal, higher frame rates result in higher noise levels and lower signal-to-noise ratios. High-resolution scientific imaging cameras, such as those employed in life science laboratories and used in astronomy, maximize signal quality by digitizing camera data slowly: commercially available astronomy grade cameras can take several minutes to output a single high-resolution image. Recently developed CMOS camera architectures can reduce read noise by using parallelization strategies (see Section 7.2.2.1), but as a general rule of thumb, high-speed imaging can't be used in low light conditions due to read noise issues.

7.2.1 HISTORICAL PERSPECTIVE

The first high-speed scientific imaging study was done over 100 years ago. Eadweard Muybridge was given a challenge by his sponsor, California Gov. Leland Stanford, to determine if a horse has all four of its hooves in the air at any point during a gallop. Muybridge came up with a strategy that is still employed by some high-speed camera designs today. He arranged 24 cameras along a segment of a race track* and set up trip wires so that the cameras would fire in sequence as the horse ran past. He achieved an effective frame rate high enough to resolve the horses motion, convincingly proved that a horse spends part of its time during a gallop completely airborne. He subsequently used multi-camera systems to perform a series of high-speed imaging studies and inspired inventors of that era to design multi-exposure cameras that employed similar ideas. Muybridge's approach of using parallelization to bypass the rate-limiting step – in his case chemical photographic plate development, but in modern cameras it is digitization – can at least in principle be used to overcome any bandwidth issue; however, system complexity and size eventually become prohibitive.

7.2.2 HIGH-SPEED IMAGING STRATEGIES

Modern high-speed scientific imaging cameras achieve high frame rates using diverse strategies. The most common approach is to trade spatial resolution for speed. Conventional CCD pixels can be binned [13]: Charge from adjacent pixels can be summed in readout registers before digitization, which both reduces bandwidth and increases signal-to-noise ratios at the expense of spatial resolution. CMOS sensors can use pixel skipping (digitizing every nth pixel) or region of interest readout (digitizing a subset of adjacent pixels) to minimize bandwidth. In principle, these schemes could give very high frame rates, but in practice, frame rates increase less than the square root of pixel count (e.g. binning 2 by 2 pixels gives a fourfold

---

* The path Muybridge took to reach his final innovation took several years (1872–1878). He initially captured a single image by setting up a system to trigger a camera remotely (which didn't convince the public), then set up an array of 12 cameras and finally a bank of 24 cameras. He also had at least two triggering mechanisms – the horse would travel through a series of trip wires that would trigger electromechanical shutters, or alternatively, the horse would pull a carriage so that its wheels would depress electrical contacts closing an electrical circuit that would open each shutter. Complete details can be found here: http://www.photo-seminars.com//Fame//muybridge.htm and http://www.victorian-cinema.net//muybridge.htm.

reduction in resolution but only a twofold increase in frame rate) in most commercially available sensors. Scientific sensors designed for life science experiments where high frame rates are needed often have modest pixel counts. For example, RedShirtImaging and SciMeasure cameras use a cooled Marconi CCD39-01 back-illuminated 80 × 80 pixel sensor with four output amplifiers to reduce bandwidth. These cameras can reach 2000 fps with a read noise of only 8 electrons. Andor and Photometrics high-speed cameras with base frame rates of just over 500 Hz use an e2V CCD60 L3Vision 128 × 128 pixel electron-multiplying CCD (EMCCD) sensor with a single digitizer, for a read noise of 35 e-. Specialized CMOS-based cameras by RedShirtImaging (128 × 128 pixel NeuroCMOS-SM128f) and SciMedia (100 × 100 pixel MiCAM Ultima) can run at 10 kHz frame rates, but here the emphasis is on large high-capacity pixels for high dynamic range imaging, and read noise is a less important parameter.

Higher-resolution cameras with noise performances suitable for life science research typically have relatively low frame rates. The popular Sony 285 interline 1.4 megapixel sensor, first released in 2002, is deployed in imaging systems sold by many manufacturers (Andor Clara, Photometrics CoolSNAP, QImaging EXi Blue) but has frame rates between 10 and 15 fps and higher read noise values (5.5–12 e⁻, depending on the manufacturer). More recently, Fairchild, Andor and PCO codeveloped a high-resolution CMOS sensor that uses an A/D converter on every pixel column to achieve very low (1.5 e⁻) read noise levels in rolling shutter mode at significantly higher frame rates: the camera can capture 5.5 megapixel images at 100 fps and a 100 × 128 pixel subregion at 1600 fps (see also sCMOS cameras by QImaging and Hamamatsu). Low-noise CMOS architectures are still an active area of research. Yasutomi et al. have developed an improved design capable of low noise in global shutter mode [14]. Commercially available sCMOS cameras are now seeing wide adoption in the life sciences, as they offer a significant step up in performance from interline CCD sensors.

7.2.2.1 Parallelization Strategies

A highly successful strategy for overcoming both read noise issues and bandwidth hurdles takes inspiration from Muybridge's multiple camera approach. Multiple cameras can be arranged so that they image roughly the same field of view from roughly the same view point, with some parallax between cameras [15], or use a single lens and optical beam splitters to direct light to several different detectors [16] (e.g. Cordin Scientific Imaging Model 222-4G). In both cases, cameras are triggered to capture images in sequence. Extremely high imaging rates (200 million fps) can be achieved in systems where camera exposures are controlled by MCP image intensifiers. Another variation on this theme are Cranz–Schardin cameras where multiple, spatially offset pulsed light sources are used to back-illuminate a given subject. A field lens is then used to reimage each light source to an individual camera, such that only light from that source reaches its corresponding camera. Because of sufficient distance between this field lens and the subject, the images of the subject in each camera are roughly equivalent, save for some parallax [17–21]: Cranz–Schardin camera frame rates are limited only by the switching speed of the light sources, but their dependence on this external back-illumination limits their use in life sciences, for example, fast fluorescence image sequences cannot be captured using this technique. Multi-camera systems have a limit to the number of cameras that can be successfully incorporated into the design. In systems where the light is distributed to the camera channels using beam splitters, the number of sensors is limited to 8 or 16 as light levels drop by a factor proportional to the number of beam splitters used to redirect the image. Multi-camera systems with individual imaging lenses have less stringent limits but cannot easily be incorporated into microscopy systems. Recent innovations on this front may overcome this issue: Brady et al. [22] demonstrated a scalable multi-sensor design consisting of 98, 14 megapixel microcameras housed in a 0.75 by 0.75 by 0.5 m enclosure.

A higher frame number can be achieved using a rotating mirror design, where a fast rotating mirror is used to project an image along a series of detectors arranged in an arc. The original design for these cameras came from a need to image atomic bomb explosions toward the end of World War II [23,24]. A high-speed motor or, more optimally, a gas turbine design drives the mirror to rotate at speeds of up to 20,000 revolutions/s. The maximum frame rate of the system is set by the mirror materials and design and has not been improved in several decades [25]. The mirror is most often made from an aluminium-coated beryllium substrate as it is one of the few materials with a stiffness-to-weight ratio high enough to result in a high enough resonant frequency to allow these speeds. Light enters the camera through an objective lens and is reflected by the spinning mirror via relay lenses to up to 100 individual detectors arranged in an arc. Rotating mirror cameras produce frame records of up to 25 million frames/s. Rotating mirror cameras are widely used in material research and military applications, where extremely high frame rates are needed to image shock waves. Rotating mirror cameras are also deployed in biomedical laboratories interested in imaging microbubble formation and destruction in living tissue and the use of microbubbles in drug delivery [26–28].

7.2.2.2 CCD and CMOS Sensor Architectures for High-Speed Imaging

CMOS image sensor designs are intrinsically faster than conventional CCDs as the digitization process can be highly parallelized. CMOS active-pixel sensor (APS) architectures allow for chip designs to be optimized for high speeds, for example, it is relatively simple to have digitizers on all pixel columns or even every pixel [29]. In contrast, the CCD architecture is based on a serial readout scheme. CCDs with a small number of readout taps are commercially available (e.g. the Kodak KAI-02050 1600 by 1200 pixel CCD with 4 taps), but highly parallel CCD chips are specialized devices due to increased off-chip circuit requirements: more power and circuitry are needed to drive the electrodes and generating timing signals each tap. At one extreme, specialized CCDs for particle physics research have been manufactured with digitization in each column (column-parallel CCDs or CP-CCDs; see ref. [30]). Although parallelization is more difficult to implement in CCD processes than in CMOS architectures, until recently, CCDs have outperformed CMOS image sensors in important metrics and developing fast CCDs was the best option.

An important parameter for CMOS pixels is fill factor. This is the proportion of the pixel area that is usable for photon detection. Since CMOS chips have more on-chip circuitry than CCDs, a greater portion of every CMOS pixel is devoted to circuitry in the form of transistors and wiring; the fill factor will always be less than that of an otherwise equivalent CCD. The fill factor directly affects the quantum efficiency (QE) of the pixel, since photons that do not reach the detector have no chance of being converted to photocurrent. The overall QE in CMOS detectors can be limited due to the structures used. Silicon absorption is low for longer wavelengths, so a deep junction photodiode is needed to get the best response. Shallow junction photodiodes are frequently used and will not provide as good red performance. Another factor is the layer stack above the photodetector. At fine fabrication processes, there are many layers of metal interconnect available. Even if these metals are not used over the photodiode itself, a stack of transparent layers will be formed with each interface between layers creating a small amount of absorption and reflection and so reducing the light that reaches the photodetector. It is possible to resolve the QE problem by using back-illuminated sensors. A back-illuminated sensor is essentially the same as a front-illuminated sensor but flipped round so that the unrestricted backside is facing the illumination. This removes the obstructions caused by wiring and circuitry, reflections due to the layer stack, and provides an effective 100% fill factor. Back-illuminated sensors must also be back-thinned to reduce the amount of light that is absorbed before being detected. This makes the wafers they are built on very fragile and so the technique is traditionally used in high-end cameras. Another approach to improve QE is the use of vertically integrated CMOS sensors, where the photosensitive elements are bonded on top of control circuitry [31], but these architectures are not yet widely available.

CMOS sensors have additional noise sources when compared to CCDs. In addition to shot noise, which is common to all image sensors, CMOS pixels display MOS device noise; flicker noise, also known as 1/f noise due to its frequency relationship; and reset noise. Reset noise comes about during the reset phase of the exposure, where the parasitic capacitance of the photodiode is charged to the voltage supply through a switch. Thermal noise present on the capacitance means that when the switch is disabled ready for exposure, the precise voltage across the photodiode is not known. The starting voltage is the reference point for the exposure, so the uncertainty is present as noise. The reset noise in volts is given by $v_n = \sqrt{k_B T/C}$, where k_B is Boltzmann's constant, T is the absolute temperature and C is the capacitance of the photodiode. In pixels with small photodiodes, C will also be small and reset noise may be the dominant source of noise. The use of per-pixel charge-to-voltage conversion and amplification results in fixed pattern noise, which manifests as different gains and offsets for each pixel. However, many noise sources can be minimized by careful design. Correlated double sampling (CDS) and bandwidth limiting of the in-pixel amplifier can reduce pixel-level noise sources. Microlenses can be used to partially offset low QE where back-illumination is not desirable, and fixed pattern noise not removed by CDS can be largely removed by post-processing techniques. Continued refinements and improvements have allowed CMOS sensors to be successfully deployed in high-end consumer and scientific camera systems. High-speed imaging functions are appearing in relatively inexpensive consumer grade cameras: the Exilim series of cameras from Casio [32] is capable of 1000 fps operation at 640 by 64 pixel resolution.

Most CMOS sensors can be classified as either passive or active-pixel arrays. The passive-pixel array consists of a grid of photodetectors, usually photodiodes and a switching MOSFET. In general, there is a single output for passive arrays, and each pixel is selected for readout by row and column multiplexers. The noise performance and speed of these devices are compromised due to the long high-capacitance wires needed to pass current to the output amplifier. Passive pixels have remained of limited interest in areas where QE is of paramount importance but are being replaced by back-thinned detectors. APSs add a simple amplifier at the pixel level. This improves the noise figure and thus the achievable dynamic range and S/N ratio of the array. Global shutters are typically used in high-speed applications as a rolling shutter adds motion distortion. The standard high-speed CMOS APS pixel consists of a photodetector (a pinned photodiode – usually an n-well in p-substrate implantation), a transfer gate, a floating diffusion node (which is sampled after reset to implement CDS), a reset switch, a selection switch and a source-follower readout transistor. In high-speed CMOS image sensors, designers must optimize readout bus line layout so that their capacitance does not limit bandwidth.

Voltages can be read out either as analogue signals (where digitization occurs off-chip) or as digitized by on-chip A/D converters and output in digital form. High-speed CMOS arrays optimize rapid readout, either by employing multiple parallel analogue readout lines [33] or by having a large number of on-chip digitizers. At one extreme, each pixel can have its own A/D converter. The large number of transistors required for A/D conversion results in a decrease in fill factor, which is partially mitigated by using smaller CMOS process sizes. High-speed imaging using per-pixel digitization has been demonstrated by Kleinfelder [29] with a 37 transistor, 15% fill factor design in a 0.18 μm CMOS process, and by Ghannoum [34] with a 57 transistor, 26% fill factor design in a 90 nm CMOS process. There are, however, trade-offs that come with smaller process sizes, as scaling effects increase leakage currents and reduce dynamic range due to the necessity of using lower voltages. Fill factor can be further increased by moving away from a monolithic CMOS process: hybrid sensor designs, where a 100% fill factor detector array is bump bonded onto a CMOS chip, have been used to implement very high transistor counts per pixel while maintaining sensitivity [35]. Another high fill factor design is based on placing an electrostatic barrier between the photodiode and supporting circuitry [36], allowing the whole pixel area to collect light, but limits pixels to using NMOS transistors, which restricts pixel complexity. A variant on this architecture, termed isolated N-wells monolithic active pixel sensors (INMAPS) by Turcheta and colleagues [37], is based on including a deep P-implant that provides screening for n-wells from the P-doped epitaxial layer. The INMAPS architecture allows the integration of both negative-channel metal oxide semiconductor (NMOS) and positive-channel metal oxide semiconductor (PMOS) transistors within these pixels, allowing for complex circuit designs.

7.2.2.3   Image Intensification

The effects of read noise on signal quality can be mitigated by using image intensification before digitization. The strategy here is to increase the signal amplitude over the noise floor of the amplification circuitry. In principle, this allows higher frame rates as the bandwidth of the amplifier and digitizer can be increased without incurring a read noise penalty. Intensified detectors either use external intensifiers (intensified CCDs (ICCDs)/CMOS devices) or amplify the signal internally (EMCCDs; electron multiplication gain CCDs). In an ICCD, an image intensifier, consisting of a photocathode, an MCP and a phosphor screen are placed in front of the CCD chip. In contrast, EMCCD adds gain by putting a solid-state multistage EM register ahead of the output amplifier. The EM register uses a high voltage at each stage to produce secondary electrons via a process called impact ionization. Although the number of extra electrons generated at each stage is fairly modest, several hundred stages can generate gains of over 1000-fold.

ICCDS typically have a lower spatial resolution than EMCCDs as intensifiers impose distortions on the image. Also, EMCCD gain adds less noise than image intensifiers (a factor 1.41 vs. >2 [38]). However, EMCCD architecture makes on-chip parallelization strategies impractical, as each digitizer would need its own high-voltage gain stage. As a result, EMCCD frame speeds are limited by the capacity of a single A/D converter, and the fastest intensified systems are CMOS devices with external intensification. An additional consideration is that all intensified devices have far lower intrascene dynamic ranges at high gains (approx. 9 bits and as low as 5 bits at maximum gain), and high-spatial-resolution devices typically have lower dynamic ranges due to smaller pixel sizes (10–12 vs. 14 bits). Although patterned illumination strategies can partially compensate for the drop in dynamic range (see reference [39]), intensified camera systems are best suited for conditions where the number of grey levels captured is not critical.

7.2.2.4 *In Situ* Image Storage Sensors

At very high speeds, even per-pixel digitization times are limiting. CMOS chips have a limited number of input and output ports – indeed, the surface area available for ports on a chip scale with the square root of pixel count, as ports must be located at the chip edge. As a result, image data must be heavily multiplexed, which causes significant delays. This bottleneck has inspired a class of sensor where light intensity data are stored in analogue form on-chip. The *in situ* image sensor (ISIS) chip design is based on having a large number of storage elements located on the imaging chip itself for each light-sensitive pixel. Storage elements are arranged so that charge can be shifted from the light-sensitive pixel to a masked storage pixel in an analogous fashion to that used by interleaved CCDs, but here each light-sensitive pixel is connected to several storage elements arranged in series. With each frame exposure, charge is shifted to the next storage element in the chain until all the storage elements are filled. The first sensor with multiple analogue storage elements was developed by Kosonocky et al. [40] in 1997. The 360 by 360 pixel sensor has a fill factor of 13.5% and is capable of capturing a burst of 30 sequential frames at 833,000 fps. Another variant was developed by Etoh et al. [41,42] and is now available commercially. The Shimadzu HyperVision HPV-2 camera has 312 by 260 pixels with a fill factor of 13% and 100 storage elements per light-sensitive pixel, allowing a burst of 100 frames to be captured at 1 million fps. A more advanced variant of the same design, with 720 by 410 pixels with a fill factor of 16.8% and 144 element storage buffer, has recently been reported [43]. CCD *in situ* sensors are thought to have an inherent top frame rate of 10 million fps, due to difficulties related to moving charge between storage wells at high speed. The voltage needed to move charge between storage elements increases with the square of the signal frequency, so eventually thermal noise caused by high driving voltages degrades the signal. CMOS sensors with *in situ* storage do not have the same inherent limit and are in principle capable of frame rates in excess of 10 million fps. A CMOS *in situ* sensor with 64 storage elements was developed by Kleinfelder [44]: the device achieves frame rates of over 10.5 million fps and can be reconfigured to use half the frames to reduce noise via CDS. El-Desouki [45] describes a similar *in situ* architecture that has the potential to reach 1 billion fps speeds, although the physical implementation of the chip was limited to 1 million fps.

In situ charge storage can be exploited for specialized applications. The modulated light camera developed by Drs. Mark Pitter and Roger Light has each photodiode connected to four storage elements that can be exposed independently. In experiments where a very small signal is superimposed on a large DC background, it is common to introduce amplitude modulation into the system and use a lock-in amplifier to extract the small signal. This technique is also used in optical experiments with a single photodiode connected to a lock-in amplifier. The modulated light camera, or optical lock-in array, recreates this feature on the pixel level by capturing four data points during a single oscillation, each integrating $90°$ of the signal, and reconstructing phase offset and frequency from the data as shown in the following equations. The camera has been demonstrated detecting signals in the range of 10^{-4} to 10^{-6} of the background level [46–48]:

$$A = \sqrt{(i_1 - i_3)^2 + (i_2 - i_4)^2}$$

$$\phi = \tan^{-1} \frac{i_2 - i_4}{i_1 - i_3}$$

Other variant of the *in situ* storage design is time-of-flight photon counting sensors. Time-of-flight sensors record the arrival time of a photon or charged particle instead of an intensity value. The MEGAFRAME sensor [49], which consists of a 128×128 element array of single-photon avalanche diodes (SPADs), can record 1 million fps with 50 ps arrival time accuracy. The pixel imaging mass spectrometry (PImMS) sensor is designed for spectroscopy applications [11,50]. It has six storage registers to record arrival times, allowing it to be used to characterize complex molecular species. The PImMS sensor was designed using an INMAPS process in collaboration with science and technology facilities council at the Rutherford Appleton laboratory (STFC-RAL).

One of the limitations with ISIS designs is that the number of pixels and storage sites is fixed at the design/manufacturing stage. This is in contrast to conventional CCD and CMOS designs, which allow user-specific trade-offs between frame rate and resolution: Conventional CCDs can increase frame rates by pixel-binning strategies, while CMOS cameras can increase frame rates by pixel skipping and subregion readout strategies. Further, in most ISIS designs, the spatial resolution and QE are low due to the large amount of on-sensor circuitry required for data storage.

7.2.2.5 Temporal Embedding Strategies

The strategies in the previous sections produce conventional (frame-by-frame) image sequences, where each frame is a complete record of a moment in time. The following sensor strategies encode high-speed information in a fundamentally different way and recover frame sequences via post-processing.

Temporal pixel multiplexing (TPM) is a relatively new imaging modality that promises the speed of *in situ* image storage with a high degree of user flexibility in both spatial resolution and frame rate [51]. A TPM sensor allows works on the principle of offsetting exposure times at the pixel level. By shifting pixel exposure times in a larger image, high-speed temporal information is embedded in high-resolution images. Figure 7.1 illustrates the general principle. Different pixel subsets are exposed at different times so that all pixels in the array are exposed once and each pixel subset covers the entire field of view. All pixels are read out once all the pixels are exposed, giving a high-resolution image. The pixels in each subset store image data from a subinterval and can be grouped to give a high-speed image sequence. Since all the data are read out once, a TPM design shares the benefits of an ISIS. However, here, the user can control the resolution and number of frames on a per-run basis. For a proposed 2k by 2k sensor, the user can choose to image either a full frame at video rates, 100 frames at 200×200 pixels or anything in between.

TPM imaging was first demonstrated using a conventional machine vision CCD and a micro-mirror array to control pixel exposure times [51], and a variant has been developed by other researchers by controlling scene illumination [52]. The authors have designed and fabricated a TPM chip based on an APS circuit with on-chip circuitry to control exposure at the pixel level. A conventional camera with a global shutter has a single shutter input to control the exposure across the entire pixel array. Each TPM pixel has a row and column input as well as a shutter input. The shutter switch is only activated when all three of the row, column and shutter inputs are enabled. This allows sparse patterns of exposure to be created across the array. The row and column inputs are controlled by programmable shift registers at the edge of the pixel array. The shift registers can be programmed with the required pattern before the beginning of the first exposure. This arrangement makes changing the exposure pattern during a frame capture extremely quick, so minimal time is required between the multiple exposures. Once the entire pixel array has been exposed, the frame data can be digitized. Since rapid charge transfer between pixels and storage elements is not necessary, many of the engineering challenges faced by the original CCD-based *in situ* image storage sensor are sidestepped, and the chip can be fabricated using standard low-voltage components. Low-level simulations of the circuitry involved demonstrate that the chip will be capable of MHz frame rates.

Compressive methods: On-chip compression has long been used to decrease the storage and transmission requirements of high-resolution imaging chips [53], but compressive method techniques can reduce signal bandwidth before digitization, which could dramatically increase throughput. An example of a compressive method that can be directed to high-speed imaging is flutter shutter photography [54]. Here, a global shutter is repeatedly applied at random times during each video frame. The resulting images can be decoded to reconstruct high-speed video. Another paradigm that shows great promise is adaptive compressive imaging [55]. This technique is based on controlling which pixels are digitized based on real-time analysis of the scene, allowing high-speed capture of dynamically repositioned regions of interest, with potentially large reductions in bandwidth.

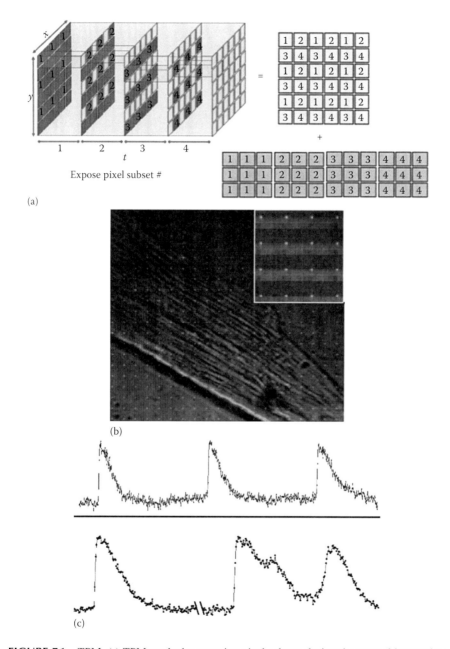

FIGURE 7.1 TPM. (a) TPM works by exposing pixel subsets during the normal integration time of the camera, resulting in a high-resolution still and a high-speed image sequence. (b) High-resolution TPM image recorded at 10 Hz of a cardiac cell loaded with a calcium-sensitive dye; inset is a zoom after a background frame subtraction, showing how time information is encoded into space during a calcium upstroke. (c) Calcium action potentials extracted from sequential frames at 250 Hz: top panel is a raw trace obtained from one 5 × 5 mirror group; bottom trace shows an average of several single traces at two time points (Nature Methods 2010).

An emerging paradigm that may further drive the evolution of high-speed imaging technology is compressive sensing [56]. Compressive sensing is based on sampling subsets of the data in an optimal way so that the entire (undersampled) signal can be reconstructed with high accuracy mathematically.* The assumption used here is that most images are compressible (i.e. the image is sparse in a known basis). Instead of analysing the image before compressing it, an image is compressed by making a small number of measurements; for example, each measurement can be the sum of a random subset of pixels of the image. Compressive sensing theory states that the number of measurements needed is significantly less than what one would expect given the Nyquist criteria. Compressive sensing imaging chips have been proposed and fabricated and work either by a combination of summation and subtraction operations on pixels before digitization [57–59] or at the digitization level [60]. So far, these compressive sensing chips have been proof-of-principle low pixel count designs that have not pushed imaging speeds beyond what is possible given the current state of the art. Compressive sensing has, however, been used to greatly increase speeds in life science imaging applications where a single image must be reconstructed from multiple sequentially captured data sets: Magnetic resonance imaging scan times [61] and superresolution microscopy [62] data set capture times have been reduced by more than an order of magnitude.

7.3   CONCLUSIONS

An overview of the different strategies employed to overcome bandwidth limits is shown in Table 7.1. It is clear that the high-speed imaging field is constantly evolving, with new approaches discovered on a regular basis. It is striking that innovations from decades (and even a century) past continue to have impact: For example, Muybridge's multiple camera concept is reapplied in CMOS form [15] and in TPM technology [51], and Cranz–Schardin cameras are being refreshed given advances in illumination techniques [19].

The challenges faced by life science applications, where record duration and frame rate are often limiting factors, may be addressed by combining the newer compressive methods with high-resolution detectors. Life science data can be highly compressible: For example, arrhythmia initiation events are statistically rare and the activation patterns eventually evolve into macroscopic, low-resolution waves, but the events need to be recorded at high resolution in order to resolve initial trigger activity at the cell level. Similarly, fast fluorescence lifetime decay occurs at a far different timescale than the biological events it is used to document. Sensors capable of efficiently capturing data across widely divergent space and timescales would lead to a revolution in life science imaging.

---

* The usual method is that randomly selected samples from the complete signal are reduced in some way (e.g. summed) to a single data point or each sample is transformed randomly (e.g. multiplied by -1 or 1) and then summed to a single data point, and this process is repeated N times. Compressive sensing theory states that it is possible to accurately reconstruct the complete signal given knowledge of which random samples were taken and if the original signal is compressible for $N < M$, where M is the number of data points needed to represent the original signal using the Nyquist criteria. These techniques are similar to those that use (non-random) Hadamard transforms to both compress and boost signals in spectroscopy applications [63,64]. A large compressive sensing resource page can be found here: http://dsp.rice.edu/cs.

TABLE 7.1
Different Imaging Strategies as Discussed in the Text

	Strategy	Notes
Optical methods		Multi-sensor approaches use individual cameras with shutters or optical methods (rotating prism or beam splitters). In contrast, the rotating mirror approach is not affected by parallax (Cranz–Schardin and image converter cameras not shown).
Bandwidth reduction		CCDs can boost speed by using frame binning (which also increases S/N) and region of interest (ROI) readout (which requires optical masking). CMOS sensors can perform ROI readout without masking and pixel skipping.
Intensification		Image intensification boosts signals before amplification and digitization to overcome read noise limits but lowers dynamic range and adds noise. EMCCDs use a high-voltage gain register (ICCDs not shown).
Parallelization		This strategy reduces bandwidth by lowering the number of pixels per A/D converter. Multi-tap CCDs usually are limited to four taps (but see CP-CCDs), while CMOS architectures can support pixel-level A/D.
ISIS		ISISs: CCD-based approach with masked wells shifts charge sequentially to masked storage pixels. CMOS approach (e.g. MODCAMS) uses switching to move charge to multiple storage capacitors.
Temporal embedding		Embedding temporal data in images. TPM embeds time information into images by controlling pixel digitization times; compressive sensing reduces bandwidth by grouping pixels and solving for the original image offline.

ACKNOWLEDGMENT

Gil Bub acknowledges support from the BHF Centre of Research Excellence, Oxford.

REFERENCES

1. H. Himel IV, J. Savarese, and N. El-Sherif, The photodiode array: A critical corner-stone in cardiac optical mapping, in *Photodiodes – Communications, Bio-Sensings, Measurements and High-Energy Physics*, ISBN 978–953–307–277–7, InTech, Rijeka, Croatia, 2011., J.-W. Shi, Ed., 2011, pp. 137–160.
2. I. Efimov, V. Nikolski, and G. Salama, Optical imaging of the heart, *Circulation Research*, 95, 21–33, 2004.
3. S. Mironov, F. Vetter, and A. Pertsov, Fluorescence imaging of cardiac propagation: Spectral properties and filtering of optical action potentials, *AJP Heart*, 291(1), H327–H335, 2006.
4. E. Entcheva and H. Bien, Macroscopic optical mapping of excitation in cardiac cell net-works with ultra-high spatiotemporal resolution, *Progress in Biophysics and Molecular Biology*, 92(2), 232–57, October 2006.
5. G. Aistrup, Y. Sherifaw, S. Kapur, A. Kadish, and J. Wasserstrom, Mechanisms underly-ing the formation and dynamics of subcellular calcium alternans in the intact rat heart, *Circulation Research*, 104, 639–649, 2009.
6. G. Aistrup, J. Kelly, S. Kapur, M. Kowalszyk, I. Sysman-Wolpin, A. Kadish, and J. Wasserstrom, Pacing-induced heterogeneities in intracellular Ca2+ signaling, car-diac alternans, and ventricular arrhythmias in intact rat heart, *Circulation Research*, 99, E65–E73, 2006.
7. S. Gaeta, G. Bub, G. Abbott, and D. Christini, Dynamical mechanism for subcellular alternans in cardiac myocytes, *Circulation Research*, 105, 335–342, 2009.
8. T. Kodama and K. Takayama, Dynamic behaviour of bubbles during extracorporeal shock-wave lithotripsy, *Ultrasound in Medicine and Biology*, 24(5), 723–738, 1998.
9. K. Suhling, P. French, and D. Philips, Time-resolved fluorescence microscopy, *Photochemical and Photobiological Science*, 4, 13–22, 2005.
10. G. Zanda, N. Sergent, M. Green, J. Levitt, Z. Petrásek, and K. Suhling, Wide-field single photon counting imaging with an ultrafast camera and an image intensifier, *Nuclear Instruments and Methods A*, 638(1), 127–133, ISSN 0168–9002, 2011.
11. M. Brouard, A. Johnsen, A. Nomerotski, C. S. Slater, C. Vallance, and W. H. Yuen, Application of fast sensors to microscope mode spatial imaging mass spectrometry, *Journal of Instrumentation*, 6, C01044, 2011.
12. A. Nomerotski, M. Brouard, E. Campbell, A. Clark, J. Crooks, J. Fopma, J. J. John, A. J. Johnsen, C. Slater, R. Turchetta, C. Vallance, E. Wilman, and W. H. Yuen, Pixel imaging mass spectrometry with fast and intelligent pixel detectors, *Journal of Instrumentation*, 5(07), C07007–C07007, July 2010.
13. J. R. Janesick, *Scientific Charge-Coupled Devices*. 2001, SPIE Press, Bellingham, WA. p. section 2.2.10.1.
14. K. Yasutomi, S. Itoh, and S. Kawahito, A two-stage charge transfer active pixel CMOS image sensor with low-noise global shuttering and a dual shuttering mode, *IEEE Transactions on Electron Devices*, 58(3), 740–747, 2011.
15. B. Wilburn, N. Joshi, V. Viash, M. Levoy, and M. Horowitz, High-speed videography using a dense camera array, in *Proceedings of Computer Vision and Pattern Recognition*, Washington, DC, 2004, pp. 1–8.
16. J. Honour, Electronic camera systems take the measure of high-speed events, *Laser Focus World*, 30(10), 121, 1994.

17. H. Schardin, The development of high-speed photography in Europe, *Journal of the Society of Motion Picture and Television Engineers*, 61(3 pt 1), 273–285, 1953.

18. B. Bretthauer, An electronic Cranz-Schardin camera, *Review of Scientific Instruments*, 62(2), 364–368, 1991.

19. Y. Deblock, High speed single charge coupled device Cranz-Schardin camera, *Review of Scientific Instruments*, 78(3), 035111–035111-4, 2007.

20. A. Hijazi and V. Madhavan, A novel ultra-high speed camera for digital image processing applications, *Measurement Science and Technology*, 19(8), 085503–085503–11, 2008.

21. F. K. Lu and X. Liu, Optical design of Cranz-Schardin cameras, *Optical Engineering*, 36(7), 1935, 1997.

22. D. J. Brady, M. E. Gehm, R. A. Stack, D. L. Marks, D. S. Kittle, D. R. Golish, E. M. Vera, and S. D. Feller, Multiscale gigapixel photography, *Nature*, 486(7403), 386–389, June 2012.

23. C. Miller, Half-million stationary images per second with refocused revolving beams, *Journal of the Society of Motion Picture and Television Engineers*, 53(5), 479–488, 1949.

24. C. Miller, Origin of the framing camera, *Journal of the Society of Motion Picture and Television Engineers*, 75(12), 1158–1160, 1966.

25. A. Frank and J. Bartolick, Solid-state replacement of rotating mirror cameras, *Proceedings of SPIE*, 6279, 62791U, 2007.

26. C. T. Chin, C. Lancée, J. Borsboom, F. Mastik, M. E. Frijlink, N. de Jong, M. Versluis, and D. Lohse, Brandaris 128: A digital 25 million frames per second camera with 128 highly sensitive frames, *Review of Scientific Instruments*, 74(12), 5026–5034, 2003.

27. P. Prentice, A. Cushieri, K. Dholakia, M. Prausnitz, and P. Campbel, Membrane disruption by optically controlled microbubble cavitation, *Nature Physics*, 1, 107–110, 2005.

28. A. van Wamel, K. Kooiman, M. Harteveld, M. Emmer, F. J. ten Cate, M. Versluis, and N. de Jong, Vibrating microbubbles poking individual cells: Drug transfer into cells via sonoporation, *Journal of Controlled Release*, 112(2), 149–155, May 2006.

29. S. Kleinfelder, S. Lim, X. Liu, and A. E. Gamal, CMOS digital pixel sensor with pixel-level memory, in *Solid-State Circuits Conference, 2001. Digest of Technical Papers. ISSCC. 2001 IEEE International*, San Franciso, CA, 2001, vol. 435, pp. 88–89.

30. K. Stefanov, CCD-based vertex detector for the future linear collider, *Nuclear Instruments and Methods in Physics Research Section A: Accelerators, Spectrometers, Detectors and Associated Equipment*, 549(1–3), 93–98, 2005.

31. O. Skorka and D. Joseph, Design and fabrication of vertically-integrated CMOS sensors, *Sensors*, 11, 4512–4538, 2011.

32. O. Nojima, Development of high speed digital camera: EXILIM EX-F1, *Journal of The Society of Photographic Science and Technology of Japan*, 72(3), 195–198, 2011.

33. S. Lauxtermann, G. Israel, P. Seitz, H. Bloss, J. Ernst, H. Firla, and S. Gick, A megapixel high speed CMOS imager with sustainable gigapixel/sec readout rate, in *IEEE Workshop on Charge-Coupled Devices and Advanced Image Sensors*, Lake Tahoe, NV, 2001, pp. 1–4.

34. R. Ghannoum, A 90nm CMOS multimode image sensor intended for a visual cortical stimulator, in *International Conference on Microelectronics, 2007*, Cairo, Egypt, 2007, pp. 179–182.

35. R. K. Reich, D. D. Rathman, D. M. O'Mara, D. J. Young, A. H. Loomis, R. M. Osgood, R. A. Murphy, M. Rose, R. Berger, B. M. Tyrrell, S. A. Watson, M. D. Ulibarri, T. Perry, F. Weber, and H. Robey, Lincoln Laboratory high-speed solid-state imager technology, *Proc. SPIE 6279*, 27th International Congress on High-speed Photography and Photonics, 62791 K, January, 2007.

36. B. Dierickx, G. Meynants, and D. Scheffer, Near 100% fill factor CMOS active pixel, in *Extended Programme of the 1997 IEEE Workshop on CCD and Advanced Image Sensors*, Bruges, Belgium, 1997, pp. 89–93.

37. J. A. Ballin, J. P. Crooks, P. D. Dauncey, A.-M. Magnan, Y. Mikami, O. D. Miller, M. Noy et al., Monolithic active pixel sensors (MAPS) in a quadruple well technology for nearly 100% fill factor and full CMOS pixels, *Sensors*, 8(9), 5336–5351, September 2008.

38. C. Coates, D. Denvir, N. McHale, K. Thornbury, and M. Hollywood, Ultrasensitivity, speed, and resolution: Optimizing low-light microscopy with the back-illuminated electron-multiplying CCD, in *Confocal, Multiphoton, and Nonlinear Microscopic Imaging*, ed. T. Wilson, *Proceedings of SPIE*, vol. 5139, Bellingham, WA, 2003, pp. 56–66.

39. R. Hoebe, C. van Oven, W. J. Gadella, P. Dhonukshe, C. van Noorden, and E. Manders, Controlled light exposure microscopy reduces photobleaching and phototoxicity in fluorescence live-cell imaging, *Nature Biotechnology*, 25, 249–253, 2007.

40. W. F. Kosonocky, G. Yang, R. K. Kabra, C. Ye, Z. Pektas, J. L. Lowrance, L. S. Member, V. J. Mastrocola, F. V. Shallcross, and V. Patel, 360 Element three-phase very high frame rate burst image sensor, *IEEE Transactions on Electron Devices*, 44(10), 1617–1624, 1997.

41. T. Etoh, D. Poggerman, A. Ruckelshausen, A. Theuwissen, G. Kreider, H. Folkerts, H. Mutoh, Y. Kondo, H. Maruno, K. Takubo, H. Soya, K. Takehara, T. Okinaka, Y. Takano, T. Reisinger, and C. Lohman, No title, in *Presented at the IEEE International Solid-State Circuits Conference (ISSCC), Digest of Technical Papers*, San Francisco, CA, 2002, pp. 46–47.

42. T. G. Etoh, D. Poggemann, G. Kreider, H. Mutoh, A. J. P. Theuwissen, A. Ruckelshausen, Y. Kondo, H. Maruno, K. Takubo, H. Soya, K. Takehara, T. Okinaka, and Y. Takano, An image sensor which captures 100 consecutive frames at 1,000,000 frames/s, *IEEE Transactions on Electron Devices*, 50(1), 144–151, January 2003.

43. H. Ohtake, T. Hayashida, and K. Kitamura, Development of a 300,000-pixel ultrahigh-speed high-sensitivity CCD, *Proceedings of SPIE*, 6119, 61190E, 2006.

44. S. Kleinfelder, Y. Chen, K. Kwiatkowski, and A. Shah, High-speed CMOS image sensor circuits with in-situ frame storage, *IEEE Transactions on Nuclear Science*, 51, 1648–1656, 2004.

45. M. M. El-Desouki, O. Marinov, M. J. Deen, and Q. Fang, CMOS active-pixel sensor with in-situ memory for ultrahigh-speed imaging, *IEEE Sensors Journal*, 11(6), 1375–1379, June 2011.

46. R. J. Smith, R. A. Light, S. D. Sharples, N. S. Johnston, M. C. Pitter, and M. G. Somekh, Multichannel, time-resolved picosecond laser ultrasound imaging and spectroscopy with custom complementary metal-oxide-semiconductor detector, *Review of Scientific Instruments*, 81(2), 024901-1–024901-6, 2010.

47. M. C. Pitter, R. A. Light, M. G. Somekh, M. Clark, and B. R. Hayes-Gill, Dual-phase synchronous light detection with 64 × 64 CMOS modulated light camera, *Electronics Letters*, 40(22), 1404–1405, 2004.

48. R. A. Light, R. J. Smith, N. S. Johnston, S. D. Sharples, M. G. Somekh, and M. C. Pitter, Highly parallel CMOS lock-in optical sensor array for hyperspectral recording in scanned imaging systems, in *Three-Dimensional and Multidimensional Microscopy: Image Acquisition and Processing XVII*, 7570th edn., J.-A. Conchello, C. J. Cogswell, T. Wilson, and T. G. Brown, Eds., *Proceedings of the SPIE*, Bellingham, WA, 2010, pp. 75700U–75700U–10.

49. M. Gersbach, R. Trimananda, Y. Maruyama, M. Fishburn, D. Stoppa, J. Richardson, R. Walker, R. Henderson, and E. Charbon, High frame-rate TCSPC-FLIM using a novel SPAD-based image sensor, in *Proceeding SPIE Optics+Photonics Single-Photon Imaging*, San Diego, CA, 2010.

50. E. S. Wilman, S. H. Gardiner, A. Nomerotski, R. Turchetta, M. Brouard, and C. Vallance, A new detector for mass spectrometry: Direct detection of low energy ions using a multi-pixel photon counter, *Review of Scientific Instruments*, 83(1), 013304, January 2012.

51. G. Bub, M. Tecza, M. Helmes, P. Lee, and P. Kohl, Temporal pixel multiplexing for simultaneous high-speed, high-resolution imaging, *Nature Methods*, 7, 209–211, 2010.

52. M. Gupta, A. Agarwal, A. Veeraraghavan, and G. Narasimhan, Flexible voxels for motion-aware videography, *Lecture Notes in Computer Science*, 6311/2010, 100–114, 2010.

53. M. Zhang and A. Bermak, CMOS image sensor with on-chip image compression: A review and performance analysis, *Journal of Sensors*, 2010, Article ID 920693, 1–17, 2010.

54. J. Holloway, A. Sankaranarayanan, A. Veeraraghavan, and S. Tambe, Flutter shutter video camera for compressive sensing of videos, in *2012 IEEE International Conference on Computational Photography (ICCP)*, Seattle, WA, 2012, pp. 28–29.

55. B. Fu, M. C. Pitter, and N. A. Russel, A reconfigurable real-time compressive-sampling camera for biological applications, *PloS one*, 6(10), e26306-1–e26306-12, 2011.

56. R. G. Baraniuk, Compressive sensing, *IEEE Signal Processing Magazine*, 24(4), 118–124, 2007.

57. R. Robucci, J. Gray, K. Leung, J. Romberg, and P. Hasler, Compressive sensing on a cmos separable-transform image sensor, *Proceedings of the IEEE*, 98(6), 1089–1101, 2010.

58. D. Reddy, A. Veeraraghavan, and R. Chellappa, P2C2: Programmable pixel compressive camera for high speed imaging, in *IEEE Conference Computer Vision and Pattern Recognition*, Colorado Springs, CO, 2011.

59. L. Jacques, P. Vandergheynst, A. Bibet, V. Majidzadeh, A. Schmid, and Y. Leblebici, CMOS compressed imaging by random convolution, in *2009 IEEE International Conference on Acoustics, Speech and Signal Processing*, Taipei, Taiwan, 2009, pp. 1113–1116.

60. Y. Oike and A. Gamal, A 256×256 CMOS image sensor delta-sigma based single-shot compressed sensing, in *IEEE International Solid-State Circuits Conference*, 60(1), 386–388, 2012.

61. M. Lustig, D. Donoho, J. Santos, and J. Pauly, Compressed sensing MRI, *Signal Processing Magazine, IEEE*, 25(2), 72–82, 2008.

62. L. Zhu, W. Zhang, D. Elnatan, and B. Huang, Faster STORM using compressed sensing, *Nature Methods*, 9, 721–723, 2012.

63. E. D. Nelson and M. L. Fredman, Hadamard spectroscopy, *Journal of the Optical Society of America*, 60(12), 1664, December 1970.

64. J. J. Decker, Experimental realization of the multiplex advantage with a Hadamard-transform spectrometer, *Applied Optics*, 10(3), 510–514, 1971.

8 Terahertz Communications for Terabit Wireless Imaging

Kao-Cheng Huang and Zhaocheng Wang

CONTENTS

8.1 INTRODUCTION

Recently, terahertz (THz) technology has attracted a great deal of interests from academia and industry. THz imaging has been used for medical and security applications. It is not only an accessible and lower-cost medical imaging tool comparable to X-ray but also an excellent real-time on-the-fly screening method of detecting explosives and narcotics.

The high-resolution full-body computed tomography scan is a high-definition 3D imaging technology that is used in airports and hospitals. A 3D image of a person or an imaged target has been presented in [1] by scanning an inward-directed vertical array around the person or imaged target. The array achieves a full 360 mechanical scan in 2–10 s with optimized illumination. Signal loss is minimized due to specular reflection away from the array. To deliver such imaging wirelessly in real time in a hospital or in an airport, THz communication is a promising solution with the following advantages:

- Terabit (Tb) data rate (point-to-point delivery to ensure data security).
- Synergistic with current semiconductor trends.
- Solid-state electronic devices have entered the sub-THz regime.

THz waves have a number of interesting features, including the tens and hundreds of gigahertz bandwidths available and the fact that this frequency band poses only a minor health threat. Also, as millimetre-wave communication systems become mature [2], the focus of research is, naturally, moving to the THz range. According to Shannon theory [3], the broad bandwidth of the THz frequency bands can be used for Tb-per-second (Tb/s) wireless indoor communication systems (see Appendix I). This enables several new applications, such as cordless phones with 360° autostereoscopic displays, optic-fibre replacement and wireless Tb/s file transferring. All of these applications provide higher quality and a better user experience. Although THz technology could satisfy the demand for an extremely high data rate, a number of technical challenges need to be overcome or better understood before its practical deployment.

This chapter provides an overview of the state of the art in THz wireless communication, and it is also a tutorial for an emerging application in Tb radio systems. The objective is to construct robust, low-cost wireless systems for THz Tb communications. The chapter is divided into five parts: The propagation of THz waves is presented in Section 8.2, Section 8.3 reviews various types of signal generation used in transmitters (Txs), Section 8.4 focuses on the detection of THz waves, Section 8.5 reviews the advancement of THz transistor that is an enabling technology for THz communications and Section 8.6 discusses some solutions with respect to suitable THz air interfaces and highlights some related system issues. The conclusion of this chapter is followed by an appendix.

8.2   TERAHERTZ WAVE PROPAGATION

To design THz wireless communications, the first step is to investigate THz wave propagation characteristics in free space and then to conduct channel measurements. The main issue for THz wave propagation is atmospheric attenuation, which is dominated by water vapour absorption in the THz frequency band. From experiments on the propagation of THz waves in air, it is observed that there are a number of THz transmission windows [4]. (The frequency that ranges between the two water vapour absorption peaks is the THz 'transmission windows'.) Nine major transmission windows throughout the 0.1–3 THz frequency range are indicated and can be used for wireless communications. Those THz transmission bands are

1. 0.1–0.55 THz
2. 0.56–0.75 THz
3. 0.76–0.98 THz
4. 0.99–1.09 THz

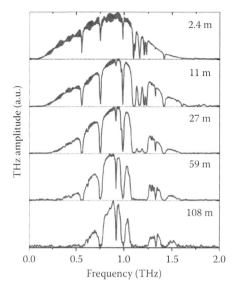

FIGURE 8.1   THz spectra at different propagation distances under free-space, LOS conditions. The temperature of measurements is 22°C, and the relative humidity is about 10% (reproduced with permission of IEEE). (From Liu et al., *Proc. IEEE*, 95, 1514, 2007.)

5. 1.21–1.41 THz
6. 1.42–1.59 THz
7. 1.92–2.04 THz
8. 2.05–2.15 THz
9. 2.47–2.62 THz [5]

Figure 8.1 illustrates the THz spectra after different propagation distances in the atmosphere. The transmission bands become narrower as the propagation distance increases. Reasonable bandwidths remain, however, even when the propagation distance reaches 108 m. The attenuation below 0.65 THz is from beam divergence due to diffraction, not from the atmosphere [6].

8.3   TERAHERTZ WAVE TRANSMITTERS

When a THz Tx that utilizes binary frequency-shift keying (FSK) is considered, it is vital to have a controllable THz source. Various methods for generating THz waves have been thoroughly studied throughout the last decade. In general, there are two types of THz signalling protocols that can be used for wireless communications: continuous-wave (CW) and pulsed signals.

This chapter could not exhaustively cover all of the detailed aspects of THz signal generation and detection techniques such as p-Ge emitter, black-body radiation and free-electron laser [7]. Instead, the following material only reviews THz techniques that are affordable and feasible for THz communication systems.

8.3.1 CONTINUOUS-WAVE SIGNAL GENERATION FROM PHOTOMIXERS

Photomixing process is commonly used to generate THz CWs from two CW laser beams. We define that these lasers have angular frequencies ω_1 and ω_2, respectively. Both beams with the same polarization are mixed and then focused onto an ultrafast semiconductor material such as GaAs. Due to the photonic absorption and the short charge carrier lifetime in the material, we obtain the modulation of the conductivity at the expected THz frequency $\omega_{THz} = \omega_1 - \omega_2$. Figure 8.2a schematically shows the two beams photomixing with a photomixer coupled to a hemispherical substrate lens. Figure 8.2b shows a schematic diagram of the photoconductive emitter and an identical detector.

Low-temperature-grown GaAs (LT-GaAs) is a widely used substrate for the photomixer. Its unique properties include a short carrier lifetime (around 0.5 ps), a large breakdown-field threshold (greater than 300 kV/cm) and a relatively high carrier mobility (approximately 200 cm^2 V^{-1} s^{-1}), each of which is vital for the efficiency of the photomixing process. The photomixing efficiency can be enhanced by reducing the transit time of majority carriers to much less than 1 ps in photomixers and photodetectors [8]. Output CW power is in the range of 0.1–5 μW, depending on pump power and operating frequency [9].

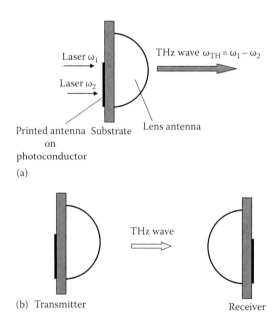

FIGURE 8.2 (a) Schematic view of two-beam photomixing with a photomixer. Two beams photomixing with a photomixer, which couples to a silicon hemispherical lens. The incident angular frequencies are ω_1 and ω_2, respectively. The output THz frequency ω_{THz} is the difference between ω_1 and ω_2. (b) A schematic diagram of the photoconductive emitter and identical detector along with a typical free-space THz system incorporating lenses. The same arrangement can be applied to both pulsed- and CW-based THz communication systems.

8.3.2 Pulse Generation (Optical Method)

The optical generation method uses a laser with a pulse width of 10–200 fs, which is in the visible or infrared spectrum. This laser is pointed at a THz-generating medium (e.g. dielectric crystals, semiconductors and organic materials). Semiconductors can be advantageous for generating THz pulses with high spectral intensity at higher THz frequencies. $LiNbO_3$ is better suited to generate THz pulses having a very large relative spectral width.

Optical rectification is a second-order phenomenon (different from frequency mixing) in a nonlinear medium with large second-order susceptibility. The difference between optical rectification and photoconduction is that the visible exciting beam creates virtual carriers rather than real ones. Thus, the transmission results in rectification and produces THz radiation, while both reflection and transmission results in photoconduction generate THz radiation. The output power of the photoconduction process can be increased with higher pump power. The photoconductive sources are capable of generating relatively large average THz powers of 100 μW [10]. The average power of optical rectification is of the order of μW range (e.g. 35–60 μW THz power generation with the use of a pulsed laser oscillator at 160–260 mW output power). As optical rectification relies on relatively low-efficiency coupling of the incident optical power to THz frequencies, it generally provides lower output powers than photoconductive antennas. On the other hand, optical rectification attracts much attention to generate 0.1–5 THz wave due to its simplicity and spectral broadness [11].

8.3.3 Pulse Generation (Electronic Method)

An all-electronic system uses monolithic *nonlinear transmission lines* as ultrafast voltage-step generators. The nonlinear transmission lines may be used for both generating picosecond pulses and for driving monolithically integrated diode samplers for the detection of these pulses.

The *nonlinear transmission lines* method, shown in Figure 8.3, is basically a transmission line. The line is periodically loaded with reverse-biased Schottky varactor diodes, serving as voltage-variable capacitors. This causes a wave travelling along the line to experience a voltage-dependent propagation velocity, resulting in a shock wave whose main features resemble a step function.

The advantages of this approach lie in its low phase noise, low cost, relative simplicity and its robustness. The disadvantage is its limited available frequency

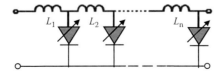

FIGURE 8.3   Nonlinear transmission lines method. The configuration is a transmission line (with inductors L_1, L_2, ...) periodically loaded with reverse-biased Schottky varactor diodes. The diodes, serving as voltage-variable capacitors (C_1, C_2, ...), cause a wave travelling along the line.

range, which is below 1 THz. The average output power level is on the order of −10 dBm [12]. Based on the (previous) wave propagation discussion in Section 8.2, nonlinear transmission lines can be applied to the first four transmission windows (0.1–0.55 THz, 0.56–0.75 THz, 0.76–0.98 THz and 0.99–1.09 THz) for wireless communications.

8.4 TERAHERTZ WAVE DETECTORS/RECEIVERS

To detect THz pulse signals, the waves first are focused on a detector or a photoconductive antenna. The double refraction of a silicon lens (Figure 8.2a) or an electrooptic crystal can be read using a probe pulse, usually split up from the laser used in THz generation. The probe beam is then measured with a quarter wavelength plate. Finally, the time delay between the probe pulse and the THz signal can be measured to obtain the electric field in the time domain.

8.4.1 PHOTOCONDUCTIVE ANTENNAS

Photoconductive antennas (see Figure 8.4) are widely used for effective THz detection, as well as for emission, and are commonly made from either silicon or GaAs. Laser pulses, separated from the pump laser, probe the antenna. To determine the THz waveform, the time delay from pump and probe pulses is measured, as well as the change in DC current from the antenna [13].

The antenna in Figure 8.4 is built with a GaAs substrate, but other semiconductors, such as GaSe, may also be used. The gap in the middle is biased by a DC voltage, A, and is probed by laser pulses split off from the laser pump in the case of detection. The gap is about 8 μm. The width of the antenna is 15 μm approximately, and the length is about 40 μm (see Figure 8.4). The THz waves strike the antenna that,

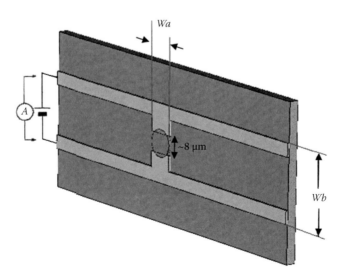

FIGURE 8.4 Photoconductive antenna structure (*Wa* 10 ~ 20 μm, *Wb* 30 ~ 50 μm, not to scale).

in turn, causes slight transient current changes [13]. The bandwidth of the antenna is affected by several key factors including the duration of the laser pulse, the physical structure of the antenna, the carrier scattering time and the THz beam optics [13].

8.4.2 MONOLITHIC MICROWAVE INTEGRATED CIRCUIT METHOD

The design and characterization of a 0.22 THz microstrip single-chip receiver (Rx) has been demonstrated with an integrated antenna in a 100 nm GaAs metamorphic high electron mobility transistor (HEMT) using metamorphic HEMT (mHEMT) technology [14]. This mHEMT THz Rx has two advantages: broad operating frequency bandwidth and low noise figure. Its performance is superior to the latest results presented using CMOS or SiGe technology. Further, this technology enables low-cost solutions when compared to pure indium phosphide (InP)-substrate-based designs. The THz Rx monolithic microwave integrated circuit (MMIC) consists of a silicon substrate lens-fed antenna, a low-noise amplifier and a sub-harmonically pumped resistive mixer. The double-sideband noise figure of this quasi-optical Rx is as low as 8.4 dB (1750 K) at 220 GHz, including the losses in the antenna and in the silicon lens [14,15].

8.4.3 LOW-POWER DESIGN WITHOUT MIXER AND LOCAL OSCILLATOR

For pulse-generated signals, a simple diode detector is preferred due to cost considerations. Alternatively, to demodulate CW-based binary FSK transmitted signals, a differential detector without a local oscillator (LO) is detailed in Figure 8.5 [1]. This Rx consists of a phase shifter, mixer and low-pass filter. The input signal, $i_0(t)$, is supplied to both the phase shifter and the mixer. The phase shifter shifts the phase at the carrier frequency of the input signal $i_0(t)$ by $-\dfrac{\pi}{2} \pm 2N\pi$, where N is an integer, and outputs a phase-shifted signal $i(t)$ to the mixer. The mixer multiplies the input signal $i_0(t)$ by the phase-shifted signal $i(t)$ and outputs a down-converted signal $r(t)$ to the low-pass filter. In the THz frequency range though, it is challenging to implement the mixer module in Figure 8.6 as the wavelength is very short, so its phase tuning in small scale is difficult and unstable using conventional methods.

A balanced detector without a complex mixer has been proposed, as depicted in Figure 8.6a. It is comprised of three adders, two square-law detectors, a phase shifter and a low-pass filter. An input signal $i_0(t)$ is supplied to the phase shifter and the adders A and B. The phase shifter shifts the phase of the input signal and outputs a shifted signal $i(t)$ to the adders A and B. The adder A adds the input signal $i_0(t)$ and

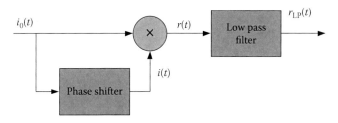

FIGURE 8.5 Basic model of differential detector.

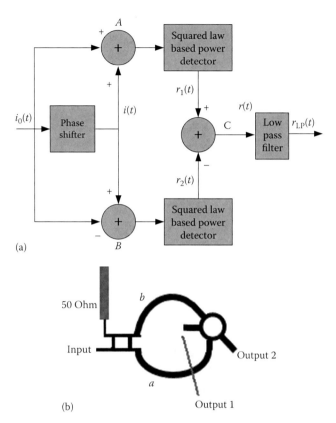

FIGURE 8.6 (a) Balanced differential detector without using a mixer and a local oscillator. (b) An example of layout (not to scale). Two outputs connect to the square-law detectors in (a).

the shifted signal $i(t)$ generated by the phase shifter and outputs the resulting signal to the first square-law detector. By squaring the output signal from the adder A, the first square-law detector presents the first squared signal $r_1(t)$ to the adder C. In contrast, the adder B subtracts the input signal $i_0(t)$ from the shifted signal $i(t)$ generated by the phase shifter and outputs the resulting signal to the second square-law detector. Similarly, by squaring the output signal from the adder B, the second square-law detector presents the second squared signal $r_2(t)$ to the adder C. The adder C then subtracts the second squared signal $r_2(t)$ from the first squared signal $r_1(t)$, respectively, and passes the combined signal $r(t)$ to the low-pass filter.

The layout of the balanced differential detector as shown in Figure 8.6b is designed by scaling down all aspects of microstrip technology. The length difference between curves a and b operates as the phase shifter in Figure 8.6a. They are designed to adjust phase difference to be $-\dfrac{\pi}{2} \pm 2N\pi$. The outputs connect to the square-law (squared law) detectors as shown in Figure 8.6a.

It is clear that both detectors from Figures 8.5 and 8.6 are mathematically identical. Since simple diodes are adopted instead of the complex mixer, the balanced differential detector is low profile for THz wireless communications.

8.5 TERAHERTZ ANTENNAS

THz frequencies correspond to wavelengths between 300 nm and 1 mm including the visible spectrum between 400 and 700 nm. With techniques now available, very small antenna structures can be fabricated making THz antennas feasible. At frequencies above about 0.02 THz, atmospheric attenuation may be as high as 20 dB/km [16]. This attenuation provides a natural shield for radio waves that is useful in many wireless applications. Thus, the same frequency may be employed by many different units in close proximity provided the attenuation is sufficient to prevent mutual interference [16].

- Terahertz waveguide structures
  Waveguide-based mixers with feed horns are relatively low loss and are excellent beam launchers, with up to 98% of the emergent power in the fundamental Gaussian beam mode. The problem at THz frequencies is achieving the required high machining tolerances (a few microns) and integrating the mixing devices. High-quality milling machines and several photolithographic techniques for etching silicon and photoresist have been developed to push the usefulness of waveguide into the THz regime. However, these techniques are often limited to producing relatively simple structures. One alternative technique is to use a galvo mirror–steered laser beam to etch split-block waveguide structures into a silicon wafer. The wafer is then coated with metal and assembled into waveguide devices, such as feed horns. The excellent performance and repeatability of waveguide feed horns make them an excellent choice for focal plane arrays. For such applications, the mixing devices can be fabricated on a matrix of thin (~1 μm) dielectric membranes that can then be readily integrated into a feed-horn array. An 810 GHz mixer utilizing a membrane-mounted SIS junction in a waveguide mount has been successfully built and tested [28].
- Off-axis scanning
  Off-axis scanning in a lens antenna may also be exploited to steer a single beam. The lens was illuminated by a fixed conical horn–type primary feed (Figure 8.7). Lenses with both a single refracting surface and two refracting surfaces were investigated, the latter offering more scope for optimization. In a demonstration, polyethylene was used for the lens material. A cited advantage is the ability to scan the antenna beam while leaving the feed fixed and so eliminating rotating waveguide joints that tend to be costly or troublesome at these frequencies.

  In Figure 8.7, a simple gimbal supports scanning in the two axes: θ increases at the lens are tilted with respect to the on-axis beam, while rotation of ϕ is around the feed axis. θ introduces a scanning loss, designed here to be minimized over ±45°, while from symmetry, ϕ is any arbitrary angle between −180° and +180° and does not bring about scan loss.

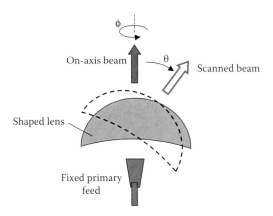

FIGURE 8.7   Beam scanning where primary feed is fixed.

- Terahertz nanotube structures
  Bundled carbon nanotubes (BCNTs) can be used as a conducting material for the fabrication of antennas. They are especially suitable for the THz applications. Typically, a gold film is used for THz antenna fabrication. It is shown that the radiation efficiency of a BCNT antenna is consistently lower than the efficiency of a gold film antenna for BCNT equivalent density values up to 104 CNTs/μm.

  Thus, the equivalent number density of BCNT antenna should be greater than to achieve acceptable values of radiation efficiency. The fact suggests the necessity of fabricating much more densely aligned single-wall carbon nanotubes. Therefore, to be utilized as an effective antenna radiator in the THz frequency range, the density of BCNTs should be approximately 10,000 CNTs/μm [29].

- Terahertz nano-antennas
  The use of novel nanomaterials as the building block of a new generation of nano-electronic components is envisioned to solve a part of the main shortcomings of current technologies. Amongst others, CNT and graphene nanoribbons (GNR) are the major candidates to make nano-antennas. From the communication perspective, the electromagnetic properties observed in these materials will determine the specific frequency bands for emission of electromagnetic radiation, the time lag of the emission or the magnitude of the emitted power for a given input energy. Amongst others, the possibility to manufacture resonant structures in the nanoscale enables the development of novel nano-antennas.

In Ref. [17], the mathematical framework for the analysis of carbon nanotubes as potential dipole antennas was developed (Figure 8.8a). In Ref. [18], a novel nano-antenna design (Figure 8.8b) based on a suspended GNR and resembling a nano-patch antenna is proposed, modelled and initially analysed combining the classical antenna theory with quantum mechanics.

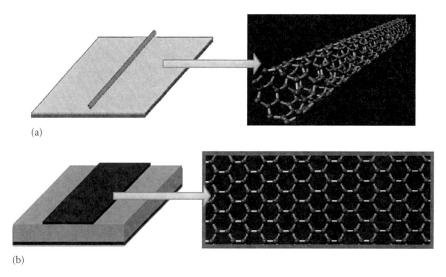

FIGURE 8.8 (a) Nano-dipole based on a carbon nanotubes and (b) a nano-patch antenna based on a GNR.

8.6 TERAHERTZ TRANSISTORS

In the perspective of cost-effectiveness and communication quality, transistors operating at THz frequency are desirable for constructing practical communication systems (e.g. balanced differential detector). On the other hand, the most fundamental tool that has been used to increase the operating frequency of electron device integrated circuits has been the progressive improvement in transistor f_T and f_{MAX}. With the f_{MAX} of current generation InP transistors pushing above 1 THz and new transistor scaling in progress, the operational frequency of solid-state amplifiers is being pushed towards THz frequencies. We thus briefly review the present situation of THz transistors in this section. The development of THz transistor includes the following approaches:

1. Heterojunction bipolar transistor (HBT) with transit frequency (f_T) ~ 0.8 THz was reported in Ref. [19]. The electrical properties of InP materials clearly offer performance possibilities that greatly exceed those of SiGe. The InP HBT technology focuses on a scaled emitter width of 300 nm tailored for high-frequency gain. Measurements indicate that the technology is capable of achieving maximum oscillation frequency (f_{max}) > 0.5 THz, making it suitable for switches, phase shifters, amplifiers and oscillators at THz frequencies [20]. Take amplifier as an example; it is expected to be reached 1 THz within a decade. For the same cutoff frequency, InP-based HBTs offer higher breakdown voltages than InP-based HEMTs. Some fabrication challenges still need to be overcome before the full potential of this material system can be harnessed to achieve 1 mW at 1 THz from a single device.
2. HEMTs with f_{max} > 1.0 THz were reported in Ref. [21]. A table of transistor performance is shown in Table 8.1 [22]. A packaged solid-state

TABLE 8.1
Summary of THz Transistor Performance

	InP HBT		InP HEMT	
Technology	Year 2010	Year 2011	Year 2010	Year 2011
Feature size	250 nm emitter	150 nm emitter	50 nm gate	30 nm gate
f_T	0.53 THz	0.64 THz	0.55 THz	0.69 THz
f_{MAX}	>0.63 THz	>1.2 THz	>1 THz	>1.2 THz

amplifier has demonstrated greater than 11 dB gain at 480 GHz using this technology. Going forwards, researchers will attempt to drive InP HEMT technology to 20 nm gate lengths and demonstrate f_T > 1.2 THz and f_{max} > 2.25 THz. These frequencies will be accessible through the reduction of internode capacitances C_{gs} to below 0.4 pF/mm and source resistance R_{source} < 0.1 Ω-mm.

3. Ballistic deflection transistor (BDT) is a six-terminal coplanar structure etched into a 2D electron gas. In the BDT, the steering voltage is much smaller than that required for gate pinch-off, which, combined with the reduced gate capacitances of the 2DEG, results in an estimated f_T in the THz range with low-noise and low-power consumption at room temperature. Experimental study confirms that the performance of BDT can be enhanced by using a complete gate configuration where gate runs along the channel. Simulations are used to study the effect of variation in the dielectric constant filled in the trenches between channel and gate. It is observed that high-permittivity dielectrics enhance the output [23].

4. Nanowire field-effect transistors (NWFETs): Nanowire building blocks can be configured as p-type and n-type FETs and also used to 'build' small logic circuits. NWFETs were firstly reported in the sub-100 nm Ge/Si channel length regime. Analysis of the intrinsic switching delay shows that 2 THz intrinsic operation speed is possible when channel length is reduced to 70 nm and that an intrinsic delay of 0.5 ps is achievable in 40 nm device. In addition to transistors, it is also possible to use crossbar nanowire circuits to make nonvolatile switch devices and synthetically coded nanowire for multiplexing/demultiplexing [24].

8.7   CHALLENGES IN SYSTEM ISSUES

Modern THz communication systems are dedicated to consumer electronics applications and, therefore, must meet strict requirements including small size, light weight and low production cost. Moreover, there are numerous challenging constraints on antenna design, such as high gain (>20 dBi) and wide bandwidth (>90° half-power beam width) while maintaining a small size [2].

To build a THz link, a single microstrip antenna is not enough, as it has low antenna gain and narrow impedance bandwidth. Also, because of the high water/air attenuation at THz frequencies, the transmitting power should be increased and the receiving noise should be minimized. We can either improve the power amplifier or use an antenna with 20 dBi gain or above. In general, THz power amplifiers consume high DC power, so a high-gain antenna is more cost effective. It is strongly recommended to apply lens antennas or phased-array antennas to amplify line-of-sight (LOS) received signals and minimize interference [25].

When a narrow-beam antenna is adopted for both the Tx and the Rx, beam directions from both sides are rotated either mechanically or electrically in order to find the strongest wireless link based on the measured channel quality information. During the rotation, the half-power beam width of the main beam or the main radiation pattern is kept unchanged. The algorithm to control the rotation and to handle the coordination between the Tx and the Rx, and successfully exchange the control signalling, is called beam steering. After beam steering, the strong LOS/non-LOS (NLOS) paths can be aligned and the link budget requirements can be satisfied due to the inherent antenna gain from both narrow-beam antennas.

If a phased-array antenna is used, one should consider the trade-off between the antenna size and the losses due to the long feed network. One example of phased-array antennas is shown in Figure 8.9. From the Tx side, the baseband and RF signal are identical for all the patch antenna elements. The coefficients W_i for $i = 1, ..., M$ are adjusted to realize beam forming, where the coefficient amplitude for each patch element can be changed and its corresponding phase can be shifted. For THz

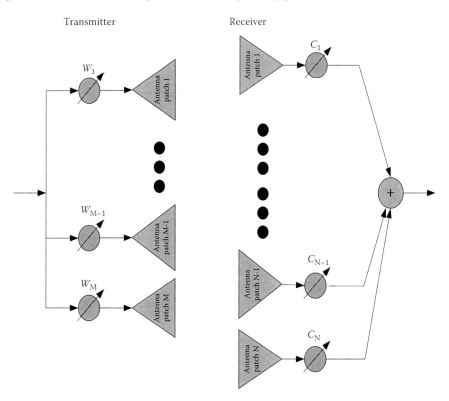

FIGURE 8.9 System model of phased-array antennas.

circuits, the feed loss of phased-array antennas is relatively high. To improve the performance, each patch antenna could have its own embedded power amplifier to reduce the effect of the feed loss.

From the Rx side, the multiple received RF signals are then combined from all of the patch antenna elements. The coefficients C_i for $i = 1, \ldots N$ are adjusted to obtain the maximum received signal-to-noise ratio (SNR), whereby the amplitude of the coefficient for each patch element can be changed and its corresponding phase can be shifted. Similar to the Tx, each patch element could have its own embedded low-noise amplifier before the combiner to reduce the effect of the feed loss and fulfil the strict link budget requirements.

At the Tx side, multiple phase shifters receive the output from the mixer that is working at the THz range. A demultiplexer is included to control which phase shifters receive the signal. These phase shifters are either quantized phase shifters or complex multipliers. The phase and/or magnitude of the currents in each patch element are adjusted to produce a desired beam pattern, as illustrated in Figure 8.10. The beam pattern may neither be symmetric nor have only one main lobe (i.e. regular beam); an irregular beam pattern with multiple main lobes is produced in order to maximize the received SNR under dynamic wireless conditions.

With respect to the Rx side, a phased-array antenna with multiple patch elements receives the signals and then sent them to the phase shifters having different coefficients. As discussed previously, phase shifters are comprised of either quantized phase shifters or complex multipliers. The phase shifters receive the signals from antennas, which are then combined to form a single output, as shown in Figure 8.9. A multiplexer is used to combine the signals from the different elements and to output the combined signal through the single feed line. As mentioned previously, the phases and magnitudes of the patch elements in the phased-array antennas are adjusted to produce a desired beam pattern, as illustrated in Figure 8.10. Additionally,

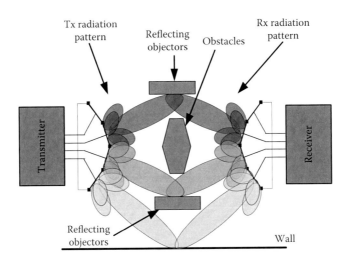

FIGURE 8.10   One example of a phased-array antenna with an irregular antenna radiation pattern.

the beam pattern may have more than one strong beam direction. An irregular beam pattern is generated to maximize the received SNR.

From a design point of view, complex multipliers are difficult to implement for THz circuits. Quantized phase shifters with very limited amplitude or phase selections are preferred, especially for THz wave-based applications [26]. As we have emphasized earlier, the idea of beam forming is not concentrated on the search for an antenna radiation pattern having a good shape but instead concentrated on maximizing the received SNR under a dynamic wireless environment. The corresponding antenna radiation patterns from both Tx and Rx sides might have strange beam shapes, which are illustrated in Figure 8.10 [27].

8.8 CONCLUSION

Different aspects of Tb/s radios working in the THz frequency range have been discussed. It is encouraging to see the advancements in technology development with respect to these radios. The characteristics of THz propagation in free space were firstly summarized. Provided that the requisite bandwidth is available, FSK can satisfy the required performance with a simpler, lower-cost transceiver configuration. Technically, the success of the THz radio will largely depend on the advancement of THz front end and THz transistors and the development of advanced system concepts. Different THz-generating and THz-detecting methods were reviewed and compared. In particular, a novel FSK Rx with no mixer and no LO was described for THz wireless applications. It has the benefits of minimizing power consumption and simplifying the circuit configuration. An example of layout is given. Finally, a new system concept was discussed with irregular antenna radiation patterns from both the Tx and Rx sides, to establish an adaptive Tb/s wireless link. The techniques discussed are intended to provide an efficient means to complete a first-order assessment of the options available to system designers and also to narrow the choices to those that provide adequate performance and less cost. This chapter provides a preliminary study on the state of the art of the THz Tb communications. The conceptual framework of these systems is still in the research phase, not yet ready for practical use.

8.A APPENDIX TERABIT VS. TERAHERTZ

Shannon's ideal communication theorem is expressed as

$$C = B \cdot log_2(1 + S/N) \text{ unit: bps}$$

where C is called the channel capacity in bits per second (bps). This represents the maximum error-free communication rate for a communication channel having bandwidth B and signal-to-noise power ratio of S/N. Hence, in theory, it is possible to convey information at a rate of up to C with no errors in the received message at all. In order to support 1 Tbps transmission under the constraint of a 40 dB SNR limit, the minimum required bandwidth is around 0.2 THz. This indicates that only THz frequency range could provide this kind of available spectrum and enable high-rate communications beyond 1 Tbps.

ACKNOWLEDGMENT

The authors thank Hsiang-Jung Huang for her assistance.

REFERENCES

1. Sheen, D. M., McMakin, D. L., and Hall, T. E., Cylindrical millimeter-wave imaging technique and applications, in *Proc. SPIE—Int. Soc. Opt. Eng.*, 6211, *Passive Millim. Wave Imaging Technol.*, IX, pp. 6357–6365, 2006.
2. Huang, K. and Wang, Z., Millimeter wave communication systems (*IEEE Series on Digital & Mobile Communication*). ISBN: 978-0470404621, Wiley-IEEE Press, Hoboken, NJ, 2011.
3. Miles, R. E., Harrison, P., and Lippens, D., *Terahertz Sources and Systems*, 1st edn., Springer, Berlin, Germany, 2001.
4. James, F. O'Bryon, *Assessment of Millimeter-Wave and Terahertz Technology for Detection and Identification of Concealed Explosives and Weapons*, The National Academies Press, Washington, DC, 2007.
5. Seta, T., Mendrok, J., and Kasai, Y., Laboratory spectroscopic measurement of water vapor for the terahertz-wave propagation model, *URSI Chicago General Assembly*, 1–4, August 2008.
6. Liu, H., Zhong, H., Karpowicz, N., Chen, Y., and Zhang, X., Terahertz spectroscopy and imaging for defense and security applications, *Proc. IEEE*, 95, 1514–1527, 2007.
7. Saldin, E. L., Schneidmiller, E. V., and Yurkov, M. V., *The Physics of Free Electron Lasers,* Springer, Berlin, Germany, 2000.
8. Pilla, S., Enhancing the photomixing efficiency of optoelectronic devices in the terahertz regime, *Appl. Phys. Lett.*, 90, 16, 161119, 1–3, 2007.
9. Sakai, K., *Terahertz Optoelectronics*, 1st edn., Springer, Verlag, New York, 2005.
10. Creeden, D., McCarthy, J. C., Ketteridge, P. A., Schunemann, P. G., Southward, T., Komiak, J. J., and Chicklis, E. P., Compact, high average power, fiber-pumped terahertz source for active real-time imaging of concealed objects, *Opt. Express*, 15, 6478–6483, 2007.
11. Andrei, G., Stepanov, J. H., and Jürgen, K., Efficient generation of subpicosecond tera-hertz radiation by phase-matched optical rectification using ultrashort laser pulses with tilted pulse fronts, *Appl. Phys. Lett.*, 83, 3000–3002, 1983.
12. Weide, D., Applications and outlook for electronic terahertz technology, *Opt. Photon. News*, 14, 4, 48–53, 2003.
13. Tani, M., Herrmann, M., and Sakai, K., Generation and detection of terahertz pulsed radiation with photoconductive antennas and its applications to imaging, *Meas. Sci. Technol.*, 13, 1739–1745, 2002.
14. Gunnarsson, S. E., Wadefalk, N., Svedin, J., Cherednichenko, S., Angelov, I., Zirath, H., Kallfass, I., and Leuther, A., A 220 GHz single-chip receiver MMIC with integrated antenna, *IEEE Microw. Guided Wave Lett.,* 18, 284–286, April 2008.
15. Koch, S., Guthoerl, M., Kallfass, I., Leuther, A., and Saito, S., A 120–145 GHz hetero-dyne receiver chipset utilizing the 140 GHz atmospheric window for passive millimeter-wave imaging applications, *IEEE J. Solid-State Circuits*, 45, 10, October 2010.
16. Huang, K. and Edwards, D., *Millimetre Wave Antennas for Gigabit Wireless Communications,* ISBN: 978-0470712139, John Wiley & Sons Ltd., London, U.K., 2008.
17. Burke, P., Li, S., and Yu, Z., Quantitative theory of nanowire and nanotube antenna performance, *IEEE Trans. Nanotechnol.*, 5, 314–334, July 2006.

18. Jornet, J. M. and Akyildiz, I. F., A nano-patch antenna for electromagnetic nanocommunications in the terahertz band, broadband wireless networking, Lab, Tech. Rep., August 2009.

19. Snodgrass, W., Hafez, W., Harff, N., and Feng, M., Pseudomorphic InP/lnGaAs heterojunction bipolar transistors (PHBTs) experimentally demonstrating ft = 765 GHz at 25°C increasing to ft = 845 GHz at -55°C, in *Proc. IEEE Int. Electron Devices Meeting*, pp. 1–4, December 2006.

20. Albrecht, J. D., Rosker, M. J., Wallacet, H. B., and Chang, T., THz electronics projects at DARPA: Transistors, TMICs, and amplifiers, *IEEE Microw. Symp. Dig. (MTT)*, 1118–1121, May 2010.

21. Lai, R., Mei, X. B., Deal, W. R., Yoshida, W., Kim, Y. M., Liu, P. H., Lee, J et al., Sub 50 nm InP HEMT Device with Fmax greater than 1 THz, *IEEE 2007 IEDM Conf. Dig.*, 609–611, December 2007.

22. Deal, W. R. Solid-state amplifiers for terahertz electronics, *IEEE MTT-S Int. Microw. Symp. Dig. (MTT)*, 1122–1125, 2010.

23. Kaushal, V. K., Iñiguez-de-la-Torre, I., and Margala, M., Topology impact on the room temperature performance of THz-range ballistic deflection transistors, in *Proceeding GLSVLSI '10 Proceedings of the 20th symposium on Great lakes symposium on VLSI*, pp. 159–162, 2010.

24. Hu, Y., Xiang, J., Liang, G., Yan, H., and Lieber, C. M., Sub-100 nanometer channel length Ge/Si nanowire transistors with potential for 2 THz switching speed, *Nano Lett.*, 8, 3, 925–930, 2008.

25. Thornton, J. and Huang, K., Modern lens antennas for communications engineering, *IEEE Ser. Electromagn. Wave Theory.* ISBN: 978-1118010655, Wiley-IEEE Press, Hoboken, NJ, 2013.

26. Dai, L., Wang, Z., and Yang, Z., Spectrally efficient time-frequency training OFDM for mobile large-scale MIMO systems, *IEEE J. Sel. Areas Commun.*, 31, 251–263, February 2013.

27. Huang, K. and Wang, Z., Terahertz terabit wireless communications, *IEEE Microw. Mag.*, 12, 4, 108–116, June 2011.

28. Zmuidzinas, J. and Richards, P. L., Superconducting detectors and mixers for millimeter and submillimeter astrophysics, *Proceedings of the IEEE*, 92, 1597–1616, October 2004.

29. Sangjo, C. Radiation efficiency assessment of bundled carbon nanotube antenna at terahertz frequency range, *General Assembly and Scientific Symposium*, 2011 XXXth URSI, pp. 1–4, August 2011.

9 Increasing Projector Contrast and Brightness through Light Redirection

Reynald Hoskinson and Boris Stoeber

CONTENTS

9.1 INTRODUCTION

A projection display creates an image on an auxiliary surface, separate from the projector itself. For this reason, the images can be made very large, able to be viewed by groups of people, for instance, at a cinema. Typically, projectors are structured with a light source separate from the image-forming element, or *light valve*. As much of the light as possible from the light source is channelled to the light valve, which in turn relays the formed image through the projection optics to the screen.

How efficient the projector is at doing this has major repercussions. Brightness is the primary characteristic determining projector price and quality, and projector efficiency is one key in determining the final brightness of the projected image. Simply increasing the brightness of the lamp to make the image brighter is not always an option. The lamp is typically the most expensive piece of the projector, even more so than the light valve itself. Also, since the light that is not directed to the screen ends up as heat, a brighter lamp carries with it the need for bulkier and noisier fans and lamp electronics. The light source in a projector supplies a uniform brightness distribution on the light valve. For most images, however, only a fraction of the total area is illuminated at peak brightness. A conventional projector simply blocks the light that is not necessary for a scene, thereby wasting this fraction of light, while it could be used to further illuminate the bright parts of the image. Furthermore, the currently available light valves are 'leaky' and cannot block all the light for black image areas (Dewald et al., 2004). The contrast or dynamic range of a projector is the ratio between the brightest and darkest levels it can display. Merely increasing the illumination with a more powerful projector lamp does not necessarily increase the dynamic range, as both the darkest and brightest levels rise by the same relative amount. Current projector technology could benefit from an improvement in both dynamic range and peak brightness of the projected image.

The dynamic range of projectors is not sufficient to display many real-world scenes, which can have up to eight orders of magnitude of luminance range (Reinhard et al., 2005). Scenes such as those that include a sunset, fireworks or daylight must have their luminances tone mapped to the limited range of current projectors. The peak brightness of a projector also limits the type of environment in which it can be used to its full potential; the brighter the room illumination, the more lower-end detail will be lost, resulting in a 'washed-out' image.

The emerging popularity of 3D movies even further strains a projector's ability to display an image. Typically, the efficiency of a theatre projector drops to 14% of what it is for a 2D content (Brennesholtz, 2009). Current digital cinema projectors 'strain' to achieve 4.5 foot-lamberts (fl) of screen luminance for a 3D film, especially on larger screens, rather than the 14 fl recommended by the Society of Motion Picture Television Engineers in the standard ANSI/SMPTE 196M (SMPTE, 2003). Even at 4.5 fl, lamps must be run at maximum power, leading to high electric bills and short lamp life. These limitations are felt by the studios as well. Because the displayed luminances are so different, they affect the perception of colours (Brennesholtz, 2009), and so studios must perform two separate expensive colour-correction processes, one for 2D and one for 3D.

9.2   CONVENTIONAL PROJECTION DISPLAY SYSTEMS

To explain our proposal for a high-contrast projector using an auxiliary mirror array, it helps to first outline the architecture of a modern projector system. Most projection displays available today are based on either Digital Light Projection (DLP) or Liquid Crystal on Silicon (LCoS). They differ in the type of light valve used, the element that selectively blocks light. A lamp provides uniform illumination to the light valve, and the liquid crystals in the LCoS, for example, selectively reduce illumination of a pixel on the screen in order to form the dark parts of the image. The digital micro-mirror device (DMD) inside a DLP projector functions in a similar manner. A DMD is an array of micromirrors, one for each pixel, each one of which can be tilted in one direction so that incident light reflects towards the projection lens and then out onto the screen, or another direction, so the light is reflected to a heat sink and that spot on the screen remains dark.

All projectors can be broken down into a number of functional subsections. Figure 9.1 shows the subsections of single-chip DLP projectors. For purposes of clarity, only the light reaching one pixel of the DMD is shown; in reality there is a ray bundle that reaches each pixel of the DMD. First, in the light path, a reflector

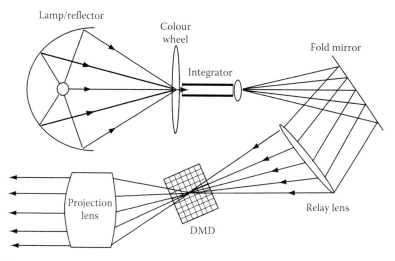

FIGURE 9.1 One ray bundle traversing a projection system.

collects the light from a small lamp or LED and directs it into the illumination optics. In a single-chip DLP projector, the lamp reflector minimizes the spot size of the light at the colour wheel. After, the colour wheel is the integrator, which spatially redistributes the image of the light source from a highly peaked to a more uniform distribution with an aspect ratio that matches that of the light valve. This affects the final distribution of the light on the screen. For DLP projectors, the integrator is usually a rod, made of hollow mirrored tunnels. From the integrator rod, the light travels through relay/folding optics, which form an image of the integrator rod face on the DMD. The image of the DMD is then transmitted to the screen using a projection lens system.

It is interesting to note how the DMD creates the image we see on the screen. Each DMD micromirror has two active tilt positions: plus or minus the maximum tilt angle α_m, with $\alpha_m = 12°$ in current devices. In one state, light is reflected to the projection lens, while a tilt in the other direction sends light to the 'light dump' where it is emitted as heat. This places particular constraints on the light incoming to the DMD mirrors. To fulfil its function as a light modulator, the incoming/outgoing light has to be limited to an angle less than $2\alpha_m \pm \alpha_m$. That way, the light going to the heat sink can be fully separated from that going to the projection lens pupil. Another way of saying this is that the numerical aperture (NA) of the projection pupil has to be small enough to prevent overlapping flat and on-state light.

9.3 OTHER ADAPTIVE DISPLAY MECHANISMS FOR PROJECTORS

There have been other proposed methods for dynamically changing portions of the display chain in a projector in response to the image being displayed. The addition of a dynamic iris that is adjusted per frame of video is a limited way of adapting the projector's light source to the content (Iisaka et al., 2003). The dynamic iris is a physical aperture near the lamp that partially closes during dark scenes and opens during bright scenes. While this method can decrease the black level of certain images, it is only a global adjustment, not allowing the improvement of contrast in a scene with several localized bright or dark areas. Also, while this approach decreases the black level for certain select images, it can't increase the maximum brightness of a scene like the analogue micromirror array (AMA) can.

There have been several methods proposed that increase contrast solely by decreasing the dark level of the projector, at the expense of brightness (Pavlovych and Stuerzlinger, 2005; Damberg et al., 2007; Kusakabe et al., 2009). Each uses two light valves in series within a projector system, which reduces not only the dark level but also the overall brightness and efficiency of the system.

9.4   CONCEPT FOR AN IMPROVED PROJECTION DISPLAY

To address the contrast shortcomings of currently available projectors, we have proposed adding a low-resolution intermediate mirror array to provide a non-homogeneous light source (Hoskinson and Stoeber, 2008). This intermediate mirror device is capable of directing the uniform light from the projector lamp incident on its surface to different areas on the light valve, in effect projecting a low-resolution version of the original image onto the light valve as shown in Figure 9.2. Adding this intermediate mirror device improves the dynamic range in two ways: by directing the light to the bright parts of the image, the achievable peak brightness will be increased. Simultaneously, the amount of light that needs to be blocked in the dark regions of the image is reduced, thus decreasing the brightness of the black level.

This intermediate device can be realized with a low-resolution AMA, made using microelectromechanical system (MEMS) technology. The tip and tilt angle (two degrees of freedom) of the micromirrors in the array can be set continuously in order to direct light to an arbitrary location on the light valve.

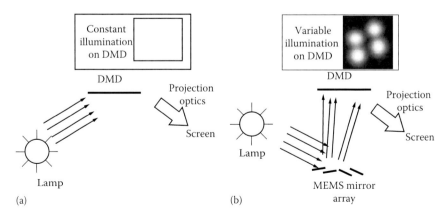

FIGURE 9.2   (a) Schematic of a conventional DLP projector and (b) schematic of an enhanced DLP projector with the second MEMS mirror array (AMA).

9.5   AMA PROJECTOR OPTICAL DESIGN

The AMA must be inserted between the lamp and the DMD in order to change the illumination distribution on the DMD. The problem is then how to achieve the optimal spot size of light from each AMA mirror onto the DMD while maintaining adequate light coverage of the DMD overall. The image of the DMD with this variable illumination will then be projected to the screen by the projection lens. As well as minimizing the AMA spot size, we would like to maximize the spot displacement for a given mirror tilt angle. A lens between the AMA and the DMD should optimally relay the light at the proper magnification onto the DMD. To determine the optical lens focal length and placement, we start with the simple lens formula (Hecht, 2002)

$$\frac{1}{f} = \frac{1}{u} + \frac{1}{v},$$ (9.1)

which gives a relationship between distances of the focal length of the lens f, the distance between the object and the lens u and the distance between the lens and the image v, as shown in Figure 9.3a. If we put the AMA at the object plane u and

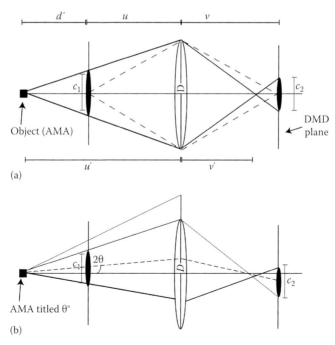

FIGURE 9.3   Illustration of circle of confusion. (a) A point on the AMA spreads to a region c_1 that in turn is imaged onto the DMD as the circle c_2. In (b), the mirror is tilted by θ, causing the light cone to be shifted by 2θ. The dotted line in (b) represents the shifted principle ray of the cone.

the DMD at the image plane v, we will get a perfectly in-focus image of the AMA on the screen, minimizing the spot size of the AMA mirror. However, since every point on the object plane is mapped to a corresponding point on the image plane, tilting the mirrors will not move the light from one region to another when the AMA is in focus on the DMD.

To achieve the desired effect of redirecting light from one region to another, the AMA is placed at a distance d' from the object plane of the lens, as shown in Figure 9.3b. By the time the light from a point on the AMA reaches a distance u from the lens, it describes a circle of diameter c_1, which in turn is imaged onto the DMD plane, forming the circle of diameter c_2. This has the effect of blurring the image of the AMA at the DMD plane.

Using plane geometry and knowing the magnification m of the system, we can show that

$$c_2 = \left| Dm \frac{d'}{u'} \right| \tag{9.2}$$

where D is in effect the lens aperture, limiting the angle of incoming light from the AMA. To lower the rate of increase of blur diameter as the disparity increases, we could reduce the aperture of the lens. However, this would also decrease the efficiency of the projector. To maintain system efficiency, the minimum size of D is that which captures the entire beam from the lamp that can be used by the DMD.

When an AMA mirror is tilted as shown in Figure 9.3b, the incident light is redirected for a distance d' before it reaches the object plane. Points on the plane at u will be imaged onto corresponding points on the plane at v, so the tilt angle of the AMA only displaces light up until d'. The displacement d_t at the DMD plane can thus be calculated as

$$d_t = md' \tan(2\theta) \tag{9.3}$$

From Equations 9.2 and 9.3 it is evident that as we increase the separation d' to increase the displacement of the ML on the DMD, the circle of confusion also grows, so we are blurring the light from the AMA mirror. This relationship is explored in more detail in Hoskinson et al. (2010). A key trade-off of the system is between the blur kernel on one side and the system efficiency and ML range on the other. A smaller blur kernel corresponds to a combination of smaller range and/or reduced system efficiency.

The distribution of light energy with the circle of confusion can be referred to as the optical point spread function (PSF). Nagahara et al. (2008) provide a model of a PSF that includes an approximation of the effect of aberrations using a Gaussian function

$$p(r,b) = \frac{2}{\pi (gc)^2} e^{-2r^2/(gc)^2} \tag{9.4}$$

where g is constant between 0 and 1. g has a large effect over the shape of the Gaussian, with smaller values giving a more narrow peak. This parameter should be determined through an optical simulation for a given light source or empirically through optical measurements.

Once we have come up with an estimate of the PSF of one spot on the AMA, we can estimate the distribution of light from one mirror in the array on the DMD, referred to here as the mobile light (ML), by convolving the area of the mirror with the PSF. We can then use the ML to get an estimate of the light distribution from the entire AMA as it appears on the DMD.

Let $g(x, y)$ be the ML, the luminance distribution on the DMD from one AMA micromirror, and $f(x, y)$ be the distribution if the ML was in focus (although still with the displacement due to mirror tilt) and $p(x, y)$ be the PSF estimated according to Equation 9.4. Then

$$g = p*f \qquad (9.5)$$

where $*$ is the convolution operator.

So to obtain the final light distribution from the AMA, we calculate the displacement and magnification of each AMA mirror, place each mirror's displaced and magnified spot into an image and then convolve (or multiply, in the Fourier domain) the entire image by the PSF.

9.6 MIRROR ALLOCATION

In an AMA approach, the intensity of light of each ML source cannot be changed, but the location on the high-resolution light valve is variable. This means that every ML can be targeted towards the region it is needed. The combination of all the virtual light sources must be sufficient to correctly display the same ratio between intensities as specified in the original image data, but the entire image can be made brighter through reallocation. This section discusses an allocation scheme to take advantage of the flexibility afforded by these MLs to increase the peak brightness of the image. Other schemes are possible; see, for example, Hoskinson et al. (2010), but this one is particularly efficient and useful for real-time implementation.

Finding the optimal light distribution for the AMA corresponds to choosing the locations for each of the n MLs from the array of n analogue micromirrors. We can approximate the representation of the light incident on the DMD, coming from the entire AMA, as a block of pixels in DMD space. For the sake of example, we will illustrate with an 800×600 pixel DMD. We initially represent the light as a white image of size 1000×800 pixels, centred on the DMD. The extra space around the edges approximates the overfill that is a common technique in projector design to ensure image uniformity.

The image is subdivided into n rectangles, each of which represents 1 ML. The position of the centre of each rectangle in image (DMD) coordinates is recorded. Each rectangle is then blurred using the convolution kernel specified in Equation 9.4, to get the shape and relative intensity distribution of each ML. The convolution kernel used to blur the light is calculated from the physical parameters of the optical system, as detailed earlier.

Initially, the mirrors of the AMA are in a non-actuated, flat state, so all MLs are equally distributed on the DMD. The summation of the MLs in this state should give a relatively uniform distribution on the DMD so that any image can be displayed in a conventional manner. We must then search for the position of each of the n MLs that provides the best improvement to the displayed image. It is assumed that there is adequate granularity in the mirror control to position the MLs at any pixel in the image.

To guide the allocation algorithm, it needs to be established what constitutes an improvement. Here, we define improvement as the ability to boost the entire range of pixel brightness values, corresponding to a projector with a higher effective lumen output.

Our approach for allocating the light from all AMA mirrors is to divide the original image into equal energy zones and allocate one ML for each zone. The median cut algorithm described in Debevec (2005) is an efficient method of subdividing an image into zones of approximately equal energy. We have modified the original algorithm so that it can divide an image into an arbitrary number of regions d. The image is first added to the region list as a single region, along with the number of desired divisions. As long as $d > 2$, subdivide the region into two parts as equally as possible, adding the two new subregions back to the region list, as well as the new divided d for each. In order to maintain the constraint that all regions at the final level be of approximately equal energy, choose the cutting line so that it divides the region into portions of energy as equal as possible, along the largest dimension of the parent region. Repeat until all d's in the region list equal 1.

Figure 9.4 shows an image cut into 28 regions. The centroids of each region are marked with squares. An ML is placed with its centre at the centroid of each region. The main advantage of dividing the image in this manner is that the image can be divided quickly into as many regions as there are MLs.

FIGURE 9.4 **(See colour insert.)** Image divided into 28 regions using the median cut algorithm, with the centroids in each region represented as dots.

One potential drawback of using this scheme for ML allocation is that the size of each region is a function only of its summed light energy, without regard to the size of the ML. Very small regions might only fit a small fraction of the total light energy of an ML, and very large regions might be larger than a single ML. Also, an equal aspect ratio of a region is not guaranteed, so some regions might be of much different shape than an ML.

Another drawback is that the MLs could have a limited range of movement from their original positions due to a limited tilt angle of the micromirrors, which is not taken into account in the aforementioned algorithm. In the Euclidean bipartite minimum matching problem (Agarwal and Varadarajan, 2004), we are given an equal number of points from two sets and would like to match a point from one set with a distinct point from the other set, so that the sum of distances between the paired points is minimized. We can use an algorithm that solves this problem to match each centroid from the median cut step to a location of a ML in its rest (non-tilted) state and thus minimize the sum total distance between the pairs in the two groups. This will minimize the sum total angle that the mirrors must tilt to achieve the points specified in the median cut solution set. The Euclidean bipartite matching problem can be done in polynomial time using the Hungarian approach (Kuhn, 1955).

The bipartite minimum matching problem does not limit any one pair to be less than a given amount, only that the sum total of all pairs is minimized. This means that there might be some pairs that exceed the range of motion of an ML for a given maximum mirror tilt angle. If any of the distances between the non-tilted ML and the centroid are larger than the range, they are placed at the furthest point along the line that connects the two points that they can reach.

9.7 DESIGN CONSIDERATIONS FOR ELECTROSTATICALLY ACTUATED MICROMIRRORS

In the preceding sections, it has become apparent that an AMA suitable for this application must have specific capabilities. Each micromirror must have adequate range of motion in order to move the light from one region to another. The array must be as efficient as possible so that the losses it introduces do not outweigh the gains. In this section, we discuss these requirements and others that must be addressed when designing a micromirror array.

The primary mode of actuation of micromirrors is electrostatic, due to its scalability and low power consumption. Other actuation methods include magnetic (Judy and Muller, 1997) and thermal actuation (Tuantranont et al., 2000), which are more difficult to confine or require more power, respectively, compared to electrostatic actuation.

A voltage applied between two separated surfaces creates an attractive electrostatic force (Senturia, 2001). Typically, the mobile surface is attached to a spring system that provides a restoring force when the mobile surface (the mirror) approaches the fixed surface.

The electrostatic torque τ_e causes the micromirror to rotate, which in turn causes an opposing mechanical spring torque in the torsion beams suspending the mirror above the substrate. The equilibrium tilt angle is reached when the electrostatic torque equals the opposing mechanical restoring moment

$$\tau_e = M_t \tag{9.6}$$

One difficulty of designing micromirrors that are fully controllable over a range of tilt angles is that at 1/3 of the range between the mirror surface and the electrode, the mirror will 'pull in', snapping towards the electrode surface. Therefore, care must be taken not to exceed the pull-in value when continuous tilt range is desired. Since gap distance h between the mirror layer and the electrode limits the maximum tilt angle α_1 of a given size mirror, increasing h will not only extend the tilt angle range but also increases the voltage required to actuate the mirror.

One other way of increasing α_1 is to decrease the mirror size L_1. Using multiple smaller mirrors ($L_2 > L_1$) in place of large mirrors allows for greater angles ($\alpha_2 > \alpha_1$). However, this also requires many more electrodes to control the direction of each of the small mirrors. It also compromises the fill factor of the array, as the ratio of space between mirrors rises as the mirror size shrinks. Also, more space has to be allocated between mirrors to the electric traces feeding voltages to all of the extra electrodes.

The main example of MEMS micromirrors available in the consumer market is the DMD, which has been discussed earlier. The DMD chip cannot be used as an AMA in this application because it has no intermediate positions between its two discrete states. However, both before and after the DMD was invented, there has been a substantial amount of research into variable-angle micromirrors for use in optical switching and other applications (Dutta et al., 2000; Bishop et al., 2002; Dokmeci et al., 2004; Taylor et al., 2004; Tsai et al., 2004b).

Applications that require just one mirror, such as laser scanning, offer high tilt angles for single mirrors (Tsang and Parameswaran, 2005). To achieve this, extensive use is made of the chip area around the mirror. This makes such designs unsuitable for applications where multiple closely packed mirrors are required, such as this one.

Some mirror arrays have a high fill factor in one direction only, such as in Hah et al. (2004), Taylor et al. (2004) and Tsai et al. (2004a). The mirrors in these configurations can be stacked tightly in one dimension, but extended components to the sides of the mirrors prevent them being stacked tightly in two dimensions. Dagel et al. (2006) describe a hexagonal tip/tilt/piston mirror array with an array fill factor of 95%. The micromirrors tilt using a novel leverage mechanism underneath the mirror.

Many tip/tilt mirror systems use gimbals to suspend the mirrors, such as in Lin et al. (2001), Bishop et al. (2002) and Wen et al. (2004). Usually, a frame surrounds the mirror and is attached to it by two torsional springs, forming an axis of rotation. The frame itself is then attached to the surrounding material by two springs in orthogonal directions, allowing the mirror to tip and tilt on torsion springs.

9.7.1 ANALOGUE MICROMIRROR ARRAY DESIGN

While the aforementioned designs meet one or several of the characteristics needed for an AMA, none meet all of them, and in any case, all but the DMD are prototypes, not available to the public. From calculations detailed in Hoskinson et al. (2010),

FIGURE 9.5   Array of composite mirrors in Micragem row design.

a 7 × 4 pixel AMA would provide sufficient resolution, and micromirrors with a tilt angle of ±3.5° would allow for sufficient light reallocation in any direction. The area of the mirror array should be close to that of the light valve selected, 11.17 mm × 8.38 mm for an 800 × 600 DMD array.

We have designed MEMS micromirrors to use in an AMA projector (Hoskinson et al., 2012) by machining into a 10 μm thick silicon plate that is suspended 12 μm above electrodes on a glass substrate. They employ a thin gimbal system, which suspends square mirrors over the 12 μm cavity. Multiple mirrors are controlled simultaneously to form one AMA composite mirror by linking their electrodes. In our design, mirrors within a row in the same composite mirror share the same gimbal frame, which improves fill factor and promotes homogeneity between submirror deflections. Multiple mirrors within one frame also minimize the ratio between frame area and mirror area. Instead of the loss of reflective area from four frame sides and two frame hinges per mirror, the loss from the frame area and outer hinges are amortized over a larger number of mirrors. The two gimbal springs and two of the four sides are only needed once per group rather than once per mirror.

Figure 9.5 shows a photograph of several of the composite mirrors in the array. Placing a row of mirrors within a gimbal frame also allows the electrodes beneath to be routed in daisy-chain form within the rows, minimizing the electrode paths. Thus, there is a series of rectangular rows of mirrors within a composite mirror, with corresponding rows in the electrode layer.

9.8   OPTICAL SYSTEM

A prototype to demonstrate dual-light modulation using an AMA and DMD has been built using two conventional projectors and a custom-made AMA chip. The light from one projector lamp is collected with a 60 mm lens and directed to the AMA. A 45 mm lens relays the light from the AMA to a second projector, a Mitsubishi PK20 pico-projector that is relatively easy to open up to allow access to the DMD. The light incident to the DMD is reflected normally through a prism to the PK20 projection lens onto a screen. Figure 9.6 shows a photograph of the prototype. Physically, the AMA array is 9 mm × 5 mm, while the DMD is 11.17 mm × 8.38 mm.

A custom-made driver provides 84 independent analog voltages to the AMA. A static image was displayed with the AMA projector by directing a VGA signal from a computer to both the light source and the DMD. Actuating the AMA

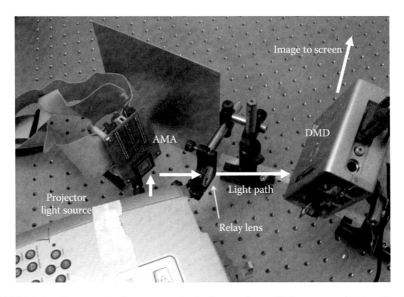

FIGURE 9.6   Photograph of prototype, including projector light source, AMA, relay lens and DMD.

mirrors via the 84-channel D/A converter changes the final intensity distribution of the projected image.

In the prototype, the light reflecting from the AMA to the DMD and finally out of the projection lens was the sole source of illumination for the images. The disparity can be adjusted by moving the relay lens between AMA and DMD. As the blur increases, details of the AMA disappear, and the resulting ML becomes more and more disperse as the displacement of the ML increases, as shown in Figure 9.7. The grid lines are the image sent to the DMD and are in 50 pixel increments. They serve as a reference to estimate the size of the AMA MLs that get larger as the disparity increases because of the increasing size of the blur kernel.

Figure 9.7a shows the MLs nearly in focus, with a slight bias in light to the top right. At such low disparities, the MLs hardly moves at all. The size of the MLs is approximately 75 DMD pixels per side. If the DMD had been exactly on the focal plane, the MLs would not have been displaced at all. Figure 9.7b shows how the MLs has become blurred as the disparity increases to 17 mm. The size of the MLs has grown to 100 pixels per side. The MLs has been displaced 60 pixels in a diagonal direction towards the bottom left. The angle of tilt is not uniform over the composite mirror, causing the MLs to distort as it is tilted.

At 37 mm disparity, the size of the MLs is now 120 pixels on its longest side, as shown in Figure 9.7c. The MLs is spread over an area of 160 pixels per side. In this case, the MLs has become so diffuse that the benefit from the AMA is much less apparent. That is even more the case with Figure 9.7d, which shows roughly the same light distribution at 49 mm.

Overall, these measurements show that the approach of using an AMA is successful at redistributing light from one region of the projected image to another. We also demonstrated that the AMA does not geometrically distort the image from the DMD

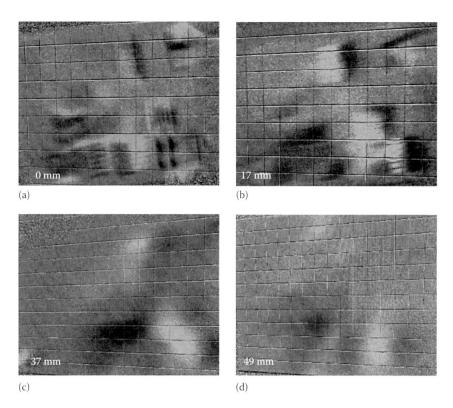

(a)   (b)

(c)   (d)

FIGURE 9.7   Relative change for four different disparity settings (a) 0 mm, (b) 17 mm, (c) 37 mm, and (d) 49 mm, showing multiple mirrors actuated.

in any way. The effect that disparity has on both range and blur was shown, and it was demonstrated that the AMA can make areas of the image both brighter and darker as the mirrors are actuated.

9.9   CONCLUSION

By channelling the light to where it is needed and away from where it is not, an AMA projector makes better use of its light source, which is one of the most expensive components of today's projectors. After the addition of an AMA, the light reaching the primary image modulator, such as a DMD, is not considered to be uniform. Instead, the distribution from the AMA mirrors is simulated, and the compensation for the non-homogeneity is applied to the original image before it is sent to the DMD. The result is an image of higher contrast and peak brightness than would otherwise be possible with the same projection lamp.

This work has demonstrated that an AMA-enhanced projector can make projectors intelligent allocators of their light sources. This ability opens up several interesting research directions, as the projector could conceivably take into account the ambient lighting in the room, the psychophysical limitations of the viewers and the content being presented to display the best possible image over a wide range of conditions.

REFERENCES

Agarwal, P.K. and K.R. Varadarajan. A near-linear constant-factor approximation for euclidean bipartite matching? In *Proceedings of the Twentieth Annual Symposium on Computational Geometry*. ACM, New York, 2004, pp. 247–252.

Brennesholtz, M.S. 3D: Brighter is Better, http://displaydaily.com/2009/05/27/3d-brighter-is-better. (2009).

Bishop, D.J., C.R. Giles, and G.P. Austin. The Lucent LambdaRouter: MEMS technology of the future here today. *IEEE Communications Magazine*. 40(3): (2002) 75–79.

Dagel, D.J., W.D. Cowan, O.B. Spahn, G.D. Grossetete, A.J. Grine, M.J. Shaw, P.J. Resnick, and B. Jokiel Jr. Large-stroke MEMS deformable mirrors for adaptive optics. *Journal of Microelectromechanical Systems*. 15(3): (2006) 572–583.

Damberg, G., H. Seetzen, G. Ward, W. Heidrich, and L. Whitehead. High dynamic range projection systems. *SID Symposium Digest of Technical Papers*. 38(1): (2007) 4–7.

Debevec, P. A median cut algorithm for light probe sampling. In *SIGGRAPH '05: ACM SIGGRAPH 2005 Posters*. ACM, New York, 2005.

Dewald, D.S., D.J. Segler, and S.M. Penn. Advances in contrast enhancement for DLP projection displays. *Journal of the Society for Information Display*. 11: (2004) 177–181.

Dokmeci, M.R., A. Pareek, S. Bakshi, M. Waelti, C.D. Fung, K.H. Heng, and C.H. Mastrangelo. Two-axis single-crystal silicon micromirror arrays. *Journal of Microelectromechanical Systems*. 13(6): (2004) 1006–1017.

Dutta, S.B., A.J. Ewin, M.D. Jhabvala, C.A. Kotecki, J.L. Kuhn, and D.B. Mott. Development of individually addressable micromirror arrays for space applications. In *Proceedings of SPIE*. 4178: (2000) 365–371.

Hah, D., S.T.-Y. Huang, J.-C.Tsai, H. Toshiyoshi, and M.C. Wu. Low-voltage, large-scan angle MEMS analog micromirror arrays with hidden vertical comb-drive actuators. *Journal of Microelectromechanical Systems*. 13(2): (2004) 279–289.

Hecht, E. *Optics*, 4th edn., Addison Wesley, San Francisco, CA, 2002.

Hoskinson, R., S. Hampl, and B. Stoeber. Arrays of large-area, tip/tilt micromirrors for use in a high-contrast projector. *Sensors and Actuators A: Physical*. 173(1): (2012) 172–179.

Hoskinson, R. and B. Stoeber. High-dynamic range image projection using an auxiliary MEMS mirror array. *Optics Express*. 16: (2008) 7361–7368.

Hoskinson, R., B. Stoeber, W. Heidrich, and S. Fels. Light reallocation for high contrast projection using an analog micromirror array. *ACM Transactions on Graphics (TOG)*. 29(6): (2010) 165.

Iisaka, H., T. Toyooka, S. Yoshida, and M. Nagata. Novel projection system based on an adaptive dynamic range control concept. In *Proceedings of the International Display Workshops*. 10: (2003) 1553–1556.

Judy, J.W. and R.S. Muller. Magnetically actuated, addressable microstructures. *Journal of Microelectromechanical Systems*. 6(3): (1997) 249–256.

Kuhn, H.W. The Hungarian method for the assignment and transportation problems. *Naval Research Logistics Quarterly*. 2: (1955) 83–97.

Kusakabe, Y., M. Kanazawa, Y. Nojiri, M. Furuya, and M. Yoshimura. A high-dynamic-range and high-resolution projector with dual modulation. In *Proceedings of SPIE*. 7241: (2009).

Lin, J.E., F.S.J. Michael, and A.G. Kirk. Investigation of improved designs for rotational micromirrors using multi-user MEMS processes. *MEMS Design, Fabrication, Characterization, and Packaging SPIE*. 4407: (2001) 202–213.

Nagahara, H., S. Kuthirummal, C. Zhou, and S.K. Nayar. Flexible Depth of Field Photography. In *European Conference on Computer Vision (ECCV)*. 4: (2008) 60–73.

Pavlovych, A. and W. Stuerzlinger. A high-dynamic range projection system. *Progress in Biomedical Optics and Imaging*. 6: (2005) 39.

Reinhard, E., G. Ward, S. Pattanaik, and P. Debevec. *High Dynamic Range Imaging: Acquisition, Display, and Image-Based Lighting (The Morgan Kaufmann Series in Computer Graphics)*. Morgan Kaufmann Publishers Inc., San Francisco, CA, (2005).

Senturia, S.D. *Microsystem Design*. Kluwer Academic Publishers, Norwell, MA, (2001).

SMPTE. *196M Indoor Theater and Review Room Projection – Screen Luminance and Viewing Conditions*. (2003).

Taylor, W.P., J.D. Brazzle, A.B. Osenar, C.J. Corcoran, I.H. Jafri, D. Keating, G. Kirkos, M. Lockwood, A. Pareek, and J.J. Bernstein. A high fill factor linear mirror array for a wavelength selective switch. *Journal of Micromechanics and Microengineering*. 14, 1: (2004) 147–152.

Tsai, J.-C., L. Fan, D. Hah, and M.C. Wu. A high fill-factor, large scan-angle, two-axis analog micromirror array driven by leverage mechanism. In *International Conference on Optical MEMS and Their Applications*. Takamatsu, Japan, 22–26 August (2004a).

Tsai, J.-C., S. Huang, and M.C. Wu. High fill-factor two-axis analog micromirror array for $1xN^2$ wavelength-selective switch. In *IEEE International Conference on Micro Electro Mechanical Systems*. (2004b) 101–104.

Tsang, S.-H. and M. Parameswaran. Self-locking vertical operation single crystal silicon micromirrors using silicon-on-insulator technology. In *Canadian Conference on Electrical and Computer Engineering*. (2005) 429–432.

Tuantranont, A., V.M. Bright, L.A. Liew, W. Zhang, and Y.C. Lee. Smart phase-only micromirror array fabricated by standard CMOS process. In *The Thirteenth Annual IEEE International Conference on Micro Electro Mechanical Systems (MEMS)*. (2000) 455–460.

Wen, J., X.D. Hoa, A.G. Kirk, and D.A. Lowther. Analysis of the performance of a MEMS micromirror. In *IEEE Transactions on Magnetics*. 40: (2004) 1410–1413.

Part III

Sensors and Microdevices

10 Theoretical Electromagnetic Survey

Application to a CMOS-MEMS Planar Electrodynamic Microphone

Fares Tounsi

CONTENTS

10.1 INTRODUCTION

Undoubtedly, one of the most exciting technological evolutions over the last decade of the twentieth century is in the field of microsystems (microelectromechanical systems (MEMS)). MEMS technology has been able to profit from the benefits and innovations created during the integrated circuit (IC) technology revolution in terms of processes, equipments and materials. The measurement of the pressure is certainly one of the most mature applications of MEMS technology. The following chapter will present an electromagnetic comprehensive study of a new micromachined microphone performed in MEMS technology. However, the main purpose of this chapter is not to introduce MEMS technology, but it will serve to develop an electromagnetic analysis theory that can serve as a model in several MEMS magnetic sensors based, for example, on the Hall effect or Lorentz force. Through different approaches, the magnetic field expression and shape, then subsequently the induced voltage, will be investigated and determined. Developed magnetic modelling techniques will be widely applied on a new electrodynamic MEMS microphone concept to prove its feasibility and interest in a purely arithmetical term. To our knowledge, the proposed electrodynamic principle has never been exploited in practice to realize silicon micromachined microphones. Therefore, the integration of the electrodynamic principle on a single chip and the realization technology choice are our two major pre-eminences.

10.2 STATE OF THE ART OF MEMS MICROPHONES

10.2.1 Contribution of MEMS Technology

The evolution of ICs in the end of the twentieth century was dominated by the miniaturization, providing more functionality and reliability. This trend is consistent with a steady reduction in unit cost and with a variety of available functions offered to users while having low tolerances and high sensitivities. A diversion of microelectronics has led to microsystems (MEMS) that combine semiconductor microelectronics and micromachining technology, allowing the realization of complete systems on chip (SoC) [1,2]. This marriage gave birth to microsystem devices, which are supposed to be more 'intelligent' by providing them with analytical skills, decision making and communication with the outside environment. The scope of microsystems is very wide, but certainly telecommunication and inertial sensor sectors have the strongest application potentials. Microsystems are miniature components including electronic, mechanical and occasionally optical functions on the same chip. Microsystems technologies combine developed microelectronic semiconductor technology and new techniques of micromachining, allowing the realization of an entire SoC. In addition, this integration will miniaturize the system, improve performance and increase the sensitivity and especially the noise reduction by reducing the size of components and thereafter the parasitic capacitances due to interconnections.

A microphone is an electro-acoustic transducer that converts pressure inputs into an electrical signal and is widely employed in numerous applications from different fields as consumer, industrial [3], health [4] and others [5]. During the last decade, microphones and other sensors have been miniaturized by micromachining of silicon. The microphones have a particular type of pressure sensor designed to detect acoustic signals by converting them into electrical signals. Miniaturization is necessary to make them competitive with existing conventional devices. Using MEMS technology, it is possible to miniaturize the microphone and reduce parasitic capacitances due to interconnections. The main advantages of MEMS technology introduction in the manufacture of microphones are as follows: (i) a high degree of control on dimensions, (ii) the miniaturization of devices and mechanical elements, (iii) the possibility of manufacturing by lots and therefore reducing costs and (iv) the integration of acoustic transducers with ICs, for example, CMOS to design an entire SoC. All these factors contribute to improve the cost/performance factor of these acoustic devices.

10.2.2 Different Principles of Micromachined Microphones

Many techniques have been used in MEMS microphones to detect pressure fluctuations. The available principles are mainly based on the membrane vibration detection using piezoelectric, piezoresistive, capacitive and optical techniques. These principles were already known since the twentieth century, but the introduction of the photolithographic manufacturing methods over the past two decades has provided strength for miniaturization and performance. Subsequently, an overview of different MEMS microphone principles will be summarizing.

10.2.2.1 Piezoresistive Micromachined Microphones

Among the various transduction principles, the piezoresistive microphones have the advantage of using a simple manufacturing process and are easily integrated into semiconductor devices [6]. The piezoresistive property of a material is defined as the change of resistivity due to deformation or mechanical stress. The semiconductor piezoresistivity has been demonstrated for many materials such as germanium and polycrystalline or amorphous silicon. A piezoresistive microphone consists of a thin membrane that is usually equipped by four piezoresistive also called piezoresistive gauges in a Wheatstone bridge configuration (see Figure 10.1). The first MEMS piezoresistive sensor [7] had a 1 µm thick silicon membrane, highly boron doped, with an area of 1 mm^2. The piezoresistive microphones have many advantages such as robustness, ease of micromachining and the lack of a need for integrated electronics, thanks to a low output impedance [8]. However, piezoresistive microphones have some drawbacks, such as high noise floor, high power consumption and mainly

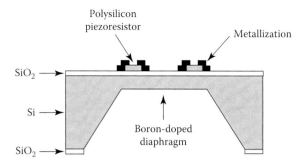

FIGURE 10.1   Sectional view of a silicon micromachined piezoresistive microphone. (From Schellin, R. and Hess, G., *Sensors Actuat. A*, 32, 555, 1992.)

thermal degradation of piezoresistive material due to heating by Joule effect and by constraint [9]. Unfortunately, this leads to a strong temperature dependence of the piezoresistive sensor.

10.2.2.2   Piezoelectric Micromachined Microphones

The piezoelectric microphones may have the same design as piezoresistive ones except that the materials used generate electrical charge on its surface instead of a resistive change. In a piezoelectric microphone, the electro-acoustic detection consists of a piezoelectric layer deposited on a thin membrane (cantilever [10], rectangular plate [11,12] or circular [13]). The piezoelectric material is placed in the region that will undergo the maximum deformation caused by the acoustic signal. A region of high stress is localized near the perimeter of the membrane [13] (see Figure 10.2). Cantilevered structures are more flexible than the membranes and are capable of achieving broader movement [14]. The most used materials include zinc oxide (ZnO), aluminium nitride (AlN), aromatic polyurea, polyvinylidene fluoride (PVDF) and lead zirconate titanate (PZT). Each one provides different mechanical and electrical properties and also transduction capabilities. In addition, some materials are CMOS compatible, and some are not, which leads to many different approaches of piezoelectric microphones implementations. The AlN material, for example, is compatible with CMOS technology, while the PZT one contains lead and is manufactured by sol-gel deposition or by chemical solution deposition (CSD) and is subjected to a polarization process. The ZnO and PVDF materials, while not being fully CMOS

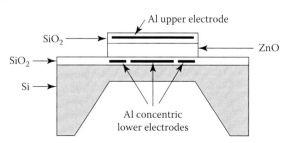

FIGURE 10.2   Sectional view of a monolithic micromachined piezoelectric microphone. (From Royer, M. et al., *Phys. Sensors Actuat. A*, 4, 357, 1983.)

compatible, can be more easily integrated with conventional treatment. The first micromachined piezoelectric microphones are based on inorganic materials (such as ZnO [11] and AlN [15]) and organic materials (such as aromatic polyurea [16]). Moreover, the interesting piezoelectric characteristics of PZT have motivated the search of deposition techniques, integration and design of devices using ferroelectric materials. Indeed, the PZT was considered a promising material for all piezoelectric applications due to its extremely high electromechanical coupling coefficients and their excellent piezoelectric actuation [17]. The use of the PZT thin films in MEMS occurred with significant delays due to difficulties of integration [18]. Piezoelectric microphones have many advantages including intrinsically low power consumption and the possibility of monolithic integration with electronic proximity. The arrival of new ferroelectric materials allowed a sensitivity of about 40 mV/Pa [17], which is comparable to other types of microphones and mainly capacitive ones. The major advantage of this type of microphone consists in the possibility of monolithic integration with processing electronics without the need for an external supply to operate. The disadvantages of piezoelectric microphones reside in the relatively high background noise, which is in the order of 50–72 dB(A) sound pressure level (SPL) [12] and aging of piezoelectric crystal while fearing heat and humidity [13].

10.2.2.3 Capacitive Micromachined Microphones

The basic structure of the capacitive microphone is composed mainly of a fixed back plate, which represents the reference, in parallel with a flexible membrane. These plates are separated by a dielectric, generally an airgap. When sound pressure acts on the membrane, the capacitance change between the diaphragm and the back plate is detected and amplified through various possible types of interface circuits [19] (see Figure 10.3). The operation of the capacitive microphone is conditioned by the application of a bias voltage to generate a constant electrical field between the two capacitor electrodes. In a micromachined microphone, the membrane should be as thin as possible to maximize its response to the sound with a narrow airgap from below, typically of the order of 1–2 μm [16]. The back plate, thick and rigid, is generally perforated so that air between the two plates can circulate freely during the membrane vibration, which allows the damping optimization and ensures appropriate dynamic characteristic. The electrical field, necessary for the operation of the microphone, can be either provided by an external source or by a layer deposition of a material that can hold a quasi-permanent electrical charge between the two electrodes.

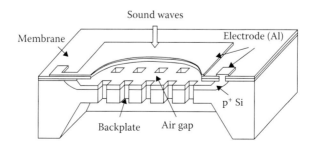

FIGURE 10.3   Structure of a typical condenser microphone.

This material, called electret, provides energy allowing the use of these microphones in portable applications [20]. The principle of capacitive sensing offers many advantages as a relatively high sensitivity, a wide bandwidth, an inherently low power consumption and a low noise [21]. However, specific problems are still present in the capacitive microphones such as instability in the electrostatic pull, the output signal attenuation due to parasitic capacitance and the sensitivity reducing in high frequencies due to viscous damping of the perforated back plate [22].

10.2.2.4 Optical Micromachined Microphones

Optical detection is another alternative for detecting the sound pressure. The idea is based on the modulation of light rather than converting one form of energy into light energy. An optical microphone converts an acoustic signal into an electrical signal by modulating a reference light signal [23]. Three transduction principles are used: (i) the intensity modulation, (ii) the phase modulation and (iii) the polarization modulation [23]. Unlike other types of microphones, an audio signal is first converted into an optical signal before it is converted into an electrical signal. Figure 10.4 shows the intensity modulation basic operating principle. In this configuration, the top electrode is the microphone diaphragm, which is used for light reflection; the bottom electrode is the diffraction grating fabricated on the silicon substrate. Both these electrodes are electrically conductive and optically reflective. The silicon substrate is etched from the backside side to dig an optical path to reach the reflective diffraction grating. When illuminated from the backside with an integrated semiconductor laser such as vertical-cavity surface-emitting laser (VCSEL) as shown in Figure 10.4, some of the light is reflected from the diffraction gratings and some reflected from the microphone diaphragm. As a consequence, a diffracted field consisting of a zero and higher orders results, whose angles remain fixed but whose intensities are modulated by the diaphragm deflection (induced by the sound pressure) to produce the interference curves of a regular Michelson interferometer [24]. Using scalar diffraction theory and knowing the gap height, the beam intensity of the zero and first diffraction

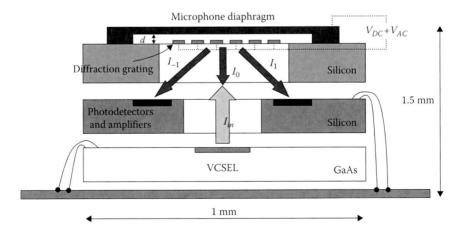

FIGURE 10.4 Operating principle of the intensity modulation detection scheme. (From Hall, N.A. et al., *J. Acoust. Soc. Am.*, 118(5), 3000, 2005.)

orders can be calculated and thereafter the diaphragm fluctuation. Optical techniques are less vulnerable to electromagnetic interference and can tolerate higher temperatures. The disadvantages of optical microphones reside in the requirement of a stable optical reference source and encapsulation of all system components, such as light sources, optical sensor and photodetector, since they must be properly aligned and positioned [25]. In addition, they generally require complex optoelectronic interface circuits to convert optical signal into electrical signal.

10.3   ELECTRODYNAMIC MICROPHONE DESIGN

Currently, the development of next microphone generation focuses on the miniaturization of all components using silicon micromachining. Miniaturization can be achieved through the development accomplished in the manufacturing technology of ICs, which can increase performance and reduce cost. The new microphones integrate monolithically on the same silicon substrate, the microstructures forming the vibrating part and the analogue/digital electronic processing. To overcome problems related to microphone integration, a more flexible approach is presented in this chapter. This monolithic integration will increase performances, miniaturize the system, increase the sensitivity and in particular decrease noise, due to the reduction of interconnections parasitic capacitance. Furthermore, our approach focuses on a low-cost manufacturing process obtained by processing the chips issued from an industrial standard CMOS process. In this approach, we use silicon and other layers issued from a CMOS process as basic materials for the mechanical part of the system.

10.3.1   PRINCIPLE OF ELECTRODYNAMIC LINKAGE

The electrodynamic microphone is based on the principles of electricity and magnetism from the nineteenth century. These microphones were not too prevalent, because of their low yields, until the advent of the electrical amplification that allowed them to find some positions in business and broadcasting. However, currently there are relatively few electrodynamic microphones used on a large scale; most of the market is taken by electret microphones [26]. The electrodynamic microphone is also referred as dynamic, inductive or moving-coil microphone. It is based on the electromagnetic induction principle wherein a conductor or a wire moves in a magnetic field inducing a voltage proportional to the force of the magnetic field, the speed of movement and the length of the conductor. The equation governing the operation is given by $e(t) = Blv(t)$ where $e(t)[V]$ is the instantaneous output voltage, $B[T]$ is the magnetic flux density, $l[m]$ is the length of the conductor emerged in the magnetic field and $v(t)$ $[m/s]$ is the instantaneous conductor movement velocity. Since B and l are constant, the output voltage is directly proportional to the instantaneous conductor movement velocity. The basic principle of the magnetic induction is shown in Figure 10.5a, indicating the relationship between the flux density vector, the direction of current in the conductor and the speed of the conductor. In a microphone, the typical implementation is in the form of a multi-turn coil placed in a radial magnetic field. A sectional view of a typical electrodynamic pressure microphone is shown in Figure 10.5b. In this configuration, the action is the reciprocal of a dynamic speaker.

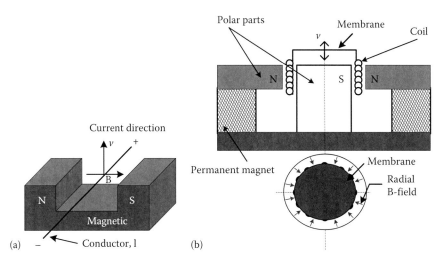

FIGURE 10.5 (a) Basic principle of the magnetic induction and (b) cross-sectional view of the conventional electrodynamic microphone.

10.3.2 Structure of the Electrodynamic Microphone

The proposed structure of the electrodynamic microphone is shown in Figure 10.6. The microphone consists mainly of two concentric planar inductors that occupy separate regions. The first one, a stationary outer inductor B_1, is placed on top of the substrate and an inner inductor B_2 is fabricated on top of a flexible plate suspended over the micromachined cavity [27]. The membrane is attached to the substrate with four arms, one at each corner. The electromagnetic field necessary for the proper functioning of the microphone can be produced by an electrical current flowing through one or both inductors. This biasing current can be either direct current (DC) or alternating current (AC). Thus, a variable magnetic field in space and/or in time will be generated in the vicinity of the inner inductor B_2. This inductor will move up and down due to the vibration of the suspended membrane under an incident acoustic pressure. According to Faraday's law, the vibration of the inductor B_2 in a magnetic field will generate, at its bounds, an electromotive force (*emf*), proportional to the amplitude of the incident acoustic pressure. This voltage will be processed by the electronic circuitry integrated on the same chip.

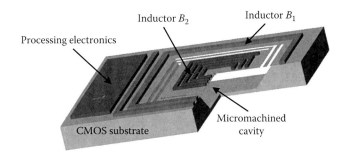

FIGURE 10.6 Basic representation of the electrodynamic microphone structure.

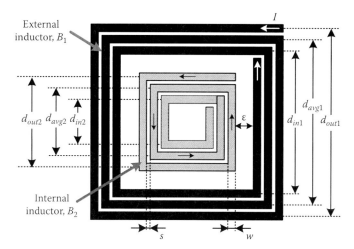

FIGURE 10.7 Disposition and dimensions of the two coplanar concentric inductances.

TABLE 10.1

Geometrical Parameters Used in Graphical Representations Illustrated Thereafter

Name[a]	Description	Value	Name[a]	Description	Value
n_1	Number of turns	50	n_2	Number of turns	50
w	Width of the metal turns	0.8 μm	s	Spacing between metal turns	0.8 μm
d_{out2}	Outside inductor diameter	1.5 mm	d_{out1}	Outside inductor diameter	1.73 mm
d_{avg2}	Average inductor diameter	1.42 mm	ε	Separation between the inductors	35 μm

[a] The 1 index (respectively 2) indicates that the parameter belongs to the outer inductor B_1 (respectively the inner inductor B_2).

In our study, throughout this chapter, we will focus on the theoretical magnetic modelling resulting from supplying one or both planar inductors. In simulations, the most common geometrical parameters to be used (see Figure 10.7) are given in Table 10.1. The other parameters can be easily deduced using geometrical properties of an inductor. The shown dimensions conceive a microphone with the same size proportions with that cited in references.

Generally, the incident sound wave is a complex signal resulting from the superposition of several sine waves having different frequencies. Fourier's theorem allows us to decompose the complex signal into a sum of pure tones. For simplicity, we will consider in our calculations a pure incident sound wave which leads to consider a sinusoidal fluctuation of the membrane given by $\xi = h\sin(\omega_p t)$, where ω_p is the angular velocity of the incident sound pressure, h is the vibration amplitude and t is time.

It is also assumed that the movement of the membrane is considered as a piston; hence, the deflection of the centre relative to the periphery is negligible.

10.4 FUNDAMENTAL RELATIONSHIPS IN TRANSFORMERS

The structure of the micromachined electrodynamic microphone is based on the use of two square planar spiral inductors. Since the transformer is formed by two coupled inductors, a solid knowledge of the integrated inductors, characteristics and limitations will be an excellent starting point in the analysis of integrated transformers.

10.4.1 EVALUATION OF THE INDUCED VOLTAGE

When two inductors are placed at a short distance and flown by i_1 and i_2 currents, as shown in Figure 10.8, this gives an ideal transformer. A magnetic interaction between two circuits and four flow components Φ_{11}, Φ_{12}, Φ_{22} and Φ_{21} will take place. The magnetic flux produced by the current i_1 flowing in the primary winding induces a current i_2 flowing through the secondary winding, and vice versa, according to the following equations:

$$\begin{cases} \Phi_1 = \Phi_{11} + \Phi_{21} = L_1 i_1 + M_{21} i_2 \\ \Phi_2 = \Phi_{22} + \Phi_{12} = L_2 i_2 + M_{12} i_1 \end{cases} \tag{10.1}$$

where L_1 and L_2 are, respectively, the primary and secondary self-inductances. Indeed, the current i_1 produces a magnetic field flux Φ_1 in which a part is lost as flow coupling, Φ_{12}, and the other part, Φ_{11}, passes through the circuit itself. The mutual inductance, M_{12}, between the two circuits 1 and 2 is defined as the ratio between the flux generated by circuit 1 through circuit 2 (Φ_{12}) and the current flowing in circuit 1 (i_1). According to Faraday's law, the change of flux generated by the magnetic field through one of the two inductors induces an electromotive force at the terminals of the secondary giving the following pair of equations:

$$\begin{cases} v_1 = -\dfrac{d\Phi_1}{dt} = -\dfrac{dL_1 i_1}{dt} - \dfrac{dM i_2}{dt} \\[2mm] v_2 = -\dfrac{d\Phi_2}{dt} = -\dfrac{dL_2 i_2}{dt} - \dfrac{dM i_1}{dt} \end{cases} \tag{10.2}$$

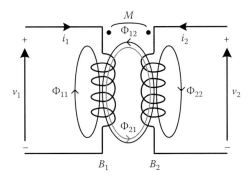

FIGURE 10.8 Magnetic interaction between two ideal inductors.

To characterize the coupling importance between the primary and the secondary, we introduce the magnetic (or mutual) coupling coefficient k, which is defined by

$$k = \frac{M}{\sqrt{L_1 L_2}} \tag{10.3}$$

For an ideal transformer, $k = 1$, while for most integrated transformers, k is between 0.3 and 0.9 due to magnetic flux leakage [28]. In practice, the reasons of these non-idealities include parasitic capacitance, resistance due to ohmic losses, skin effect, proximity effect and eddy current in the substrate. In the case of an integrated transformer, its model is achieved by representing the primary and secondary inductors by their equivalent localized π models, then adding the mutual inductance and coupling capacitances between the primary and the secondary [28,29].

10.4.2 DIFFERENT REALIZATIONS OF MONOLITHIC TRANSFORMERS

Monolithic transformers have been widely used in RF circuits. Integrated transformer structures are mainly divided into two categories: planar or stacked, depending on the used magnetic coupling, lateral or vertical [30]. The compromise between these two categories is the greater terminal substrate capacitance and a smallest self-resonant frequency. Planar transformers occupy generally a larger area than stacked ones, with many varieties in their achievements as shown in Figure 10.9. Tapped transformer, shown in Figure 10.9a, is composed of a 2-terminal spiral inductor connected on its middle. The parasitic capacitance between the spiral and the substrate can be minimized by an implementation of the inductance in the upper level of metal in order to achieve a high resonant frequency. This transformer allows a broader range of relationships between the number of turn of its primary and the number of turn of its secondary, although its coupling coefficient k is usually small (0.4 ~ 0.6) [29].

Figure 10.9b shows an interleaved transformer; its primary and secondary are wound (roll up) in parallel. This type of transformer is best suited for inductors and allows a moderate coupling ($k \approx 0.7$) reached at the cost of a decrease in the self-inductance. This coupling may be increased by a greater series resistance achieved by reducing the width of the inductor, w, and spacing, s.

The stacked transformer, shown in Figure 10.9c, uses multiple metal layers and exploits both the vertical and the lateral magnetic coupling to permit best efficiency and surface occupancy, and it has a higher coupling coefficient ($k \approx 0.9$). The main disadvantage is the high terminal–terminal capacity, that is, a low self-resonant frequency. In modern multilevel manufacturing processes, the capacity can be reduced by increasing the oxide thickness between spirals.

Table 10.2 shows a performance comparison between different types of the presented manufactured transformers. Other types of transformers are presented in the literature, but they are inspired from these past three models [31].

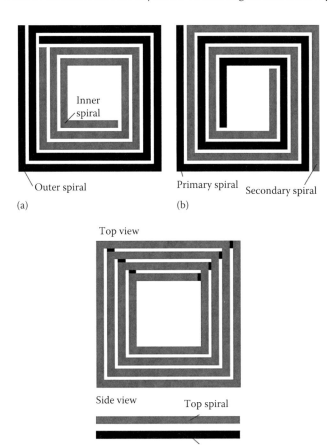

FIGURE 10.9 Various structures of transformers: (a) concentric, (b) interleaved and (c) stacked.

TABLE 10.2
Comparison between Different Types of Transformers Manufactured

Type of Transformer	Surface	Coupling Coefficient, k	Self-Inductance	Self-Resonant Frequency
Concentric	High	Low	Medium	High
Interleaved	High	Medium	Low	High
Stacked	Low	High	High	Low

10.4.3 EVALUATION OF THE TOTAL INDUCTANCE IN A PLANAR INDUCTANCE

There are two numerical methods for assessing the inductance value of an inductor: the methods of Grover and Wheeler [32]. Based on the Grover study, which showed results more consistent with the practice [33], Greenhouse developed a theory to calculate the total inductance of a planar square spiral [37]. Indeed, a square spiral can be carved into a discrete set of straight lines and the total inductance of the inductor can be evaluated as the sum of the self-inductances L_i of each line. In addition to the self-inductance of each conductor, the interaction between the parallel segments must also be included in the calculation of the total inductance. This interaction is expressed by the mutual inductance $M_{i,j}$ (positive or negative) between the ith and the jth (with $i \neq j$) parallel segments. Thus, the total inductance expression can be calculated using the following equation [35]:

$$L = \sum L_i + \sum M_{i,j}^+ - \sum M_{i,j}^-$$

(10.4)

The expression of the rectangular conductor self-inductance L_i was given by Grover as follows [35]:

$$L_i = \frac{\mu_0 l}{2\pi}\left[\ln\left(\frac{2l}{w+t}\right) + 0.50049 + \frac{w+t}{3l}\right]$$

(10.5)

where
  μ_0 is the air magnetic permeability
  w, t and l represent the conductor width, thickness and the inductor length, respectively

This expression is valid as long as the width or thickness of the conductor does not double its length, which is difficult to reproduce. A simpler approximate expression using the total length of the spiral can be found in [36]. The mutual inductance between two conductors depends on their intersection angles, length and separation. Two orthogonal conductors do not have mutual inductance since their magnetic fluxes are not coupled together. For a square inductance, the mutual inductance is constituted by two components: the positive component that contributes to increasing the magnetic coupling between the spirals and the negative that contributes to decrease it. Parallel segments, with currents flowing in the same phase, produce a positive contribution to the mutual inductance M^+, such as the segments located on the same side of the square inductance. Antiparallel segments, where the currents are in opposite phase, have a negative contribution to the mutual inductance M^-, such as currents on one side and the other of the inductor (see Figure 10.10) [34,36].

   In the simple case of two parallel conductors, having the same length and rectangular section of width w and thickness t (Figure 10.11), placed opposite at a distance between the two centres equal to d (for planar inductance $d = w + s$), the mutual inductance is defined by the following equation [37]:

$$M_l = \frac{\mu_0 l}{2\pi}Q$$

(10.6)

where l is the length of the conductor and Q is the coefficient that depends only on the geometry and is given by the following equation [38]:

$$Q = \ln\left[\frac{l}{GMD} + \sqrt{1 + \frac{l^2}{GMD^2}}\right] - \sqrt{1 + \frac{GMD^2}{l^2}} + \frac{GMD}{l}$$

(10.7)

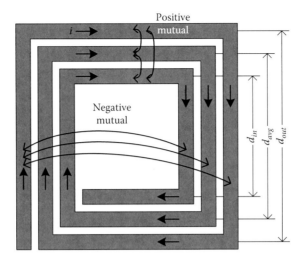

FIGURE 10.10 Illustration of the positive and negative mutual inductance in a square planar spiral.

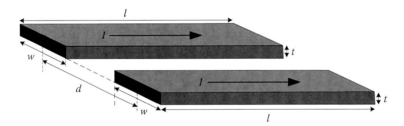

FIGURE 10.11 Parallel arrangement of two equal-length conductors.

The GMD (acronym for 'Geometric Mean Distance') is the geometric mean of the distance between the surfaces of two conductors. In the case of two rectangular shape conductors (Figure 10.11), the exact value of the GMD can be obtained from the Grover table [37] and can be estimated with an error less than 3% by the following equation [29]:

$$\ln(\text{GMD}) = \ln(d) - \left\{ \frac{w^2}{12d^2} + \frac{w^4}{60d^4} + \frac{w^6}{168d^6} + \frac{w^8}{360d^8} + \frac{w^{10}}{660d^{10}} + \cdots \right\} \quad (10.8)$$

As expected, the GMD is slightly smaller than the distance d between the two-conductor centre. The mutual inductance, M_l, varies relatively slightly depending on the width of the conductor when the distance between the two centres d remains fixed. This implies that the mutual inductance value between two inductors, with similar distances between spirals, barely changes when the width of these spirals is modified [39].

In a planar inductor, structure used for the microphone, we can identify interactions between segments of different lengths as shown in Figure 10.12a. The mutual inductance between two parallel conductors of different lengths, l and m, can be

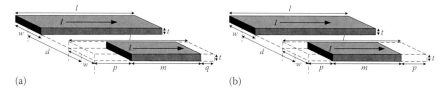

(a)   (b)

FIGURE 10.12 Arrangement: (a) arbitrary and (b) symmetrical with respect to the middle of two different-length parallel conductors.

expressed as a difference in mutual between segments of the same length, given by the following equation [37]:

$$M_{l \leftrightarrow m} = \frac{1}{2}[M_{m+p} - M_p + M_{m+q} - M_q] \qquad (10.9)$$

In the case where each segment of length m is located symmetrically in the middle of the segments of length l ($p = q$) (Figure 10.12b), the expression of the mutual inductance can be written in a simplified form as

$$M_{l \leftrightarrow m} = M_{m+p} - M_p \qquad (10.10)$$

10.5 EVALUATION OF THE THEORETICAL INDUCED VOLTAGE, THROUGH THE MUTUAL INDUCTANCE

The coupling between the two inductors represents a concentric planar transformer with the internal inductance occupying a distinct region than the external. When only the primary inductor is crossed by a DC (see Figure 10.13), the flow through the inductor B_2 is restricted to $\Phi_{12} = M \cdot I_1$ and the voltage in the secondary is given by the following equation:

$$v_2 = -I_1 \frac{dM}{dt} \qquad (10.11)$$

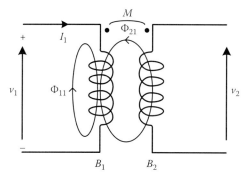

FIGURE 10.13 Ideal magnetic interaction between the two inductors when only the primary is crossed by a *DC*.

where M is the mutual inductance (or the magnetic coupling) between the segments of the outer primary inductor, B_1, and the internal secondary, B_2. The induced voltage is even more important when the mutual inductance variation with respect to time is larger, that is, faster movement of the inner inductor. The concentric transformers are based solely on the lateral magnetic coupling (Figure 10.13). Thus, the total mutual inductance between two inductors, M, will be calculated using the Greenhouse algorithm that is based on the magnetic coupling interactions, sum of each segment from B_1 with that from B_2 [37]. As mentioned previously, the mutual inductance is the superposition of two terms, a positive term that contributes to the magnetic flux coupling increase minus the negative magnetic couplings that reduce it. According to Equations 10.6 and 10.8, the mutual inductance is a purely geometrical parameter, which is proportional to the conductor's length, inversely proportional to the distance between the two centres and varies slightly with the conductor width. There are two approaches to calculate the mutual inductance between spirals of two inductors, the recursive and the average method.

10.5.1 EVALUATION OF THE MUTUAL INDUCTANCE

10.5.1.1 Evaluation through the Accurate Recursive Approach

This approach is based on a recursive principle that consists of computing the effect of each individual segment constituting the inductor on all other segments. This method is divided into two parts; the first part calculates the positive mutual inductance for segments where currents have the same direction. The other part is used to calculate the negative mutual between the segments with opposite currents (see Figure 10.10). The total mutual inductance will be equal to the difference between positive and negative. The mutual inductance between the two spiral inductors depends also on the distance between the two planes of B_1 and B_2. So we must consider the instantaneous vertical distance between the planes of the two inductors, already called ξ. So, to calculate the distance between the two segments that belong to two different inductors, we must first calculate the square root of $\varepsilon^2 + \xi^2$ as shown in Figure 10.14.

In the case of a square inductor, where each spiral turn consists of four line segments, we can work on one side, then multiply by four factors to find the total

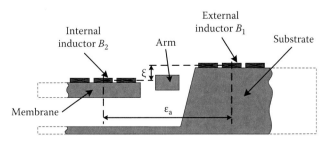

FIGURE 10.14 Sectional view of the relative movement of the internal inductor B_2 relative to B_1.

mutual inductance. Subsequently, the positive mutual inductance M^+ is calculated using the following algorithm:

$Mi_1 = 0$;

for $\quad i = 1 \quad$ to $\quad n_2 - 1$

$\quad\quad$ for $\quad j = 1 \quad$ to $\quad n_1 - 1$

$$a_1 = \frac{d_{out1} + d_{out2}}{2} - j \times (w + s) - i \times (w + s);$$

$$a_2 = \frac{d_{out1} - d_{out2}}{2} - j \times (w + s) + i \times (w + s);$$

$$d = \sqrt{a_2^2 + h^2}; \; M_{a_1} = \frac{\mu_0 a_1}{2\pi} Q;$$

$$Mi_1 = Mi_1 + M_{a_1};$$

$$M_{a_2} = \frac{\mu_0 a_2}{2\pi} Q;$$

$$Mi_1 = Mi_1 - M_{a_2};$$

$\quad\quad$ end;

$\quad$ end;

$$M^+ = 4 \times Mi_1;$$

The execution of this algorithm with the values initially given in Table 10.1 was done in Figure 10.15a. We note that the mutual inductance is inversely proportional to the distance ξ between the two inductor planes. Following the same reasoning, the negative mutual inductance M^- is determined using the following algorithm:

$Mi_2 = 0$;

for $\quad i = 1 \quad$ to $\quad n_2 - 1$

$\quad\quad$ for $\quad j = 1 \quad$ to $\quad n_1 - 1$

$$a_1 = \frac{d_{out1} + d_{out2}}{2} - j \times (w + s) - i \times (w + s);$$

$$a_2 = \frac{d_{out1} - d_{out2}}{2} - j \times (w + s) + i \times (w + s);$$

$$d = \sqrt{a_1^2 + h^2}; \; M_{a_1} = \frac{\mu_0 a_1}{2\pi} Q;$$

$$Mi_2 = Mi_2 + M_{a_1};$$

$$M_{a_2} = \frac{\mu_0 a_2}{2\pi} Q;$$

$$Mi_2 = Mi_2 - M_{a_2};$$

$\quad\quad$ end;

$\quad$ end;

$$M^- = 4 \times Mi_2;$$

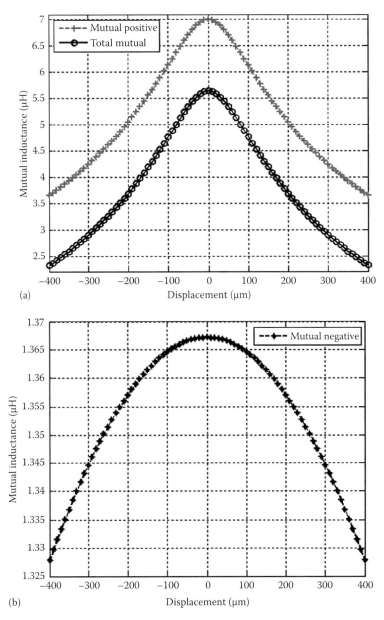

(a)

(b)

FIGURE 10.15   Variation of the (a) positive and total mutual inductance and (b) negative mutual inductance, for a displacement amplitude less than 400 μm.

Similarly, the mutual negative is also inversely proportional to the distance ξ. However, its variation and amplitude are significantly lower than the positive mutual since the distance between the antiparallel current segments is greater than that between the segments with parallel flow (Figure 10.15b). Finally, the variation $M_T = M^+ - M^-$ is shown in Figure 10.15a. This algorithm provides a very accurate assessment of the mutual inductance variation; however, it is unable to give a simple expression usable for the induced voltage estimation. The following approach tries to apply some approximations on the aforementioned theory to give a fairly accurate estimate of the mutual inductance expression generated between the two planar inductors. Among the approximations is to neglect the component produced by the negative mutual inductance because of the large diameter of the internal inductance (mm range).

10.5.1.2 Evaluation through the Approximately Average Approach

The second approach, by the average, for calculating the mutual inductance, consists of replacing the two inductors B_1 and B_2 by two single turns formed by n_1 superposed spirals for the first and n_2 spirals for the second (see Figure 10.16). The length of the resulting spiral is given by the average diameter of the inductors, d_{avg}, of each inductor. Subsequently, the problem is reduced to calculate the mutual inductance between two parallel conductors of different lengths and multiply the result by $n_1 \times n_2$. Using the symmetry and noting that the segments perpendicular to each other have a zero mutual, the total value of positive interactions of the mutual inductance in one side may be replaced by an equivalent of $4n_1 \times n_2$ average interactions between two parallel conductors of different lengths (see Figure 10.16).

In light of these considerations and by combining Equations 10.6, 10.8 and 10.10, the total mutual inductance between the two concentric inductors is a parabolic function that can be defined by the following equation:

$$
M = 2n_1n_2 \frac{\mu_0}{\pi} \left\{ (a + \varepsilon_a) \left(\ln \left[\frac{a + \varepsilon_a}{GMD} + \sqrt{1 + \frac{(a + \varepsilon_a)^2}{GMD^2}} \right] - \sqrt{1 + \frac{GMD^2}{(a + \varepsilon_a)^2}} + \frac{GMD}{a + \varepsilon_a} \right) \right.
$$

$$
\left. - \varepsilon_a \left(\ln \left[\frac{\varepsilon_a}{GMD} + \sqrt{1 + \frac{\varepsilon_a^2}{GMD^2}} \right] - \sqrt{1 + \frac{GMD^2}{\varepsilon_a^2}} + \frac{GMD}{\varepsilon_a} \right) \right\} \tag{10.12}
$$

where a and ε_a are, respectively, the length of the average outer spiral and the distance between the two average outer and inner spirals (see Figure 10.16). The mutual inductance can be normalized by writing that it is equal to the sum of two terms as

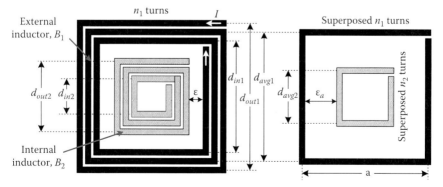

FIGURE 10.16 Considered equivalent scheme of the two inductors.

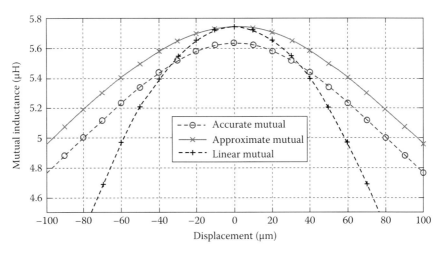

FIGURE 10.17 Variation of the mutual inductance M versus displacement ξ, for the accurate, approximate and linear expressions.

follows: $M = M_0 + \Delta M(\xi)$. The term M_0 represents the permanent and constant magnetic link, with respect to time, between the turns of the two inductors corresponding to a separation equal to ε_a. The second term $\Delta M(\xi)$ represents the mutual caused by the relative vibration, ξ, between the plane of the internal inductor and that of the external one (see Figure 10.17). For small displacements, the mutual inductance components can be estimated using a second-order Taylor expansion of M when ξ varies in the neighbourhood of zero, which gives the following equation:

$$
M_0 \approx n_1 n_2 \frac{2\mu_0}{\pi} \left[(a - \varepsilon_a) \cdot \ln \left(\frac{(a - \varepsilon_a) + \sqrt{\varepsilon_a^2 + (a - \varepsilon_a)^2}}{\varepsilon_a} \right) \right.
$$

$$
\left. + \varepsilon_a \ln \left(\frac{\varepsilon_a + \sqrt{\varepsilon_a^2 + (a - \varepsilon_a)^2}}{(a - \varepsilon_a)} \right) - 2\sqrt{\varepsilon_a^2 + (a - \varepsilon_a)^2} - a\left(\ln(1 + \sqrt{2}) - \sqrt{2} \right) \right] \quad (10.13)
$$

$$
\Delta M(\xi) \approx n_1 n_2 \frac{\mu_0}{\pi} \left(\frac{\sqrt{2}\varepsilon_a - \sqrt{\varepsilon_a^2 + (a - \varepsilon_a)^2}}{\varepsilon_a^2} \right) \xi^2 = -\frac{A_2}{2} \xi^2
$$

where A_2 is a purely constant geometrical factor.

10.5.1.3 Comparison between the Two Approaches

For the same values used previously, a graphical representation of the linear and approximate equations of the mutual inductance is given in Figure 10.17. We clearly see that for low vibration amplitudes, the two curves overlap.

Figure 10.18 shows the error of the average method compared to the recursive method; we can see that the error is less than 5% for a number of turns equal to 50.

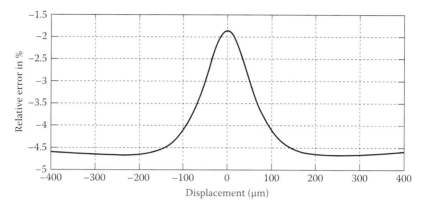

FIGURE 10.18 Percentage of the relative error between the accurate mutual and the approximate mutual.

So both methods lead to similar results with an error less than 5% (Figure 10.18). The average method allows us to calculate the mutual inductance value with a simulation time much less than the recursive method especially for a high number of turns.

10.5.2 EVALUATION OF THE THEORETICAL INDUCED VOLTAGE

We note that the magnetic coupling coefficient is at its maximum when the two inductors are located on the same plane (Figure 10.19a) and when the distance between them, ε_a, is minimal (see Figure 10.19b). We can notice also that the mutual coupling coefficient, k, is relatively small due to the large spatial separation, ε, between the two inductors (see Figure 10.19b). Stacked transformers have a higher coupling coefficient due to the small distance between the layers forming the two inductors; nevertheless, we are limited in our case by the operation principle and the use of industry standard technology for manufacturing the electrodynamic microphone.

10.5.2.1 External Inductor Biasing by a DC

By combining Equations 10.11 and 10.13, a simple expression of the induced voltage in the open-circuit secondary output, around the linear range, can be obtained by differentiating $\Delta M(\xi)$, as shown in the following equations:

$$v_2 = -I_1 \frac{d\Delta M}{dt} = A_2 I_1 \xi v \tag{10.14}$$

10.5.2.2 External Inductor Biasing by an AC

By considering that the two inductors are a concentric planar transformer [34], the flux through the inductor B_2 is given by $\Phi_{12} = M \cdot i_1$, and the voltage in the secondary is given by

$$v_2 = -\frac{dMi_1}{dt} = A_2 i_1 \xi v - M \frac{di_1}{dt} \tag{10.15}$$

where M is defined given by two terms as shown in Equation 10.13, a constant term, and the other is variable.

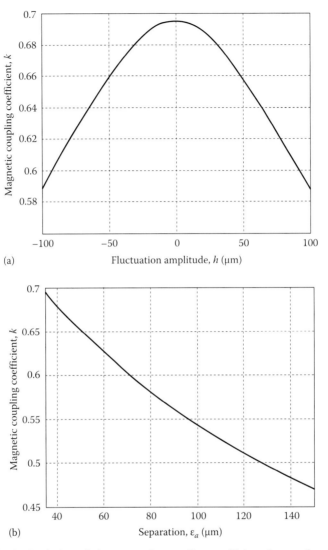

(a)

(b)

FIGURE 10.19 Variation of the magnetic coupling coefficient k as a function of the (a) membrane fluctuation amplitude and (b) separation between the two inductors.

10.5.2.3 Both Inductors Biasing by an AC

In this latter polarization case, we consider both inductances biasing by an AC. Hence, a third factor will appear in the total induced voltage expression. This coefficient represents the flux produced by the secondary inductance on itself. Therefore, the expression of the induced voltage is given by the following expression:

$$v_2 = -\frac{dMi_1}{dt} - L_2\frac{di_2}{dt} = A_2 i_1 \xi v - M\frac{di_1}{dt} - L_2\frac{di_2}{dt} \tag{10.16}$$

If we consider the same bias current pulsation for the secondary, as well as the primary, the additional term will be added to the carrier, so it will not provide additional useful data about the incident sound wave (see Section 10.6.3 for further explanation).

10.6 EVALUATION OF THE THEORETICAL INDUCED VOLTAGE, THROUGH THE MAGNETIC FIELD

The following section aims to evaluate the magnetic field expression produced by the outer inductor within the internal. This second method allows us to find other alternative to validate and understand the induced voltage expression already found using the mutual formulas.

10.6.1 EVALUATION OF THE MAGNETIC FIELD PRODUCED BY A SLENDER SPIRE

The simplified considered spirals configuration schematic of the MEMS electro-dynamic microphone is shown in Figure 10.20a. The considered structure of the microphone consists of two concentric square spirals. The internal inductor vibrates vertically along the z-axis under the influence of sound pressure, while the outer inductor is always stationary. To calculate the magnetic field produced by the outer inductor within the inner, we can apply the principle of superposition. In fact, circuits having a rectangular shape are made by a juxtaposition of several conductors together; hence, according to the principle of superposition, the resulting magnetic field, created at a point M, is the vector sum of magnetic fields created by the contribution of each conductor segment. Consequently, we can start initially by calculating the magnetic field produced by the spiral segment $[AB]$ (see Figure 10.20b).

The Cartesian coordinate system (x, y, z), shown in Figure 10.20a, will be used throughout the rest of the chapter. Initially, we consider an electrical current flow I and $\overrightarrow{dl}$ denotes an elementary displacement along this conductor oriented in the same direction as the current (Figure 10.20b). The Biot and Savart law teaches us that the elementary magnetic field $\overrightarrow{dB}$ at point $M(x, y, z)$ inside the spire, by the wire element $\overrightarrow{dl}$ located at the point P, carrying the current i is written by

$$\overrightarrow{dB}(M) = \frac{\mu_0}{4\pi} \frac{I\overrightarrow{dl} \wedge \overrightarrow{PM}}{|PM|^3} \tag{10.17}$$

where μ_0 is the magnetic permeability of vacuum and $\overrightarrow{PM}$ is the displacement vector from the wire element to the point at which the magnetic field will be calculated at. The magnetic field produced by a finite length straight wire at a point M of space can be evaluated using the following equation:

$$B_z(M) = \frac{\mu_0}{4\pi} \frac{I}{r}(\sin\theta_1 - \sin\theta_2) \tag{10.18}$$

where
  r is the perpendicular distance between the point M and the conductive wire
  θ_1 and θ_2 are the two oriented angles indicated in Figure 10.20b

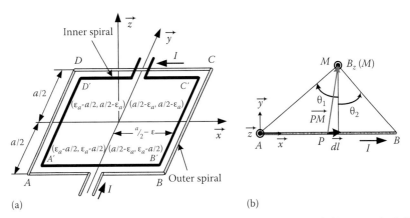

(a)   (b)

FIGURE 10.20   (a) Geometrical arrangement of the two spirals and (b) magnetic field B produced by a straight conductor wire.

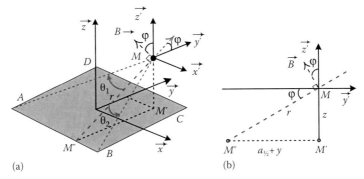

(a)   (b)

FIGURE 10.21   (a) 3D (b) y–z plane, illustration of the produced B magnetic field at an arbitrary point M in the space inside the inductor.

To find the magnetic field expression produced by the rectangular conductor $[AB]$, we will project the produced magnetic field vector on the Cartesian coordinate axes (see Figure 10.21a); we can write that

$$\vec{B}_{[AB]}(M) = \frac{\mu_0}{4\pi} \frac{I}{r} \left[\sin(\alpha_{11}) - \sin(\alpha_{12}) \right] \times \left[\cos(\phi)\vec{z} - \sin(\phi)\vec{y} \right] \qquad (10.19)$$

From Figure 10.21a and b, we can deduce that

$$r = \sqrt{z^2 + \left(\frac{a}{2} + y \right)^2}$$

$$\sin(\theta_1) = \frac{\frac{a}{2} - x}{\sqrt{r^2 + \left(\frac{a}{2} - x \right)^2}} \quad \text{and} \quad \sin(\theta_2) = -\frac{\frac{a}{2} + x}{\sqrt{r^2 + \left(\frac{a}{2} + x \right)^2}}$$

$$\cos(\phi) = \frac{\left(\frac{a}{2} + y \right)}{r} \quad \text{and} \quad \sin(\phi) = \frac{z}{r} \qquad (10.20)$$

Thus, by substituting Equations 10.20 into Equation 10.19, the y-magnetic field components produced by $[AB]$ wire can be written as

$$B_y(M) = \frac{\mu_0 I}{4\pi} \frac{z}{\left(\dfrac{a}{2}+y\right)^2 + z^2}$$

$$\left[\frac{\dfrac{a}{2}-x}{\sqrt{\left(\dfrac{a}{2}-x\right)^2 + \left(\dfrac{a}{2}+y\right)^2 + z^2}} + \frac{\dfrac{a}{2}+x}{\sqrt{\left(\dfrac{a}{2}+x\right)^2 + \left(\dfrac{a}{2}+y\right)^2 + z^2}} \right] \quad (10.21)$$

Similarly, we can find that the z magnetic field component produced by $[AB]$ is given by

$$B_z(M) = \frac{\mu_0 I}{4\pi} \frac{\dfrac{a}{2}+y}{\left(\dfrac{a}{2}+y\right)^2 + z^2}$$

$$\left[\frac{\dfrac{a}{2}-x}{\sqrt{\left(\dfrac{a}{2}-x\right)^2 + \left(\dfrac{a}{2}+y\right)^2 + z^2}} + \frac{\dfrac{a}{2}+x}{\sqrt{\left(\dfrac{a}{2}+x\right)^2 + \left(\dfrac{a}{2}+y\right)^2 + z^2}} \right] \quad (10.22)$$

Thus, according to the principle of superposition applied to the magnetic fields, the total magnetic field created inside the square planar spiral of side a, placed in the x–y plane (see Figure 23a), flown by an I current density, will produce three components given by the following set of equations [40]:

$$B_x(M) = \frac{\mu_0 I z}{4\pi} \left[\frac{1}{\left(\dfrac{a}{2}-x\right)^2 + z^2}\left(\frac{\dfrac{a}{2}-y}{c_1} + \frac{\dfrac{a}{2}+y}{c_2} \right) - \frac{1}{\left(\dfrac{a}{2}+x\right)^2 + z^2}\left(\frac{\dfrac{a}{2}-y}{c_3} + \frac{\dfrac{a}{2}+y}{c_4} \right) \right]$$

$$B_y(M) = \frac{\mu_0 I z}{4\pi} \left[\frac{1}{\left(\dfrac{a}{2}-y\right)^2 + z^2}\left(\frac{\dfrac{a}{2}-x}{c_1} + \frac{\dfrac{a}{2}+x}{c_3} \right) - \frac{1}{\left(\dfrac{a}{2}+y\right)^2 + z^2}\left(\frac{\dfrac{a}{2}-x}{c_2} + \frac{\dfrac{a}{2}+x}{c_4} \right) \right]$$

$$B_z(M) = \frac{\mu_0 I}{4\pi} \left[\frac{\frac{a}{2} - x}{\left(\frac{a}{2} - x\right)^2 + z^2} \left(\frac{\frac{a}{2} - y}{c_1} + \frac{\frac{a}{2} + y}{c_2} \right) + \frac{\left(\frac{a}{2} + x\right)}{\left(\frac{a}{2} + x\right)^2 + z^2} \left(\frac{\frac{a}{2} - y}{c_3} + \frac{\frac{a}{2} + y}{c_4} \right) \right.$$

$$\left. + \frac{\frac{a}{2} - y}{\left(\frac{a}{2} - y\right)^2 + z^2} \left(\frac{\frac{a}{2} - x}{c_1} + \frac{\frac{a}{2} + x}{c_3} \right) + \frac{\frac{a}{2} + y}{\left(\frac{a}{2} + y\right)^2 + z^2} \left(\frac{\frac{a}{2} - x}{c_2} + \frac{\frac{a}{2} + x}{c_4} \right) \right]$$

The constants c_1 to c_4 are given by

$$c_1 = \sqrt{\left(\frac{a}{2} - x\right)^2 + \left(\frac{a}{2} - y\right)^2 + z^2}, c_2 = \sqrt{\left(\frac{a}{2} - x\right)^2 + \left(\frac{a}{2} + y\right)^2 + z^2}$$

$$c_3 = \sqrt{\left(\frac{a}{2} + x\right)^2 + \left(\frac{a}{2} - y\right)^2 + z^2}, c_4 = \sqrt{\left(\frac{a}{2} + x\right)^2 + \left(\frac{a}{2} + y\right)^2 + z^2}$$

$$(10.23)$$

Therefore, both magnetic field radial components, B_x and B_y, are null on the substrate plane. For a displacement amplitude of 20 μm, the magnetic field B_x shape is shown in Figure 10.22a. We note that the radial components B_x as a function of x (also for B_y as a function of y) increase exponentially when approaching to the spiral. However, its shape variation depending on the z position, for x slightly lower than $a/2$ and $y = 0$, is shown in Figure 10.22b. To extend previous results in the case of two rectangular loops into two concentric planar inductors, case of the microphone, we need to apply some approximations. We assume that each inductor B_i will be represented by n_i superimposed turns. This means that the magnitudes of the three magnetic field components will be amplified by n_1 factor.

10.6.2 External Inductor Biasing Using a DC

When the microphone is biased by a *DC*, an induced electromotive force, according to Faraday's law, will be produced inside the closed circuit by the Lorentz electromagnetic induction phenomenon. This electromotive force is called motional electromotive force (emf_m) and is generated by the movement of the inner inductor in a magnetic field variable in space and time independent. The expression of the emf_m is proportional to the time variation of the magnetic flux, Φ, as indicated by the following [40, 41]:

$$emf_m = \oint_{loop} \overrightarrow{E_m} \overrightarrow{dl} = \oint_{loop} (\vec{v} \wedge \vec{B}) \overrightarrow{dl} \qquad (10.24)$$

where
  $\overrightarrow{E_m}$ is the motional electrical field
  $\overrightarrow{dl}$ is the elementary displacement vector (see Figure 10.23) on the inner inductor
  $\vec{v}$ is the membrane vibration velocity obtained by deriving the membrane instantaneous displacement equation

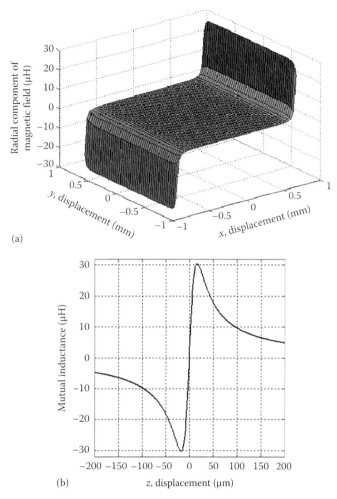

(a)

(b)

FIGURE 10.22 Shape of the radial component of the magnetic field B_x (a) as a function of (x, y) for $z = 20$ µm and (b) as a function of z for $x \approx a/2$ et $y = 0$.

Equation 10.24 means that the circulation of an electromotive field on the closed curve described by the internal inductor is nonzero and is equal to the induced emf_m value. These produced electromotive forces will be, as well as the B field, amplified by an n_2 factor since we have supposed n_2 superimposed turns in the inner inductor. Thus, for a time-invariant magnetic field, the accurate expression of the emf_m versus displacement ξ is given when deriving Equation 10.23 by the following [42]:

$$emf_m = n_1 n_2 \frac{2\mu_0}{\pi} \frac{\left(\sqrt{\xi^2 + (a - \varepsilon_a)^2 + \varepsilon_a^2} - \sqrt{\xi^2 + 2\varepsilon_a^2} \right)}{\xi^2 + \varepsilon_a^2} I_1 \xi v = A_1 I_1 \xi v \quad (10.25)$$

The A_1 factor, which is an even function of the displacement ξ, is a nonlinear coefficient. As a result, the induced voltage remains sinusoidal only for small values of ξ.

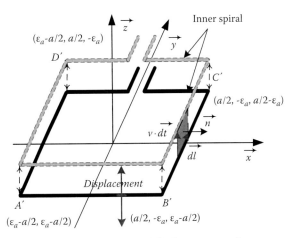

FIGURE 10.23   Illustration of the inner spiral displacement mode.

A first-order Taylor expansion of the emf_m neighbouring zero (in the linear region) can be estimated by the following:

$$emf_m \approx n_1 n_2 \frac{2\mu_0}{\pi} \frac{\left(\sqrt{(a-\varepsilon_a)^2 + \varepsilon_a^2} - \sqrt{2}\varepsilon_a\right)}{\varepsilon_a^2} I_1 \xi v = A_2 I_1 \xi v \qquad (10.26)$$

It is noteworthy that the derived Equation 10.26 is identical to that found through the mutual inductance approach in paragraph 5.2.1. The A_2 is a purely constant geometrical factor representing a close linear approximation of A_1 neighbouring zero. The behaviour of the two factors A_1 and A_2 multiplied by the displacement ξ is plotted in Figure 10.24a. For a relative error between A_1 and A_2 less than k (in%), the vibration amplitude should be restricted to the linear region and bounded by h_{max}, which can be estimated by Equation 10.27. For instance, if we need an error k less than 5%, h_{max} is found to be 25 μm (see Figure 10.24b):

$$h_{max} \leq 2\varepsilon_a \sqrt{\frac{k\left(a - \sqrt{2}\varepsilon_a\right)}{4a - 3\sqrt{2}\varepsilon_a - 2\varepsilon_a^2/a}} \qquad (10.27)$$

For small membrane movements, when substituting ξ with $h \cdot \sin(\omega_p t)$ in Equation 10.26, the open-circuit total output voltage in the linear range will be given by the following equation:

$$emf_m \approx A_2 I_1 \xi v = \frac{1}{2} A_2 I_1 h^2 \omega_p \sin(2\omega_p t) \qquad (10.28)$$

As a conclusion, the emf_m produced is inversely proportional to the square of the separation distance ε_a between B_1 and B_2. Consequently, the two inductors should

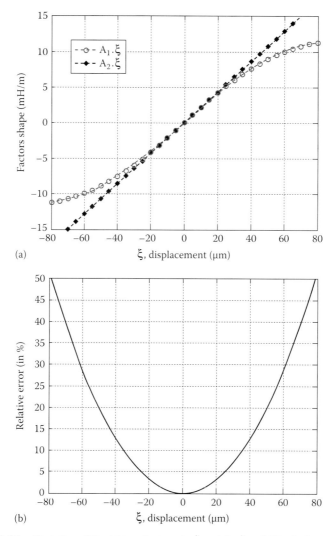

FIGURE 10.24 Variation of the (a) two factors $A_1 \cdot \xi$ and $A_2 \cdot \xi$ and (b) relative error between A_1 and A_2, versus the displacement amplitude.

be placed as close as possible to produce the maximum output voltage. We can also note that the frequency of the induced voltage is doubled compared to that of the incident pressure wave. This is due to the product of speed, v, by the displacement, ξ. This effect is a drawback of the microphone operation and will have a direct influence on the sensor linearity.

For $f_p = 1$ kHz, the curves of the estimated and accurate emf_m are shown for two different vibration amplitudes in Figure 10.25. For an amplitude of 20 μm located in the linear region, we can acquire two superimposed sinusoidal signals in the range of μV. We can reach higher-induced voltage amplitudes if the fluctuation

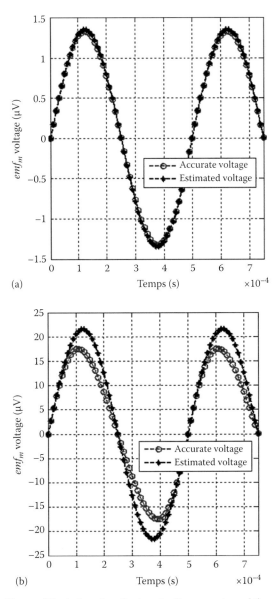

(a)

(b)

FIGURE 10.25 Shape of the induced emf_m given by the accurate and the estimated equation for displacement amplitudes (a) $h = 20$ μm and (b) $h = 80$ μm.

amplitude is larger. Unfortunately, this will be to the detriment of the signal distortion (Figure 10.25). Subsequently, we will need to use a more complicated electronic processing to filter and recover the signal.

Additional biasing of the internal inductor by a *DC* does not add a new component to the induced voltage, since the self-inductance and current are invariable with respect to time.

10.6.3 EXTERNAL INDUCTOR BIASING USING AN AC

An electromotive force, as defined by Faraday's law, is merely a voltage that results from a moving conductor in a constant magnetic field or a stagnant conductor submerged in a time-variable magnetic field or a combination of both. Thus, to increase the induced voltage, we can generate a time-varying variable magnetic field component that will be added to the original motional induction. This can be achieved by an *AC* applied to the external inductance. According to Faraday's law, the phenomenon of electromagnetic induction occurs once there is a magnetic flux time variation through the armature. This new produced voltage is called transformer induction (*emf$_t$*) and is defined by [41]

$$emf_t = \oint_{loop} \vec{E} \cdot \vec{dl} = -\frac{d}{dt} \iint_S \vec{B} \cdot \vec{dS} \qquad (10.29)$$

where

S is the surface based on the contour of the closed internal inductor A′B′C′D′ directed outwards in the z direction

$\vec{E}$ is the electromotive field

Eventually, when the external inductor is driven by an *AC*, the total induced voltage value emf_T is given by the superposition of both contributions defined in Equations 10.29 and 10.24. The superposition of these two voltages is expressed by the following equation:

$$emf_T = \oint_{loop} \left(\vec{v} \wedge \vec{B} + \vec{E} \right) . \vec{dl} = \underbrace{\oint_{loop} (\vec{v} \wedge \vec{B}) . \vec{dl}}_{\text{Motional induction}} - \underbrace{\frac{d}{dt} \iint_S \vec{B} . \vec{dS}}_{\text{Transformer induction}} \qquad (10.30)$$

This equation can be expressed differently by considering concentric planar transformer; thus, we can write that the induced total voltage is described by

$$emf_T = -i_1 \frac{dM}{dt} - i_1 \frac{dM}{dt} = \underbrace{A_2 i_1 \xi v}_{\text{Motional induction}} - \underbrace{(M_0 + \Delta M) \frac{di_1}{dt}}_{\text{Transformer induction}} \qquad (10.31)$$

In the *AC* configuration, the value of the current flowing through the outer inductor will be given by $i_1 = I_0 \cdot \sin(\omega_c \cdot t)$, where ω_c is the angular velocity. Thus, the new value of the emf_m will be the same as found in Equation 10.26, after substituting the *DC* expression of the current by the *AC* one. The result will be an amplitude-modulated signal, given by the following equation:

$$emf_m \approx A_2 i_1 \xi v \tag{10.32}$$

Thus, for small displacement, Equation 10.32 can be simplified as

$$emf_m \approx \frac{A_2}{4} I_0 h^2 \omega_p \left[\sin(\omega_c t + 2\omega_p t) - \sin(\omega_c t - 2\omega_p t) \right] \tag{10.33}$$

The emf_m describes an amplitude-modulated signal; subsequently, the largest value among ω_c and $2\omega_p$ indicates the carrier angular frequency and the other that of the modulating signal [43]. If the modulated signal is the multiplication of two pure sinusoids, the modulation is called suppressed-carrier signal ('double-sideband suppressed carrier' [DSB-SC]). Thus, the frequency spectrum has only two frequency frames, symmetrically spaced above and below the carrier frequency, at $\omega_c - 2\omega_p$ and $\omega_c + 2\omega_p$. However, it's necessary to distinguish the carrier frequency from that of the incident sound wave, which scans the audio band [20 Hz–20 kHz]. The requirement to respect in the choice of the carrier pulse ω_c, coming from the bias current, is that it should be off the sound band in order to be able finally to separate them, which means that ω_c must be ≪20 Hz or ≫20 kHz. The choice to keep is the second ($\omega_c \gg 20$ kHz), since the first case does not present a significant gain in the induced voltage [42].

Regarding the second term in Equation 10.30 that designates the transformer induction, the elementary surface $\overline{dS}$ being oriented towards the normal, only the z-component magnetic field is useful in the assessment, which leads after calculation of the double derivative to the following:

$$emf_t = -\frac{d}{dt} \iint_S B_z . dx . dy = -n_1 n_2 \frac{2\mu_0}{\pi} \frac{d}{dt} \left[i_1 f(\xi) \right] \tag{10.34}$$

where $f(\xi)$ is a function that depends on the geometry of the inductor and the motion ξ also; it is evaluated by this expression:

$$f(\xi) = \left[(a - \varepsilon_a) \left(\operatorname{arctanh} \left(\frac{\sqrt{\xi^2 + \varepsilon_a^2 + (a - \varepsilon_a)^2}}{(a - \varepsilon_a)} \right) - \operatorname{arctanh} \left(\frac{\sqrt{\xi^2 + 2(a - \varepsilon_a)^2}}{(a - \varepsilon_a)} \right) \right) \right.$$

$$+ \varepsilon \left(\operatorname{arctanh} \left(\frac{\sqrt{\xi^2 + (a - \varepsilon_a)^2 + \varepsilon_a^2}}{\varepsilon_a} \right) - \operatorname{arctanh} \left(\frac{\sqrt{\xi^2 + 2\varepsilon_a^2}}{\varepsilon_a} \right) \right)$$

$$\left. + \sqrt{\xi^2 + 2(a - \varepsilon_a)^2} + \sqrt{\xi^2 + 2\varepsilon_a^2} - 2\sqrt{\xi^2 + \varepsilon_a^2 + (a - \varepsilon_a)^2} \right] \tag{10.35}$$

By differentiating with respect to time the double integral, the emf_t will be equal to

$$emf_t = -n_1 n_2 \frac{2\mu_0}{\pi} \left[i_1 v \frac{df}{d\xi}(\xi) + f(\xi) \frac{d}{dt}(i_1) \right] \tag{10.36}$$

where $df(\xi)/d\xi$ is the derivative of the f function with respect to ξ; it is given by the expression

$$\frac{df}{d\xi} = \xi \left[\frac{\sqrt{\xi^2 + 2(a - \varepsilon_a)^2}}{\xi^2 + (a - \varepsilon_a)^2} - \frac{\left(2\xi^2 + (a - \varepsilon_a)^2 + \varepsilon_a^2\right)\sqrt{\xi^2 + \varepsilon_a^2 + (a - \varepsilon_a)^2}}{\left(\xi^2 + \varepsilon_a^2\right)\left(\xi^2 + (a - \varepsilon_a)^2\right)} + \frac{\sqrt{\xi^2 + 2\varepsilon_a^2}}{\xi^2 + \varepsilon_a^2} \right]$$

$$\tag{10.37}$$

For low vibration amplitudes, the emf_t presents an amplitude-modulated signal with an amplitude spectrum composed of three frequency frames located at $\omega_c - 2\omega_p$, ω_c and $\omega_c + 2\omega_p$. Because of the nonzero component of the mutual inductance when the two inductors are coplanar (specified by $f(0) \neq 0$ when $\xi = 0$ in Equation 10.35), the carrier magnitude will be greater than of the modulating signal. Then, the modulation depth is virtually zero ($m \approx 0$) and the recovered voltage will be a purely constant magnitude sinusoidal signal regardless of the incident acoustic wave frequency [41,42]. Therefore, since the carrier provides no useful data on the incident wave and the image of the acoustic incident wave is included only in the two lateral frames, so we can proceed with its elimination or attenuation. The elimination of the carrier can be achieved through balanced transistor structures (structure of Gilbert [44]) or diode structure such as ring modulator. Consequently, the modulation trapezoid pattern is triangular instead of rectangular ($m \approx 1$). To validate the developed results, we use Equation 10.31 to evaluate the transformer induction, which results to

$$emf_t \approx \frac{A_2}{8} h^2 I_0 \omega_c \left[\cos(\omega_c t + 2\omega_p t) + \cos(2\omega_p t - \omega_c t) \right] + \omega_c I_0 \left(M_0 - \frac{A_2}{4} h^2 \right) \cos(\omega_c t)$$

$$\tag{10.38}$$

For the following simulations, we set $f_p = 1$ kHz and $f_c = 40$ kHz. The total induced voltage emf_T with the motional induction, given by Equation 10.32, and the transformer induced, given by Equation 10.36, are plotted in Figure 10.26a. The emf_T presents an amplitude-modulated signal since it is the sum of the two modulated signals, with the sidebands are the sum of the emf_m and the emf_t sideband's (Figure 10.26b). By increasing the vibration amplitude, other n-rank frequency harmonics ($\omega_c \pm 2n\omega_p$) will appear in the emf_t frequency spectrum. For low fluctuation amplitude, the suppressed-carrier accurate emf_T fits perfectly with the results found by Equation 10.31 and then can be simplified as [42]

$$emf_T \approx \frac{A_2}{8} h^2 I_0 \omega_c \left[\sin(\omega_c t - 2\omega_p t) + \sin(\omega_c t + 2\omega_p t) \right] \approx emf_t \tag{10.39}$$

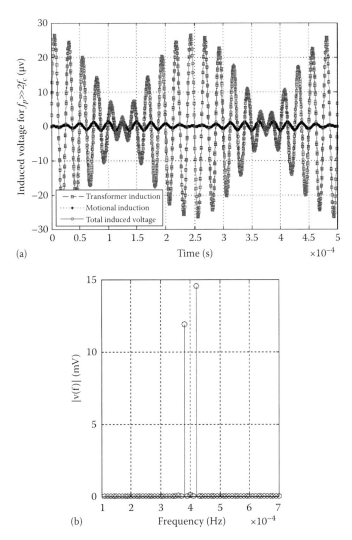

FIGURE 10.26 Suppressed-carrier (a) Induced voltage shapes and (b) total induced voltage single-sided amplitude spectrum given by the accurate equation, for a displacement amplitude $h = 20~\mu m$.

For the choice $\omega_c \gg 2\omega_p$, we can affirm that the total induced voltage emf_T, with suppressed carrier, has an amplitude independent of the incident acoustic wave frequency. Therefore, we find that the emf_T is almost equal to emf_t. This means that the electromagnetic induction is provided, therefore, only by varying the magnetic flux over time. Moreover, there could be a factor between the emf_T and the recovered voltage with a DC bias, emf_L, adjustable and above the unit defined by the coefficient $(\omega_c/2\omega_p)$. Keeping the same geometrical parameters, we can reach an emf_T amplitude, 20 times higher than the emf_L voltage with two angular pulses located at $\omega_c - 2\omega_p$ and $\omega_c + 2\omega_p$ (see Figure 10.26a).

10.7 MODELLING OF THE MICROPHONE WITH VERTICAL OFFSET MEMBRANE

10.7.1 EXTERNAL INDUCTOR BIASING USING A DC

In different polarization modes, the induced voltage recovered through the secondary is proportional either to the product of the displacement by the velocity (in *DC* case) or to the square of the displacement (*AC* case). However, a conventional electrodynamic microphone consists of a coil oscillating in a constant magnetic field produced by a permanent magnet. Thus, the induced voltage across the coil is an electromotive force given by Faraday's law by $B \cdot l \cdot v$. From Figure 10.22b, we note that the radial components of the magnetic field (B_x and B_y) increase linearly up to an optimum value before decreasing. If we consider that the membrane is positioned at this optimum altitude, the radial magnetic field will be almost constant and maximum. Once the radial component of the magnetic field is at its maximum, the induced voltage will also be maximized. On the other hand, having a constant component will induce a voltage depending solely on the displacement speed of the membrane. Several technological solutions to act on the membrane offset can be employed. One solution is to place the microphone structure on a metal electrode and exploit the electrostatic force, or we can take advantage of the inherent residual stresses after the structure liberation along with the auto-generated Lorentz force in the inner inductor. For this consideration, with a *DC* bias, when the membrane is driven by a v velocity, Equation 10.24 becomes

$$emf_{m\_off} = 4 \int_{-\frac{a}{2}+\varepsilon_z}^{\frac{a}{2}-\varepsilon_z} B_{x,max} v \, dy = 4(a - 2\varepsilon_a)vB_{x,max} \qquad (10.40)$$

Since the radial magnetic field is almost constant throughout the conductor *AB* (see Figure 10.22a), the altitude d_0 in which the vertical field amplitude is at its maximum can be given by

$$d_0 = \pm\frac{1}{4}\sqrt{\sqrt{(12\varepsilon_a^2 + a^2)^2 + (4\varepsilon_a a)^2} - (4\varepsilon_a^2 + a^2)} \approx \pm\varepsilon_a \qquad (10.41)$$

Subsequently, the maximum radial magnetic field $B_{x,max}$ calculated at this amplitude will be expressed by the following equation:

$$B_{x,max} = n_1 n_2 \frac{\mu_0}{4\pi}\left(\frac{a}{\varepsilon_a\sqrt{8\varepsilon_a^2 + a^2}}\right)I_0 = A_3 I_0 \qquad (10.42)$$

Therefore, in the case of vertical offset, the final induced voltage will have a magnitude that depends only on the geometry of the microphone, the bias current and the membrane vibration velocity, given by

$$emf_{m\_off} = n_1 n_2 \frac{\mu_0}{\pi}\left(\frac{a(a - 2\varepsilon_a)}{\varepsilon_a\sqrt{8\varepsilon_a^2 + a^2}}\right)I_0 v = 4(a - 2\varepsilon_a)B_{x,max}v \qquad (10.43)$$

We notice that the approximated emf_{m_off} in Equation 10.43 preserves the same form as that of the conventional microphone given by $B \cdot l \cdot v$. To confirm this result, we plot the exact voltage shape given by Equation 10.25 but substituting ξ by $h \cdot \sin(\omega_p t) + d_0$. For the same geometry and spectral parameter simulation used previously, a representation of the exact and estimated induced voltage equations, for two different vibration amplitudes, is given in Figure 10.27. For pure sinusoidal vibration amplitude of 20 μm, both accurate and estimated induced voltages coincide largely without doubling the frequency (Figure 10.28a). The amplitude of the acquired voltage is in the tens of

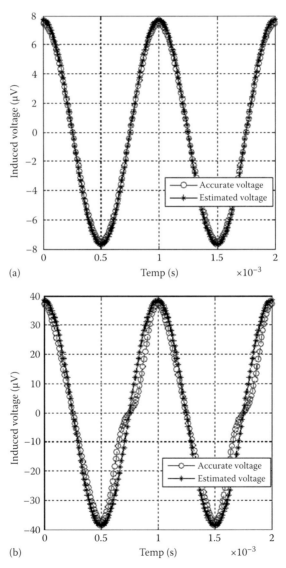

FIGURE 10.27 Shape of the induced voltage given by the accurate and estimated equation for displacement amplitude equal to (a) $h = 20$ μm and (b) $h = 100$ μm.

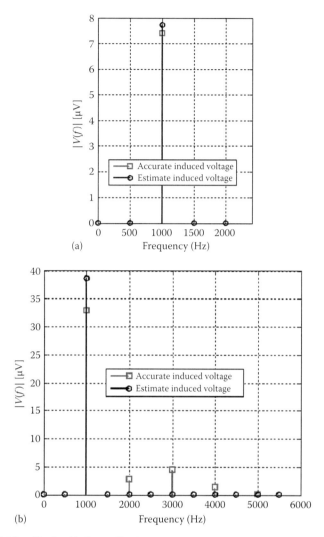

FIGURE 10.28 Single-sided amplitude spectrum of the induced voltage given by the estimated and the accurate equation for displacement amplitude equal to (a) $h = 20$ μm and (b) $h = 100$ μm.

microvolts, that is, five times larger than the case without offset, within the used dimensions. We can reach higher-induced voltage amplitudes if the membrane performs higher oscillations but at the detriment of increased signal distortion (Figure 10.28b).

10.7.2 EXTERNAL INDUCTOR BIASING USING AN AC

The emf_i equation includes a derived term of the f function by the current product (Equation 10.34). This derivative diverges from the accurate value when assuming that B_x is constant. Subsequently, we must use the original equation (10.25) by replacing ξ by $h \cdot \sin(\omega_p t) + d_0$. Similarly as in the first case, we must also eliminate the carrier for the same reason mentioned earlier. The induced voltage variation for two different fluctuation amplitudes is shown in Figure 10.29 and Figure 10.30. For a displacement amplitude $h = 20$ μm, the amplitude of the voltage gained is in the range of hundreds of μV, which gives a coefficient between the amplitudes with and without offset located in the vicinity of 10, in the case of used dimensions. By increasing the vibration amplitude, other n-rank frequency harmonics $(\omega_c \pm 2n \times \omega_p)$ will appear in the emf_t frequency spectrum (see Figure 10.29b and Figure 10.30b).

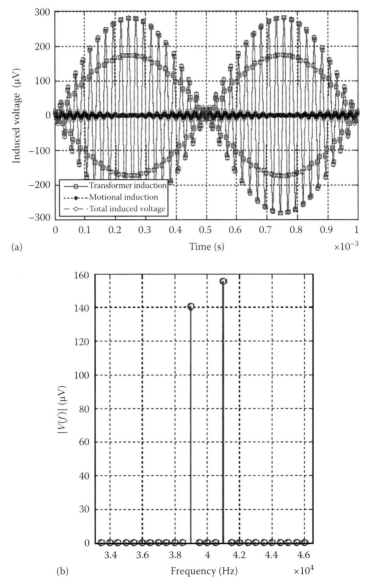

(a)

(b)

FIGURE 10.29 Total induced voltage with suppressed-carrier (a) shape and (b) amplitude spectrum, for a displacement amplitude $h = 20$ μm.

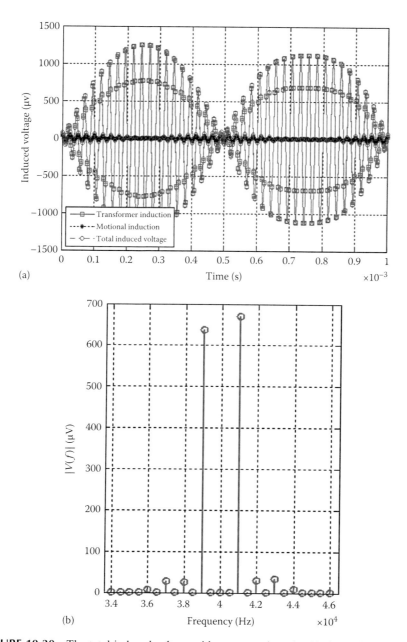

FIGURE 10.30 The total induced voltage with suppressed carrier (a) shape (b) amplitude spectrum, for a displacement amplitude $h = 100\ \mu m$.

10.8   CONCLUSION

In this chapter, we presented a detailed theoretical magnetic modelling of a new monolithic electrodynamic CMOS-MEMS microphone structure. Several approaches have been detailed, leading to determine the induced voltage expression in different biasing modes. This theory allowed us to estimate the theoretical expected output voltages range obtained in the MEMS electrodynamic microphone output. It is found that this induced voltage output extends from a few μV to hundreds of μV. In fact, the use of an *AC* produces an induced voltage greater and independent of the incident wave frequency as was the case for a *DC* bias. The expected increase of the induced voltage between the *AC* and the *DC* can be estimated in the order of $\omega_c/2\omega_p$. The study of the magnetic field according to an offset membrane displacement has shown that the induced voltage equation is similar to a conventional microphone under a *DC* biasing. In addition, there is a nonlinearity relationship between the output voltage and the maximum displacement of the membrane. For small amplitude fluctuation, the problem of nonlinearity can be solved electronically to the detriment of complicating the proximity electronics.

REFERENCES

1. G.T.A. Kovacs, N.I. Maluf, and K.E. Petersen, Bulk micromachining of silicon, *Proceedings of the IEEE* 86, 1998, 1536–1551.
2. J.M. Bustillo, R.T. Howe, and R.S. Muller, Surface micromachining for microelectromechanical systems, *Proceedings of the IEEE*, 86, 1998, 1552–1574.
3. P.R. Scheeper, B. Nordstrand, J.O. Gullov, B. Liu, T. Clausen, L. Midjord, and T. Storgaard-Larsen, A new measurement based on MEMS technology, *Journal of Microelectromechanical Systems*, 12, 2003, 880–891.
4. J. Bay, O. Hansen, and S. Bouwstra, Design of a silicon microphone with differential read-out of a sealed double parallel-plate capacitor, *Sensors and Actuators A*, 53, 1996, 232–236.
5. D.P. Arnold, T. Nishida, L.N. Cattafesta, and M. Sheplak, A directional acoustic array using silicon micromachined piezoresistive microphones, *Journal of the Acoustical Society of America*, 113(1), 2003, 289–298.
6. G. Li, Y. Zohar, and M. Wong, Piezoresistive microphone with integrated amplifier realized using metal-induced laterally crystallized polycrystalline silicon, in *17th IEEE International Conference on MicroElectro Mechanical Systems(MEMS'04)*, Maastricht, the Netherlands, 2004, pp. 548–551.
7. R. Schellin and G. Hess, A silicon subminiature microphone based on piezoresistive polysilicon strain gauges, *Sensors and Actuators A*, 32, 1992, 555–559.
8. M. Sheplak, J.M. Seiner, K.S. Breuer, and M.A. Schmidt, A MEMS microphone for aeroacoustics measurements, in *Proceedings of 37th AIAA Aerospace Sciences Meetings and Exhibit*, paper no. AIAA99-0606, Reno, NV, 1999.
9. E.C. Wente, A condenser transmitter as a uniformly sensitive instrument for the absolute measurement of sound intensity, *Physical Review, the American Physical Society*, Series II 10, 1917, 39–63.
10. S.S. Lee, R.P. Ried, and R.M. White, Piezoelectric cantilever microphone and microspeaker, *Journal of Microelectromechanical Systems*, 5(4), 1996, 238–242.
11. M. Royer, J.O. Holmen, M.A. Wurm, and O.S. Aadland, ZnO on Si integrated acoustic sensor, *Physical Sensors and Actuators A*, 4, 1983, 357–362.

12. E.S. Kim, J.R. Kim, and R.S. Muller, Improved IC-compatible piezoelectric microphone and CMOS process, in *IEEE International Conference Solid State Sensors Actuators*, San Francisco, CA, 1991, pp. 270–273.

13. S.B. Horowitz, T. Nishida, L.N. Cattafesta, and M. Sheplak, Design and characterization of a micromachined piezoelectric microphone, *11th AIAA/CEAS Aeroacoustics Conference (26th AIAA Aeroacoustics Conference)*, Monterey, CA, May 23–25, 2005, pp. 1–10.

14. J. Franz, Piezoelektrische Sensoren auf Siliziumbasis, PhD thesis, Darmstadt, Germany, 1988.

15. J. Franz, Aufbau Funktionsweise und technische Realisierung eines piezoelektrischen Siliciumsensors für akustische Grössen, *VDI-Berichte*, 667, 1988, 299–302.

16. R. Schellin, G. Hess, W. Kuehnel, G.M. Sessler, and E. Fukada, Silicon subminiature microphones with organic piezoelectric layers: Fabrication and acoustical behaviour, in *7th IEEE International Symposium on Electrets*, Berlin, Germany, 1991, pp. 929–934.

17. T.L. Ren, L.T. Zhang, J.S. Liu, L.T. Liu, and Z.J. Li, A novel ferroelectric based microphone, *Microelectronic Engineering*, 66, 2003, 683–687.

18. K. Suzuki, K. Higuchi, and H. Tanigawa, A silicon electrostatic ultrasonic transducer, *IEEE Transactions on Ultrasonics Ferroelectrics and Frequency Control*, 36(6), 1989, 620–627.

19. S.D. Senturia, *Microsystems Design*, Kluwer Academic Publishers, Norwell, MA, 2001.

20. R. Kessmann, M. Klaiber, and G. Hess, Silicon condenser microphones with corrugated silicon oxide/nitride electret membranes, *Physical Sensors and Actuators A*, 100(2/3), September 2002, 301–309.

21. S. Bouwstra, T. Storgaard-Larsen, P.R. Scheeper, J.O. Gullov, J. Bay, M. Mullenborg, and P. Rombach, Silicon microphones – A danish perspective, *Journal of Micromechanics and Microengineering*, 8, 1998, 64–68.

22. P.R. Scheeper, A.G.H. Van der Donk, W. Olthuis, and P. Bergved, A review of silicon microphones, *Sensors and Actuators A*, 44, 1994, 1–11.

23. N. Bilaniuk, Optical microphone transduction techniques, *Applied Acoustics*, 50, 1997, 35–63.

24. N.A. Hall, B. Bicen, M.K. Jeelani, W. Lee, S. Qureshi, F.L. Degertekin, and M. Okandan, Micromachined microphones with diffraction-based optical displacement detection, *Journal of the Acoustical Society of America*, 118(5), 2005, 3000–3009.

25. K. Kadirvel, Design and characterization of a MEMS optical microphone for aeroacoustic measurement, Master thesis, Gainesville, FL: University of Florida, 2002.

26. J. Eargle, *The Microphone Book*, 2nd edn., Focal Press, Burlington, MA, June 2004.

27. F. Tounsi, L. Rufer, B. Mezghani, M. Masmoudi, and S. Mir, Highly flexible membrane systems for micromachined microphones–modeling and simulation, in *3rd IEEE International Conference on Signals, Circuits and Systems (SCS'09)*, Djerba, Tunisia, November 6–8, 2009.

28. S.S. Mohan, C.P. Yue, M.del.M. Hershenson, S.S. Wong and T.H. Lee, Modeling and characterization of on-chip transformers, in *IEEE International Electron Devices Meeting (IEDM)*, San Francisco, CA, December 6–9, 1998.

29. S.S. Mohan, The design, modeling and optimization of on-chip inductor and transformer circuits, Ph.D Dissertation, Stanford University, Stanford, CA, 1999.

30. J.R. Long, Monolithic transformers for silicon RF IC design, *IEEE Journal of Solid-State Circuits*, 35, 1368–1382, September 2000.

31. H. Gan, On-chip transformer modeling, characterization, and applications in power and low noise amplifiers, Stanford University, Stanford, CA, March 2006.

32. H.-A. Wheeler, Simple inductance formulas for radio coils, *Proceedings of IRE*, 16(10), 1398–1400, 1928.

33. B. Estibals, Conception, Réalisation et Caractérisation de micro-miroirs à déflexion localisée appliqués aux télécommunications optiques, Ph.D dissertation, Institut National Polytechnique de Toulouse, Toulouse, France, 2002.

34. F. Tounsi, B. Mezghani, L. Rufer, M. Masmoudi, and S. Mir, Electromagnetic investigation of a CMOS MEMS inductive microphone, *Sensors and Transducers Journal* (ISSN 1726–5479), 108(9), September 2009, 40–53.

35. F.W. Grover, *Inductance Calculations*, Dover, New York, 1946.

36. S. Jenei, B.K.J.C. Nauwelaers, and S. Decoutere, Physics-based closed-form inductance expression for compact modeling of integrated spiral inductors, *IEEE Journal of Solid-State Circuits*, 37(1), January 2002, 77–80.

37. H.M. Greenhouse, Design of planar rectangular microelectronic inductors, *IEEE Transactions on Parts, Hybrids, Packaging*, PHP-10, June 1974, 101–109.

38. P.L. Alan Ling, On-chip planar spiral inductor induced substrate effects on radio frequency integrated circuits in CMOS technology, The Hong Kong University of Science and Technology, Hong Kong, China, January 1998, PhD work.

39. J. Aguilera and R. Berenguer, *Design and Test of Integrated Inductors for RF Applications*, Kluwer Academic Publishers, Dordrecht, the Netherlands, 2003.

40. F. Tounsi, L. Rufer, B. Mezghani, M. Masmoudi, and S. Mir, Electromagnetic modeling of an integrated micromachined inductive microphone, in *4th IEEE International Conference on Design and Test of Integrated Systems in Nanoscale Technology (IEEE DTIS 2009)*, Cairo, Egypt, April 6–10, 2009.

41. W.H. Hayt and J.A. Buck, *Engineering Electromagnetics*, 6th edn., *Mcgraw Hill*, Chapter 10, pp. 320–347, 2001.

42. Farès Tounsi, Microphone électrodynamique MEMS en technologie CMOS: étude, modélisation et réalisation, Editions universitaires europeennes, July 2011.

43. T.L. Chow, *Introduction to Electromagnetic Theory: A Modern Perspective*, Chapter 4, vol. 132, Jones & Bartlett Publishers, Boston, MA, pp. 126–170, 2006.

44. D. Dubac, Contribution a la conception de convertisseurs de fréquence intégration en technologie arséniure de gallium et silicium germanium, PhD dissertation, Université Paul Sabatier de Toulouse, Toulouse, France, December 2001.

11 Thin-Film Sensors

Mutsumi Kimura

CONTENTS

11.1 INTRODUCTION

Thin-film transistors (TFTs) [1,2] have been widely utilized for flat-panel displays (FPDs) [3], such as liquid-crystal displays (LCDs) [4], organic light-emitting diode displays (OLEDs) [5] and electronic papers (EPs) [6]. They have been recently applied to driver circuits, system-on-panel units [7] and general electronics [8], such as information processors [9], integrated beside the FPDs. However, the essential features of the thin-film technology are that functional devices, such as active-matrix circuits, amplifying circuits and general circuits, can be fabricated on large and flexible substrates at low temperature and low cost. These features can be employed not only in the FPDs but also sensor applications because the active-matrix circuits are required for some kinds of area sensors, amplifying circuits are necessary to amplify weak signals from sensor devices, large areas are sometimes needed to improve the sensitivities, and flexible substrates are convenient when they are put anywhere on demand. Therefore, we have studied potential possibilities of the thin-film sensors based on TFT technologies [10].

In this article, as examples of the thin-film sensors, we introduce a photosensor and temperature sensor. First, we propose a p/i/n thin-film phototransistor (TFPT) [11], compare it with other photodevices [12], characterize it from the viewpoint of the device behaviour [13] and conclude that the p/i/n TFPT is an excellent thin-film photodevice. Moreover, we develop an artificial retina with the p/i/n TFPTs as

247

photodevices and TFTs as amplifying circuits [14]. The artificial retina can be driven using wireless power supply and can be implanted into eyeballs of the blind people of retina pigmentary degeneration and age-related macular degeneration to recover their eyesight [15]. Next, we also propose a temperature sensor [16]. It is known that the TFTs have temperature dependence of the current–voltage characteristic. We employ the temperature dependence to realize the temperature sensor and invent a sensing circuit and sensing scheme. It is confirmed that the temperature sensitivity of this temperature sensor is less than 1°C. We think that it is promising to integrate this temperature sensor in some applications using TFTs. An example is to integrate it in FPDs such as LCDs and OLEDs in order to optimize driving conditions and compensate temperature dependences of display characteristics of liquid crystals or OLEDs.

11.2 p/i/n THIN-FILM PHOTOTRANSISTOR

11.2.1 Fabrication Processes and Device Structures

Figure 11.1 shows the device structure of the p/i/n TFPT. The p/i/n TFPT is fabricated on a glass substrate using the same fabrication processes as low-temperature poly-Si (LTPS) TFTs [17–19]. First, an amorphous-Si film is deposited using low-pressure chemical-vapour deposition (LPCVD) of Si_2H_6 and subsequently crystallized using XeCl excimer laser to form a 50 nm thick poly-Si film. Next, a SiO_2 film is deposited using plasma-enhanced chemical-vapour deposition (PECVD) of tetraethylorthosilicate (TEOS) to form a 75 nm thick control-insulator film. A metal film is deposited and patterned to form a control electrode. Afterwards, phosphorous ions are implanted through a photoresist mask at 55 keV with a dose of 2×10^{15} cm^{-2} to form an n-type anode region, and boron ions are implanted through a photoresist mask at 25 keV with a dose of 1.5×10^{15} cm^{-2} to form a p-type cathode region. Finally, a water-vapour heat treatment is performed at 400°C for 1 h to thermally activate the dopant ions and simultaneously improve the poly-Si film, the control-insulator film and their interfaces.

11.2.2 Comparison of Thin-Film Photodevices

Thin-film photodevices generate the detected current (I_{detect}) caused by the generation mechanism of electron–hole pairs at semiconductor regions where the

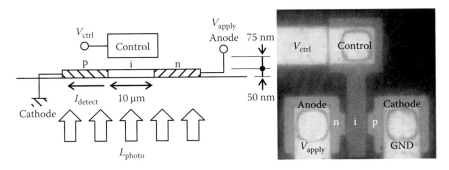

FIGURE 11.1   Device structure of the p/i/n TFPT.

photo-illuminance (L_{photo}) is irradiated, the depletion layer is formed and the electric field is applied. Since thin-film photodevices are mostly used to detect the L_{photo} and the applied voltage (V_{apply}) often unintentionally changes in the circuit configuration, it is preferable that the I_{detect} is dependent only on the L_{photo} but independent of the V_{apply}.

First, the characteristic of the p/i/n thin-film photodiode (TFPD) is shown in Figure 11.2a. It is found that the I_{detect} is proportional to the L_{photo} and relatively high but dependent on the V_{apply}. This is because the depletion layer is formed in the whole intrinsic region, which means that the generation volume of electron–hole pairs is large, and the I_{detect} is relatively high. However, the electric field increases as the V_{apply} increases, which means that the generation rate increases, and the I_{detect} is dependent on the V_{apply}.

Next, the characteristic of the p/n TFPD is shown in Figure 11.2b. It is found that the I_{detect} is not high but independent of the V_{apply}. This is because the depletion layer is formed only near the p/n junction due to a built-in potential, which means that the generation volume of electron–hole pairs is small, and the I_{detect} is not high. However, the electric field is always high, which means the generation rate is constant, and the I_{detect} is independent of the V_{apply}.

Finally, the characteristic of the p/i/n TFPT is shown in Figure 11.2c, where the control voltage (V_{ctrl}) is optimized. It is found that the I_{detect} is simultaneously relatively high and independent of the V_{apply}. This is because the depletion layer is formed in the whole intrinsic region, which means that the generation volume of electron–hole pairs is large, and the I_{detect} is relatively high. Moreover, the electric field is always high because it is mainly determined by the V_{ctrl}, which means that the generation rate is constant, and the I_{detect} is independent of the V_{apply}. Therefore, the p/i/n TFPT is recommended.

11.2.3 Device Characterization of the p/i/n Thin-Film Phototransistor

Figure 11.3a shows the optoelectronic characteristic of the actual device, namely, the measured dependence of I_{detect} on V_{ctrl} with a constant V_{apply} while varying L_{photo}. First, it is found that the dark current, I_{detect} when $I_{photo} = 0$, is sufficiently small except when V_{ctrl} is large. This is because the p/i junction and i/n junction act as insulators, preventing any significant electric current under the reverse bias. This characteristic is useful to improve the S/N ratio of the p/i/n TFPT for photosensor applications. Next, I_{detect} increases as L_{photo} increases. This characteristic should be useful in quantitatively determining L_{photo}. Finally, I_{detect} is maximized when $V_{ctrl} \cong V_{apply}$. This relation is valid also for other V_{apply} [20]. This reason is discussed as follows by comparing an actual device with device simulation.

The same device structure as the actual device is constructed in the 2D device simulator [21]. The uniform photo-induced carrier generation rate (R_{gen}) is set to 2.6×10^{17} cm^{-3} s^{-1} lx^{-1} as a fitting parameter by fitting the simulated I_{detect} to the experimental one. Shockley–Read–Hall (SRH) recombination [22,23], phonon-assisted tunnelling (PAT) with the Poole–Frenkel effect [24] and band-to-band tunnelling (BBT) [25] are included as generation–recombination models as these models sufficiently simulate poly-Si thin-film devices [26]. The electron and hole mobilities are respectively set to 165 and 55 cm^2 V^{-1} s^{-1}, which are determined from

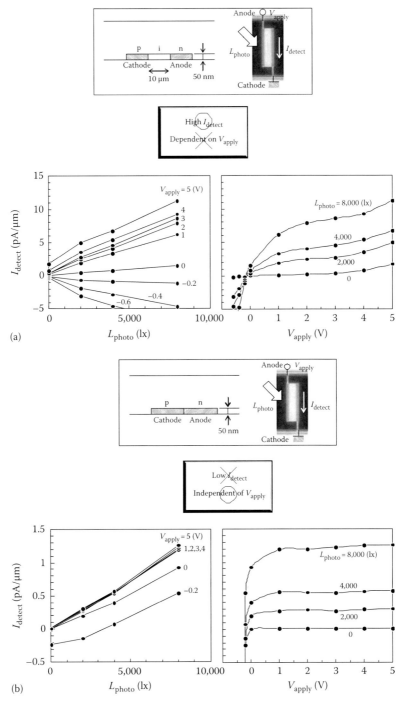

FIGURE 11.2 Comparison of the thin-film photodevices. (a) p/i/n thin-film photodiode. (b) p/n thin-film photodiode.

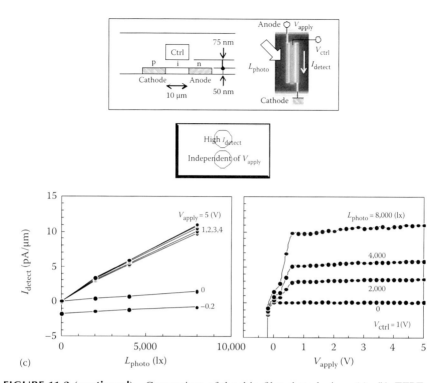

FIGURE 11.2 (continued)   Comparison of the thin-film photodevices. (c) p/i/n TFPT.

a transistor characteristic of a LTPS-TFT fabricated at the same time. I_{detect} is calculated by sweeping V_{ctrl} while maintaining V_{apply} but changing R_{gen}.

Figure 11.3b shows the optoelectronic characteristic from the device simula-
tion, namely, the calculated dependence of I_{detect} on V_{ctrl} with a constant V_{apply} while
varying R_{gen}. The qualitative features of the experimental results are reproduced
in the simulation results. In particular, I_{detect} is maximized when $V_{ctrl} \cong V_{apply}$. This
reason is discussed as follows by considering the detailed behaviour of the depletion layer and the difference between the electron and hole mobilities.

Figure 11.4a shows the electric potentials (ϕ) in the p/i/n TFPT, namely, the calculated distributions of ϕ in both the poly-Si film and the control-insulator film in
a cross section perpendicular to p/i/n TFPT as 3D images. Figure 11.5 shows ϕ at
the top poly-Si surface. ϕ is zero in the cathode electrode, which is grounded as
shown in Figure 11.1. Thus, ϕ in the anode region and ϕ in the cathode region are
not V_{apply} and zero owing to the built-in potentials in the n-type anode and p-type
cathode regions, respectively. Figure 11.4b shows the free electron density (e⁻) and
the free hole density (h⁺) in the poly-Si film, namely, the calculated distributions of
e⁻ and h⁺ in a cross section perpendicular to the poly-Si film. The depletion layers
are defined as regions where both free carrier densities are below 10^{12} cm⁻³. Figure
11.4c shows the carrier recombination rates (R_{rec}) in the poly-Si film, namely, the
calculated distributions of R_{rec} in a cross section perpendicular to the poly-Si film.

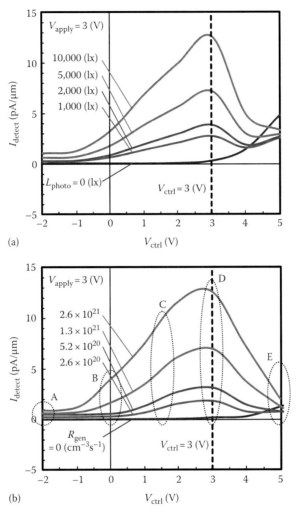

(a)

(b)

FIGURE 11.3 Optoelectronic characteristics of the p/i/n TFPT. (a) Actual device. (b) Device simulation.

Although L_{photo} is assumed to be 10,000 lx in Figures 11.4 and 11.5, the values of ϕ, e^-, h^+ and R_{rec} are qualitatively similar for all L_{photo}. The bias points A, B, C, D and E in Figure 11.3b correspond to the graphs and figures A, B, C, D and E in Figures 11.4 and 11.5, respectively.

I_{detect} depends on the volume of the depletion layer. R_{gen} is uniform in the poly-Si film, whereas R_{rec} is dependent on e^- and h^+. In the depletion layer, the electrons and holes are generated owing to light irradiation, and they do not recombine but are transported and contribute to I_{detect} because both e^- and h^+ are low and the chance of the recombination is low. However, electrons and holes not in the depletion layer immediately recombine owing to the aforementioned generation–recombination models and do not contribute I_{detect} because either e^- or h^+ is high and the chance

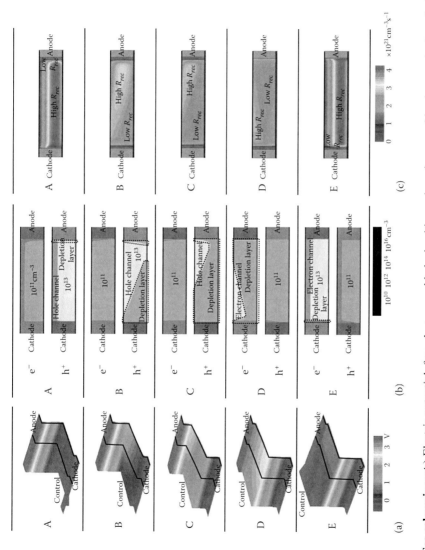

FIGURE 11.4 **(See colour insert.)** Electric potential, free electron and hole densities and carrier recombination rate in the poly-Si film. (a) Electric potential. (b) Free electron and hole densities. (c) Carrier recombination rate.

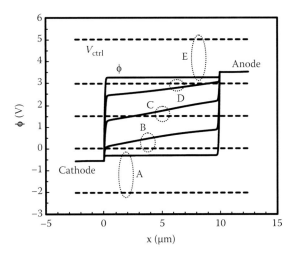

FIGURE 11.5 Electric potential at the top poly-Si surface.

of the recombination is high. This is consistent with the presence of a depletion layer in Figure 11.4b that corresponds to the area where R_{rec} is low in Figure 11.4c. Consequently, the value of I_{detect} depends on the volume of the depletion layer.

A. When $V_{ctrl} < 0$, because $V_{ctrl} < \phi$ throughout the intrinsic region, a hole channel is induced, and a pseudo p/n junction appears near the anode region. Because a depletion layer is narrowly formed, I_{detect} is small.

B. When $V_{ctrl} \cong 0$, similar to the pinch-off phenomena in the saturation region of metal-oxide-semiconductor field-effect transistors (MOSFETs), although the hole channel is induced, because $V_{ctrl} \cong \phi$ near the cathode region, h$^+$ decreases. Besides the depletion layer at the pseudo p/n junction near the anode region, because another depletion layer is widely formed from the cathode region to the back surface, I_{detect} increases.

C. When $0 < V_{ctrl} < V_{apply}$, because $V_{ctrl} > \phi$ on the side of the cathode region, h$^+$ further decreases, but because V_{ctrl} remains below the threshold voltage, an electron channel is not observed. At the same time, because $V_{ctrl} < \phi$ on the side of the anode region, the hole channel is induced. Because the depletion layer is widely formed from the cathode region to the underside of the hole channel, I_{detect} further increases.

D. When $V_{ctrl} \cong V_{apply}$, although an electron channel is induced, because $V_{ctrl} \cong \phi$ near the anode region, e$^-$ is low. Because the depletion layer is widely formed from the anode region to the underside of the electron channel, I_{detect} further increases. Incidentally, I_{detect} for B is smaller than I_{detect} for D. In the case of B, the generated holes cannot be rapidly transported because the hole mobility is less than the electron mobility; hence, the holes accumulate in the poly-Si film. On the other hand, in the case of D, the generated electrons can be rapidly transported because the electron mobility is high; hence, the electrons do not accumulate much in the poly-Si film. As a result, the depletion layer in B is less than that in D. Moreover, the value of I_{detect} depends on the volume of the depletion layer as well as e⁻ and h⁺. Consequently, I_{detect} for B is smaller than I_{detect} for D.

E. When $V_{ctrl} > V_{apply}$, because $V_{ctrl} > \phi$ throughout the intrinsic region, an electron channel is induced, and a pseudo p/n junction appears near the cathode region. Because another depletion layer is narrowly formed, I_{detect} is small.

11.3 ARTIFICIAL RETINA

11.3.1 ARTIFICIAL RETINA WITH THE p/i/n THIN-FILM PHOTOTRANSISTORS

Artificial retinas are promising because they may provide sight for the blind [27] and new applications of information electronics [28]. Until now, bulk Si MOSFETs have been commonly used for artificial retinas. We will propose an artificial retina using the p/i/n TFPTs and LTPS-TFTs, which is expected to be suitable for human bodies because it can potentially be fabricated on a plastic substrate [29–32].

The circuit configurations, planar photographs and measured characteristics of the retina pixel and retina array are shown in Figure 11.6. Although this artificial

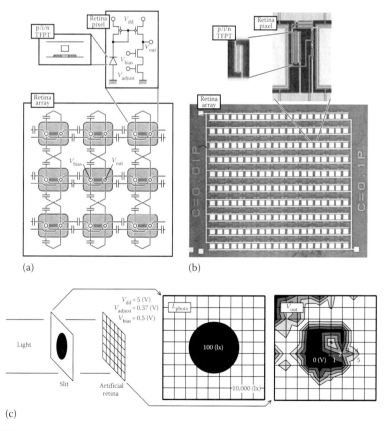

FIGURE 11.6 Artificial retina with the p/i/n TFPTs. (a) Circuit configuration. (b) Planar photograph. (c) Measured characteristic.

retina has its peculiar advanced function, namely, edge enhancement function, it is switched off in this measurement. It is found that this artificial retina can detect the L_{photo} profile and output the output voltage (V_{out}) profile. Although some noises occur owing to the characteristic deviations of the p/i/n TFPTs and LTPS-TFTs, the shape of the V_{out} profile is roughly the same as that of the L_{photo} profile.

An advantage of the artificial retina in comparison with the other photosensor applications is that it can output the V_{out} in DC signal. This advantage is achieved because the I_{photo} can be both relatively large and independent of the V_{apply} using the p/i/n TFPT. This advantage will be useful for the in- and inter-pixel information processing because it is not necessary to sample transient signals with precise timing control and the in- and inter-pixel functional circuits can freely utilize the DC signals.

11.3.2 Artificial Retina Driven Using Wireless Power Supply

The concept model of the artificial retina fabricated on a transparent and flexible substrate and implanted using epiretinal implant is shown in Figure 11.7a. Electronic photodevices and circuits are integrated on the artificial retina, which is implanted on the inside surface of the living retina at the back part of the human eyeballs. Since the irradiated light comes from one side of the artificial retina and the stimulus signal goes out of the other side, the transparent substrate is preferable. Moreover, since the human eyeballs are curved, the flexible substrate is also preferable. It is possible to make spherical shape by designing a petal-like pattern. As a result, the artificial retinas using TFPTs and TFTs are suitable for the epiretinal implant on the curved human eyeballs.

The wireless power supply using inductive coupling is shown in Figure 11.7b. This system includes a power transmitter, power receiver, diode bridge and Zener diodes. The power transmitter consists of an AC voltage source and induction coil. The V_{pp} of the AC voltage source is 10 V, and the frequency is 34 kHz, which is the resonance frequency of this system. The material of the induction coil is an enamelled copper wire, the diameter is 1.8 cm, and the winding number is 370 times. The power receiver also consists of an induction coil, which is the same as the power transmitter and located face to face. The diode bridge rectifies the AC voltage to the DC voltage, and the Zener diodes regulate the voltage value. The diode bridge and Zener diodes are discrete devices and encapsulated in epoxy resin. Although the current system should be downsized and bio-compatibility has to be inspected, the supply system is in principle very simple to implant into the human eyeballs. As a result, the generated power is not so stable, which may be because the artificial retina is fabricated on an insulator substrate, has little parasitic capacitance and is subject to the influence of noise. Therefore, it is necessary to confirm whether the artificial retina can be correctly operated even using the unstable power source.

The detected result of L_{photo} profile versus V_{out} profile is shown in Figure 11.7c. It is found that the L_{photo} profile can be correctly detected as the V_{out} profile even if it is driven using the unstable power source, although shape distortion is slightly observed, which is due to the misalignment of the optical system or characteristic variation of TFPTs and TFTs.

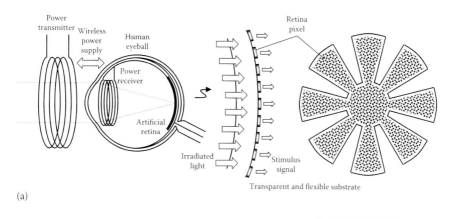

(a)

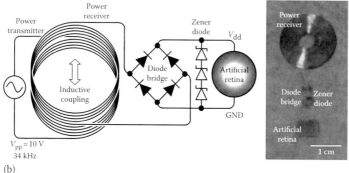

(b)

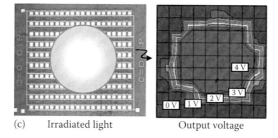

(c)   Irradiated light   Output voltage

FIGURE 11.7 **(See colour insert.)** Artificial retina driven using wireless power supply. (a) Concept model of the artificial retina implanted using epiretinal implant. (b) Wireless power supply using inductive coupling. (c) Measured characteristic.

11.4 TEMPERATURE SENSOR

11.4.1 FABRICATION PROCESSES AND DEVICE STRUCTURES

The fabrication processes for the top-gate, coplanar and n-type poly-Si TFT are as follows [33–38]. First, an amorphous-Si film is deposited using LPCVD of Si_2H_6 and crystallized using XeCl excimer pulse laser to form a poly-Si film, whose thickness (t_{Si}) is 50 nm. Next, a SiO_2 film is deposited using PECVD of TEOS to prepare a gate-insulator film, whose thickness (t_{SiO_2}) is 75 nm. Afterwards, a gate-metal film is deposited and patterned, and phosphorus ions are implanted and thermally activated to form source and drain regions. Subsequently, a SiO_2 film is deposited and patterned to prepare an interlayer-insulator film, and a source and drain-metal film is deposited and patterned. Finally, water-vapour heat treatment is performed to improve the poly-Si film, the SiO_2 film and their interfaces. The gate width (W) and length (L) are 50 and 4.5 μm, respectively. The field-effect mobility (μ) and threshold voltage (V_{th}) are 93 $cm^2 \cdot V^{-1} \cdot s^{-1}$ and 3.6 V, respectively.

11.4.2 Temperature Dependences of Transistor Characteristics

We measure the temperature dependence of the transistor characteristic. The poly-Si TFT is located on a chuck stage of a manual prober in a shield chamber, whose temperature is controlled by heating or cooling the chuck stage and measuring the temperature by a thermometer inserted into the chuck stage. Figure 11.8 shows the temperature dependence of the transistor characteristic of the poly-Si TFT. Although the transfer characteristic for the drain voltage (V_{ds}) = 0.1 V is only shown here, that for higher V_{ds} also has similar features. The temperature dependence of the off current is much larger than that of the on current. Moreover, although it is known that off current of poly-Si TFTs is caused by multiple complicated mechanisms such as SRH generation, PAT with Poole–Frenkel effect and BBT [39], the temperature dependence is monotonic increasing from −30°C to 200°C. Furthermore, while the on current may affect device temperature owing to self-heating effect, the off current does not so. Therefore, we determine that it is suitable to employ the off current for the temperature sensor.

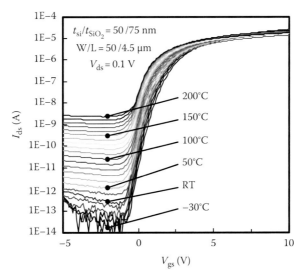

FIGURE 11.8 **(See colour insert.)** Temperature dependence of the transistor characteristic of the poly-Si TFT.

11.4.3 CELL CIRCUIT AND DRIVING METHOD

We suggest a cell circuit consisting of 1-transistor and 1-capacitor and driving method composed of initializing, holding and detecting periods. This sensing scheme can effectively utilize the temperature dependence of the off current. Figure 11.9 shows the cell circuit and driving method of this sensing scheme. Although the cell circuit is similar to the pixel circuit of LCDs, the cell capacitor (C_{cell}) has to have much larger capacitance by considering the actual working written in the next section. Such large capacitance can be located because the cell circuit does not have to be fine unlike LCDs.

First, during an initializing period, scan voltage (V_{scan}) is applied in order to switch on the TFT, and signal voltage (V_{sig}) is applied. Then, cell voltage (V_{cell}) is charged and becomes equal to V_{sig}. Next, during a holding period, V_{scan} is grounded in order to switch off the TFT, and V_{sig} is also grounded. Then, V_{cell} is discharged and decreases towards V_{sig} due to the off current. Since the discharging speed depends on the temperature due to the temperature dependence of the off current, V_{cell} at the end of the holding period (V_{out}) also depends on the temperature. Finally, during a detecting period, V_{scan} is again applied, then V_{out} is detected through V_{sig}. As a result, it is possible to detect the temperature by measuring V_{out}.

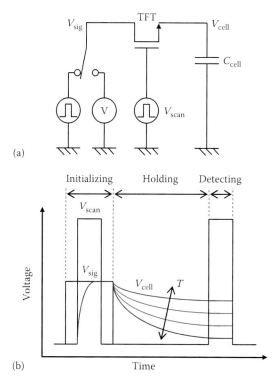

FIGURE 11.9   Cell circuit and driving method. (a) Cell circuit. (b) Driving method.

11.4.4 EXPERIMENTAL RESULTS

We find that it is possible to detect the temperature using this sensing scheme, for example, from the room temperature to 70°C. Figure 11.10a shows the voltage waveform of this sensing scheme. Here, the TFT is the same as that written in the previous

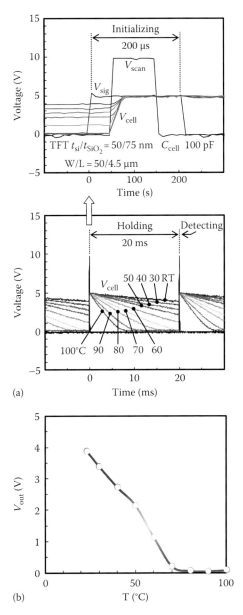

FIGURE 11.10 **(See colour insert.)** Experimental results. (a) Voltage waveform. (b) Relationship between the temperature and V_{out}.

section, and C_{cell} is 100 pF. The initializing and holding periods are 200 μs and 20 ms, respectively. Pulse voltages of 10 and 5 V are applied as V_{scan} and V_{sig}, respectively, and V_{cell} is measured using a high-impedance FET probe. The temperature is varied from the room temperature to 100°C by considering temperature range for regular working of the FET probe although the temperature controller is the same as written in the previous section. It is confirmed that the voltage waveform does not change even after the actual operation during several days, which suggests that the characteristic degradation of the poly-Si TFT does not occur even at the high temperature. However, further AC and DC stress tests are required to check long-time reliability, although the characteristic degradation of the poly-Si TFT hardly occurs at the room temperature [39,40]. Figure 11.10b shows the relationship between the temperature and V_{out}. It is possible to detect the temperature from the room temperature to 70°C by measuring V_{out}.

Although this relationship might be uneven owing to the non-uniformity of the transistor characteristic of the poly-Si TFT, the temperature can accurately be detected by preparing look-up tables between the temperature and V_{out} for each cell circuit. Although the look-up tables must be calibrated, only several points such as every 10°C are necessary because the relationship between the temperature and V_{out} is nearly straight as shown in Figure 11.10b. Even if multiple cell circuits are integrated, they can be simultaneously calibrated by putting it all in temperature chambers. Since some automatic calibration circuits can be also integrated, such additional steps are not unpractical. Although the need for the calibration should be discussed after the evaluation of the non-uniformity of the transistor characteristic, since the look-up table must be certainly prepared owing to the non-linearity of the temperature dependence, we think that the additional cost for the calibration can be within a permitted extent.

Consequently, the detection accuracy will be less than 1°C because the relationship is nearly straight as shown in Figure 11.10b. The sensitivity of V_{out} on the temperature is roughly 80 mV/°C, which is sufficiently large to detect by common readout circuits including those using TFTs. Moreover, it is expected to extend the temperature range by controlling the holding period, which we will report in the near future.

11.5   CONCLUSION

We have studied potential possibilities of the thin-film sensors based on TFT technologies. In this chapter, as examples of the thin-film sensors, we introduced a photosensor and temperature sensor. First, we proposed a p/i/n TFPT, compared it with other photodevices, characterized it from the viewpoint of the device behaviour and concluded that the p/i/n TFPT is an excellent thin-film photodevice. Moreover, we developed an artificial retina with the p/i/n TFPTs as photodevices and TFTs as amplifying circuits. The artificial retina can be driven using wireless power supply and can be implanted into eyeballs of the blind people of retina pigmentary degeneration and age-related macular degeneration to recover their eyesight. Next, we also proposed a temperature sensor. It is known that the TFTs have temperature dependence of the current–voltage characteristic. We employed the temperature dependence to realize the temperature sensor and invented a sensing circuit and sensing scheme. It was confirmed that the temperature sensitivity of this temperature sensor is less than 1°C. We thought that it is promising to integrate this temperature sensor in some applications using TFTs. An example is to integrate it in FPDs such as LCDs and OLEDs in order to optimize driving conditions and compensate temperature dependences of display characteristics of liquid crystals or OLEDs.

REFERENCES

1. Y. Kuo, *Thin Film Transistors, Materials and Processes, 1: Amorphous Silicon Thin Film Transistors*, Kluwer Academic Publishers, Boston, MA (2004).
2. Y. Kuo, *Thin Film Transistors, Materials and Processes, 2: Polycrystalline Silicon Thin Film Transistors*, Kluwer Academic Publishers, Boston, MA (2004).
3. W. den Boer, *Active Matrix Liquid Crystal Displays: Fundamentals and Applications*, Newnes, Oxford, U.K. (2005).
4. S. Morozumi, K. Oguchi, S. Yazawa, T. Kodaira, H. Ohshima, and T. Mano, *SID*, 83, 156 (1983).
5. M. Kimura, I. Yudasaka, S. Kanbe, H. Kobayashi, H. Kiguchi, S. Seki, S. Miyashita, T. Shimoda, T. Ozawa, K. Kitawada, T. Nakazawa, W. Miyazawa, and H. Ohshima, *IEEE Trans. Electron Devices*, 46, 2282 (1999).
6. S. Inoue, H. Kawai, S. Kanbe, T. Saeki, and T. Shimoda, *IEEE Trans. Electron Devices*, 49, 1532 (2002).
7. Y. Aoki and H. Kimura, *IDW'09*, 239 (2009).
8. Y. Kuo, *Jpn. J. Appl. Phys.*, 47, 1845 (2008).
9. N. Karaki, T. Nanmoto, H. Ebihara, S. Utsunomiya, S. Inoue, and T. Shimoda, *ISSCC '05*, 272 (2005).
10. M. Kimura, T. Hachida, Y. Nishizaki, T. Yamashita, T. Shima, T. Ogura, Y. Miura, H. Hashimoto, M. Hirako, T. Yamaoka, S. Tani, Y. Yamaguchi, Y. Sagawa, K. Setsu, Y. Imuro, and K. Bundo, *IMID '09*, 957 (2009).
11. T. Yamashita, T. Shima, Y. Nishizaki, M. Kimura, H. Hara, and S. Inoue, *Jpn. J. Appl. Phys.*, 47, 1924 (2008).
12. M. Kimura, T. Shima, T. Yamashita, Y. Nishizaki, and H. Hara, *IMID '07*, 7, 1745 (2007).
13. M. Kimura and Y. Miura, *IEEE Trans. Electron Devices*, 58, 3472 (2011).
14. T. Yamashita, T. Shima, Y. Nishizaki, M. Kimura, H. Hara, and S. Inoue, *Jpn. J. Appl. Phys.*, 47, 1924 (2008).
15. Y. Miura, T. Hachida, and M. Kimura, *IEEE Sensors J.*, 11, 1564 (2011).
16. A. Nakashima, Y. Sagawa, and M. Kimura, *IEEE Sensors J.*, 11, 995 (2011).
17. S. Inoue, M. Matsuo, T. Hashizume, H. Ishiguro, T. Nakazawa, and H. Ohshima, *IEDM '91*, 555 (1991).
18. T. Sameshima, S. Usui, and M. Sekiya, *IEEE Electron Device Lett.*, 7, 276 (1986).
19. N. Sano, M. Sekiya, M. Hara, A. Kohno, and T. Sameshima, *IEEE Electron Device Lett.*, 16, 157 (1995).
20. M. Kimura, Y. Miura, T. Ogura, S. Ohno, T. Hachida, Y. Nishizaki, T. Yamashita, and T. Shima, *IEEE Electron Device Lett.*, 31, 984 (2010).
21. Silvaco International, Atlas, Device Simulator, http://www.silvaco.com/products/device_simulation/atlas.html.
22. W. Shockley and W. T. Read, Jr., *Phys. Rev.*, 87, 835 (1952).
23. R. N. Hall, *Phys. Rev.*, 87, 387 (1952).

24. O. K. B. Lui and P. Migliorato, *Solid-State Electron.*, 41, 575 (1997).
25. G. A. M. Hurkx, D. B. M. Klaassen, and M. P. G. Knuvers, *IEEE Trans. Electron Devices*, 39, 331 (1992).
26. M. Kimura, A. Nakashima, and Y. Sagawa, *Electrochem. Solid-State Lett.*, 13, H409 (2010).
27. T. Yagi, *Oyo Buturi*, 73, 1095 (2004) [in Japanese].
28. Y. Nakagawa, K.-W. Lee, T. Nakamura, Y. Yamada, K.-T. Park, H. Kurino, and M. Koyanagi, *ICONIP '00*, 636 (2000).
29. T. Noguchi, J. Y. Kwon, J. S. Jung, J. M. Kim, K. B. Park, H. Lim, D. Y. Kim, H. S. Cho, H. X. Yin, and W. Xianyu, *Jpn. J. Appl. Phys.*, 45, 4321 (2007).
30. J. Y. Kwon, H. Lim, K. B. Park, J. S. Jung, D. Y. Kim, H. S. Cho, S. P. Kim, Y. S. Park, J. M. Kim, and T. Noguchi, *Jpn. J. Appl. Phys.*, 45, 4362 (2007).
31. M. He, R. Ishihara, E. J. J. Neihof, Y. Andel, H. Schellevis, W. Metselaar, and Kees Beenakker, *Jpn. J. Appl. Phys.*, 46, 1245 (2007).
32. H. Ueno, Y. Sugawara, H. Yano, T. Hatayama, Y. Uraoka, T. Fuyuki, J. S. Jung, K. B. Park, J. M. Kim, J. Y. Kwon, and T. Noguchi, *Jpn. J. Appl. Phys.*, 46, 1303 (2007).
33. H. Ohshima and S. Morozumi, *IEDM '89*, 157 (1989).
34. S. Inoue, M. Matsuo, T. Hashizume, H. Ishiguro, T. Nakazawa, and H. Ohshima, *IEDM '91*, 555 (1991).
35. T. Sameshima, S. Usui, and M. Sekiya, *IEEE Electron Device Lett.*, 7, 276 (1986).
36. H. Watakabe and T. Sameshima, *IEEE Trans. Electron Devices*, 49, 2217 (2002).
37. H. Watakabe, Y. Tsunoda, N. Andoh, and T. Sameshima, *J. Non-Cryst. Solids*, 299/302 B, 1321 (2002).
38. N. Sano, M. Sekiya, M. Hara, A. Kohno, and T. Sameshima, *IEEE Electron Device Lett.*, 16, 157 (1995).
39. M. Kimura and T. Tsujino, *IDW '07*, 1841 (2007).
39. N. Morosawa, T. Nakayama, T. Arai, Y. Inagaki, K. Tatsuki, and T. Urabe, *IDW '07*, 71 (2007).
40. T. Kasakawa, H. Tabata, R. Onodera, H. Kojima, M. Kimura, H. Hara, and S. Inoue, *IDW '09*, 1829 (2009).

12 Technologies for Electric Current Sensors

*G. Velasco-Quesada, A. Conesa-Roca
and M. Román-Lumbreras*

CONTENTS

12.1 INTRODUCTION

Within the field of study of classical physics, we can find the branch concerned with electromagnetic phenomena: electromagnetism. Electromagnetism describes the interaction between electric charges through the concepts of electric and magnetic fields. The formulation of the electromagnetic theory principles has enabled the engineering sciences, such as electricity and electronics, to develop applications related to the treatment of energy and information using electric charges as physical support.

Most operations performed on the processing of energy or information supported by electric charges involve the movement of these charges, and there are two basic

265

electric magnitudes to represent this movement: electric voltage (also called potential difference or voltage) and electric current.

12.1.1 DEFINITION OF VOLTAGE AND ELECTRIC CURRENT

The electric voltage measures the energy involved in the movement of electric charges between two points. Its unit is the volt [V] and represents the energy required (expressed in joules [J]) to move one unit of electric charge (expressed in coulombs [C]) between the two considered points.

The electric current corresponds to the physical phenomenon caused by the motion of electric charges, being the electric current intensity the magnitude used to describe it. The current intensity measures the amount of charge that circulates through an enclosed space section per unit of time. The unit of electric current intensity is the ampere [A] and corresponds to the charge unit (coulomb [C]) divided by the time unit (second [s]).

Regardless of the scope of the developed application, many electric or electronic systems need to measure the value of either of these two magnitudes. This chapter is devoted to reviewing the various techniques currently available to measure or sense the electric current intensity.

12.1.2 DIRECT AND INDIRECT MEASURE OF CURRENT INTENSITY

An instrument that allows the direct measurement of electric current intensity is the galvanometer. Its working principle is based on the deflection of a magnetic needle by current circulating in a wire, and it was first discovered by Hans Oersted around 1820. Magnetic needle or movable coil galvanometers are suitable for the measurement of the average value of pulsating direct currents (DCs), but they are not suitable for alternating current (AC) measurement. However, it is possible to construct galvanometers suitable for this purpose.

The lack of similar devices for measuring voltage has led to that for a long time, these measurements were made as current measurements. This implied the need to convert the voltage into current.

Nowadays, the use of data acquisition systems in electronic instrumentation or control applications, in both analogue and digital versions, means that most electric current measurements are done as voltage measurements. In this regard, electric current transducers provide at its output a voltage proportional to the instantaneous value of the measured current.

Accordingly, this chapter addresses the review of techniques that allow the measurement of the electric current intensity of an electric current flowing through electric conductors and providing at its output a voltage proportional to the instantaneous value of the measured current.

12.2 KEY PARAMETERS OF ELECTRIC CURRENT

From the definition of electric current conducted in the previous section, it follows that the volume of moving charge per time unit and the direction of this movement are two important parameters in the current intensity measurement. This measurement involves the determination of both the magnitude and the direction or sign of the electric current.

These two parameters are not completely independent and allow the classification of electric currents into two major groups: AC and DC. These two kinds of currents can be characterized according to parameters such as current magnitude, bandwidth, frequency and phase.

12.2.1 MAGNITUDE

The classification of electric currents in AC or DC is done according to the evolution of its instantaneous values $i(t)$. From this evolution, the current average value (I_{AVG}) can be determined using Equation 12.1, depending on whether the current waveform analysed is periodic (with period T_P) or not:

$$I_{AVG} = \frac{1}{T_P} \cdot \int_0^{T_P} i(t) \cdot dt \quad I_{AVG} = \frac{1}{T_2 - T_1} \cdot \int_{T_1}^{T_2} i(t) \cdot dt \tag{12.1}$$

An electric current is defined as AC when its average value is zero. This means that during the period under consideration, their instantaneous values have been changing in direction or sign.

To characterize ACs, the root-mean-square (RMS) value is often used (I_{RMS}). This value is also determined from the current instantaneous values and it is calculated using the expression for 'square root of the mean (average) squared' value calculation (2):

$$I_{RMS} = \sqrt{\frac{1}{T_P} \cdot \int_0^{T_P} i(t)^2 \cdot dt} \tag{12.2}$$

An electric current is defined as DC when its average value is non-zero. In this case, two new kinds of currents can be differentiated in accordance with, once again, its instantaneous values evolution.

A DC is also defined as constant if all of its instantaneous values are equal: $i(t) = I_0$. On the other hand, it is defined as pulsating current when its instantaneous values are different between them.

It is possible to find constant DCs in applications such as battery charge or photovoltaic systems management, pulsating DCs are typical in the filter inductors present in static power converters, and ACs appear in systems devoted to generation and distribution of electric energy.

To characterize electric currents, in addition to the average and RMS values, their maximum, minimum and/or peak-to-peak values are also commonly used.

12.2.2 BANDWIDTH, FREQUENCY AND PHASE

Another important parameter to characterize electric currents is the bandwidth. Periodic currents are usually characterized by the total harmonic distortion (THD) value. This parameter is calculated using the current waveform fundamental frequency and its harmonic content. For non-periodic currents, the maximum rate of change ($di(t)/dt$) is commonly used for its characterization.

Electric current phase measurement is interesting in applications where the determination of the phase difference between the measured current and other electric magnitude is required. It is very common to measure the phase difference between voltage and current supplied by a power source, because this kind of measurements allow to determine the power factor (PF) imposed by loads in power systems. This kind of measurement is crucial in grid power quality applications, where power factor compensation (PFC) systems and active harmonics filters are commonly used.

According to the preceding explanation, this chapter is devoted to reviewing the techniques for sensing instantaneous values of electric currents $i(t)$, and the determination of key parameters that can be useful for the considered application is left to subsequent processing systems.

12.3 FUNDAMENTAL PHYSICAL PRINCIPLES FOR ELECTRIC CURRENT INTENSITY MEASUREMENT

It is usual to classify the electric current meters according to the basic principles or physical properties on which their operating principle is based. In this sense, we commonly speak about current meters based on Ohm's law, on Faraday's law of electromagnetic induction, on Faraday effect or on magnetic field measurements.

This section is devoted to a basic description of these principles or physical properties and what are the main features of the current measuring devices based on them.

12.3.1 CURRENT SENSORS BASED ON OHM'S LAW

Ohm's law states that the voltage drop across a resistor is proportional to the flowing current intensity. This relationship between these electric magnitudes can be used to sense currents. The component utilized to exploit this principle is known as current shunt resistor (or shunt) and it is, basically, a low-resistance precision resistor. The main disadvantages of shunts can be summarized as follows:

- The circuit under measurement must be interrupted to insert the shunt.
- The circuit under measurement is electrically connected to the measuring circuit through the shunt.
- As a resistive component, shunts present power losses that can lead to great heat dissipation.
- Shunts for large current measurement are normally bulky and heavy: for example, a shunt able to measure 6000 A has dimensions of 30 × 15 × 15 cm and weighs about 5.5 kg.

On the other hand, shunts are low-cost and robust components, and they allow the measurement of DC and AC.

Practical current shunts are usually specified with a variety of electric specifications:

- *Current rating* indicates the maximum current through the shunt without damaging it.
- *Output voltage* indicates the voltage across the shunt produced by the rated current. Typical values for output voltage are 50, 60, 100, 120 or 150 mV at rated current and rated power.
- *Resistance accuracy* indicates the variation margin, expressed as a percentage, of the shunt nominal resistance value. Common accuracy values are ±0.1%, 0.25% and 0.5%.
- *Resistance drift* indicates the shunt's resistance changes, expressed as a percentage or in parts per million (ppm), per°C of temperature change. This specification is directly related to the material utilized on shunt construction.
- *Power rating* indicates the maximum power that can continuously dissipate the shunt without damaging it or adversely affecting its resistance. It is directly related to the shunt continuous current value and, in general, the value of the shunt continuous current can be calculated as two-thirds of the shunt rated current value.
- *Power derating* is a graphic specification that indicates how the shunt continuous current value is derated as a function of the ambient temperature. Typically, a shunt can be used at full continuous current if the temperature around the shunt is 0°C but only at 40% of full rating with a 100°C ambient temperature.

The frequency response of shunts is directly related to its self-inductance and the mutual inductance between loops built by sense wires and the main current circuit. Although this parasitic inductance affects the magnitude of the shunt impedance at relatively high frequencies, its effect on the current phase at medium or low frequencies is sufficient to cause significant errors. Another important factor affecting negatively the frequency response of shunts is the skin effect. This effect increases the effective shunt resistance for high-frequency currents and can be modelled as a frequency-dependent resistance connected in series with the shunt resistance. In order to minimize this drawback, many topologies have been used for shunt design and construction. Normally, these types of designs can be found referenced as coaxial shunt, squirrel shunt, tubular shunt and disc shunt.

The value of shunt resistance should be chosen considering the trade-off between output voltage and power dissipation. A large value of resistance can produce a reasonable output voltage ($v(t) = R \cdot i(t)$) and high power dissipation ($p(t) = R \cdot i^2(t)$). By contrast, small resistance values imply small losses and small output voltage values, meaning low-sensibility devices.

Shunts can be found with two, four or five terminals. Two-terminal shunts (Figure 12.1a), also known as milliohm resistors, are often utilized in applications for the measurement of currents lower than 100 A, such as portable multimeters.

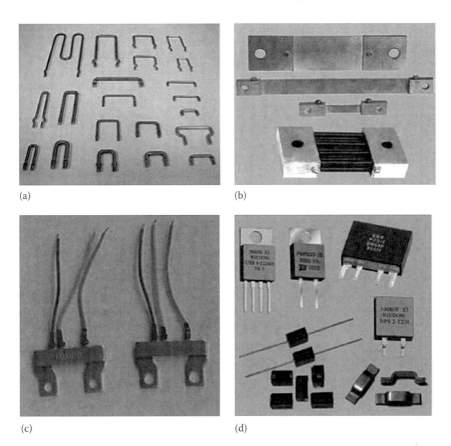

(a)   (b)

(c)   (d)

FIGURE 12.1   Several types of shunt resistors (a) two-terminal shunts, (b) four-terminal shunts, (c) five-terminal shunts and (d) shunts intended for PCB mounting.

Four-terminal shunts (Figure 12.1b) can be utilized to measure currents ranging from 1 A to 100 kA. Two terminals are utilized to intercalate the shunt into the main current circuit and the other two terminals are used for connecting the sense wires from the shunt to the measuring circuit. Using this approach, the effects of contact resistance between the wires of the main current circuit and the shunt can be eliminated. The contacts between the sense wires and the shunt also have a contact resistance, but this resistance can be neglected. This is a valid approach because these terminals are connected to the measuring voltage circuit, and the input resistance of this circuit is sufficiently high to neglect the current through these terminals. By using this type of shunts, it is possible to measure pulsed current of 650 kA peak value. The typical operating cycle of these shunts is 1 s on and 30 min off, if natural cooling is used.

Five-terminal shunts (Figure 12.1c) are normally utilized in power meter applications. In this application, it is necessary to measure the current through one electric phase and the voltage between this phase and the neutral conductor. Two of the three sense wires are devoted to measure the phase current and the other sense wire is dedicated to the phase voltage measurement. The second wire needed for the phase voltage measurement is the neutral conductor.

In many applications, it is possible to incorporate the current sensor on the same printed circuit board (PCB) used for placing the electronic control system. For these uses, it is possible to find many shunt resistors specially intended for PCB mounting (Figure 12.1d). These shunts are commonly used to sense currents up to 100–200 A and are designed to be mounted using through-hole technology (THT) or surface-mount technology (SMT).

Also, a shunt resistor can be created from the intrinsic resistance of a copper trace on a PCB. This realization is a low-cost substitute to discrete shunt resistors with no additional power losses. However, their resistance thermal drift is poor when compared with a shunt resistor. The resistance drift coefficient of a copper PCB trace is equal to approximately 3930 ppm/°C, while other materials such as Aluchrom, Manganin or Constantan (very common in shunt construction) have thermal drift coefficients equal to 50, 100 and 40 ppm/°C, respectively. This large temperature dependence makes this approach only suitable for low-accuracy applications.

Shunt resistors can be inserted in the main current circuit, remaining in series with the power supply and the main circuit load. This connection offers two possibilities: low-side and high-side shunt placement. Low-side placement refers to placing the shunt in return path from the load to the power supply, and high-side location refers to placing the shunt in the path from the supply source to the load. Both locations offer certain advantages and drawbacks.

If the shunt is utilized in the low-side placement, its output voltage is referred to the ground potential. As a consequence, this voltage can be easily amplified using well-known and low-cost techniques and presents a low common mode voltage. But this shunt location also has some important drawbacks, which include the following:

- The load is not directly connected to ground. The shunt voltage drop may create problems in the control circuits, with unwanted electromagnetic interference (EMI) emissions as a result.
- The assumption that common mode voltage will be close to zero cannot prevail in some applications or conditions. The potential difference between the ground of the main current circuit power supply and the ground of the measuring circuit may modify the measurement and even damage the circuitry.
- Low-side current monitoring is not possible in systems powered via a single wire connection, when the ground connection is made using the load chassis or metal frame.
- Only the current returned to the power supply is measured. Current derived to ground through other possible paths (load chassis, control circuit, etc.) is not measured. These alternative paths are normally related to electric fault conditions, conditions that could not be correctly detected and then can lead to erroneous diagnostic results.

High-side current measurement solves most of these drawbacks, but when the shunt resistor is placed in this location, the common mode voltage present on these measurements complicates the design of the amplification stage. Differential amplifiers, instrumentation amplifiers or integrated high-side current-sense amplifiers are commonly utilized for this kind of measurements.

In both cases of shunt location, the lack of isolation between the main current circuit and the sensing circuit can be overcome using an isolation amplifier. This is an expensive component that raises the cost of the measurement system and also adversely affects the bandwidth and precision of the measurement principle. For this reason, the use of shunt resistors in applications where electric isolation is required is not suggested.

12.3.2 Current Sensors Based on Faraday's Law of Electromagnetic Induction

The operating principle of these current sensors must be exposed starting with Ampere–Maxwell law. This law relates magnetic flux density with electric currents, and in its original form, Ampere's law relates these two magnitudes in stationary situations, that is when a static magnetic field is originated by continuous electric current.

In the case of ACs, Ampere's law needs to be corrected introducing the effect of the displacement current. From this concept, introduced by Maxwell in 1861, it follows that a magnetic field also can be generated by a variable electric field.

For practical purposes, the relationship between magnetic fields due to electric charges acceleration (B_A) and electric charges velocity (B_V) on sinusoidal currents is given by Equation 12.3:

$$\frac{B_A}{B_V} = \frac{2\pi \cdot f \cdot r}{c} \qquad (12.3)$$

where
  f is the electric current frequency
  r is the distance between the current conductor and the spatial point considered for the magnetic field evaluation
  c is the speed of light in vacuum

In accordance with this expression, the contribution of the displacement current to the magnetic flux density, B_A, is negligible for low-frequency currents. For this kind of currents, also called quasi-stationary currents, the relationship between magnetic flux density and electric current is properly defined by Ampere's law in its original form.

On the other hand, Faraday's law of induction relates the electromotive force (EMF) induced in a closed electric circuit with the variation of the magnetic flux that crosses the surface formed by this circuit. Finally, the sign of this relation is given by Lentz's law.

The combination of these three fundamental laws of the electromagnetic theory describes the operation of two devices largely used as current sensors; they are current transformers (CTs) and Rogowski coils. The main feature of these transducers is the inherent electric isolation provided between the measured current circuit and the output signal circuit. Other important features of these transducers are described as follows.

12.3.2.1 Current Transformer

The construction of CTs is based on a high-permeability ring core with two independent windings. In general, the primary winding is a single turn through the central opening of the core, but more turns can be used to measure small currents. Usually, the secondary winding is connected to a small sense resistor or load impedance, also referred to as 'burden', although ideally this winding should be short-circuited. In an ideal CT, the ratio of the number of secondary turns (N_S) to the number of primary turns (N_P if it is different from 1) determines the amplitude of the secondary or output current, and then the CT operates as controlled current source.

By contrast, a practical CT is affected by several factors that result in errors in amplitude and phase measurements. These factors can be summarized as follows:

- Voltage drops due to the CT windings resistance. These resistances are responsible for the usually called copper losses, and these losses are manifested on a rise in the device temperature.
- Voltage drops due to leakage or non-concatenated magnetic flux between the two CT windings. This effect is modelled using two leakage inductors, one per winding. A careful construction of these devices guarantees a very small value of leakage flux, making this effect negligible if it is compared with the windings resistance.
- Power losses in the core of the magnetic circuit due to hysteresis effect and Foucault currents. This effect is modelled by a resistor. The core-loss current circulates through this resistor, being this current one component of the CT no-load current.
- Magnetizing current needed to produce the magnetic flux for the CT operation. This effect is modelled by an inductor, usually named magnetizing inductance. Magnetizing current is the second component of the CT no-load current.

On well-designed CT, errors related to the output current amplitude are function of the core-loss current and errors related to the output current phase are function of the magnetizing current. As a consequence, a reduction in the core-loss current value implies a decrease in amplitude errors, and similarly, a reduction in the phase error is achieved by decreasing the magnetizing current. Strategies generally used to reduce these errors in low-frequency applications are the utilization of high-permeability materials with large cross-sectional area on core construction and high number of turns in the secondary coil construction.

CTs are very simple and robust components that allow the measurement of ACs (DC measurement is not possible) with high galvanic isolation between primary current and output signal circuits, they do not require external power sources and the output signal usually does not require additional amplification, if adequate burden is utilized.

Commercial CTs are usually characterized by a set of electric specifications:

- *Application*: Protection or metering. Protective CTs are designed to quantify currents in distribution power systems, and the device output current is used as input to the protective relays devoted to automatic isolation of one area of the distribution grid in abnormal operating conditions. Moreover, measuring CTs are used to adapt the level of measured currents to the span of utilized instruments. According to applications, CTs also can be classified as devices for indoor or outdoor purpose.

- *Rated frequency*: Indicates the frequency or range of frequencies in which the CT is designed to operate.
- *Rated primary voltage*: Indicates the maximum RMS voltage value of the circuit where the CT primary winding can be connected.
- *Insulation voltage:* Indicates the maximum voltage that the CT can withstand between the primary and secondary circuits without permanent damage.
- *Rated primary current*: Indicates the CT input current for continuous and normal operation.
- *Rated continuous thermal current*: Indicates the primary current value for continuous operation without thermal overloads, and unless otherwise specified, it is equal to rated primary current.
- *Rated thermal short-circuit current*: Indicates the RMS value of primary current that the CT can withstand for a specified time (usually 1 s is assumed) without damage on its operation, and it is specified when the secondary winding is short-circuited. This current is indicated as multiple of the rated primary current, 80 being a commonly used factor.
- *Rated dynamic short-circuit current*: Indicates the peak value of primary current that a CT can withstand without thermal, electrical or mechanical damage when the secondary winding is shorted. This current is proportional to rated thermal short-circuit current and 2.5 is a usual factor.
- *Rated secondary current*: Indicates the value of output current due to rated primary current and values as 1, 2 or 5 A are usual for protective CTs. A clear trend towards 1 A is identified, because the lower burden needed for this output current implies a significant reduction in size and cost of CT cores.
- *Rated current ratio*: Ratio of primary and secondary rated currents. This ratio is not expressed as a number and normally the values of the two currents are indicated, such as 1000/5.
- *Accuracy class*: For metering CTs, typical values for accuracy class are 0.1, 0.2, 0.5, 1 or 3, and this value indicates the maximum percentage of error in the rated current ratio at the rated primary current. For protective CTs, the accuracy class is usually 5P or 10P. This value indicates the maximum percentage of composite error (both ratio and phase errors) in the rated current ratio.
- *Rated burden*: Indicates the maximum load that may be connected on the CT secondary preserving the specified accuracy class. For metering CTs, burden is expressed as ohms impedance, and for protective CTs, burden is expressed as apparent power of the secondary circuit (in volt-amperes), at rated secondary current and at a specific PF value.
- *Rated security factor*: Indicates the value of the ratio between the primary current and the rated primary current up to which the CT will operate complying with the accuracy class requirements. For metering CTs, the value of this factor is up to 1.5. For protective CTs, this characteristic is known as *accuracy limit factor* (ALF) and indicates the maximum number of times that the primary current can exceed the rated primary current. The ALF value can be up to 30, but 10 or 20 are the most typical values.

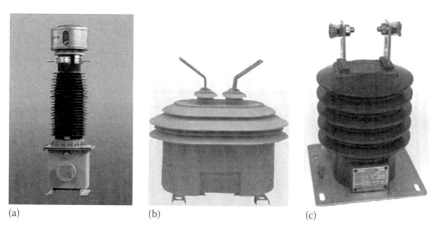

(a)   (b)   (c)

FIGURE 12.2   Several types of protective CTs for outdoor (a and b) and indoor (c) purposes.

The first classification of CTs is according to their application, and as mentioned previously, CTs can be designed for protection or for metering applications. Protective CTs are used in power transmission and distribution networks monitoring, detecting fault conditions or for load survey purpose, in order to guarantee the proper operation of the different grid sections. Using this kind of CTs, it is possible to measure rated primary currents up to 5 kA; rated thermal and dynamic short-circuit currents up to 80–200 kA, respectively; and with rated primary voltages from 1 to 765 kV. Dimensions and weight of these CTs are directly related to the rated primary voltage value and can reach heights about 7.5 m, diameters close to 1.2 m and weights up to 3800 kg. Figure 12.2 shows several protective CTs for outdoor and indoor purposes.

Measuring CTs are used in a wide range of applications, which include current monitoring and metering for visualization in instrumentation devices, for calculating energy consumption in electric power supply billing and for generating error signals in automatic control systems. The currents that can be measured using these devices are from mA to kA, with rated primary voltage lower than 600 V and rated secondary currents among mA and A.

The second possibility for CT classification is according to the used way of construction. In this sense, there are several variations, but they can be reduced to four basic types:

- *Window type*: This type has no primary winding and the primary current wire is passed through the window of the CT. It has electric insulation in its window, so conductor wires can be passed through the windows without special precautions (Figure 12.3a).
- *Bushing type*: Essentially, this CT is similar to window type, but it is for use with a fully insulated conductor acting as primary winding, because it has no electric insulations in its window (Figure 12.3b).
- *Bar type*: In these CTs, the primary conductor (usually a bar of suitable size and material) passes through the CT core forming an integral part of the device. Connectors located at the ends of this bar allow the easy installation of the CT in the primary current circuit (Figure 12.3c).

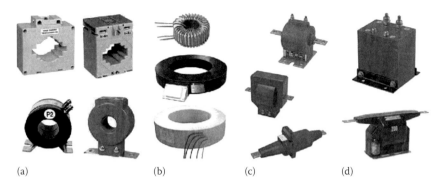

FIGURE 12.3   Several types of metering CTs (a) window type, (b) bushing type, (c) bar type and (d) wound type.

- *Wound type*: These CTs have an internal primary winding, which usually consist of more than one turn. The primary and secondary windings are insulated from each other and from the core and all these parts are assembled as an integral structure (Figure 12.3d).

The high-frequency response of CTs is directly related to the secondary winding resistance and the parasitic capacitances between its turns and layers. As a consequence, and contrary to the case of low-frequency operation, the increasing of the secondary turns number leads to substantial increase of errors in high-frequency operation. Precise CTs at high frequencies use high-permeability cores and careful design for the construction of this winding. For example, one way to reduce the parasitic capacitance on coils is by means of a progressive winding technique, rather than the usual continuous winding technique. Protective CTs and a large variety of measuring CTs are designed for line frequency (50–60 Hz), but measuring CTs with bandwidths from 0.1 Hz to 500 MHz can be found in the market.

Current transformers based on window-type and bushing-type have in common its solid core, as is shown in Figure 12.3. This feature can be a drawback since it implies that the primary current wire must be disconnected or interrupted for CT installation. This trouble is solved in split-core CTs. This core configuration allows fast and easy CT assembly without significant disturbances in the primary circuit. Split-core configuration is commonly used on CTs for portable instruments as current clamps or current probes.

Split cores present an undesirable effect over the CT accuracy because the leakage flux increases near the core closure. To avoid this inconvenience, it is recommended that the primary circuit wire passes through the core window as far as possible from the core closure.

It is possible to find CTs intended for PCB mounting and designed for THT and SMT. These CTs can measure currents up to 100 A, and their bandwidth is strongly dependent to mounting technology. The bandwidth of CTs based on THT is around 500 kHz, but if SMT is used, the bandwidth can be extended up to 10 or 15 MHz.

12.3.2.2   Rogowski Coil

One important drawback of CTs is related to hysteresis and saturation features that the material used for core construction exhibits. When the primary current exceeds

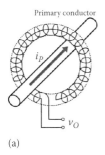

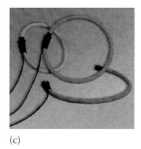

(a)   (b)   (c)

FIGURE 12.4 (a) Rogowski coil construction, (b) wound on rigid core and (c) wound on flexible core.

the value established by the rated security factor, the saturation of the CT core material is reached, and then the CT linearity and accuracy class are lost.

This apparent disadvantage of CTs is not present in Rogowski coils. The way to construct a Rogowski coil is essentially the same as the one used for a CT, but no ferromagnetic materials are used as core, and for this reason, the Rogowski coils are also called air-cored coils. As a consequence, they do not exhibit saturation features and are inherently linear devices.

The simplest design of a Rogowski coil is a solenoid placed around the primary current-carrying conductor. As shown in Figure 12.4a, an important feature of the solenoid construction is that the end of the wire used for the coil implementation is returned to the beginning *via* the central axis of the solenoid. If the Rogowski coil design does not incorporate this return loop, the sensor would be sensitive to magnetic fields perpendicular to the coil plane and, therefore, measurement errors may appear.

Figure 12.4 also outlines the two typical designs of Rogowski coils, wound on a rigid core (b) and wound on a flexible and opened core (c).

A Rogowski coil is a kind of CT, and consequently, its operating principle is defined by Ampere's and Faraday's laws. In this way, the mutual inductance of the coil (M) relates the induced output voltage ($v_O(t)$) to the rate of change of the current to be measured ($i_P(t)$), as shown in Equation 12.4:

$$v_O(t) = -M \cdot \frac{d i_P(t)}{dt} \qquad (12.4)$$

Equation 12.4 is valid if the coil core shape is not circular or if the primary current conductor is not centred on the coil; nevertheless, a non-uniform density of turns and a variation in the area of the turns along the coil lead to an increase on measurement errors. This effect is greater at the junction of the beginning and the end of the coil, where a short gap of turns appears. Some manufacturers compensate this gap error increasing the turn density around the coil junction; however, this effect can be minimized if the primary current wire is installed as far as possible from the coil junction. Finally, this drawback can be more significant in flexible core coils, since bending the core during installation can lead to deformations, thus affecting the winding characteristics.

In order to obtain an output signal proportional to the waveform of measured currents, the integration of the coil output voltage is required. Among the possible methods for implementing this operation, electronic integrators are commonly used. In this sense, Rogowski coils using digital or analogue integrators, in both their active and passive configurations, are described on specialized literature.

The high input impedance typical of the integrators connected as load to Rogowski coils implies that this kind of transducers presents very low insertion impedance when compared with CTs, making this an interesting feature according to the designed application.

Although in theory Rogowski coils can measure DCs, practical devices are not suitable for low-frequency current measurement. This behaviour is related to the impossibility of knowing the initial current value at the beginning of the integration process, making the DC component reconstruction practically impossible.

The following summarizes the most common technical specifications that the Rogowski coil manufacturers usually indicate on datasheets.

- *Rated peak input (or primary) current*: Rogowski coil has extremely large measurement range, because it does not exhibit saturation effects. The limitation conditions on measurement range are normally imposed by the integrator saturation. Transducers with measurement ranges from a few mA up to hundreds of kA, as rated peak input current, can be found.
- *Maximum input current*: Only a few manufacturers indicate the value of this parameter, and as indicated previously, the value of this current depends on the integrator saturation conditions.
- *Peak di/dt*: This specification indicates the maximum primary current change ratio above which the transducer fails to measure correctly. This value is not usually provided by manufacturers, but commercially available transducers can provide capability from hundreds of A/s to thousands of kA/s.
- *Maximum peak di/dt*: This value indicates the absolute maximum rating for primary current change ratio that must not be exceeded because otherwise, transducers can be damaged due to the voltage generated between adjacent turns in the coil. For commercial transducers, this specification is in the range of 0.1–100 kA/μs.
- *Maximum RMS di/dt*: Even if the maximum peak *di/dt* is not exceeded, the current transducer can be damaged by primary currents with sufficiently high repetitive *di/dt*. A high repetitive *di/dt* causes excessive power dissipation in the damping resistor of the Rogowski coil. The typical value for this parameter is around one $A_{RMS}/\mu s$.
- *Output voltage or sensitivity*: Indicates the ratio between the input current and the transducer output voltage signal. It can be specified as the output voltage at rated input current (i.e. ±6 V for ±6 kA), as the current to voltage ratio value (i.e. 1 mV/A) or as the current value to give 1 V output voltage (i.e. 100 A/V).
- *Current linearity*: The linearity of Rogowski current transducers typically includes the effects of both Rogowski coil and integrator, and typical value ranges are from 0.2% to 0.05%.

- *Measurement accuracy*: Accuracy values around 1% are typical on flexible core devices, while values around 0.2% are usual on rigid core Rogowski coils. As mentioned previously, the transducer accuracy varies slightly according to the position of the primary current wire within the measurement window. This 'positional accuracy' leads to errors that can reach 2% or 3%.
- *Frequency response or bandwidth*: Indicates the range of primary current frequencies in which the stated accuracy applies. The low-frequency bandwidth is set by the practical limitation included on the integrator gain at low frequencies (in theory, infinite for DCs). This frequency is several tens of Hz for most devices, but it can be 0.1 Hz for high-end devices. Besides, the high-frequency bandwidth is basically related to the parasitic inductance and capacitance of the Rogowski coil and with the gain-frequency characteristic of the amplifier used for the integrator implementation. This frequency is up to 5–50 kHz for most devices and can reach values around tens of MHz for high-performance transducers.
- *Phase error*: Output phase error is intrinsically zero in Rogowski coils, but it can reach high values when a load, as an integrator circuit, is connected to the coil. Nevertheless, this error can be quantified and externally compensated by a phase-shift compensation stage. Since this correction depends on the primary current frequency, it is necessary to define the frequency that has been used to design the compensation stage in order to minimize the phase error (50, 60 or 400 Hz are typical values). The phase error reported by many manufacturers is around 1° at 50–60 Hz.
- *Thermal drift*: It is basically determined by two factors: the thermal drift on integrator characteristics and the variation of the cross-sectional area of the coil due to thermal expansions/contractions. Few manufacturers indicate this parameter value, but typically it ranges from +0.03%/°C to +0.08%/°C.
- *Isolation voltage*: As CTs, Rogowski coils provide galvanic isolation between the primary current conductor and the measurement circuit. The working voltage for most transducers is typically compressed between 60 V and 1 kV. However, specialized transducers are available with isolation voltages of more than 10 kV.
- *External power requirements*: Many commercial Rogowski current transducers include batteries to power the internal electronic parts, such as the integrator and compensation circuits. Some transducers can be powered by external AC adaptor or external power supply. In any case, typical power requirements are from 3 V_{DC} to 24 V_{DC}. Obviously, transducers based on passive integrators do not need external power supply.
- *Battery life*: This parameter means the estimated time that the battery can power the current transducer. When batteries are used, lifetimes from 50 to 100 h are typical, but this value is reduced from 25 to 50 h when rechargeable batteries are utilized. When the current transducer incorporates the battery charger, the recharge time is also specified by manufacturer. In any case, most manufacturers provide visual signal as battery state-of-charge indicator.

In order to minimize the influence of constructive aspects in the accuracy of these transducers, PCB technology was introduced for designing the coil windings. The use of this technology practically eliminates variations in density and area of the turns along the coil due to construction inaccuracies, deformations on installation, thermal deformations or sliding by vibrations.

The measurement of current transients is the area of applications where Rogowski coils are especially valuable. Some examples of this application are current monitoring in precision welding systems, in arc melting furnaces, in electromagnetic launchers, in short-circuit test of electric generators, in electric plants for protection purposes and in fusion experiments for measuring the plasma current. The harmonic components measurement in electric currents is also another area where these transducers are used due to their excellent linearity. In this sense, they are used in applications where the current presents complex waveforms such as in high-power rectifiers.

12.3.3 MAGNETIC FIELD SENSORS

According to Ampere–Maxwell law, magnetic fields and electric currents are closely related, and as a direct consequence, magnetic field sensing devices can be used for sensing electric currents. The capacity of magnetic field sensors to measure static and dynamic magnetic fields makes them an interesting option for measuring AC and DC.

When a magnetic field sensor is used, current sensors can be constructed using two basic configurations: open-loop configuration and closed-loop configuration.

In first approximation of the open-loop configuration, the magnetic field sensing device is located near the current-carrying conductor. This configuration assumes that if the distance between the sensing device and the conductor is constant, the magnetic field measured is proportional to the circulating current. This is a very simple and cheap principle, but its performance is directly related to the physical geometry of the sensor.

Some problems of this approximation can be solved if a magnetic core is used in order to concentrate the magnetic field generated by the primary current onto the magnetic sensing device, as shown in Figure 12.5a. This second approximation reduces the influence of the external magnetic fields, the influence of the relative position between current conductor and magnetic sensing device and the skin effect on the measurement accuracy and increases the sensor sensitivity. On the other hand, the performance of these sensors is also related to the properties of the core material. In this sense, core losses due to hysteresis and eddy currents and the possible saturation of the magnetic core can lead to the sensor overheating and important measurement errors.

In the current sensors based on closed-loop configuration, a secondary winding is incorporated to the magnetic core and the output signal of the magnetic field sensor is used for generating the secondary current (I_S), as is illustrated in Figure 12.5b. The secondary current generates a magnetic field that is opposed to the field generated by the primary current (I_P), and when the secondary current compensates the magnetic flux inside the sensor core ($\Phi = 0$), this current is proportional to the primary current. The ratio between primary and secondary currents is fixed by the number of turns of

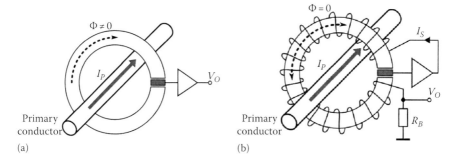

FIGURE 12.5 Current sensors using magnetic field sensing devices in (a) open-loop configuration and (b) closed-loop configuration.

the secondary winding, and the ratio between the secondary current and the output voltage (V_O) is fixed by the value of the used burden (R_B).

This configuration reduces the influence of the thermal behaviour and characteristics of the magnetic field sensing devices on the measurement accuracy and linearity. Core losses due to eddy currents and hysteresis are greatly reduced if the current sensor operates very closely to zero-flux conditions. Also, the combination of the secondary winding and the magnetic core can operate as a CT, thereby increasing the current sensor bandwidth.

On the other hand, the more complicated construction of this configuration implies larger sizes and costs, and since the magnetic flux on magnetic core must be compensated, the required performance of the power supply (in output current capability) is higher.

In the following, the three technologies more commonly utilized for magnetic field sensing are introduced and briefly explained.

12.3.3.1 Hall Effect Current Sensor

The Hall effect was discovered by Edwin Hall in 1879 and it originates from the Lorentz force experienced by a charged particle moving in a magnetic field. When a current circulates through a slab of semiconductor or metal that is penetrated by a magnetic field, the charge carriers are deflected to one side of the slab, and as a consequence of the difference on the conductive carrier density, a voltage is generated perpendicular to both current and magnetic flux density directions.

The slab of commercial Hall elements is manufactured using semiconductor materials instead of metals because the sensitivity of these elements is related to the carrier density in the material utilized for their construction. These elements are constructed in different shapes as rectangles or symmetrical crosses, and semiconductor materials as indium antimonide (InSb), gallium arsenide (GaAs) or doped silicon (Si) are commonly used on commercial Hall elements manufacturing.

In practice, the behaviour of Hall elements also depends on other physical factors such as the pressure and the temperature. The output voltage dependence on the mechanical pressure (piezoresistive effect) is not an important factor to be considered by users of these devices, because it is considered and compensated by manufacturers when the device is encapsulated. In contrast, temperature affects the Hall element accuracy in two different ways. An increase in temperature means, on the one hand, an increase of the Hall element resistance and, on the other hand, an increase of the mobility of charge carriers. Although these two factors affect the accuracy with opposite sign and a partial compensation between them is possible, Hall current sensors have high values of thermal drift in accuracy specifications.

Another important drawback of these sensors is the presence of offset voltages at the sensor output. This is an output voltage different from zero in the absence of magnetic field and is related to material non-uniformities, physical inaccuracies in the slab construction and misalignments in the position of sensing electrodes around the Hall element. The practical current sensors based on Hall elements include additional electronic circuitry devoted to compensate the offset voltage and differences on thermal drifts.

The most common technical specifications of Hall current sensors are summarized as follows.

- *Rated current*: Indicates the input current for continuous and normal operation. Devices with rated currents from 1 A up to 5 kA can be found in the market.
- *Saturation current*: Indicates the input current value for which the output deviates from the estimate output voltage by more than 10%.
- *Output type:* The output signal of these sensors may be a voltage or a current.
- *Rated output*: Denotes the output signal value when the input to the sensor is the rated current.
- *Residual output*: This specification is also called output offset and indicates the output value when the input current to the sensor is zero.
- *Bandwidth*: Indicates the range of primary current frequencies in which the stated accuracy applies. Hall effect transducers operate from DC to several hundreds of kHz.
- *Output linearity*: Denotes the error between the value of the sensor output and the estimated output value. It is calculated by the least mean squares method from the output and residual output when the input to the sensor is the rated current and one-half of the rated current.
- *Linearity limits*: Indicates the range of the input current value for which the output is within 1% of the estimated output voltage.
- *Output temperature characteristic*: Indicates the rate of output change due to temperature change when the rated current is input, within the working temperature range. The rate of change is shown as output temperature coefficient and is evaluated per 1°C with the output at 25°C as the reference.
- *Residual output temperature characteristic*: Denotes the change of the residual output due to temperature change when the input is the rate current, within the working temperature range. The change per 1°C is shown as residual output temperature coefficient.
- *Response time*: Indicates the output response time when a pulse current is the input current. This time is evaluated as the time difference when the input and output waveforms drop to a specified level of their initial values.

(a)   (b)   (c)

FIGURE 12.6 Hall effect current sensors: (a) SMT device [www.melexis.com], (b) THT devices for onboard-mount [www.lem.com] and (c) screw-mount device.

- *External power supply*: The internal electronic parts of Hall Effect current transducers can be powered by external power supply. Typical power requirements are from 1 V_{DC} to 24 V_{DC}, and it is usual that the bidirectional sensors need a bipolar power supply, being ±5 V_{DC}, ±12 V_{DC} or ±15 V_{DC} typical values.
- *Isolation voltage*: Hall effect current sensors provide galvanic isolation between the primary current conductor and the measurement circuit. The isolation voltage for most transducers is typically compressed between 100 V and 2.5 kV.

As was presented previously, the accuracy is fair for open-loop devices (values between 0.5% and 2% are typical) and high when the closed-loop configuration is implemented (reaching values around 0.02% and 0.05%). Both types of configurations are commonly used in this kind of transducers and it is possible to find Hall effect current sensors intended for screw-mount and onboard-mount types, as shown in Figure 12.6. Also, PCB devices are designed for both THT and SMT.

Typical applications for Hall effect current sensors are related to DC measurement. This is the case of electric power conversion systems, motor drives and electric traction, uninterruptible power supplies (UPS), electric welding and electrolyzing equipments, electric power network monitoring and electrowinning or smelter industry. These sensors are also utilized in current clamps used for DC and low-frequency current measurement.

An example of high current metering system is the LKP-500 manufactured by DynAmp LLC. The LKP-500 is a high-accuracy closed-loop Hall effect system designed to accurately and reliably measure unidirectional DC bus currents up to 500 kA, with 500 mV rated output voltage and 0.1% of accuracy. The device core shape is basically rectangular, has a perimeter of approximately 4×5 m and a depth of 30 cm and is weighing slightly over 2500 kg.

12.3.3.2 Fluxgate Current Sensor

Fluxgate sensors appeared in the early 1930s and are still used today in many applications that require high precision, because this technology is one of the most accurate for sensing magnetic fields.

The basic operating principle of these transducers is based on the detection of saturation states in a magnetic circuit. These states are related to the non-linear relationship between the magnetic field and the magnetic flux density shown by magnetic materials.

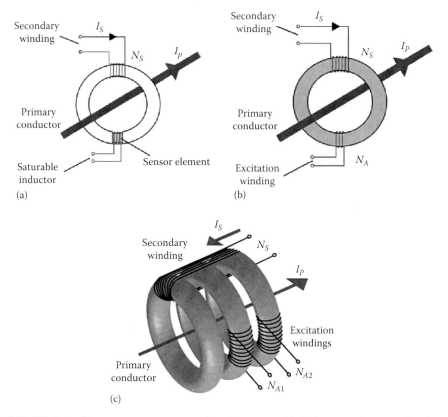

FIGURE 12.7   Fluxgate current sensors: (a) using a saturable inductor as detector, (b) using a closed toroid core as detector and (c) using multiple cores for voltage disturbances reduction and extend the device bandwidth.

The three basic designs used for the construction of fluxgate sensors are depicted in Figure 12.7.

The so-called 'standard' fluxgate transducer (Figure 12.7a) uses a toroidal magnetic circuit that includes an airgap with a saturable inductor as the element for magnetic field detection. This inductor is led to saturation periodically between positive and negative values by effect of the signal applied to its excitation winding. In the design shown in Figure 12.7b, the sensor uses its own ring core as a magnetic field detector and includes no gap on the magnetic path. An auxiliary winding N_A (normally called excitation winding) is added to the core, which is used as a saturable inductor for flux detection purposes.

In the absence of the primary current (I_P), the flux through the magnetic field detector is zero. Under these conditions, if an adequate square voltage is applied to the excitation winding, a growing magnetic flux appears in the magnetic circuit until the core saturation state. The primary current (I_P) effect on the excitation current implies a waveform deformation, and this phenomenon can be utilized to primary current value estimation.

In open-loop fluxgate sensors, the primary current value (I_P) is obtained measuring the excitation current second harmonic component or average value. In closed-loop devices, one of these signals serves as error signal for generating an additional magnetic field by means of a secondary current (I_S) applied to an additional secondary winding (N_S). When a zero-flux condition is achieved, the secondary current is a replica of the primary current and the ratio between these two currents is fixed by the number of turns used in the secondary winding.

Fluxgates based on one-core designs present two important drawbacks. The first one is related to the possibility of noise injection from the excitation winding circuit into the primary current circuit. This noise can be coupled to the primary current circuit due to the transformer effect present in the magnetic core of the transducer. The solution usually adopted is the use of a second core with a new excitation auxiliary winding. Under ideal conditions, these two cores and their winding must be identical. If these two auxiliary windings (N_{A1} and N_{A2}) are excited with the same current but in opposite directions, then the two induced currents on the primary conductor will be equal and with opposite directions, cancelling their effects.

The small bandwidth (only a few hundred of Hz) is the other significant problem of these devices. The bandwidth of these fluxgate configurations is limited by the frequency of the excitation signal or, in other words, by the time spent in driving the magnetic core between positive and negative saturation states. To increase the device bandwidth and improve its dynamic response, a third core is included in the transducer design (Figure 12.7c). The combination of this new core, the secondary winding and the primary current wire operates exclusively as a conventional CT, and this feature guarantees medium- and high-frequency current measurement.

Fluxgate technology offers several advantages over other technologies based on magnetic field measurement, such as low offset, low offset thermal drift and excellent overcurrent recovery because any permanent magnetization or offset on the fluxgate core is reset with subsequent saturation cycles. Furthermore, its large dynamic range allows measurements of small and large currents with the same device and also they present a large temperature range of accurate operation limited only by used materials and electronic components.

Both open-loop and closed-loop configurations are commonly used for these transducers and it is possible to find fluxgate current sensors intended for screw-mount or onboard-mount types (Figure 12.8).

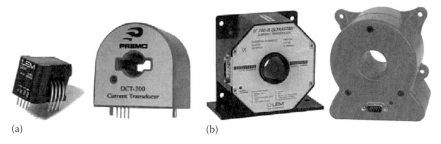

(a)   (b)

FIGURE 12.8   Fluxgate current sensors: (a) THT devices for onboard-mount and (b) screw-mount devices [www.grupopremo.com and www.lem.com].

The measuring range of a typical commercial 'standard' fluxgate is from 0 to 500 A, accuracy 0.2%, linearity 0.1%, response time <1 µs and bandwidth from 0 to 200 kHz. Moreover, the measuring range of three core fluxgate sensors is from 0 up to 1 kA, accuracy 0.0002%, linearity 0.0001%, response time <1 µs and bandwidth from 0 to 100 kHz.

Finally, these sensors are commonly used in high-precision measurement applications due to its high cost and size. Typical applications are calibration and diagnosis systems, medical equipment, precise energy metering, high-performance laboratory equipment or feedback element in high-precision systems.

12.3.3.3 Magnetoresistance Current Sensor

Magnetoresistance is defined as the property of a material to change the value of its electric resistance when an external magnetic field is applied to it, and consequently, devices based on these materials can be used as sensors of magnetic fields. Their main application has been as read head in magnetic storage devices, but nowadays, it is possible to find commercial sensors devoted to electric current intensity measurement.

The sensing direction of these sensors is in the same plane as the magnetoresistive element, so they cannot be located into narrow airgap in magnetic cores. The use of large airgaps implies increasing the sensitivity to external magnetic fields and external currents, thus these sensors are commonly coreless.

The magnetoresistive elements are normally configured in a Wheatstone bridge to compensate the homogeneous external magnetic fields and thermal drifts due to variations of the electric resistivity of the magnetoresistive elements. Also, these sensors can be found in the market in both open-loop and closed-loop configurations, but they are only intended for onboard-mount type.

There are several known physical effects that cause variations in the resistance of certain materials when a magnetic field is present, but only two are currently applied to the measurement of electric currents and they are discussed as follows.

12.3.3.3.1 Anisotropic Magnetoresistance Sensor

The discovery of the anisotropic magnetoresistance (AMR) effect is attributed to William Thomson (or Lord Kelvin) in 1856. This effect is associated with Hall effect elements. The deflection of current carriers when the sensor slab is penetrated by magnetic field has as a consequence the increase of the path length followed by some charge carriers and consequently an increase in the device resistance. In most conductors, this effect is less important when compared with the Hall effect, but in anisotropic materials, such as permalloy (an iron nickel alloy), where its electric resistance is influenced by their magnetization state, the application of an external magnetic field may produce resistance variations from 2% to 5%. These variations are dependent on the angle between the current flow direction and the magnetization field direction.

In this regard, the minimum resistance variation occurs when the magnetization field and the current directions are perpendicular (90°), and the maximum variation is reached when the current flow in the same direction as the magnetization field (0°). If the directions of the magnetic field and the circulating current are separated by 45° angle, then the AMR effect is also sensitive to the direction (or sign) of the magnetization field. This operating condition is obtained using a series of permalloy strips separated by a thin film of vapourized aluminium deposited on the permalloy surfaces. This structure is commonly known as barber pole.

The magnetization field is obtained as the addition of the external applied magnetic field and the initial magnetization direction of the permalloy strips. In this regard, and if the external field is applied perpendicular to the initial magnetization direction, the orientation of the resulting magnetization field depends on the intensity of the applied magnetic field. Therefore, the angle between the resulting magnetization field and the current across the barber pole changes, and consequently, the barber pole resistance also changes.

One of the major drawbacks of AMR sensors is that a strong external magnetic field can permanently change the direction of the permalloy initial magnetization, introducing a permanent measurement error until a new permalloy re-magnetization in the right orientation has been performed.

The bandwidth of these sensors is from DC to several MHz and commonly is limited by the amplifiers used in the signal conditioning circuitry.

AMR current sensors are available in both open-loop and closed-loop configurations but always, as mentioned earlier, configured in a Wheatstone bridge as shown in Figure 12.9.

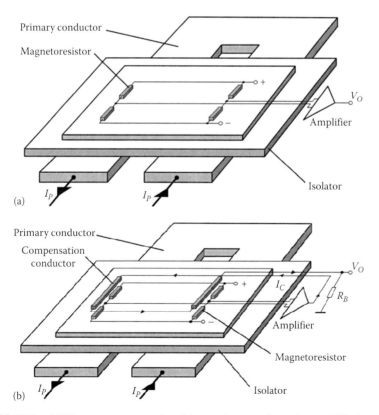

FIGURE 12.9 AMR current sensor using: (a) open-loop configuration and (b) closed-loop configuration.

FIGURE 12.10 Commercial AMR current sensors [www.fwbell.com, www.sensitec.com, sensing.honeywell.com and www.diodes.com].

In the closed-loop configuration, depicted in Figure 12.9b, the Wheatstone bridge output voltage is used as error signal in order to generate the compensation current and the associated magnetic field. The value of the compensation current is adjusted until its magnetic field cancels the magnetic field associated with the primary current. In this condition, the compensation current is proportional to the primary current.

Commercial AMR current sensors are manufactured by Sensitec, F.W. Bell (part of Meggitt PLC group), Zetex Semiconductors (part of Diodes Incorporated) and Honeywell. They offer sensors up to 150 A of primary rated current, accuracy lower than 0.3%, linearity errors lower than 0.1% and bandwidths up to 2 MHz with isolation voltage of 3.5 kV. Figure 12.10 shows some of these devices.

12.3.3.3.2 Giant Magnetoresistance Sensor

In 1988, two research groups independently discover the giant magnetoresistance (GMR) effect measured on Fe/Cr thin multilayers and described it as a large change in electric resistance that occurs when thin stacked layers of ferromagnetic and non-magnetic materials are exposed to a magnetic field. The cause of this huge variation of the resistance (about 50%) is attributed to the scattering of the electrons at the thin layers interfaces. In 2007, the Royal Swedish Academy of Sciences awarded the Nobel Prize in Physics jointly to Albert Fert and Peter Grünberg (the leaders of these two research groups) for the discovery of the GMR effect.

The GMR effect allows the detection of very small currents due its high sensitivity, and it is a cheaper technology because the massive production of small devices is possible using standard semiconductor technology. However, GMR technology has important drawbacks as high non-linear behaviour and distinct thermal drift. Furthermore, the GMR sensors behaviour may be permanently altered after a magnetic shock due to a very strong external field or overcurrent.

Current sensors based on this technology are constructed using open-loop configuration and, in order to increase their immunity to external magnetic fields and temperature variations, are configured in a Wheatstone bridge setup. The half-bridge configuration with two active resistances and two shielded ones is commonly used because the manufacturing process of this GMR structures is cheaper if it is compared with the two-step deposition process needed for the full-bridge configuration implementation.

Commercial GMR current sensors can be found in onboard-mount types and are designed for SMT. NVE Corporation is the most important manufacturer of GMR

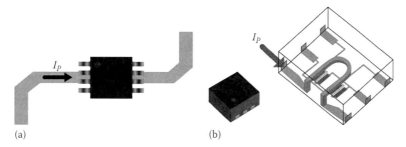

(a)   (b)

FIGURE 12.11 Commercial GMR by NVE Corporation: (a) the AA00X-02 series current sensor and (b) the AAV003-10E device [www.nve.com].

current sensors and offers devices for linear measurement of electric currents in the range of ±80 mA and with frequency response of 100 kHz. Figure 12.11 depicted some of these devices.

The main application field of GMR current sensors is related to the measurement of low currents using very small size and low-power consumption devices. This is the case of portable equipments and battery-powered systems, as mobile phones, laptops or battery chargers.

12.3.4 CURRENT SENSORS BASED ON FARADAY EFFECT

This effect was discovered in 1845 by Michael Faraday and it was the first evidence that electromagnetism and light are related. This effect describes the interaction between magnetic fields and light and explains the rotation in the light polarization plane proportional to the component of the magnetic field in the same direction of light propagation. A medium that changes the state of light polarization is said to be birefringent and Faraday's discovery was that a certain type of birefringence can be induced into a medium by a magnetic field.

A linear polarized light wave can be considered as the superposition of two orthogonal circular polarized light waves, the right-hand circular polarized (RHCP) wave and the left-hand circular polarized (LHCP) wave. In a circular birefringent medium, the RHCP wave velocity decreases if the applied magnetic field has the same direction as the light beam and increases if it has the opposite direction. In contrast, the LHCP light wave has an opposite behaviour. The rotation of the polarization plane is the result of the phase difference induced between the two circular polarized components of a linear polarized light.

Faraday effect, as Hall and magnetoresistive effects, can be exploited to design devices dedicated to the electric current measurement. These devices provide exceptional electric isolation and enable the measurement of DC and AC up to 500 kA using a negligible amount of energy and space if they are compared with other technology current sensors.

Current sensors based on Faraday effect can be constructed using two different configurations usually referred to as extrinsic and intrinsic fibre sensor classes. These two classes of architectures are briefly described as follows.

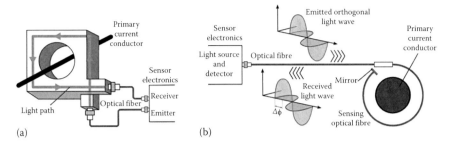

FIGURE 12.12 Optical current sensor types commercialized by ABB: (a) MOCT type and (b) FOCS type.

12.3.4.1 Extrinsic Fibre Optic Sensor

This kind of sensors, named magneto-optic current transducer (MOCT) by ABB, uses a piece of glass wrapping the current-carrying conductor, as is shown in Figure 12.12a, and optical fibres are used for light transmission and not for sensing purposes. The sensitivity of these transducers can be increased using multiple light reflections and creating a 3D path with several turns around the current conductor.

ABB perform these sensors for metering and for protective (MOCT-P) applications in 72.5–800 kV and 50 or 60 Hz electric energy distribution systems. The MOCT system is suitable for outdoor application and has an accurate (0.2 class) metering current range from less than 1 A to 4000 A using the same sensor. The MOCT-P system is also suitable for outdoor applications with a 5TPE accuracy class for a continuous current rating up to 3150 A_{RMS}.

12.3.4.2 Intrinsic Fibre Optic Sensor

These sensors, named fibre optic current sensors (FOCS) by ABB, use an optical fibre wrapping the current-carrying conductor, as is depicted in Figure 12.12b. The sensitivity of these transducers can be easily increased using multiple fibre turns, creating a fibre optic coil around the current conductor. Moreover, the accuracy may be deteriorated due to birefringence induced by bending the fibre optic cable. For this reason extrinsic fibre optic sensors was firstly introduced, since the use of a solid piece of glass avoids the optical fibre bending stress. However, the use of high birefringent materials with low sensitivity to stress for optical fibre manufacturing makes the intrinsic sensors an interesting alternative to the larger and expensive extrinsic sensors.

In order to reduce the sensor sensitivity to mechanical stress and vibrations, back-and-forth propagation through the sensing optical fibre can be used. This approach exploits the fact that the birefringence induced into the optical fibre by bending stress is reciprocal, in contrast with the birefringence due to Faraday effect. This feature is implemented reflecting by a mirror the light beam at one side of the optical fibre and detecting the differential rotation in the light polarization at the other end. Furthermore, these one-side open sensors can be wound around current conductors in existing installations.

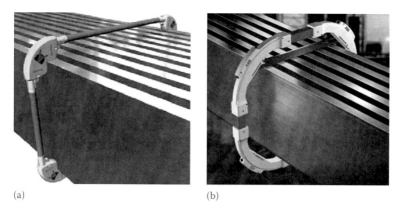

(a)   (b)

FIGURE 12.13 Optical current sensor types commercialized by (a) DynAmp LLC and (b) ABB [www.dynamp.com and www.abb.com].

Some manufacturers of commercial intrinsic FOCS are ABB, DynAmp LLC, Ariak Inc. and ALSTOM. They offer sensors in the range of 1 A–600 kA with measurement accuracy and linearity error around 0.1%. These sensors are designed for DC and from 50–60 Hz to 5 kHz applications with isolation voltages of several kV. Figure 12.13 shows some of these devices.

As is easily deduced from its specifications, the main fields of application of these sensors are the current metering for protection and measurement on electric energy distribution and production systems and very high DC metering for control systems on electrochemical industries.

12.3.5 Do You Want to Learn More?

This chapter only describes the current sensors technologies actually applied to commercial devices, but there are still several physical effects or properties applicable to measurement of magnetic fields or electric currents. Some examples are the tunnelling magnetoresistance (TMR) effect, the colossal magnetoresistance (CMR) property, the giant magneto-impedance (GMI) effect, devices based on superconducting quantum interference detectors or SQUID (used for measuring extremely small currents), devices based on magnetically sensitive complementary metal-oxide-semiconductor (CMOS) chapter transistors (MAGFETS) or sensors that exploit Lorentz force.

In order to learn more about these future technologies or to increase our knowledge on the current sensors described in this chapter, we can access various information sources and technical documents.

Two very interesting reviews dedicated to electric current sensors are

- Silvio Ziegler et al, 'Current Sensing Techniques: A Review', published in April 2009 (volume 9, number 4) in IEEE Sensors Journal (IEEE Sensors Council). Doi: 10.1109/JSEN.2009.2013914
- Pavel Ripka, 'Electric current sensors: a review', published in November 2010 (volume 21, number 11) in Measurement Science and Technology journal (IOP Publishing). Doi: 10.1088/0957-0233/21/11/112001

Other interesting documents are published by the companies that manufacture these devices. Normally this kind of documents includes an overview about the technology used for their products and a detailed set of electric and physical specifications and characteristics. Some examples of these documents are

- 'Isolated current and voltage transducers. Characteristics–Applications–Calculations', published by LEM. This document exposes the operating principles of Hall effect, fluxgate and Rogowski coil current sensors developed by this company.
- 'General Catalogue. Current Sensors', published by PREMO. This document is devoted to explain the operating principles and main features of CTs and fluxgate current sensors developed by this company.
- 'Instrument Transformers. Application Guide', published by ABB and devoted to explain the general principles of measuring current and voltage, how to specify and design a CT and how the optical current transducers developed by this company work.

Finally, it is possible to find a large number of application notes and technical reviews published by manufacturers and their research and technical staffs. This kind of documents usually includes a technical overview about the operating principle of the described devices and some ideas or hints on how to use these devices in a particular application. A few examples of these documents are

- Bill Drafts, P.E., 'Magnetoresistive Current Sensor Improves Motor Drive Performance', published by Pacific Scientific-OECO
- 'Application Notes', published by Power Electronic Measurements Ltd. and devoted to explain some limitations of Rogowski coils and to help the engineer maximize the many advantages of these current transducers
- 'Resistors for Energy Metering–Application Note', published by TT electronics
- 'Introduction to Current Transformers', published by Elkor Technologies Inc
- 'Current Transformers. The Basics', published by NK Technologies
- 'High-Accuracy Current Sensors for Hybrid Vehicles and Electric Vehicles', published by TDK Corporation

All these documents can be easily found on the net through the manufacturers' website.

13 Power Density Limits and Benchmarking of Resonant Piezoelectric Vibration Energy Harvesters

A. Dompierre, S. Vengallatore and L.G. Fréchette

CONTENTS

13.1 INTRODUCTION

Recent progress in low power electronics, wireless technologies and miniaturiza-tion of electromechanical devices using complementary metal oxide semiconductor (CMOS) and micro-electromechanical systems (MEMS) technologies suggests that many intelligent embedded devices will soon operate on very small power budgets, in the range of 1–100 µW [1]. Closely associated with these developments is the challenge of developing portable sources of electric power. Electrochemical batteries are widely used but often restrict the performance and autonomy of microdevices in many applications. Furthermore, for important emerging applications such as wire-less sensor networks employing a large number of nodes, it is not practical to use batteries. In fact, the task of replacing batteries is a major obstacle towards the wide-spread deployment of sensor networks for health monitoring applications and the Internet of things [2]. Hence, there is great interest in developing miniaturized energy harvesters that can convert ambient energy – thermal fluctuations and gradients, solar radiation, mechanical vibrations, fluid flow and radio-frequency waves – into electric power. Over the past 20 years, numerous device concepts have been proposed rang-ing from miniaturized engines and fuel cells to thermoelectric, thermophotovoltaic and piezoelectric devices. Many crucial questions remain open in this field. What are the performance limits of different types of harvesters? Can we develop effective strategies for selecting harvesters for specific applications and environments? Can we develop rational design methodologies to optimize the performance, reliability and manufacturability?

In this chapter, we seek answers to some of these questions for one particular type of harvester, namely, piezoelectric vibration energy harvesters (PVEHs). Vibrations are of interest for energy harvesting because of their ubiquitous nature and they do not require direct access to ambient wind or sunlight; harvesting usually works by using a suspended mass to inertially create a force from the vibration at one point. For PVEH, the stresses induced in a piezoelectric material due to this force are used as the mean of energy conversion. MEMS harvesters typically use cantilever or clamped–clamped beam geometries in a way very similar to accelerometers. In addition, a large proof mass is usually attached to the structure to tune its resonance frequency as well as increase the applied force, thus its output power. Thin-film deposition processes are often used for the integration of the piezoelectric material into these small devices.

Many research groups in academic and industrial research laboratories have designed and built miniaturized PVEH devices using different configurations, geometries, materials and electric circuits and then tested their performance under different operating conditions. Typically, power densities of the order of µW/cm^3 have been reported, which is significantly lower than the early estimates that moti-vated efforts to build PVEH [3,4]. Although increasing the power density is a major focus of current research, it remains unclear whether there is indeed any significant potential for increasing these values. Recognizing the limits of VEH is also important to better understand which wireless sensor applications are compatible with this technology. To this end, we identify and discuss the various aspects that can limit the power density of PVEH and quantify their impact using a simple model. Figures of merit (FOM) are proposed for benchmarking different designs, thus enabling a close comparison of different devices that have been reported in the literature.

13.2   ENERGY TRANSFER IN PVEH DEVICES

The energy chain in typical piezoelectric energy harvesters is depicted in a simplified manner in Figure 13.1. The ambient mechanical energy source first acts on the harvesting device, either inertially (e.g. vibration, shocks) or directly (e.g. impacts, fluid flows, pressure gradients), and transfers some of its energy. Part of the transferred energy is present as kinetic energy (from the mass motion) and potential energy (from the elastic energy in the strained spring and the electric energy in the charged capacitance), with a continuous exchange between these two forms as the harvester vibrates. An electric potential is generated on the electrodes of the piezoelectric element while it is stressed. This enables the conversion of some of the mechanical energy into electric energy. Losses usually occur during this process (e.g. mechanical damping and dielectric leakage), while electric energy is also extracted to be consumed, processed or stored. If some electric energy remains, it will be restored as mechanical kinetic energy. Once extracted and stored, the electric energy can be used by the electronic device after it has been processed by the power distribution and control interfaces.

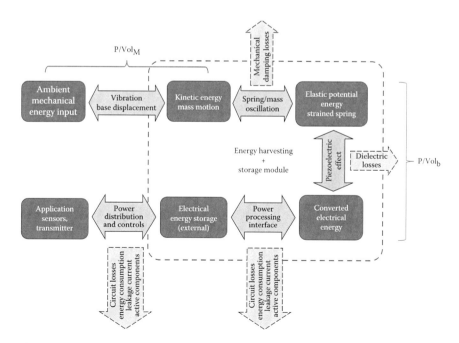

FIGURE 13.1   General energy chain of a PVEH device.

Note that bidirectional arrows are used in this chain to show that the flow of energy can indeed go both ways. For instance, the control interface could be used to electrically tune the harvester by injecting previously stored energy into the piezo-electric element [5]. This tuning could also affect the capture of ambient energy by the harvester. For the miniaturized devices under consideration, the effects of the harvester on the source are negligible. The performance is typically reported in terms of the power density (i.e. power per unit volume of the device). Several limiting aspects can be identified as potential bottlenecks in terms of net power output and power density:

- How much energy is transmitted to the structure? This depends on the vibration source and the quality of its mechanical coupling with the device. Is the excitation due to inertia or contact forces applied on the device? Is transfer of energy based on resonance, impact or plucking? This limit for linear resonator is discussed in Section 13.4.
- How much energy can the device collect and sustain? Stress concerns (i.e. distribution and concentration), fatigue and materials degradation will limit how much energy can be stored in the material before it fails. In theory, this value will represent the upper bound for power density, but one must assume that enough energy can be captured by the harvesting device in the first place. These issues are discussed in Section 13.5.
- How much mechanical energy can be converted into electric energy? For piezoelectric transducers, the generalized effective piezoelectric coupling factor, k_e^2, gives the fraction of the applied energy that is transformed in the electric domain by the piezoelectric structure under a specific strain distribution. As we discuss in Section 13.6, this coupling factor depends on material properties, mainly the material coupling factor expressed by its k_{ij} constants, and the device geometry.
- How much electric power can be extracted and stored? Even if a large part of the work applied on the device is converted into electric energy, only a fraction of it might be extractable, depending on the load connected to the transducer. Many harvesting interfaces have been proposed to improve the extraction effectiveness and reduce losses (i.e. improve efficiency). The impact of the extraction approach is presented in Section 13.7.

All these aspects can be more or less interrelated depending on the design of the device. For instance, typical inertial, resonant PVEH usually consist of a multiple layered suspended structure clamped to the vibration source and a suspended mass attached to this structure to inertially induce stresses. By designing this structure to match one of its resonant mode with the vibration source frequency, more energy can be captured, in turn inducing larger stresses as well. Hence, these influences are particularly strong in this case because a single heterogeneous structure captures, receives and converts the energy, while energy injection and extraction also occur simultaneously. This chapter will explore each of these limits from a high-level theoretical framework with the objective of introducing or reviewing relevant FOMs.

The selection of an optimal piezoelectric material and electric interface will also be discussed. In Section 13.8, we will use the FOMs to compare the performance of various devices reported in the literature and paint a global picture of the current state of the art.

13.3   SINGLE DEGREE OF FREEDOM MODEL OF PVEH

Piezoelectric resonant structures can be modelled accurately using distributed parameter methods [6], but a simpler, lumped element approach is often sufficient and more convenient for design around a single resonant mode. Such a model will be presented here to introduce the dynamics. Like mechanical resonators, the lumped representation of the piezoelectric resonator consists of a mechanical spring, K_m; an equivalent mass, M_{eq} and a mechanical damper, C_m. However, an electrically coupled spring, K_{el}, and an electric damper, C_{el}, also change the behaviour of the system [7]. These two elements are affected by the piezoelectric coupling, frequency and the electric load that is connected to the piezoelectric element. Figure 13.2 illustrates the modelling approach for a piezoelectric unimorph cantilever beam with a tip mass, M_t, and connected to a resistive load, R_{eq} [8].

We shall consider the transverse vibration mode of a thin beam of length L and width b. The thicknesses of the piezoelectric and support layers are respectively denoted by h_p and h_s, with their modulus of elasticity similarly denoted by

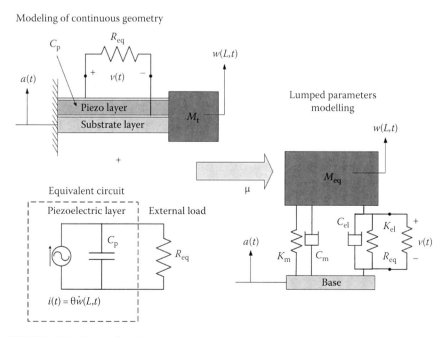

FIGURE 13.2   Equivalent lumped parameter model of a PVEH beam.

Y_p and Y_s. It is worth mentioning that for thin beams, the elasticity is derived from the compliance of the material, such as [9]

$$Y = (s_{11}^E)^{-1} \tag{13.1}$$

where s_{11}^E is the first term of the compliance tensor of the material (the superscript E indicates that the property is measured with a constant or null electric field, i.e. in short-circuit condition). Let us further assume that the electrodes cover a length $L_1 \le L$, starting from the base of the cantilever beam because this region will be the most stressed. However, the electrodes are assumed to be much thinner than the beam; hence, their effects on stiffness are negligible. The beam displacement is

$$w(x,t) = W\hat{w}(x)\sin(\omega t) \tag{13.2}$$

where
  W is the phaser related to the tip displacement (amplitude and phase)
  $\hat{w}$ is the normalized deformed shape of the beam

The energy source is a harmonic acceleration $a(t) = A\sin(\omega t)$ applied to the base of the beam. The mechanical domain equation is based on Newton's second law and can be reduced to Equation 13.3. The electric behaviour is captured by Equation 13.4, obtained from the Kirchhoff circuit laws, which is equivalent to an RC circuit connected in parallel with a variable current source. Hence, in the electric domain, the piezoelectric layer is equivalent to a current source and a capacitor, C_p:

$$M_{eq}\ddot{w}(L,t) + C_m\dot{w}(L,t) + K_m w(L,t) - \theta v(t) = F_{in} \tag{13.3}$$

$$\theta\dot{w}(L,t) + C_p\dot{v}(t) + \frac{v(t)}{R_{eq}} = 0 \tag{13.4}$$

The piezoelectric effect is therefore completely captured through the coupling force, θv, and piezoelectric current, $\theta\dot{w}(L,t)$, where $v(t)$ is the voltage and θ is the beam coupling coefficient. The equivalent lumped parameters are extracted from [6] and can be expressed as

$$\theta = Y_p d_{31}\left(b\bar{h}_p\int_0^{L_1}\frac{d^2\hat{w}(x)}{dx^2}dx\right) \tag{13.5}$$

$$C_p = \varepsilon_{33}^S\left(\frac{bL_1}{h_p}\right) \tag{13.6}$$

$$K_m = \frac{b}{3}\left[Y_s(\bar{h}_b^3 - \bar{h}_a^3) + Y_p(\bar{h}_c^3 - \bar{h}_b^3)\right]\int_0^L\left(\frac{d^2\hat{w}(x)}{dx^2}\right)^2 dx \tag{13.7}$$

with

$$\bar{h}_a = -\frac{\frac{h_s}{2}Y_s h_s + \left(h_s + \frac{h_p}{2}\right)Y_p h_p}{Y_s h_s + Y_p h_p} \tag{13.8}$$

$$\bar{h}_b = h_s + \bar{h}_a \tag{13.9}$$

$$\bar{h}_c = h_p + \bar{h}_b \tag{13.10}$$

$$\bar{h}_p = \frac{\bar{h}_b + \bar{h}_c}{2} \tag{13.11}$$

In those expressions, the terms $\bar{h}_a$, $\bar{h}_b$ and $\bar{h}_c$ are, respectively, the distance between the lower surface, the piezo/substrate interface and the top surface from the neutral axis, while $\bar{h}_p$ is the distance between the neutral axis and the mid-plane of the piezoelectric layer (see Figure 13.3). The 31-mode piezoelectric constant is denoted by d_{31} and the permittivity of the clamped capacitive piezoelectric layer is denoted by ε_{33}^S (the superscript S indicates that the property is measured with a constant strain).

As we can see, θ is proportional to d_{31}, while both θ and K_m depend on geometry and strain distribution. This is an important point that will be addressed in more detail in Sections 13.5 and 13.6, but for now, we will complete the presentation of the model by describing the right-hand portion of Equation 13.3, which is the inertial forcing term, F_{in}:

$$F_{in} = -M_{eq}a(t)\mu \tag{13.12}$$

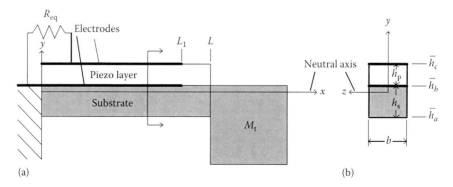

(a)   **(b)**

FIGURE 13.3 Considered beam geometry and relevant dimensions. (a) Side view and, (b) cross section.

The correction factor, μ, is used here to express the force as a function of the equivalent mass, M_{eq}, by accounting for the effect of the mass distribution of the continuous beam and the tip mass. For typical configurations, the correction factor ranges from 1 to 1.5, and the equivalent mass, M_{eq}, in the lumped parameter model will be smaller than the sum of the actual mass of the beam, M_b, and the tip mass, M_t [10].

We can express the electric damping and stiffness explicitly using Laplace transforms to combine these relationships. First, let us isolate the voltage in Equation 13.4 to obtain

$$v(t) = -\left(\frac{j\omega R_{eq}}{1 + j\omega R_{eq} C_p}\right)\theta w(L, t) \qquad (13.13)$$

then multiply both the numerator and the denominator by the complex conjugate of the denominator $(1 - j\omega R_{eq} C_p)$ to bring the complex term at the numerator:

$$v(t) = -\left(\frac{j\omega R_{eq} + \omega^2 R_{eq}^2 C_p}{1 + \omega^2 R_{eq}^2 C_p^2}\right)\theta w(L, t) \qquad (13.14)$$

By substitution of the voltage in Equation 13.3 we obtain

$$\left\{-\omega^2 M_{eq} + j\omega\left[C_m + \theta^2\left(\frac{R_{eq}}{1 + \omega^2 R_{eq}^2 C_p^2}\right)\right]\right.$$

$$\left.+\left[K_m + \theta^2\left(\frac{\omega^2 R_{eq}^2 C_p}{1 + \omega^2 R_{eq}^2 C_p^2}\right)\right]\right\} w(L, t) = F_{in} \qquad (13.15)$$

or alternatively

$$\left[-\omega^2 M_{eq} + j\omega\left(C_m + C_{el}\right) + \left(K_m + K_{el}\right)\right] w(L, t) = F_{in} \qquad (13.16)$$

where [7]

$$K_{el} = \frac{\left(\omega R_{eq} C_p\right)^2}{1 + \left(\omega R_{eq} C_p\right)^2}\left(\frac{\theta^2}{C_p}\right) \qquad (13.17)$$

$$C_{el} = \frac{\left(R_{eq} C_p\right)}{1 + \left(\omega R_{eq} C_p\right)^2}\left(\frac{\theta^2}{C_p}\right) \qquad (13.18)$$

Equations 13.17 and 13.18 show how the electric stiffness and damping are affected by the relative impedance difference of the external load, R_{eq}, with respect to the capacitive layer $(\omega C_p)^{-1}$. In fact, K_{el} increases as the impedance ratio between the external load and the capacitive layer increases. This stiffening behaviour occurs because the flow of electric energy out of the electromechanical structure is progressively blocked, thereby increasing the potential energy. Hence, the term θ^2/C_p effectively represents the maximum feedback piezoelectric force acting on the structure per displacement when the load is an open circuit. Also note that the electric damping is maximized when $R_{eq} = (\omega C_p)^{-1}$, which is the impedance matching condition, but is null in both open- and short-circuit conditions. The tip displacement amplitude, W, and the coupling force amplitude, θV, can now be expressed as

$$W = \frac{M_{eq} A \mu}{\left(K_m + K_{el} - \omega^2 M_{eq} \right) + j\omega \left(C_m + C_{el} \right)} \tag{13.19}$$

$$\theta V = -\left(K_{el} + j\omega C_{el} \right) W \tag{13.20}$$

The undamped natural frequency, given by $\omega_n = \sqrt{K_m/M_{eq}}$, does not consider the effect of electric stiffness. Therefore, the system's resonant frequency is instead expressed by

$$\omega_r = \sqrt{\frac{K_m + K_{el}}{M_{eq}}} \tag{13.21}$$

For convenience, each parameter can be expressed in dimensionless form. As defined in Equations 13.23 through 13.25, Ω is the frequency ratio, ζ_m the mechanical damping factor, α the dimensionless time constant and κ^2 the dimensionless coupling ratio. This last parameter gives the ratio of electric to mechanical energy in open circuit and is related to the effective coupling factor, k_e^2, via Equation 13.26 [11]:

$$\Omega = \frac{\omega}{\omega_n} \tag{13.22}$$

$$\zeta_m = \frac{C_m}{2\omega_n M_{eq}} \tag{13.23}$$

$$\alpha = \omega_n R_{eq} C_p, \tag{13.24}$$

$$\kappa^2 = \frac{\theta^2}{K_m C_p} = \frac{\text{Electrical energy}}{\text{Mechanical energy}} \tag{13.25}$$

$$k_e^2 = \frac{\kappa^2}{1+\kappa^2} = \frac{\text{Electrical energy}}{\text{Applied energy}} \qquad (13.26)$$

Finally, by rearranging Equations 13.19 and 13.20, we obtain the partially dimensionless expressions for the output voltage and tip deflection [8]:

$$V = -\frac{K_{\mathrm{m}}}{\theta}\left(\Delta\Omega_{\mathrm{el}}^2 + 2j\Omega\zeta_{\mathrm{el}}\right)W \qquad (13.27)$$

$$W = \frac{A\mu}{\omega_n^2\left[1+\Delta\Omega_{\mathrm{el}}^2 - \Omega^2 + 2j\Omega\left(\zeta_{\mathrm{m}} + \zeta_{\mathrm{el}}\right)\right]} \qquad (13.28)$$

where $\Delta\Omega_{\mathrm{el}}^2$ and ζ_{el} are the dimensionless forms of the electric stiffness and damping, respectively, given by

$$\Delta\Omega_{\mathrm{el}}^2 = \frac{K_{\mathrm{el}}}{K_{\mathrm{m}}} = \kappa^2\left(\frac{\Omega^2\alpha^2}{\Omega^2\alpha^2 + 1}\right) \qquad (13.29)$$

$$\zeta_{\mathrm{el}} = \frac{C_{\mathrm{el}}}{2\omega_n M_{\mathrm{eq}}} = \frac{\kappa^2}{2}\left(\frac{\alpha}{\Omega^2\alpha^2 + 1}\right) \qquad (13.30)$$

Physically, $\Delta\Omega_{\mathrm{el}}^2$ represents the resonant frequency ratio shift due to the piezoelectric coupling, as shown by the dimensionless form of expression 13.21:

$$\Omega_r = \frac{\omega_r}{\omega_n} = \sqrt{1+\Delta\Omega_{\mathrm{el}}^2} \qquad (13.31)$$

Therefore, it is possible to directly measure the effective coupling factor by observing the frequency shift occurring between the short-circuit and the open-circuit conditions. Furthermore, we previously mentioned that the maximum electric damping occurs when the impedances are matched. In dimensionless form, this condition is expressed by

$$\omega_n R_{\mathrm{eq}} C_{\mathrm{p}} = \frac{\omega_n C_{\mathrm{p}}}{\omega C_{\mathrm{p}}} \qquad (13.32)$$

$$\alpha = \Omega^{-1} \qquad (13.33)$$

and with such a load, the shift and the resonant frequency become constants:

$$\Delta\Omega_{el}^2\big|_{\alpha=\Omega^{-1}} = \frac{\kappa^2}{2} \tag{13.34}$$

$$\Omega_r\big|_{\alpha=\Omega^{-1}} = \sqrt{\frac{1+\kappa^2}{2}} \tag{13.35}$$

meaning that both of those conditions are met for a singular load and frequency. Therefore, when the system resonates and the impedances are matched, the maximum achievable electric damping is

$$\zeta_{el}\big|_{\Omega=\Omega_r,\alpha=\Omega_r^{-1}} = \frac{\kappa^2}{4\sqrt{1+\kappa^2/2}} \approx \frac{\kappa^2}{4} \tag{13.36}$$

with the approximation introducing an error less than 2.5% for $\kappa^2 < 0.1$. Finally, based on Equations 13.27 through 13.30, the mean electric power output can be evaluated by

$$P_{el} = \frac{V}{2} \frac{V^*}{R_{eq}} \tag{13.37}$$

$$= \frac{K_m^2}{2\theta^2 R_{eq}} \left(\Delta\Omega_{el}^4 + 4\Omega^2\zeta_{el}^2 \right) |W|^2 \tag{13.38}$$

$$= \frac{\mu^2 A^2 K_m^2}{2\theta^2 R_{eq}\omega_n^4} \frac{\left(\Delta\Omega_{el}^4 + 4\Omega^2\zeta_{el}^2 \right)}{\left[\left(1 + \Delta\Omega_{el}^2 - \Omega^2\right)^2 + 4\Omega^2 \left(\zeta_m + \zeta_{el} \right)^2 \right]} \tag{13.39}$$

$$= \frac{\mu^2 A^2 M_{eq}}{\omega_n} \frac{K_m C_p}{2\theta^2 (\omega_n R_{eq} C_p)} \frac{\frac{\kappa^4 \left[(\Omega^2\alpha^2)^2 + \Omega^2\alpha^2 \right]}{\left(\Omega^2\alpha^2 + 1\right)^2}}{\left[\left(1 + \Delta\Omega_{el}^2 - \Omega^2\right)^2 + 4\Omega^2 \left(\zeta_m + \zeta_{el} \right)^2 \right]} \tag{13.40}$$

$$= \frac{\mu^2 A^2 M_{eq}}{\omega_n} \frac{\frac{\kappa^2}{2} \frac{\Omega^2\alpha}{\Omega^2\alpha^2 + 1}}{\left[\left(1 + \Delta\Omega_{el}^2 - \Omega^2\right)^2 + 4\Omega^2 \left(\zeta_m + \zeta_{el} \right)^2 \right]} \tag{13.41}$$

$$= \frac{\mu^2 A^2 M_{eq}}{\omega_n} \frac{\Omega^2 \zeta_{el}}{\left[(1 + \Delta\Omega_{el}^2 - \Omega^2)^2 + 4\Omega^2 \left(\zeta_m + \zeta_{el} \right)^2 \right]} \qquad (13.42)$$

with V^* the complex conjugate of the voltage output. The first term of Equation 13.42 regroups parameters such as the vibration amplitude and frequency, as well as the equivalent mass of the device. The second term provides the frequency response and the effect of damping on the output power. At resonance, the expressions for the tip displacement and the power output (Equations 13.28 and 13.42) can be simplified to

$$\left. |W| \right|_{\Omega=\Omega_r} = \frac{A\mu}{2\omega_n\omega_r \left(\zeta_m + \zeta_{el} \right)} \qquad (13.43)$$

$$\left. |P_{el}| \right|_{\Omega=\Omega_r} = \frac{\mu^2 A^2 M_{eq}}{4\omega_n} \frac{\zeta_{el}}{\left(\zeta_m + \zeta_{el} \right)^2} \qquad (13.44)$$

These equations will now be used through this chapter to evaluate the limits and the performance of a typical PVEH device with the base line parameters found in Table 13.1. The piezoelectric material considered is a common hard lead zirconate titanate (PZT) ceramic, PZT4 [12], and silicon is used as the tip mass because it can be readily micromachined. Also note that in the generator mode, the piezoelectric material coupling factor, k_{31}, is the ratio between the electric energy transformed and the energy provided (i.e. $k_{31}^2 = E_{el}/E_{applied}$). For a single-direction, 31-mode plane stress situation [9,11],

$$k_{31}^2 = \frac{d_{31}^2 Y_p^2}{\varepsilon_{33}^T} \qquad (13.45)$$

$$\varepsilon_{33}^T = \varepsilon_{33}^S + d_{31}^2 Y_p^2 \qquad (13.46)$$

where ε_{33}^T is the permittivity of the piezoelectric material measured with a constant and preferably null stress.

TABLE 13.1

Baseline Parameters Used for Estimates

Young Modulus Y_p (GPa)	Dynamic Tensile Strength σ_l (MPa)	Coupling Factor k_{31}	Quality Factor Q_m	Frequency f (Hz)	Acceleration Amplitude A (m/s²)	Density ρ_M (kg/m³)
81.3	24	0.33	250	150	1	2330

13.4 LIMIT BASED ON INERTIAL COUPLING

How much energy is transmitted to the device fundamentally restrains how much can be harvested. For PVEH, both the vibration source and the quality of the mechanical coupling limit this quantity, which we seek to evaluate in this section. First, let us assume that the piezoelectric device has a very large tip mass; hence, $M_{eq} \approx M_t$ and $\mu = 1$. Its volume, Vol_M, occupies most of the space of the device and the effect of its sweeping motion is also neglected. Finally, we assume that the frequency shift due to the piezoelectric effect is small; hence, $\omega_n \approx \omega_r$.

13.4.1 INERTIAL COUPLING LIMIT OF LINEAR RESONATOR

For steady-state oscillations at the resonant frequency, ω_r, the total energy stored in a resonating device remains constant and is exchanged between the spring and the mass. At the same time, some energy is supplied by the source and the same amount is removed via energy harvesting and other losses. The stored energy can be evaluated as $E_s = 1/2\left(K_m + K_{el}\right)W_{max}^2 = 1/2\omega_r^2 M_{eq}W_{max}^2$, where W_{max} is the maximum tip amplitude. The rate at which this energy is converted from potential to kinetic energy during every cycle of oscillation is thus given by

$$P_s = \frac{1}{2}\left(K_m + K_{el}\right)W_{max}^2 f = \frac{1}{4\pi}\omega_r^3 M_{eq}W_{max}^2 \qquad (13.47)$$

The maximum displacement at resonance under a base acceleration is

$$W_{max} = \frac{A}{2\omega_n^2\left(\zeta_m + \zeta_{el}\right)} \qquad (13.48)$$

Hence, we could define the *internal energy density exchange rate* as (in terms of mass volume, Vol_M)

$$\frac{P_s}{Vol_M} = \frac{1}{16\pi}\frac{\rho_M A^2}{\omega_n\left(\zeta_m + \zeta_{el}\right)^2} \qquad (13.49)$$

Furthermore, the electric damping, ζ_{el}, should equal the mechanical damping, ζ_m, for optimal power generation [8,13]. Considering a damping ratio $\bar{\zeta} = \zeta_{el}/\zeta_m = 1$, and recognizing that the mechanical quality factor $Q_m = (2\zeta_m)^{-1}$, Equation 13.49 can be written as

$$\frac{P_s}{Vol_M} = \frac{1}{4\pi}\frac{\rho_M A^2 Q_m^2}{\omega_n\left(1 + \bar{\zeta}\right)^2} = \frac{1}{16\pi}\frac{\rho_M A^2 Q_m^2}{\omega_n} \qquad (13.50)$$

Using the parameters found in Table 13.1, this energy density exchange rate is evaluated at 3 mW/cm³, which is about two orders of magnitude greater than the electric power densities typically achieved by inertial PVEH devices excited by vibrations of a similar magnitude.

Moreover, mechanical energy must be continuously supplied to maintain the high-amplitude steady-state oscillations, and it is the inertial force acting on the oscillating mass that provides this energy by doing work. The amplitude of this harmonically applied force is $F_{in} = M_{eq} A$, which is also perfectly in phase with the mass velocity at resonance. In this condition, the average input power is maximized and given by the product of their root mean squared values:

$$P_{in} = P_{el} + P_{m,loss} = F_{in,RMS} \left(\omega W_{RMS} \right) \tag{13.51}$$

which for $\overline{\zeta} = 1$ leads to

$$\frac{P_{in}}{Vol_M} = \frac{\rho_M A^2}{4\omega_n \zeta_m \left(1 + \overline{\zeta}\right)} = \frac{\rho_M A^2 Q_m}{4\omega_n} \tag{13.52}$$

For the baseline parameters, this input mechanical power density is 154 μW/cm^3, approximately 20 times less than the internal energy density exchange rate. This ratio is directly a function of the quality factor, as $P_s/P_{in} = Q_m/4\pi$, and simply states by definition that a large amount of energy must remain in the structure to sustain the high-amplitude oscillations.

From the expression of the electric power output (Equation 13.44), the electric power density can be estimated as [8,13]

$$\frac{P_{el}}{Vol_M} = \frac{\rho_M A^2}{4\omega_n} \frac{\zeta_{el}}{\left(\zeta_m + \zeta_{el} \right)^2} = \frac{\rho_M A^2 Q_m}{2\omega_n} \frac{\overline{\zeta}}{\left(1 + \overline{\zeta}\right)^2} \tag{13.53}$$

Again, mechanical and electric damping should be equal to maximize the net power output. In this situation, the theoretical electric power density limit due to inertial coupling is reached and is equal to [14]

$$\frac{P_{lim}}{Vol_M} = \frac{\rho_M A^2 Q_m}{8\omega_n} \tag{13.54}$$

Here, the resulting electric power density is estimated to be 77 μW/cm^3, exactly half the input mechanical power density. This ratio of input mechanical to output electric powers, which is in fact the harvesting efficiency of the resonator, can be expressed as

$$\eta = \frac{P_{el}}{P_{in}} = \frac{\overline{\zeta}}{\left(1 + \overline{\zeta}\right)} \tag{13.55}$$

For $\bar{\zeta} = 1$, we get a value of 0.5, meaning that half of the energy pumped into the resonator at each cycle is lost through mechanical damping and the other half is extracted as electric energy. Therefore, if ζ_{el} is higher than ζ_m, a larger fraction of the energy pumped into the system and converted electrically will be extracted ($0.5 < P_{el}/P_{in} < 1$). However, the electric power output will not increase since the excess damping reduces the amplitude, W_{max}. Consequently, the input mechanical power and the electric power output are both lower. The device is more *efficient*, but it produces less power from the same source. In other words, it is also less *effective* at capturing energy and producing electric power [15]. Therefore, another useful FOM, the harvesting effectiveness, ξ, is defined as

$$\xi = \frac{P_{el}}{P_{lim}} = 4\frac{\bar{\zeta}}{\left(1+\bar{\zeta}\right)^2} \tag{13.56}$$

To summarize, the energy involved in the operation of a resonator can be grouped in three categories:

1. The energy that flows into the device. The inertial force does work on the oscillating mass and the system captures energy every cycle. Resonance optimizes the capture of the energy because the applied force is in phase with the tip mass velocity.
2. The energy that flows out of the device, via various dissipation mechanisms such as mechanical damping losses, dielectric losses and energy harvesting. In the model, the electric power output is directly given by the electric damping, but we can expect dielectric losses to also introduce additional electric damping.
3. The energy that remains inside the harvester, flowing back and forth as potential and kinetic energy. During steady-state vibrations, electric power output will increase as long as the effect of the electric damping is less than or equal to the effect of the mechanical damping. If $\zeta_{el} > \zeta_m$, the power output will actually decrease, but the device will be more efficient. While the electric energy extraction rate is increased, less energy is captured, creating a backward coupling effect.

Figures 13.4 and 13.5, respectively, plot the evolution of those different energy and power categories in the time domain for three different regimes: (a) ramp up; (b) steady state, with simultaneous energy injection and extraction; and then (c) decay, with extraction only. A sinusoidal base acceleration with an amplitude of 1 m/s² is applied to the base of the system during the 4 s. Once the excitation is stopped completely, no more energy is injected and the stored energy is gradually dissipated or extracted. At this point, energy extraction would be optimized by maximizing electric damping so that most of the energy remaining in the device can be harvested.

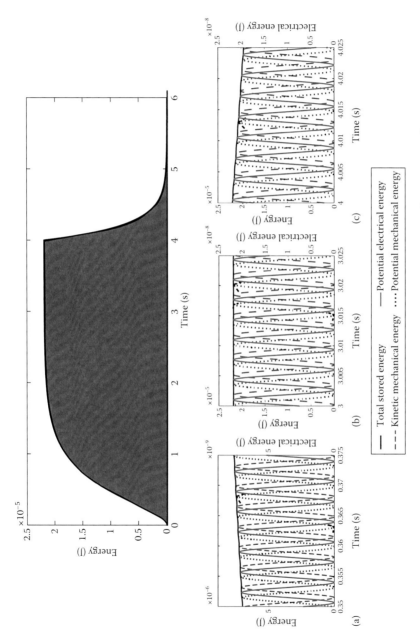

FIGURE 13.4 Evolution in the time domain of the different types of energy for different regimes. $\zeta_m = 0.002$ and $\zeta_{el} = 0.0005$. (a) Ramp-up, (b) steady state, and (c) decay.

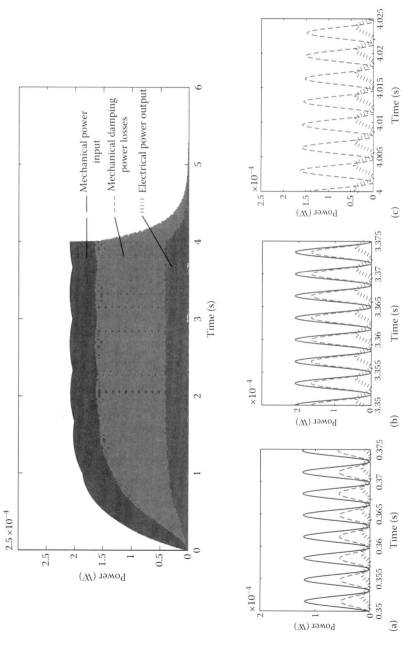

FIGURE 13.5 Evolution in the time domain of the different types of power for different regimes. $\zeta_m = 0.002$ and $\zeta_{el} = 0.0005$. (a) Transient, (b) steady state, and (c) decay.

According to Equation 13.54, high values of Q_m and A^2/ω lead to devices with high power densities, mostly because more energy can be injected in both cases for the same mass volume. However, this result does not account for the design of the strained piezoelectric structure. There is a limit on the amount of strain that can be applied and to the amount of energy that can be converted and extracted. These additional limits on performance are the focus of the next sections.

13.5 STRESS-BASED LIMITS

The ultimate tensile stress of piezoceramics can be as high as a few hundred megapascals. In fact, this value can increase further for small-scale structures because they can be prepared with higher crystalline quality and lower defect densities than bulk ceramics [16]. However, for some materials, only a fraction (20%–30%) of this value can be applied during operation to avoid degradation or failure due to depolarization and fatigue [17,18].

For this reason, the maximum stress in the spring element must be limited to ensure reliable operation for the lifetime of the device. The maximum mechanical power density is then given by the rate at which this maximum strain energy can be induced to the transducer without leading to its failure. Hence, the highest power density is achieved by stressing the material to this limit, here noted by σ_l, and using it optimally by applying this stress as uniformly as possible. To represent such a uniform stress field, we first assume that the device is simply made of a piezoelectric block of length L, thickness h and width b, clamped on one end and stressed along its axis. The stress amplitude is σ_l and it is applied in a harmonic fashion, cycling between tension and compression. Knowing that for this block, $K_m = Y_p(bh/L)$ and $W_{max} = (bh\sigma_y)/K_m$, and also supposing that all the electric energy is extracted, the average electric power density is given by [3]

$$\frac{P_{el}}{Vol_b} = \frac{1}{2}\frac{K_m W_{RMS}^2}{Vol_b}k_{31}^2 f = \frac{1}{4}\frac{K_m W_{max}^2}{Vol_b}k_{31}^2 f \tag{13.57}$$

$$= \frac{1}{4}\frac{(\sigma_l)^2}{Y_p}k_{31}^2 f \tag{13.58}$$

Here, Vol_b refers to the volume of a monolithic piezoelectric spring and this limit suggests an ultimate power density of 29 mW/cm³ using our baseline parameters. Therefore, the volume of the mass is not considered in this calculation.

However, most resonators are not subject to a uniform stress field. Stress gradients during operation are commonly encountered. This issue is considered in detail in the next subsection by considering the widely used cantilever beam geometry.

13.5.1 POWER DENSITY REDUCTION WITH STRESS DISTRIBUTION: CASE OF THE CANTILEVER

Let us assume that this same piezoelectric block is now subjected to a transverse force and behaves like a bending cantilever beam instead. The power density expression found in Equation 13.57 remains mostly valid, but two assumptions are made for simplicity:

- The coupling factor of the material, k_{31}^2, adequately represents the piezoelectric conversion effectiveness of the transducer.
- The stress distribution is that of a statically deformed cantilever beam with a constant rectangular section and a force applied at its free end. The static deflection shape is given by

$$\hat{w}(x) = \frac{1}{2}\left[-\left(\frac{x}{L}\right)^3 + 3\left(\frac{x}{L}\right)^2\right] \tag{13.59}$$

From the well-known cantilever beam equations for deformation, stress and stiffness (Equation 13.7),

$$K_m = \frac{bh^3Y}{4L^3} \tag{13.60}$$

$$\sigma_{max} = \frac{3}{2}\frac{hY}{L^2}W_{max} \tag{13.61}$$

Replacing Equations 13.60 and 13.61 into Equation 13.57, the power density can be written as

$$\frac{P_{el}}{Vol_b} = \frac{1}{36}\frac{\sigma_{max}^2}{Y}k_{31}^2f \tag{13.62}$$

Comparing Equations 13.58 and 13.62, we conclude that a power density reduction by a factor of nine (to 3.2 mW/cm³) is due to the distribution of the strain along the length and across the thickness of the beam (see Figure 13.6).

A triangular tapered beam, while stiffer, would maintain the stress over the entire length and only have a reduction factor of three coming from the linear stress distribution through the thickness [19]. Hence, strictly from the mechanical strain distribution, the power density reduction in the beam is already significant. Still, by comparison with the inertial limit established previously in Section 13.4, the mass to spring volume ratio for a similar power output remains fairly large at ≈40. This means that relatively speaking, a large mass is indeed

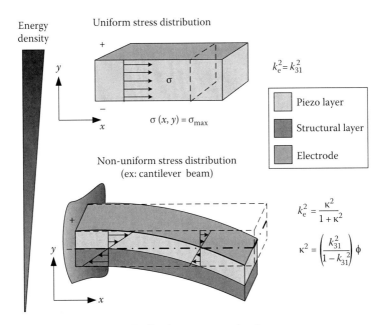

FIGURE 13.6 Effect of stress distribution on energy density.

needed to achieve such stress levels with the specified source and quality factor. In these conditions, the size of the device is therefore driven by its mass. For these two values to reach a similar level, one would have to significantly improve Q_m and the density of the mass, ρ_M, by this same ratio. Alternatively, a vibration source that is more than six times larger in magnitude would lead to the same result.

However, the influence of the non-uniform stress distribution on the electromechanical conversion factor was not considered thus far. This effect is discussed in the next section.

13.6 ELECTROMECHANICAL CONVERSION

Up to this point, we have assumed that the effective piezoelectric conversion factor, k_e^2, was only dependent of the piezoelectric material (i.e. $k_e^2 = k_{31}^2$). However, this is only true for a bulk piezoelectric specimen that is uniformly stressed in a single direction. The conversion factor can reduce and even reach small values approaching zero, when a non-uniform stress is applied. In this section, the effect of the geometry as well as the material properties is considered to better assess the conversion effectiveness.

13.6.1 Geometric Considerations

A convenient manner of expressing κ^2 consists of separating the piezoelectric properties from the geometric and mechanical parameters [8]. Such a framework is useful for comparing and optimizing different configurations. First, we assume that for a d_{31} mode device,

$$\theta = Y_p d_{31} \phi_1 \tag{13.63}$$

$$C_p = \varepsilon_{33}^S \phi_2 \tag{13.64}$$

where the variables ϕ_1 and ϕ_2 define the geometric contribution of each term. Based on Equations 13.5 and 13.6 from Section 13.3,

$$\phi_1 = \left(\overline{bh_p} \int_0^{L_1} \frac{d^2 \hat{w}(x)}{dx^2} dx \right) \tag{13.65}$$

$$\phi_2 = \varepsilon_{33}^S \left(\frac{bL_1}{h_p} \right) \tag{13.66}$$

Equation 13.25 can then be rewritten as

$$\kappa^2 = \frac{d_{31}^2 Y_p^2}{\varepsilon_{33}^S} \frac{\phi_1^2}{\phi_2 K_m} = \frac{k_{31}^2}{1 - k_{31}^2} \phi \tag{13.67}$$

with $\phi = Y_p \phi_1^2 /(\phi_2 K_m)$ being the global geometric factor. For simplicity, the static deflection shape (Equation 13.59) is again used to evaluate the deformation-dependent terms. Evaluating the integrals in Equations 13.65 and 13.67, we obtain

$$\int_0^{L_1} \frac{d^2 \hat{w}(x)}{dx^2} dx = \frac{3}{2L} \left[-\left(\frac{L_1}{L} \right)^2 + 2\left(\frac{L_1}{L} \right) \right] \tag{13.68}$$

$$\int_0^L \left(\frac{d^2 \hat{w}(x)}{dx^2} \right)^2 dx = \frac{3}{L^3} \tag{13.69}$$

After some algebraic manipulations, ϕ for the unimorph beam can finally be expressed as

$$\phi = \frac{9}{4} \frac{\tilde{Y} H (1-H)^2 \frac{L_1}{L} \left(2 - \frac{L_1}{L} \right)^2}{\left[(\tilde{Y} - 1)H + 1 \right] \left[(\tilde{Y} - 1)^2 H^4 + (\tilde{Y} - 1)H(4H^2 - 6H + 4) + 1 \right]} \tag{13.70}$$

where $\tilde{Y} = Y_p / Y_s$ is the elasticity ratio between the active and the structural layers, $H = h_p / h_{total}$ is the fraction of piezoelectric material that composes the beam and L_1 / L

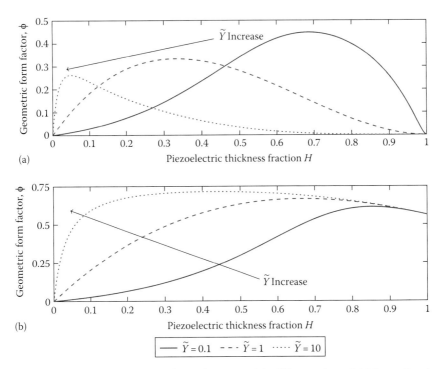

FIGURE 13.7   Form factor charts for various material stiffness ratio and thickness fraction for the (a) unimorph and (b) bimorph configurations. $L_1/L = 1$.

is the partial coverage ratio of the electrode. Therefore, ϕ is a function of the elasticity ratio and the spatial distribution of the piezoelectric material on the structural layer.

Equation 13.70 is plotted in Figure 13.7a for several elasticity ratios. The first thing noticeable is the invalidity of the previous assumption made in Section 13.5.1 regarding the conversion effectiveness of a bending beam made entirely from piezo-electric material. In fact, such a configuration would be completely ineffective, with $\phi = 0$. This occurs due to the charge cancellation between the bottom half and the top half of the cantilever that are stressed in opposite directions. Moreover, each elasticity ratio presents an optimal thickness fraction. For instance, with $\tilde{Y} = 1$ and $L_1/L = 1$, the form factor simplifies to $\phi = 9/4[H(1 - H)^2]$; the optimal thickness fraction is then $H = 0.33$, which gives $\phi_{opt} = 0.33$. This figure also demonstrates that coupling for the unimorph configuration is ideally increased by using a thick and compliant piezoelectric layer on a thin, rigid substrate. However, if we consider the thin-film deposition techniques usually used in MEMS fabrication, the low fraction region is more relevant, and in this case, the trend rather favours a stiff and thin layer on a thick and compliant substrate.

The same simple scheme can be used on a symmetric bimorph beam; similar results are illustrated in Figure 13.7b. Contrary to the unimorph configuration, the bimorph made exclusively from piezomaterial does not suffer from the same charge cancellation problem, thanks to the segregation of the regions that are stressed in opposite directions. While the form factor remains far from unity, it is much better than for the unimorph with a value of ≈ 0.56. Moreover, the bimorph configuration is generally better exploited when stiffer and relatively thinner films are used.

In both cases, it appears that using piezoelectric material near the neutral axis decreases the effective coupling. Other configurations such as air-spaced [20] and tapered [21] cantilevers have shown improved conversion factors as well. In fact, Figure 13.7b tends to support those claims in the case of air-spaced cantilevers as the curves go towards maximum asymptotic values as $\tilde{Y}$ increases. According to Yen, corrugated cantilevers can also be optimized to reach form factors that are similar to bimorph beams [22]. The advantages of this geometry are that no passive support layer is required while also being easier to process than a bimorph.

In addition, Equation 13.70 shows that the geometric factor can also be optimized by covering only a portion of the beam with the electrode instead of its whole length. With our assumptions, the optimal coverage ratio is $L_1/L = 2/3$, which increases ϕ by about 19% (to 0.39 for a unimorph). Because the bending moment in the last third of the beam is very weak, very few charges are induced there. Hence, covering this part contributes to an overall decrease of the conversion effectiveness and of the effective coupling factor. This charge redistribution phenomenon and these results were recently validated experimentally by Stewart et al. [23].

Therefore, the capacitive piezoelectric layer has to be located strategically on portions of the spring that are under large stresses if the goal is to optimize the conversion effectiveness of the device. With these results, the electromechanical conversion decreases from $k_e^2 = k_{31}^2 = 10.9\%$ (uniform stress) to $k_e^2 = 4.6\%$ for an optimized unimorph cantilever. This represents a 58% drop, giving a beam power density of 1.4 mW/cm³. This value is about 20 times less than what was first estimated for a uniformly stressed sample.

Table 13.2 presents a summary of the assessments made in this section by collecting values of the global geometric factor estimated for various configurations. Note that the estimate for the air-spaced cantilever is based on the results for the bimorph by assuming $\tilde{Y} \to \infty$ and that the beam does not bend with an S shape, which is possible for thin layers far apart from each others. In this case, the electrode has to be segmented to avoid charge cancellation [24].

TABLE 13.2

Global Geometric Factor for Various Configurations ($L_1/L = 1$)

Configuration	$\tilde{Y}$	H_{opt}	ϕ_{opt}
Unimorph	1	1/3	1/3
Symmetric bimorph	1	2/3	2/3
Tapered unimorph	1	1/3	4/9
Tapered symmetric bimorph	1	2/3	8/9
Air spaced (symmetric)	∞	—	< 0.75
Corrugated (optimized) [22]	1	—	2/3

13.6.2 PIEZOELECTRIC MATERIAL CONSIDERATIONS

Even more important than geometry, the piezoelectric material properties have the dominant influence on the conversion effectiveness. Piezoelectric materials exist in a variety of phases and crystallographic structures and can be synthesized through a large range of processes. Each material and integration process presents their own different challenges and affects the cost as well as the performances of the device [25]. While a complete treatment of the synthesis and characterization of piezoelectric material is out of the scope of this work, it is relevant to provide an overview of potential candidates in terms of their processing and resulting properties. Tables 13.3 and 13.4 contain data collected for a wide variety of bulk, thick- and thin-film piezoelectric materials.

A major difference of thin films compared to bulk materials is that their mechanical compliance is often dominated by the underlying thick substrate. The influence of residual stresses due to mismatch in lattice constant and coefficient of thermal expansion, as well as coupling with other piezoelectric modes, is also included in the measurements. Therefore, it is difficult to measure the compliance of the film and measurements rather lead to effective film properties, most notably the $e_{31,f}$ constant (in stress–charge notation), which include some coupling with the thickness mode. Equations 13.71 and 13.72 express its relation to the d_{ij} piezoelectric constants in the strain–charge notation, depending on the stress conditions [9,26,27]:

$$\text{Thin beam, main stress in one direction: } e_{31,f} = \frac{d_{31}}{s_{11}^{E}} \tag{13.71}$$

$$\text{Thin plate, main stresses in two directions: } e_{31,f} = \frac{d_{31}}{s_{11}^{E} + s_{12}^{E}} \tag{13.72}$$

For this reason, an energy harvesting FOM, which is noted by $e_{31,f}^{2}/(\varepsilon_0 \varepsilon_{33})$, is frequently used in the thin-film processing literature instead of the usual coupling factor k_{31}^{2} that is used for bulk materials [26,27].

The data points in Tables 13.3 and 13.4 show that even though barium titanate ($BaTiO_3$) has lower d_{31} constants than PZT ceramics, this material offers comparable or better coupling factors than PZT due to its higher stiffness and lower dielectric constant. This is also the case for thin-film aluminium nitride (AlN).

Finally, it is commonly assumed that a very high coupling factor, such as those of single-crystal relaxor materials, is highly desirable for high-performance energy harvesters. Equation 13.58 supports this claim. However, one must also consider the conclusions relative to the inertial coupling limit established in Section 13.4. Indeed, for a resonant device, there is a limit where increasing coupling will provide no benefits on the net power output or power density, and this limit is closely related to mechanical damping. Later in this chapter, we shall discuss the importance of the mechanical quality factor, Q_m, in the selection of a piezoelectric material, but first, we address the topic of the energy extraction.

TABLE 13.3
Reported Piezoelectric Properties for Bulk Materials

Material	Type	$(s_{11}^E)^{-1}$ (GPa)	$\varepsilon_{33}^T/\varepsilon_0$	d_{31} (pC/N)	k_{31}(%)	Q_m	Ref.
PZT–5A	Ceramic	61/71	1500/1928	−171/−190	34/37	75/80	[12,28,29]
PZT–5H	Ceramic	58/63	3200/3935	−250/−320	36/44	30/75	[12,28,29]
PZT–4	Ceramic	72/81	1135/1494	−120/−150	33/36	500/600	[12,28,29]
PZT–8	Ceramic	87/96	1000/1205	−97/−127	33/36	900/1050	[28–30]
PMN–0.25PT	Ceramic	184	1167	−74	31	283	[31]
PMN–0.25PT	Single crystals	37.3	2560	−240/−569	33/73	131/362	[31]
PMN–0.29PT	Single crystals	—	5500	−1350	87	100	[32]
PMN–0.3PT	Single crystals	17.5/50	6610/7800	−742/−1395	49/90	44	[33–35]
PMN–0.32PT	Single crystals	15/34.8	650/5700	−160/−930	32/78	31/68	[36]
PMN–0.33PT	Single crystals	14.5	8200	−1330	59	—	[37]
PMN–0.345PT	Ceramic	75.4	4952	−255	75	—	[38]
PMN–0.42PT	Single crystals	106	660	−91	39	—	[39]
Soft PMN-PZT	Single crystals	7.87	4500/8000	−1400/−2252	90/95	100	[32,40]
Hard PMN-PZT	Single crystals	—	3100/4000	−850/−1200	86/88	>500	[32]
PZN–0.045PT	Single crystals	9.35/28.6	2553/5600	−690/−1540	43/85	95/430	[30,41]
PZN–0.045PT + 0.01Mn	Single crystals	13.2/43.5	1572/3491	−542/−830	42/80	375/441	[41]
PZN–0.045PT + 0.02Mn	Single crystals	27	1626/1873	−458/−502	69	336	[41]
PZN–0.07PT	Single crystals	14.8	3180	478	35	—	[42]
PZN–0.08PT	Single crystals	11.5	7700	−1455	60	40	[35,43]
PZN–0.12PT	Single crystals	54	612.4	−148	50	—	[38]
BaTiO$_3$	Single crystals	124	168	−35	32/59	400	[29,33,44]
BaTiO$_3$	Ceramics	110/125	625/1700	−32/−78	15/21	300/1400	[12,28,29]
ZnO	—	127	11/12.64	−5.2/−5.43	18/19	N/A	[12]
LiNbO$_3$	—	173	30	−1	2.6	—	[29]

TABLE 13.4

Reported Piezoelectric Properties for Some Thin-Film Materials

Material	Type	Substrate	$\varepsilon_{33}^S/\varepsilon_0$	$e_{31,f}$ (pC/N)	$e_{31,f}^2/\varepsilon_0\varepsilon_{33}$ (GPa)	Q_m	Ref.
AlN	MOCVD epitaxial	Sapphire	9.5	−1.37	22.3	2490	[45]
AlN	Sputtering	Si	10.2	−1.3	18.7	—	[45]
AlN	Sputtering	—	10.5	−1.05	11.9	—	[26]
AlN	Sputtering	Si	—	—	—	120/500	[46,47]
ZnO	Sputtering, single crystal	Si	10.9	−1.0	10.3	1770	[26,45,48]
PZT (53/47)	Single crystal, sputtering	MgO	200	−6.2	21.7	—	[49]
PZT (53/47)	Polycrystalline, sputtering	Si	200	−7.7	33.48	—	[49]
PZT	Sol-gel	SiO₂/Si	1100/1300	−4/−6	6/18	54/237	[45,50]
PZT (48/52)	—	MgO	—	−3.98	—	114	[51]
PMnN–0.94PZT (48/52)	Single crystal, sputtering	MgO	100	−12.0	163	185	[51]
PMnN–0.94PZT (50/50)	Polycrystalline, sputtering	SiO₂/Si	834	−14.9	30	—	[51]
PMN–0.33PT	Single crystal, sputtering	MgO	500	−5	5.65	20	[52]
PMN–0.33PT	Single crystal, sputtering	SrTiO₃/Si	1600	−27	50	—	[53]

13.7 ELECTRIC ENERGY EXTRACTION

Following the conversion of the energy from the mechanical to the electric form by the transducer, it must be extracted from the piezocapacitance to be used. The amount of energy converted depends on the piezoelectric coupling of the harvesting structure, but the fraction of it that is extracted depends on the load to which it is connected. The performance of an electric energy extraction interface could therefore be considered in terms of *effectiveness* (the fraction of converted energy that is extracted) and *efficiency* (the fraction of successfully extracted energy, accounting for the Joule effect or current leakage losses, for instance). In this work, we mostly touch on the effectiveness aspect of this process.

13.7.1 RESISTIVE LOAD

It is simple to show that the impedance-matched resistance enables half of the electric energy converted to be harvested continuously at steady state. As explained in Section 13.3, the piezoelectric element can be modelled as a current source connected in parallel to a capacitance, C_p. While that actual output power, P_{el}, was derived from the power delivered to the resistive load, we can also evaluate the percentage of the energy that is extracted by considering the power delivered to the whole RC circuit. In this case, the circuit impedance is $Z = (R_{eq}^{-1} - j\omega C_p)^{-1}$ and the complex electric power, S, is given by

$$S = \frac{VV^*}{Z^*} = \frac{1}{2}\frac{K_m^2}{\theta^2}\frac{(\Delta\Omega_{el}^2 + 4\Omega^2\zeta_{el}^2)}{Z^*}WW^* \tag{13.73}$$

After manipulating these equations, S can be rewritten as

$$S = \frac{1}{2}\omega K_m \left| W \right|^2 \kappa^2 \left(\frac{\Omega^2\alpha^2}{\Omega^2\alpha^2 + 1} \right) \left(\frac{1 - j\Omega\alpha}{\Omega\alpha} \right) \tag{13.74}$$

The first part is the electric power expressed in terms of the electric stiffness (because $K_m\,\kappa^2 = (K_{el})_{max}$). The second term modulates this power as a function of operating frequency and load, and the third term provides the proportions of real and reactive powers. For a matched impedance, the modulator is 0.5, and the reactive to real power ratio is 1. For $\Omega\alpha\rightarrow\infty$, the modulator is 1, but all the generated power is reactive. Therefore, the maximum extraction effectiveness of a crude matched resistance is indeed of 50% for steady state resonance operation.

To optimize the power output of the linear resonator, one should ensure that $\zeta_{el} = \zeta_m$, meaning that the piezoelectric element should provide sufficient coupling to fulfil this condition. Considering Equation 13.36 and that $\zeta_m = (2Q_m)^{-1}$, we can rewrite these criteria as

$$\kappa^2 Q_m \approx 2 \tag{13.75}$$

$\kappa^2 Q_m$ is also called the piezoelectric resonator figure of merit [11] and is used to compare the mechanical losses to the electromechanical coupling of the structure. This FOM is also commonly used to characterize the performance of surface acoustic wave (SAW) devices. As the criterion for PVEH, it basically states that devices with low piezoelectric coupling or low mechanical quality factor ($\kappa^2 Q_m \ll 2$) may not be able to reach the optimal power generation condition. Nonetheless, once the critical coupling condition is reached ($\kappa^2 Q_m \geq 2$ for a resistive load), further increasing the coupling will not increase the maximum output power. Also, while increasing Q_m is always beneficial for power density according to Equation 13.54, the MEMS packaging must still provide enough travelling range [14]. Finally, the resonator FOM also relates to the maximum harvesting efficiency of the resonator (Equation 13.55), η_{max}. With a resistance, this efficiency is given by [54,55]

$$\eta_{max} \approx \frac{\kappa^2 Q_m}{2 + \kappa^2 Q_m} \tag{13.76}$$

In addition, the raw output of the standard interface is generally AC, meaning that to store and use this energy, the current must be rectified, filtered and regulated appropriately first. Doing so requires the use of diodes and storage capacitances, introducing losses (which affect the global efficiency) and threshold voltages blocking the energy from flowing out of the transducer (which affect the effectiveness of the interface). Furthermore, as we will see in the following section, it is also possible to use non-linear extraction schemes to improve the energy transfer and the harvesting efficiency.

13.7.2 Non-linear Extraction Interfaces

A practical energy harvester requires the implementation of an energy buffer in order to assure a reliable operation under various circumstances. Either with a capacitance, a super capacitance or a battery, such a system is required because of the fluctuation of the available ambient energy. Indeed, energy generation and consumption are not necessarily synchronized [1]. Although this complexifies the system, it allows the use of energy extraction circuits between the harvester and the storage to maximize extraction effectiveness. The storage stage may also not be directly connected to the piezoelectric transducer, which affects the extraction process [56].

Most notably, non-linearities can be used to synchronize the extraction of energy with the maximum displacement of the transducer, therefore somewhat decoupling the energy injection and extraction process. The synchronous electric charge extraction (SECE) and double synchronized switch harvesting (DSSH) architectures also allow a decoupling of the extraction and storage stages [57]. All these non-linear harvesting interfaces can increase the electric damping for a given amount of coupling. Hence, the benefits of these circuits are significant for highly damped or lightly coupled devices that have low values of $\kappa^2 Q_m$ ($\kappa^2 Q_m \ll 2$). In some

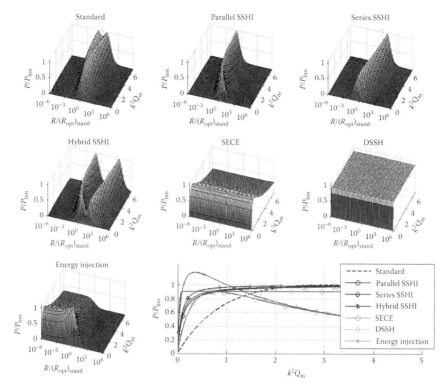

FIGURE 13.8   Normalized harvested power as a function of load and normalized maximum power for several extraction interfaces under a constant force magnitude. (From Lallart M and Guyomar D, *IOP Conf. Ser. Mater. Sci. Eng.*, 18(9), 092006, 2011.)

cases, gains of more than 400% were observed in terms of power output. The plots on Figure 13.8, adapted from [57], show that every interface tends towards the same limit, $P_{\lim}$, as $\kappa^2 Q_m$ increases. The sole exception is the injection technique, which can actually surpass the inertial limit since electric energy is added in the system [56].

Hence, even with very effective extraction interfaces, inertial PVEH devices that operate in resonance cannot overcome the inertial limit stated by Equation 13.54. Therefore, the advantages of these synchronous extraction schemes are diminished for highly coupled or weakly mechanically damped devices [58]. Despite this, these interfaces can improve both the harvesting efficiency and the bandwidth [59].

13.8   BENCHMARKING

The benchmarking of PVEH devices is not trivial and there is currently no well-established framework or procedure for doing so. While power density has so far been the usual metric used to compare performances, it only partially captures the value of a harvester design. Moreover, Equations 13.47 through 13.54 show that the power density is a function not only of the properties of the PVEH device (e.g. quality factor Q_m and density ρ_M) but also of the vibration source (described by its amplitude A and frequency ω). Therefore, it is misleading to use this value as a metric to compare designs operating under different conditions. The piezoelectric resonator FOM, $\kappa^2 Q_m$, can also be a useful metric to compare devices. Hence, we believe that the fundamental limits that have been pointed out in this chapter can serve as a basis for developing useful comparison metrics.

13.8.1 NORMALIZED POWER DENSITY

Instead of strictly using the power density, Beeby et al. [60] normalized the power density by the square of the acceleration amplitude, A^2. However, this normalization does not capture the effects of the frequency of the source. To eliminate this problem, a normalized power density FOM, P_ρ, would instead be expressed as [14,47]

$$P_\rho = \frac{P_{el}\omega}{(A^2 Vol)} = \frac{M_{eq}Q_m}{2Vol} \frac{\overline{\zeta}}{(1+\overline{\zeta})^2} \tag{13.77}$$

Here, $Vol = Vol_b + Vol_M$ is the total effective volume of the transducer with its tip mass. Moreover, A^2/ω is expressed in W/kg (or $(m/s^2)^2$/Hz), which can be seen as a specific power characterizing the quality of the source. Therefore, P_ρ is also similar to a density and scales with ρQ_m to enable a fair comparison of the power output from different vibration conditions. At the same time, normalizing by the volume provides an indication of how much of the whole transducer inertia is useful in capturing energy. The use of a similar FOM was also proposed by Mitcheson et al. [14], who suggested normalizing the output power by $Y_0^2\omega^3 M_{eq}$, where Y_0 is the source vibration displacement. However, we propose using the volume instead of the mass such that a design that integrates high-density materials will be rewarded.

13.8.2 PIEZOELECTRIC RESONATOR FOM, HARVESTING EFFICIENCY AND EFFECTIVENESS

The resonator harvesting efficiency, η (Equations 13.55 and 13.76), and its harvesting effectiveness, ξ (Equation 13.56), are two other FOMs that were discussed in Section 13.4 of this chapter. The first indicates the fraction of the energy pumped in the device used to produce power, while the second states if the device is close to the limit imposed by inertial coupling. As we have shown, both of these values are linked to the piezoelectric resonator FOM, $\kappa^2 Q_m$. More precisely, they are affected by the electric damping that also depends on the frequency and the load.

Figure 13.9 shows the general trend expected for these FOMs as a function of the piezoelectric resonator FOM in the condition of matched impedances and resonance. For $\kappa^2 Q_m < 2$, both of the FOMs increase. However, once the critical coupling condition is achieved, the resonant device can be operated to optimize either the efficiency

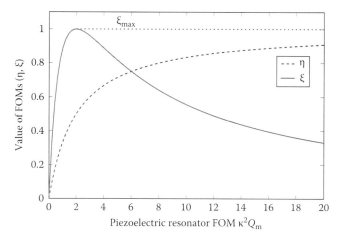

FIGURE 13.9   Variation of the FOMs with the resonator FOM, $\kappa^2 Q_m$.

(by matching the load impedance to the transducer) or the effectiveness (by adjusting the load in order to match the electric damping with the mechanical damping) [15].

13.8.3   BEAM ENERGY DENSITY

Even though a design might be capable of providing a very large power density, it might be at the expense of its reliability. Therefore, the induced stresses should be considered in the benchmarking procedure to evaluate if the piezoelectric spring element is used or sized in an appropriate manner. Based on the discussions presented in Sections 13.5 and 13.6, a fourth FOM should be introduced to evaluate the power density of the beam. To account for different operating frequencies, it is, however, more useful to express this new FOM in terms of energy density instead of power density. Defined as $E^* = P_{el}/Vol_bf$, it should scale with stress, σ, and coupling, k_e^2. We cannot directly use Equation 13.62 to evaluate the stress because it was developed for a homogeneous cantilever beam; actual devices rather use composite structures. Equation 13.57 cannot be used due to either the lack of information in the experimental literature concerning the beams stiffness or their amplitude of displacement. We rather estimated the bending modulus, YI, and deduced the strain energy from the electric power output, the coupling factor and the load. Indeed, from the definition of the damping factor, it is possible to say

$$E_{strain} = \frac{1}{2} YI \int_0^L \left(\frac{\partial^2 w}{\partial x^2} \right)^2 dx = \frac{P_{el}}{4\pi\zeta_{el}} f \qquad (13.78)$$

The strain energy is then used to evaluate an RMS curvature:

$$\left. \frac{\partial^2 w}{\partial x^2} \right|_{RMS} = \left(\frac{2E_{strain}}{YIL} \right)^{1/2} \qquad (13.79)$$

TABLE 13.5
Summary of the FOM

FOM	Expression	Meaning
Normalized power density	$P_\rho = P_{el}\,\omega/(A^2 Vol)$	Normalize the power output by the specific power of the source vibration and the device volume
Piezoelectric resonator FOM	$\kappa^2 Q_m$	Compare the mechanical losses to the electromechanical coupling of the structure (i.e. electric damping)
Harvesting efficiency	$\eta = \dfrac{P_{el}}{P_{in}} = \dfrac{\bar{\zeta}}{(1+\bar{\zeta})}$	Gives the portion of energy pumped in the device that is used to produce power
Harvesting effectiveness	$\xi = P_{el}/P_{lim} = 4\dfrac{\bar{\zeta}}{(1+\bar{\zeta})^2}$	Indicates if the device is close to the maximum power limit imposed by inertial coupling
Beam energy density	$E^* = P_{el}/Vol_b\,f$	Evaluates if the piezoelectric spring element is highly stressed

An estimate of the mean stress at the top of the piezoelectric layer is finally obtained as

$$\sigma = Y_p h_c \left.\frac{\partial^2 w}{\partial x^2}\right|_{RMS} \tag{13.80}$$

A summary of the proposed FOMs and their meaning is presented in Table 13.5.

13.8.4 Device Assessments

The proposed FOMs were used to compare different PVEH resonators reported in the literature. The results of this investigation are summarized in Table 13.6. All these devices produce power on the order of 1–50 µW.

As previously noted, P_ρ does scale with ρQ_m, motivating the choice of high-density materials and devices with high Q_m. For example, Renaud's device from Ref. [63], with its very high Q_m, has a P_ρ larger than all the others. However, a low value of Q_m can be somewhat compensated by using materials with higher density, as can be seen from the device reported in Ref. [62]. Using stainless steel as the structural layer instead of silicon, this device can capture more inertial energy due to an increased inertial force.

In terms of maximum efficiency, η_{max}, the devices made of AlN fabricated at IMEC all do very well compared to the other devices reported, and that is despite their relatively low coupling, showing the benefits of their high quality factor.

TABLE 13.6
Assessment of Several PVEH Devices Reported in the Literature

	Reference	[61]	[62]c	[63]	[40]c	IMEC 1Ab	IMEC 4Bb	IMEC 7Ab
Design	Piezomaterial	d_{33} PZT	d_{31} PZT	d_{31} AlN	d_{31} PMN-PT	d_{31} AlN	d_{31} AlN	d_{31} AlN
	Substrate	Silicon	Stainless steel	Silicon	Aluminium	Silicon	Silicon	Silicon
	Process	Sol-gel	Epitaxial	—	Bonded bulk	—	—	—
	Piezo-thickness (μm)	2	2.8	1.2	500	2	2	2
	Beam thickness (μm)	9	56	51.2	1330	52	52	52
	Thickness fraction	0.22	0.05	0.023	0.39	0.038	0.038	0.038
	Mass[a] (mg)	0.76	38	14	490	15	15	40
	Mass material	Silicon	—	Silicon	—	Silicon	Silicon	Silicon
	Vol_b (mm³)	0.0029	5.1	0.061	106	0.22	0.27	0.37
	Vol_M (mm³)	0.32	0	6	0	6	6	16.9
	f (Hz)	877	126	1082	≈1835	1178	1002	620
	A (m/s²)	19.6	5	3.14	9.81	6.28	6.28	6.28
	A^2/ω (m²/s³)	0.070	0.032	0.0015	0.0083	0.0053	0.0063	0.010
Performance	Q_m	220[a]	85	1200	22	1900	1450	865
	k_e^2	0.05	0.013	0.0025	0.18	0.0039	0.004	0.003
	$\kappa^2 Q_m$	11.6[a]	1.12	3	4.83	7.44	5.82	2.60
	σ^a (MPa)	84	7	170	92	163	158	459
	P_{el} (μW)	1.4	5.3	3	7.3	10	10	32
FOM	P_{el}/Vol (μW/cm³)	4255	1039	249	69	1608	1595	1853
	P_ρ (g/cm³)	62	33	333	8.3	297	250	186
	η_{max} (%)	85	36	60	71	79	74	57
	ξ	0.98	0.67	1	1	0.53	0.58	0.74
	E^* (nJ/mm³)	550	8.3	46	0.04	38	37	87

[a] Estimated values.
[b] Operating at maximum efficiency.
[c] Calculation corrected for lack of tip mass.

Only the device reported in [61] is better, because of the combination of its good mechanical quality factor and the fact that its interdigitated electrode design benefits from the d_{33} mode properties that leads to a high coupling factor.

As expected, highly stressed IMEC 7A and highly coupled [61] beams present higher values of E^*. Although increasing coupling will not increase the power density based on the mass volume (P_ρ), it will reduce the beam or piezoelectric volume for a desired electric power.

In addition, Table 13.6 clearly shows that devices with $\kappa^2 Q_m \geq 2$ have been frequently fabricated at the MEMS scale and even some at the mesoscale [40], which contradicts the claims of Guyomar and Lallart that realistic devices usually have $\kappa^2 Q_m$ values not much higher than 0.2 [56].

13.9 OTHER CONSIDERATIONS

This chapter focused primarily on linear resonant PVEH operating under harmonic excitation and connected to a resistive load. In this section, however, we broaden the scope with a qualitative discussion of nonresonant and non-linear approaches for vibration energy harvesting.

13.9.1 FREQUENCY DEPENDENCE OF THE INERTIAL COUPLING LIMIT

An important limit of resonant devices is that they must operate at a specific frequency, which implies a small operation bandwidth. If the mechanical energy source is broadband or time variant in nature, which is the case for most ambient vibrations, the harvester may collect only a small portion of the available energy or simply have a very low power output and a poor power density. Reducing the quality factor improves the bandwidth but at the expense of the power output. Addressing this problem is an active area of current research.

A simple solution is to use multimodal structures or even arrays of piezoelectric structures sensitive to different frequencies [64]. This approach broadens the operational bandwidth but at the expense of reducing the volumetric power density. Several papers have proposed methods to mechanically [65] or electrically [66] tune the resonant frequency to provide broadband operation while limiting the total size of the device. Nonetheless, active (continuous) or semi-active (discontinuous) tuning itself requires energy. It is not clear if the net energy balance can be improved significantly by these schemes. Hence, passive approaches are also being investigated using non-linear compliant structures [67–72], magnetic coupling [73–75] or stoppers [76–80] to introduce instabilities and other non-linearities.

13.9.2 POSSIBLE COMPROMISE CONSIDERING MATERIALS AND EXTRACTION CIRCUITS

Figure 13.10 maps the quality factor Q_m found in the literature with estimates of κ^2 for different piezoelectric materials, providing a large perspective of the

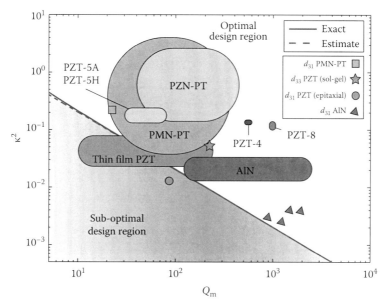

FIGURE 13.10 Critical coupling factor with respect to Q_m and mapping of materials for these criteria.

available options. Note that this figure only considers the 31-mode material properties and not the geometry of the composite structure (i.e. a geometric form factor $\phi = 1$ is assumed). The dashed line is the condition $\kappa^2 Q_m = 2$, which corresponds to the critical coupling value with a resistive load. We note that most of the materials are located in the optimal design region. Therefore, coupling should not be the only or the main criteria for PVEH material selection.

By using non-linear extraction interfaces, designs below the critical coupling threshold are however of potential interest, allowing multiple design trade-offs. For instance, with a high coupling material and a highly effective extracting interface, the size of the piezoelectric active element could be reduced significantly [31,81]. Alternatively, a wider selection of piezoelectric materials (e.g. low coupling materials) can be used to produce the same power output. Many other factors can therefore be considered for the selection of materials including

- Processing of the materials: time, costs or design requirements (e.g. integrated circuits (ICs), wafer-level packaging process compatibility)
- Toxicity (e.g. RoHS compliance, biocompatibility for bioMEMS)
- Curie temperature (no degradation during fabrication or usage)
- Dielectric losses and leakage behaviour
- Device operating principle (resonant or nonresonant) and characteristics of vibrations in the considered application (magnitude and type, e.g. random, shocks, harmonics, coloured noise)
- Fatigue strength (concerning in harsh environment)

13.10 CONCLUSION

This chapter presented a discussion of the various aspects that can limit the power density of resonant PVEHs. The motivation was to address the need for relevant metrics of comparison for PVEH system design. To this end, we investigated several ways to assess the performance of devices.

1. Compare the amount of energy captured by the device with the quality of the source.
2. Evaluate the amount of energy converted into electricity by the transducer and its energy density.
3. Evaluate the amount of electric energy extracted from the device and its global efficiency.

All these points were reviewed for linear resonators by considering the impact of geometry, material properties and electric energy extraction interface used. For typical values, a power density limit below the 100 μW/cm^3 range was estimated due to the inertial coupling limit. However, to reach this power density, the piezoelectric spring must be well designed. A good design must provide sufficient coupling to not limit the achievable power output and support the induced stress without failure. However, we've shown that for low-amplitude vibrations, capturing a significant amount of energy is the most important bottleneck, because of the large mass required compared to the size of the beam. Therefore, the power density will be first limited by the mass density of the device and then by stress. Nonetheless, if very large accelerations or forces are available in the ambient environment (e.g. in harsh environments), the strength of the materials may in this case become a concerning factor. For designs that provide low coupling, non-linear extraction interfaces can be alternatively used to improve the performance. The inertial coupling limit, however, limits the gain in performance for devices that present high Q_m.

Based on this investigation, we proposed alternative FOMs such as the normalized power density, the harvester's efficiency and the harvester's effectiveness. Interestingly, all of those depend on the piezoelectric resonator FOM, $\kappa^2 Q_m$. The beam energy density was also proposed as a new FOM to evaluate the beam stress, thus an indirect method to assess its design and guide the selection of both materials and geometry to achieve a specified power level in a compact beam volume. An assessment of several resonant devices recently published in the literature was then conducted, revealing that many have already reached the inertial limit expected for resonators. Moreover, the devices lagging behind could reach this limit with some geometric optimization or with more effective extracting interfaces. As expected, devices with a large quality factor presented the highest normalized power density values. However, the travel of the mass was not considered in our estimates. Taking this into account would reduce further the power density [14]. While these metrics provide a framework for comparing inertial piezoelectric energy harvesters, further reflections are required to define FOMs that account for broadband operation, costs and fabrication considerations. Equivalent FOMs for nonresonant devices would also be welcomed as a mean to compare those different operating principles.

This global analysis suggests that the technology for piezoelectric generator energy conversion is quite developed. Research on processing of very high coupling materials is well under way and analytical tools for the design and optimization of piezoelectric transducers have also been developed. Concerning the energy extraction interfaces, an array of circuit topologies have been proposed, but their implementation and performance at the microscale pose unresolved challenges. However, the power density of PVEH devices is inherently limited by the power input and consequently the methods that are currently used to capture the ambient energy. It is assumed that very large quantity of vibrational energy is available in the environment, but it remains challenging to harness energy from such unpredictable, low-quality sources. Linear, resonant harvesters appear to have reached their theoretical limits and significant research efforts are under way to bring the same level of performance to a broader range of operating frequencies and scenarios. Frequency-tuning methods have been proposed and novel non-linear techniques to transfer energy from one frequency band to another should also be researched. This challenge has been tackled for a couple of years, but no clear winning solution has emerged yet. In fact, it is quite probable that no single solution can address every type of solicitations.

ACKNOWLEDGEMENTS

This research was supported by General Motors of Canada and by the Natural Sciences and Engineering Research Council (NSERC) of Canada through the Canada Research Chairs, postgraduate scholarship and CREATE ISS programmes. We'd also like to acknowledge the cooperation of Ruud Vullers and Rene Elfrink at IMEC, as we are grateful for the PVEH devices data that they provided for our assessments.

REFERENCES

1. Vullers RJM, van Schaijk R, Doms I, Van Hoof C, and Mertens R. Micropower energy harvesting. *Solid-State Electronics*, 53(7):684–693, 2009.
2. Beeby SP, Tudor MJ, and White NM. Energy harvesting vibration sources for microsystems applications. *Measurement Science Technology*, 17:R175, 2006.
3. Roundy S, Wright PK, and Rabaey J-M. *Energy Scavenging for Wireless Sensor Networks: With Special Focus on Vibrations*. Kluwer Academic Publishers, Norwell, MA, 2004.
4. Roundy S. On the effectiveness of vibration-based energy harvesting. *Journal of Intelligent Material Systems and Structures*, 16(10):809–823, 2005.
5. Lallart M and Guyomar D. Piezoelectric conversion and energy harvesting enhancement by initial energy injection. *Applied Physics Letters*, 97:014104, 2010.
6. Erturk A and Inman DJ. A distributed parameter electromechanical model for cantilevered piezoelectric energy harvesters. *Journal of Vibration and Acoustics-Transactions of the ASME*, 130(4):041002, August 2008.
7. Tabesh A and Fréchette LG. On the concepts of electrical damping and stiffness in design of a piezoelectric bending beam energy harvester. *Proceedings of Power MEMS 2009*, 368–371, 2009.

8. Dompierre A, Vengallatore S, and Fréchette LG. Compact model formulation and design guidelines for piezoelectric vibration energy harvesting with geometric and material considerations. In *Proceedings of the 10th Workshop on Micro and Nanotechnology for Power Generation and Energy Conversion Applications–Power MEMS 2010 December 1–3*, Leuven, Belgium, Technical Digest Poster Sessions, pp. 99–102. Micropower Generation Group, imec/Holst Centre, December 2010.

9. Erturk A. Electromechanical modeling of piezoelectric energy harvesters. PhD thesis, Virginia Polytechnic Institute and State University, Department of Engineering Science and Mechanics, Blacksburg, Virginia, November 2009.

10. Erturk A and Inman DJ. Issues in mathematical modeling of piezoelectric energy harvesters. *Smart Materials and Structures*, 17(6):065016, December 2008.

11. Standards Committee of the IEEE Ultrasonics, Ferroelectrics, and Frequency Control Society. ANSI/IEEE Std 176–1987: Standard on piezoelectricity, January 1988.

12. Jaffe H and Berlincourt DA. Piezoelectric transducer materials. *Proceedings of the IEEE*, 53(10):1372–1386, 1965.

13. Williams CB and Yates RB. Analysis of a micro-electric generator for microsystems. *Sensors and Actuators A: Physical*, 52(1–3):8–11, 1996.

14. Mitcheson PD, Yeatman EM, Rao GK, Holmes AS, and Green TC. Energy harvesting from human and machine motion for wireless electronic devices. *Proceedings of the IEEE*, 96(9):1457–1486, 2008.

15. Renaud M, Elfrink R, Jambunathan M, de Nooijer C, Wang Z, Rovers M, Vullers R, and van Schaijk R. Optimum power and efficiency of piezoelectric vibration energy harvesters with sinusoidal and random vibrations. *Journal of Micromechanics and Microengineering*, 22(10):105030, 2012.

16. Izyumskaya N, Alivov YI, Cho SJ, Morkoc H, Lee H, and Kang YS. Processing, structure, properties, and applications of PZT thin films. *Critical Reviews in Solid State and Materials Sciences*, 32(3–4):111–202, 2007.

17. PI piezo tutorial: mechanical considerations. http://www.physikinstrumente.com/tutorial/4_22.html. Febuary 10, 2012.

18. Cain MG, Stewart M, and Gee MG. Degradation of piezoelectric materials. National Physical Laboratory Management Ltd., Teddington, Middlesex, UK, NPL Rep. SMMT (A), p. 148, 1999.

19. Goldschmidtboeing F and Woias P. Characterization of different beam shapes for piezoelectric energy harvesting. *Journal of Micromechanics and Microengineering*, 18(10):104013, 2008.

20. Wang Z and Xu Y. Vibration energy harvesting device based on air-spaced piezoelectric cantilevers. *Applied Physics Letters*, 90(26):263512–263512, 2007.

21. Halvorsen E and Dong T. Analysis of tapered beam piezoelectric energy harvesters. In *Proceedings of PowerMEMS 2008 + microEMS 2008*, Sendai, Japan, pp. 241–244, November 9–12, 2008.

22. Yen TT, Hirasawa T, Wright PK, Pisano AP, and Lin L. Corrugated aluminum nitride energy harvesters for high energy conversion effectiveness. *Journal of Micromechanics and Microengineering*, 21:085037, 2011.

23. Stewart M, Weaver PM, and Cain M. Charge redistribution in piezoelectric energy harvesters. *Applied Physics Letters*, 100(7):073901–073901, 2012.

24. Erturk A, Tarazaga PA, Farmer JR, and Inman DJ. Effect of strain nodes and electrode configuration on piezoelectric energy harvesting from cantilevered beams. *Journal of Vibration and Acoustics*, 131(1):011010, 2009.

25. Vullers RJM, van Schaijk R, Goedbloed M, Elfrink R, Wang Z, and Van Hoof C. Process challenges of MEMS harvesters and their effect on harvester performance. In *2011 IEEE International Electron Devices Meeting (IEDM)*, Washington, DC, pp. 10.2.1–10.2.4, December 2011.

26. Trolier-McKinstry S and Muralt P. Thin film piezoelectrics for MEMS. *Electroceramic-Based MEMS*, 9:199–215, 2005.

27. Trolier-McKinstry S, Griggio F, Yaeger C, Jousse P, Zhao D, Bharadwaja SSN, Jackson TN, Jesse S, Kalinin SV, and Wasa K. Designing piezoelectric films for micro electro-mechanical systems. *IEEE Transactions on Ultrasonics, Ferroelectrics and Frequency Control*, 58(9):1782–1792, 2011.

28. Online materials information resource—Matweb, http://www.matweb.com/. January 4, 2011.

29. eFunda: Piezo material data, http://www.efunda.com/materials/piezo/material_data/matdata_index.cfm. January 4, 2011.

30. Lebrun L, Sebald G, Guiffard B, Richard C, Guyomar D, and Pleska E. Investigations on ferroelectric PMN-PT and PZN-PT single crystals ability for power or resonant actuators. *Ultrasonics*, 42(1–9):501–505, 2004.

31. Badel A, Benayad A, Lefeuvre E, Lebrun L, Richard C, and Guyomar D. Single crystals and nonlinear process for outstanding vibration-powered electrical generators. *IEEE Transactions on Ultrasonics, Ferroelectrics, and Frequency Control*, 53(4):673–684, 2006.

32. Ceracomp single crystal, single crystal transducer, single crystal devices, http://www.ceracomp.com/. Febuary 8, 2012.

33. Peng J, Luo H, He T, Xu H, and Lin D. Elastic, dielectric, and piezoelectric characterization of 0.70Pb(Mg1/3Nb2/3)O3–0.30PbTiO3 single crystals. *Materials Letters*, 59(6):640–643, 2005.

34. Zhang R, Jiang W, Jiang B, and Cao W. Elastic, dielectric and piezoelectric coefficients of domain engineered 0.70Pb(Mg1/3Nb2/3) O3–0.30 PbTiO3 single crystal. In *AIP Conference Proceedings*, IOP Institute of Physics Publishing Ltd., Bristol, pp. 188–197, 2002.

35. Park SE and Shrout TR. Characteristics of relaxor-based piezoelectric single crystals for ultrasonic transducers. *IEEE Transactions on Ultrasonics, Ferroelectrics and Frequency Control*, 44(5):1140–1147, 1997.

36. PMN PT crystal | PMN-PT | APCI non-PZT crystals APC International Ltd, http://www.americanpiezo.com/product-service/pmn-pt.html. Febuary 15th 2012.

37. Zhang R, Jiang B, and Cao W. Elastic, piezoelectric, and dielectric properties of multidomain 0.67Pb(Mg1/3Nb2/3)O3–0.33PbTiO3 single crystals. *Journal of Applied Physics*, 90:3471, 2001.

38. Delaunay T, Le Clézio E, Guennou M, Dammak H, Thi MP, and Feuillard G. Full tensorial characterization of PZN−0.12PT single crystal by resonant ultrasound spectroscopy. *IEEE Transactions on Ultrasonics, Ferroelectrics and Frequency Control*, 55(2):476–488, 2008.

39. Cao H, Schmidt VH, Zhang R, Cao W, and Luo H. Elastic, piezoelectric, and dielectric properties of 0.58Pb(Mg1/3Nb2/3)O3–0.42PbTiO3 single crystal. *Journal of Applied Physics*, 96(1):549–554, 2004.

40. Erturk A, Bilgen O, and Inman DJ. Power generation and shunt damping performance of a single crystal lead magnesium niobate-lead zirconate titanate unimorph: Analysis and experiment. *Applied Physics Letters*, 93(22):224102, December 2008.

41. Kobor D. Synthèse, dopage et caractérisation de monocristaux ferroélectriques type PZN-PT par la méthode du flux. PhD thesis, Doc'INSA-INSA de Lyon, 2005.

42. Zhang R, Jiang B, Jiang W, and Cao W. Complete set of elastic, dielectric, and piezoelectric coefficients of 0.93Pb(Zn1/3Nb2/3)O3–0.07PbTiO3 single crystal poled along [011]. *Applied Physics Letters*, 89:242908, 2006.

43. Jiang W, Zhang R, Jiang B, and Cao W. Characterization of piezoelectric materials with large piezoelectric and electromechanical coupling coefficients. *Ultrasonics*, 41(2):55–63, 2003.

44. Lefeuvre E, Sebald G, Guyomar D, Lallart M, and Richard C. Materials, structures and power interfaces for efficient piezoelectric energy harvesting. *Journal of Electroceramics*, 22(1):171–179, 2009.

45. Muralt P, Antifakos J, Cantoni M, Lanz R, and Martin F. Is there a better material for thin film BAW applications than A1N? In *IEEE Ultrasonics Symposium 2005*, 1:315–320, 2005.

46. Chen Q, Quin L, and Wang QM. Property characterization of AlN thin films in composite resonator structure. *Journal of Applied Physics*, 101(8):084103, 2007.

47. Marzencki M. Conception de microgénrateurs intégrés pour systèmes sur puce autonomes. PhD thesis, Université Joseph Fourier Grenoble I, Laboratoire TIMA, Grenoble, France, March 2007.

48. Carlotti G, Socino G, Petri A, and Verona E. Elastic constants of sputtered ZnO films. In *IEEE Ultrasonics Symposium 1987*, pp. 295–300, 1987.

49. Kanno I, Kotera H, and Wasa K. Measurement of transverse piezoelectric properties of PZT thin films. *Sensors and Actuators A: Physical*, 107(1):68–74, 2003.

50. Muralt P. Piezoelectric thin films for MEMS. *Integrated Ferroelectrics*, 17(1–4):297–307, 1997.

51. Wasa K, Matsushima T, Adachi H, Kanno I, and Kotera H. Thin-film piezoelectric materials for a better energy harvesting MEMS. *Journal of Microelectromechanical Systems*, 99:1–7, 2012.

52. Wasa K, Ito S, Nakamura K, Matsunaga T, Kanno I, Suzuki T, Okino H, Yamamoto T, Seo SH, and Noh DY. Electromechanical coupling factors of single-domain 0.67Pb(Mg1/3Nb2/3)O3–0.33PbTiO3 single-crystal thin films. *Applied Physics Letters*, 88:122903, 2006.

53. Baek SH, Park J, Kim DM, Aksyuk VA, Das RR, Bu SD, Felker DA et al. Giant piezoelectricity on Si for hyperactive MEMS. *Science*, 334(6058):958–961, 2011.

54. Richards CD, Anderson MJ, Bahr DF, and Richards RF. Efficiency of energy conversion for devices containing a piezoelectric component. *Journal of Micromechanics and Microengineering*, 14:717, 2004.

55. Shu YC and Lien IC. Efficiency of energy conversion for a piezoelectric power harvesting system. *Journal of Micromechanics and Microengineering*, 16:2429, 2006.

56. Guyomar D and Lallart M. Recent progress in piezoelectric conversion and energy harvesting using nonlinear electronic interfaces and issues in small scale implementation. *Micromachines*, 2(2):274–294, 2011.

57. Lallart M and Guyomar D. Nonlinear energy harvesting. *IOP Conference Series: Materials Science and Engineering*, 18(9):092006, 2011.

58. Shu YC, Lien IC, and Wu WJ. An improved analysis of the SSHI interface in piezoelectric energy harvesting. *Smart Materials and Structures*, 16:2253, 2007.

59. Lien IC, Shu YC, Wu WJ, Shiu SM, and Lin HC. Revisit of series-sshi with comparisons to other interfacing circuits in piezoelectric energy harvesting. *Smart Materials and Structures*, 19(12):125009, 2010.

60. Beeby SP, Torah RN, Tudor MJ, Glynne-Jones P, O'Donnell T, Saha CR, and Roy S. A micro electromagnetic generator for vibration energy harvesting. *Journal of Micromechanics and Microengineering*, 17(7):1257, 2007.

61. Muralt P, Marzencki M, Belgacem B, Calame F, and Basrour S. Vibration Energy harvesting with PZT micro device. *Procedia Chemistry*, 1(1):1191–1194, 2009.

62. Morimoto K, Kanno I, Wasa K, and Kotera H. High-efficiency piezoelectric energy harvesters of C-axis-oriented epitaxial PZT films transferred onto stainless steel cantilevers. *Sensors and Actuators A: Physical*, 163(1):428–432, 2010.

63. Renaud M, Elfrink R, Op het Veld B, and van Schaijk R. Power optimization using resonance and antiresonance of MEMS piezoelectric vibration energy harvesters. In *Proceedings of the 10th Workshop on Micro and Nanotechnology for Power Generation and Energy Conversion Applications - PowerMEMS 2010 December 1–3*, Leuven, Belgium, Technical Digest Oral Sessions, pp. 23–26. Micropower Generation Group, imec/Holst Centre, December 2010.

64. Kasyap A. Development of MEMS-based piezoelectric cantilever arrays for vibrational energy harvesting. PhD thesis, University of Florida, Department of Aerospace Engineering, Mechanics and Engineering Science, Gainesville, Florida, 2007.

65. Hu Y, Xue H, and Hu H. A piezoelectric power harvester with adjustable frequency through axial preloads. *Smart Materials and Structures*, 16:1961, 2007.

66. Cammarano A, Burrow SG, Barton DAW, Carrella A, and Clare LR. Tuning a resonant energy harvester using a generalized electrical load. *Smart Materials and Structures*, 19:055003, 2010.

67. Marinkovic B and Koser H. Smart Sand - A wide bandwidth vibration energy harvesting platform. *Applied Physics Letters*, 94(10):103505, March 2009.

68. Marzencki M, Defosseux M, and Basrour S. MEMS vibration energy harvesting devices with passive resonance frequency adaptation capability. *Journal of Microelectromechanical Systems*, 18(6):1444–1453, 2009.

69. Sebald G, Kuwano H, Guyomar D, and Ducharne B. Experimental duffing oscillator for broadband piezoelectric energy harvesting. *Smart Materials and Structures*, 20:102001, 2011.

70. Sebald G, Kuwano H, Guyomar D, and Ducharne B. Simulation of a duffing oscillator for broadband piezoelectric energy harvesting. *Smart Materials and Structures*, 20:075022, 2011.

71. Hajati A, Bathurst SP, Lee HJ, and Kim SG. Design and fabrication of a nonlinear resonator for ultra wide-bandwidth energy harvesting applications. In *2011 IEEE 24th International Conference on Micro Electro Mechanical Systems (MEMS)*, pp. 1301–1304. IEEE, 2011.

72. Hajati A. Ultra wide-bandwidth micro energy harvester. PhD thesis, Massachusetts Institute of Technology, Department of Electrical Engineering and Computer Science, Boston, MA, February 2011.

73. Erturk A, Hoffmann J, and Inman DJ. A piezomagnetoelastic structure for broadband vibration energy harvesting. *Applied Physics Letters*, 94:254102, 2009.

74. Lin JT, Lee B, and Alphenaar B. The magnetic coupling of a piezoelectric cantilever for enhanced energy harvesting efficiency. *Smart Materials and Structures*, 19:045012, 2010.

75. Andò B, Baglio S, Trigona C, Dumas N, Latorre L, and Nouet P. Nonlinear mechanism in MEMS devices for energy harvesting applications. *Journal of Micromechanics and Microengineering*, 20:125020, 2010.

76. Blystad LCJ and Halvorsen E. An energy harvester driven by colored noise. *Smart Materials and Structures*, 20:025011, 2011.

77. Blystad LCJ and Halvorsen E. A piezoelectric energy harvester with a mechanical end stop on one side. *Microsystem Technologies*, 17(4):505–511, 2011.

78. Blystad LCJ, Halvorsen E, and Husa S. Piezoelectric MEMS energy harvesting systems driven by harmonic and random vibrations. *IEEE Transactions on Ultrasonics, Ferroelectrics and Frequency Control*, 57(4):908–919, 2010.

79. Liu H, Tay CJ, Quan C, Kobayashi T, and Lee C. Piezoelectric MEMS energy harvester for low-frequency vibrations with wideband operation range and steadily increased output power. *Journal of Microelectromechanical Systems*, 20(5):1131–1142, 2011.

80. Soliman MSM, Abdel-Rahman EM, El-Saadany EF, and Mansour RR. A wideband vibration-based energy harvester. *Journal of Micromechanics and Microengineering*, 18:115021, 2008.

81. Badel A. Récupération d'énergie et contrôle vibratoire par éléments piézoélectriques suivant une approche non linéaire. PhD thesis, Université de Savoie, Chambéry, France, 2005.

14 Microwave Sensors for Non-Invasive Monitoring of Industrial Processes

B. García-Baños, Jose M. Catalá-Civera,
Antoni J. Canós and Felipe L. Peñaranda-Foix

CONTENTS

The rapid development of industrial automation processes has created a growing need for improved sensors for process monitoring and control. With the introduction of ISO 9000, fast and accurate measurements of dielectric properties of materials have grown to a great importance in different applications of industry and agricultural engineering.

Microwave sensors, which allow determining the dielectric properties of materials, have found industrial and scientific applications in the nondestructive testing of materials, identification of surface defects, spectroscopy, medical diagnosis, etc.

Their applicability in industrial process monitoring tasks has allowed new control functionalities, which directly improve the efficiency in the use of resources and, thus, the overall competitiveness.

14.1 DIELECTRIC PROPERTIES OF MATERIALS

14.1.1 Interaction of Microwaves and Matter

When an electric field is applied across a dielectric material, the atomic and molecular charges in the dielectric are displaced from their equilibrium positions and the material is said to be polarized. Dielectric analysis involves the determination of this polarization in materials subjected to a time-varying electric field.

The main dielectric parameter, the permittivity, represents the measurement of the maximum dipolar polarization that can be attained by the material under specific conditions (temperature, chemical state, etc.). The permittivity is a frequency-dependent complex number with the following expression:

$$\varepsilon(f) = \varepsilon_0(\varepsilon_r'(f) - j\varepsilon_r''(f)) \tag{14.1}$$

where
   f is the frequency
   ε_0 is a constant, which represents the vacuum permittivity
   the real part $\varepsilon_r'(f)$ is known as the relative dielectric constant, which characterizes the material's ability to store and release electromagnetic energy
   the imaginary part $\varepsilon_r''(f)$ is known as the relative loss factor, which characterizes the material's ability to absorb (attenuate) electromagnetic energy to create heat

Electromagnetic fields and waves are very sensitive to the dielectric properties of the material in which they exist. Microwave sensors are based on the fact that the interaction between microwaves and the medium of propagation is completely determined by the medium dielectric properties. By monitoring the electromagnetic fields/waves that are interacting with the material, it is possible to simultaneously and independently determine the dielectric parameters ($\varepsilon_r'(f)$ and $\varepsilon_r''(f)$) and/or two independent physical parameters (e.g. moisture content and density), while all other physical properties are held constant.

14.1.2 Why Measure Dielectric Properties

Permittivity is the main factor that determines how the material interacts with an electromagnetic field. In communications and radar devices, it is necessary to know the dielectric properties of the materials involved both in the design and analysis processes [1].

In the present era of worldwide communications, international regulations require rigorous studies about how the electromagnetic fields are absorbed or affect the human body. To this end, specific absorption rate (SAR) measurements are performed using equivalent dielectric materials whose permittivities have to be previously accurately determined.

If the electromagnetic fields are applied for thermal processing of materials, dielectric characterization is necessary to model, predict, design and understand these processing systems. For example, dielectric properties can give an answer to questions such as 'Is the material loss factor high enough?', 'How does the permittivity change with temperature?' and 'Is a thermal runaway possible?'.

The range of applicability is vastly broaden because dielectric properties are also sensitive indicators of numerous physical properties of the material, such as moisture content, biomass, density, bacterial content, chemical reaction, viscosity, concentration of components and most other physical properties. Correlation between dielectric properties and other parameters of interest can be generally done through empirically modelled relationships, which is of crucial interest in industrial applications [2].

14.1.3   TECHNIQUES FOR DIELECTRIC PROPERTY MEASUREMENT

Due to the design flexibility after decades of microwave device development, there are many different fixtures available to perform dielectric characterization of materials [3,4].

The different techniques can be classified depending on many criteria: size of the measurement cell, resonant/nonresonant structures, open/closed cells, etc. One of the possible classifications distinguishes between lumped-element circuits, travelling-wave methods and resonance/cavity methods.

When the size of the measurement cell is smaller than the wavelength (frequency is sufficiently low), the measurement cell can be considered as a lumped circuit. The method usually consists of placing a slab of dielectric material between two conductors. The material permittivity can be recovered from measurements of the circuit impedance, which can be performed with an impedance analyzer.

If the measurement cell has a size comparable or larger than the wavelength, propagation parameters of the wave have to be considered. These techniques for dielectric characterization are known as travelling-wave methods. The material is inserted or placed in contact with transmission lines (rectangular waveguide, coaxial line, etc.) or between two antennas in a free-space configuration. Measurements of the complex transmission (S_{21}) and/or reflection (S_{11}) scattering parameters are performed by vector network analyzers (VNAs). From these measurements and the electromagnetic model of the measurement cell and the propagating waves, it is possible to calculate the permittivity of the dielectric material.

A widely used group of travelling-wave methods is based on open measurement cells, which are placed near or in contact with the dielectric material at one side. The waves are emitted from the open end and penetrate the material. Their propagation characteristics (transmission, reflection and attenuation) are thus modified depending on the material dielectric properties. VNA measurements of the reflection coefficient (S_{11}) allow for complex permittivity determination, which can cover several decades of frequency. Some of the most popular fixtures are the open-ended coaxial probe [5] and the coplanar waveguide [6].

If the wavelength is much shorter than the sample size, free-space methods are used in which the sample is placed between two antennas (or in front of only one antenna). Quasi-optical methods are used to model the propagation characteristics. Again, VNA measurements of complex scattering parameters allow for permittivity determination of the sample.

Finally, almost any of the previous techniques can be converted into resonance or cavity methods by modifying the propagation conditions in the measurement cell in such a way that a resonant mode is established at a certain frequency. Instead of scattering parameters, resonance frequency and quality factor of the cavity are the measured quantities. The permittivity of the sample can be recovered from these measurements only at the discrete frequencies at which the fixture resonates. The advantage of this method is the great accuracy in the determination of material permittivity, specially the loss factor for low-loss materials.

14.1.4   EXAMPLES OF MEASUREMENTS

Due to the wide range of applicability of dielectric characterization of materials, numerous studies and also deep review papers and books have been published by expert dielectric metrologists [3,4,7].

Some illustrative results obtained at Instituto de las Aplicaciones de las Comunicaciones Avanzadas (ITACA) with in-house measurement cells are shown in the following. Figure 14.1 shows several open-ended coaxial cells developed for specific applications. In [8], it is demonstrated how a careful selection of measurement frequency and coaxial dimensions can drastically affect the cell performance. Despite the fact that there are some commercially available general-purpose open-coaxial cells, it is clear that many applications require a specific design to match with the application requirements. For example, bigger coaxial cells may mitigate the error caused by sample roughness or may improve the penetration of electromagnetic fields in thick samples. This may be necessary if dielectric properties exhibit gradients at the sample surface, and it is of interest to obtain dielectric values representative of the bulk material.

FIGURE 14.1   Examples of coaxial cells developed at ITACA for dielectric characterization of materials at microwave frequencies.

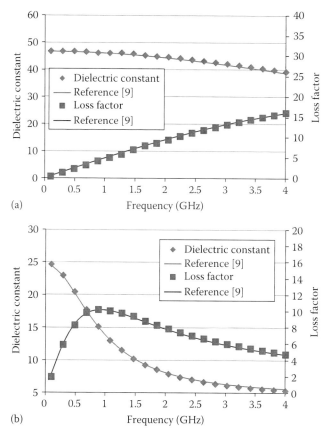

(a)

(b)

FIGURE 14.2   Dielectric characterization of some liquid samples performed with an open-ended coaxial probe at microwave frequencies. (a) Dimethyl sulphoxide (at 23.3°C) and (b) ethanol (at 23.5°C).

Figure 14.2 shows the dielectric properties of some liquid samples (ethanol and dimethyl sulphoxide) performed in the frequency range from 100 MHz to 4 GHz with an open-ended coaxial probe. This kind of measurement cell is particularly convenient for liquid characterization, because in the absence of air bubbles, a good contact between the sample and the probe surface is assured. Permittivity values obtained by other authors [9] are in good agreement with the presented results.

Table 14.1 shows permittivity results obtained for some common materials with open-ended coaxial cavity cells. In this case, the dielectric characterization is performed at discrete points of frequency, which coincide with the frequencies of the coaxial cavity resonant modes. As the table shows, the resonant coaxial structure gives accurate results for materials with low and high losses. This is possible only if an accurate model of the resonator coupling network is applied, and thus, accurate measurements of resonance frequency and quality factor are made. One of these methods will be discussed in next sections. Again, results obtained with coaxial

TABLE 14.1

Dielectric Characterization of Materials with an Open-Ended Coaxial Resonator

Material	Freq. (GHz)	ε'_r	ε''_r	ε'_r (References)	ε''_r (References)	References
PTFE	2.37	2.049	6E-4	2.050	6E-4	[10]
PVC	2.34	2.898	0.023	2.919	0.023	[11]
Nylon	2.33	3.092	0.047	3.082	0.044	[11]
Acetal	2.34	2.977	0.124	2.968	0.129	[11]
Cross-linked Polystyrene	2.35	2.532	0.002	2.540	0.002	[12]
Methanol	1.83	24.461	11.963	25.738	12.700	[13]

resonators are in good agreement with measurements performed by other authors and with other techniques.

Measurements with an open-coaxial resonator cell assume that the electromagnetic signal emitted from the probe is confined within the sample volume. However, there are applications in which, due to the cost or the material nature, only small sample volumes are available. If the sample is not big enough, a portion of the signal arrives to the sample edges and reflects back to the probe, inducing measurement errors. For these cases, a specific measurement cell has been designed at ITACA, which is based on a single-post coaxial re-entrant cavity with a partially dielectric filled gap (see Figure 14.3). With this measurement cell, liquid or solid samples of 1 mL volume can be accurately characterized at microwave frequencies.

FIGURE 14.3   Single-post coaxial re-entrant cavity sensor developed at ITACA for dielectric material characterization (1 mL samples in standard vials).

TABLE 14.2

Dielectric Characterization of Materials with a Single-Post Coaxial Re-entrant Cavity Cell

Material	Freq. (GHz)	ε'_r	ε''_r	ε'_r (References)	ε''_r (References)	References
Water	1.97	77.18	6.537	77.83	6.789	[9]
Dimethyl sulphoxide	2.10	45.39	8.820	45.02	8.966	[9]
2-Propanol	2.47	4.06	2.925	3.89	2.724	[9]
Quartz sand	2.50	2.49	0.003	2.27	0.012*	[*]
Granular paraffin	2.52	1.71	0.001	1.77	0.011*	[*]
Powdered milk	2.51	2.06	0.072	2.12	0.068	[*]
SiC	2.41	8.57	1.289	8.49	1.335	[*]
Perlite	2.49	2.68	0.018	2.73	0.011	[*]

[*] Measurements performed with Agilent coaxial probe HP-85070B (not suitable for low-loss materials).

Table 14.2 shows permittivity results obtained for some common materials with the single-post coaxial re-entrant cavity cell. It should be noted that this fixture allows convenient measurements of liquid, granular and powder samples. The results show good agreement with measurements performed by other authors and with other techniques.

14.2 MICROWAVE SENSORS FOR DIELECTRIC PROPERTY MONITORING

14.2.1 BASIC PRINCIPLES OF MICROWAVE SENSORS

Microwave measurement cells can be applied to the characterization of dielectric properties of materials but can also be used to perform a continuous monitoring of these properties if the material under test is undergoing a certain process. To this end, measurements of the cell response have to be performed at a sufficient rate to allow for real-time monitoring of changes in the material state.

This kind of application is of notable interest in industry, since it permits to implement new control functionalities, which assure the quality of final products by assessing and adapting in real time the process conditions [2].

In a typical example of a microwave sensor system, the material is subjected to a process (curing, heating, drying, etc.), the microwave sensor is placed in contact or near the material and its response is dependent on the material state. A microwave transducer must necessarily comprise a microwave source, a receiver and a network to separate the incident waves (from the source) from the reflected waves (from the sensor). Everything is controlled by a processing unit having implemented the additional control functionalities (displays, thresholds, alarms, etc.).

As explained in previous sections, microwave measurement cells have been traditionally used with VNAs, which perform broadband measurements of the complex scattering parameters. This equipment includes the functionalities of microwave transducer and control system. VNAs can be considered as commercially available sophisticated equipment to cover dielectric measurements in the whole microwave spectrum (when used in combination with microwave measurement cells). This equipment is designed to run in the laboratory environment, with specialized operators, and has low portability. On the other hand, the high cost of commercial VNAs, which might be acceptable for scientific institutions, is often prohibitive for industrialists.

The practical nature of monitoring applications has forced researchers to develop microwave equipment according to industrial conditions, simple, affordable and robust, while retaining the necessary accuracy. In this context, microwave reflectometers have been developed to tackle with this demand [14–16].

In [14,16], detailed descriptions of microwave reflectometers developed at ITACA can be found. The microwave source is designed as a phase-locked loop-based synthesizer (AD8314) with voltage-controlled oscillator (VCO) ranging from 1.5 to 2.6 GHz. The separation network has been implemented with two directional couplers. The receiver is designed from the commercial gain and phase detector (AD8302). The reflectometer system is connected to a PC through the USB port to perform the required calculations and to transform the outputs into the desired display.

The reflectometer can be integrated with different microwave measurement cells (sensors) for specific applications, giving rise to robust, compact and portable systems to perform dielectric material characterization both in industrial or laboratory environments.

14.2.2 Advantages and Drawbacks of Microwave Sensors

The main advantages of microwave sensors for material measurement can be summarized in the following points [2]:

- Contact between the material under test and the sensor is not necessary.
- Microwave energy penetrates inside dielectric materials (from mm to several cm). The resulting dielectric properties are representative of a volume of that material, not only of the surface.
- In the absence of sample, the response of a resonator sensor is only dependent on its dimensions (very stable).
- Microwave sensors are not sensitive to common industrial ambient conditions (dust, vibration, etc.) or static charges.
- At microwave frequencies, the influence of DC conductivity is small and material state is not masked by its ion content.
- The sensors are completely safe for the operator (contrary to ionizing radiation such as x-rays) and do not affect the material.
- The response of the microwave sensor to a change in the material under test is immediate. For this reason, several measurements per second are feasible.
- Manufacturing cost of microwave sensors is not high.

However, there still may be some disadvantages that have prevented a wider industrial use of microwave sensors:

- These sensors are usually designed and optimized for specific applications.
- If there are several variables changing at the same time (e.g. pressure, temperature and degree of cure), one microwave sensor is not enough to quantify the contribution of each variable to the final response. Additional sensors (microwave or other kind of sensors) may be necessary to solve the ambiguities.
- Due to the cell size and the wavelength of the microwave signals, the spatial resolution with microwave sensors is low.
- The relationship of dielectric properties with process parameters needs to be calibrated for each specific application (i.e. moisture content, degree of curing).

14.3 MICROWAVE RESONATOR SENSORS

14.3.1 BASICS OF MICROWAVE RESONATORS

A microwave resonator is formed by a section of a transmission line bounded by impedance discontinuities that cause reflections. At a frequency, where the multiple reflections are in phase, constructive interference occurs in the form of a resonance. Two main parameters are necessary to characterize a resonator response: the frequency (f) at which the resonance occurs and the quality factor (Q) of the cavity at that frequency.

The stationary fields in resonators can be considered as a superposition of forward and backward travelling waves; therefore, a dielectric brought in contact with those fields interacts continuously with them, resulting in high sensitivity of the measurement. For this reason, open resonators represent an attractive solution for non-destructive sensing of dielectric materials, since they combine the convenience of one-sided configuration with the high sensitivity of resonance structures [17].

As an example, a transmission line can be converted to a resonator if a coupling network is connected to the line at one side and the other side is left open (this end is placed in contact with the material under test). Depending on the type of transmission line, the microwave resonator is called, for example, coaxial, microstrip, stripline, slotline or coplanar resonator.

When a resonator is loaded with a material, its response is related to the permittivity of the material under test. Electromagnetic models of the resonant structure provide a quantitative relation between the resonator parameters and the material dielectric properties. Using them, a measurement of the resonant frequency and quality factor of a resonator gives the necessary information to calculate the complex permittivity of the dielectric material.

14.3.2 COUPLING OF MICROWAVE RESONATORS

In practice, any resonator is not an isolated structure. It has to be fed through a network, which couples the energy from the source. There are some examples of feeding networks: electric probes, apertures or magnetic loops. The use of a feeding network to launch energy into the resonator modifies its theoretical (unloaded) response: the resonance frequency is shifted to the loaded resonant frequency and the quality factor is lowered to the loaded quality factor [18].

The effect of the coupling network can be neglected if the energy is weakly coupled into the resonator, and then loaded resonant frequencies are practically equal to the unloaded resonances. However, the undercoupling condition involves an important limitation: materials exhibiting medium or high losses cannot be measured because they absorb and attenuate most of the energy [19]. Then, it is a challenge to develop microwave resonators to cope with any kind of material losses. One possible solution is to design a strongly coupled (overcoupled) cavity and, consequently, to develop an adequate modelling of the effect of the coupling network to correct the disturbance effect [18].

14.3.3 MEASUREMENT OF RESONANT FREQUENCY AND QUALITY FACTOR

A one-port microwave resonator, connected to a measurement system by means of a transmission line, can be represented by an equivalent circuit in the vicinity of the resonance. This representation is referred to as the first Foster's form. The coupling network is modelled by an ideal transformer of ratio n and by the complex impedance $r_e + jx_e$. Ohmic losses within the coupling mechanisms are represented by an equivalent resistance r_e. The reactance x_e represents the extra energy storage introduced by the coupling structure. A more detailed description of this representation can be found in [18].

According to this model, the detuning of a resonator due to the coupling elements, namely, the relation between loaded (f_L, Q_L) and unloaded resonator parameters (f_U, Q_U), is given by the following expressions:

$$f_L = f_U \left(1 + \frac{k\, x_e}{2Q_U} \right) \tag{14.2}$$

$$Q_L = \frac{Q_U}{1 + k} \tag{14.3}$$

where k is the coupling factor. It can be appreciated that, for highly undercoupled resonators ($k \rightarrow 0$), the loaded parameters f_L and Q_L can be approximated by the unloaded ones f_U and Q_U. The values of k, f_L, Q_L and Q_U of Equations 14.2 and 14.3 can be directly derived from the measurement of the magnitude and phase of the reflection factor of the sensor around the resonance, following the procedure described in [20]. However, the reactance x_e cannot be directly extracted from the measurements [20].

This parameter can be obtained by simulating the structure with different coupling levels with the aid of electromagnetic simulators. For example, [21] shows these results in the case of a coaxial resonator with an electric feeding probe. Once the feeding network is modelled (x_e is known for each f_L), f_U and Q_U can be calculated from measurements (f_L and Q_L) by using Equations 14.2 and 14.3. With the unloaded parameters, dielectric properties of materials can be accurately determined from electromagnetic models of the resonator.

14.3.4 EXAMPLES

The unique characteristics of microwave sensors, together with the advance of high-frequency circuits and processors, are promoting that microwave sensors can take the leap from laboratory into industrial environments with high possibilities of success. In this section, several microwave sensors developed for specific applications are shown. Each of them has been developed with the characteristics required to be used in industrial applications: robust, accurate and simple to use.

14.3.4.1 Microwave Sensor for Characterization of Liquid Mixtures

Water-in-oil (W/O) emulsions appear in the sludges of many industries. These pollutants must be treated before discharge, and microwave heating is considered as an environmentally friendly technique to separate these emulsions. The amount of heat generated in the emulsion greatly depends on the emulsion dielectric properties. Precise characterization of emulsions is essential to predict their interaction with microwave radiation.

Previous studies on emulsion dielectric properties have been conducted at RFs, which do not include the standard heating frequency (2.45 GHz). Now, with a single-post coaxial re-entrant cavity sensor (see Figure 14.4), accurate characterization of

FIGURE 14.4   Single-post coaxial re-entrant cavity sensor developed at ITACA for dielectric material characterization (8 mL samples in standard vials).

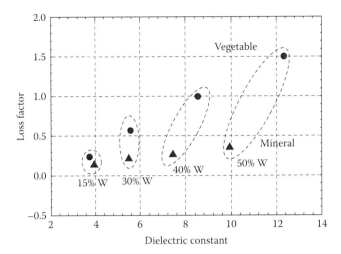

FIGURE 14.5   Dielectric characterization of W/O emulsions (with vegetable and mineral oils) performed with a single-post coaxial re-entrant cavity sensor.

different W/O emulsions (mineral and vegetable) at microwave frequencies has been performed. Also, an open-ended coaxial sensor has been designed for dielectric characterization of this type of emulsions while flowing through a pipe. These and other related results can be found in [22].

Two types of W/O emulsions were considered: mineral and vegetable oil emulsions. The concentrations of water in samples were 15%, 30%, 40% and 50% by volume. Figure 14.5 shows that dielectric properties of emulsions depend not only on the water volume fraction but also on the chemical properties of the oils. The information given by the sensor can be used to simultaneously determine the type of emulsion, influence of oil composition (vegetal or mineral) and water content. In contrast with other methods to characterize emulsions (such as conductivity measurements, gravity separation, centrifugation), dielectric measurements are a simple, convenient, nondestructive and real-time alternative.

14.3.4.2 Microwave Sensors for Monitoring Curing Processes

Cure is a chemical reaction that converts a certain mixture of liquids into a solid piece. There are many products that need some kind of cure process during their manufacturing. Typical examples of curing processes are adhesive bonding or polymer shaping into solid pieces inside a mould. Usually, the quality of products obtained after the curing process is highly dependent on variables such as component dosage, ageing or contamination, temperature or ambient conditions. In the industry, it is very difficult to control these variables and, thus, to obtain repetitive cure processes, which assure the adequate final product quality. The common practice is to apply recommended specifications or apply conservative estimations. As a result, cure is less efficient than it could be, and material waste can be up to 10%, increasing costs and environmental problems [23].

Broadband dielectric relaxation spectroscopy has been widely used in the bibliography to reveal details about curing processes, with the aid of VNAs or impedance analyzers. Most of these measurements are performed with coplanar sensors with interdigitated comb-like electrode design [24], in the frequency range between 1 KHz and 1 MHz, conductivity being the main parameter of interest. However, cure monitoring through conductivity changes presents many drawbacks [21]: lack of sensitivity, conductivity masked by ionic changes, erroneous readings due to electrode polarization, etc.

One alternative for cure monitoring is to use microwave frequencies (above 1 GHz) and follow the changes in dielectric properties instead of conductivity. At these frequencies, the mobility of molecules (indication of the material viscosity) plays the main role and it is not masked by conductivity. Despite the advantages presented by microwave sensors, industrial application of the technique has been inhibited by a lack of basic knowledge of the relationships between molecular structure and the macroscopic dielectric behaviour and by limited availability of robust sensors and instrumentation suitable for the industrial environment [25,26].

Figure 14.6 shows an example of microwave sensor system developed for the non-invasive monitoring of the curing process of thermoset materials. The thermoset material is placed inside a mould and the microwave sensor is designed with a curved surface adapted to the mould inner shape. In particular, an open-ended coaxial resonator has been designed as the microwave sensor head.

Figure 14.7 shows some measurements of the sensor response during the cure process of a polyurethane sample. The figure shows how the effect of the polyurethane formation can be seen in the change of the sensor response during time. The figure also shows the complex permittivity during the curing process. Both the dielectric constant and the loss factor decrease with the reaction progress. This means that, during cure, the increase of viscosity of the liquid components causes a drop of the molecular mobility, and thus, the permittivity of the material is drastically decreased. Thus, the rate of reaction can be followed by simple inspection of these curves. The system allows not only monitoring the evolution of the curing process but also assessing the process variables, such as component ageing or contamination [26].

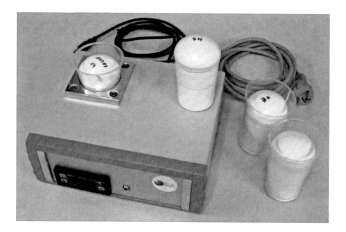

FIGURE 14.6 Open-ended coaxial resonator sensor developed at ITACA for monitoring curing process of thermoset samples.

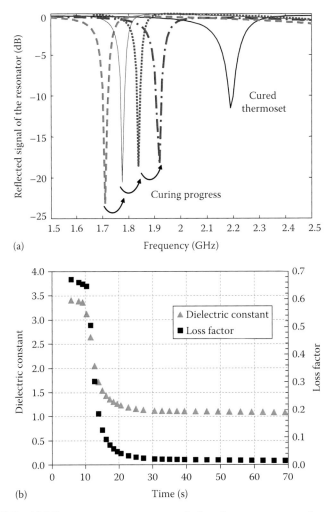

FIGURE 14.7   (a) Microwave sensor response during the cure process of a polyurethane sample. (b) Dielectric properties of the sample during cure.

Adhesives represent another interesting example of materials, which undergo cure in numerous manufacturing processes. In the case of adhesives, cure monitoring allows to reliably determine the time of cure and avoid the dependence on the apparent 'hardness' of the bond or manufacturer values of cure time, which do not take specific conditions into account.

Thermocouples and pressure transducers have been traditionally almost exclusively used for *in situ* monitoring of the adhesive cure [27]. For practical reasons, these methods are difficult to incorporate into industrial production.

In [16], a microwave sensor developed for cure monitoring of adhesive samples is shown. In this case, it is based on an open-ended coaxial resonator. The sample is placed on a protective layer, and with the aid of a metallic lid, a slight pressure is applied to get a uniform adhesive layer of the desired thickness and rid the bond of air bubbles.

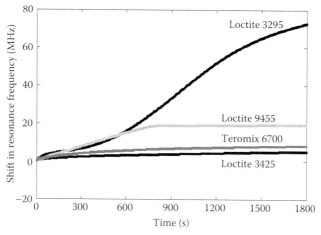

FIGURE 14.8 Microwave sensor response during the cure process of some adhesive samples. (Reproduced by B. García-Baños et al., Microwave sensor for continuous monitoring of adhesives curing, Prof. Tao J., (Ed)., *Proceedings of the 13th International Conference on Microwave and RF Heating, AMPERE*, Toulouse, France, 2011. With permission.)

Figure 14.8 shows the evolution of the shift of resonance frequency during the first 30 min of curing process for different adhesive samples. It is visible from the figure a clear dependence of the adhesive type on the curing response of the sample. As in previous case with polymer samples, the effect of the adhesive curing can be followed with the change of the resonant frequency of the sensor, showing typical curing curves. For example, Figure 14.8 shows a fast variation of the sensor response for the Loctite 9455 during the first 700 s and a practically flat response in the rest of 1800 s of monitoring. This is a clear indicator that most of the curing of Loctite 9455 falls in these first 700 s. From these measurements, more information can be obtained about the reaction, since the sensor also provides related results such as the reaction kinetics and cure time. This allows the system to be used as a production monitoring and control tool as well as for laboratory studies [28].

14.3.4.3 Microwave Moisture Sensors for Powder Materials

Knowing the moisture content has enormous economic value in the manufacture and processing of materials. Such information is useful for determining the value of raw materials, for in-process control and for output quality control.

Moisture determination has been traditionally done by analysing the loss of weight of a sample during drying, which is very time consuming and labour intensive. When fast and nondestructive moisture measurements are required, two main techniques are available: infrared (IR) methods and microwave dielectrometry. One of the differences between them is that IR methods are only sensitive to the moisture at the material surface, while microwave dielectrometry provides the *bulk* moisture of the material. This is of particular importance if moisture gradients exist in the material.

To illustrate this application, several samples of quartz sand were moistened and measured with a single-post coaxial re-entrant cavity sensor (see Figure 14.3). Quartz sand is a granular material often used as representative because its properties are

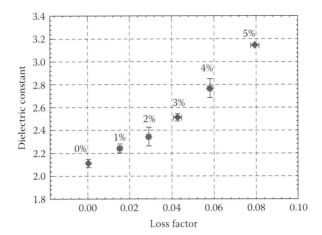

FIGURE 14.9   Dielectric characterization of quartz sand samples with different moisture contents (in% of dried weight) performed with a single-post coaxial re-entrant cavity sensor.

very similar to other materials employed in chemical or pharmaceutical industries. The moisture content of each sample was determined as the mass of water added (ranging from 0% to 5%) to the total mass of dry material.

Figure 14.9 shows the relation between dielectric properties (dielectric constant and loss factor) and moisture content of quartz sand samples at 2.45 GHz. The results show how dielectric properties are strongly influenced by water content at this frequency, as expected, due to the high dielectric constant and loss factor of water drops. Typical measurement deviations (see error bars in the figure) are more than adequate for practical moisture determination.

14.3.4.4   Microwave Sensors for Monitoring Liquid Underflow

In many industrial processes, it is very important to control the state of liquid components to ensure the final product quality. For instance, polyol is a highly reactive basic monomer commonly used for obtaining polymers in the footwear industry. It tends to absorb water, thus modifying the proper stoichiometric relationship in the reaction. Since changes in the composition of a reagent involve variations in its dielectric properties, instantaneous and nondestructive measurements with a microwave sensor can be used to monitor the purity and composition of the material.

Figure 14.10 shows the cylindrical cavity designed for this application. The liquid flows inside a pipe, passing through the resonant cavity, allowing dynamic measurements.

To follow the evolution of the degradation process in the liquid, only measurements of phase at a fixed frequency are performed. It yields to a cheaper and robust system for the industry and simplified measurements from the operator's point of view.

The designed sensor has been tested in the laboratory, where a continuous line with the liquid flowing through a PTFE pipe from and to a tank was prepared. A small pump is used to maintain constantly the movement of the liquid. The degradation process of the polyol was produced artificially by injecting water to an initially pure polyol sample.

The evolution of the sensor response in time is plotted in Figure 14.11, where each procedure of adding water to the pure material is indicated with the corresponding

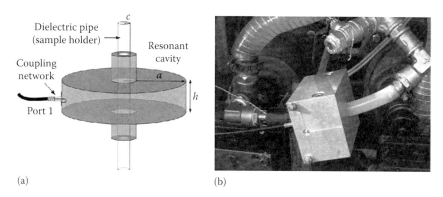

(a)   (b)

FIGURE 14.10   (a) Microwave sensor design (cylindrical cavity). (b) Nonintrusive installation of the sensor in the production line. (Reproduced with kind permission from Springer Science + Business Media: *Advances in Microwave and Radio Frequency Processing*, Microwave non-destructive evaluation of moisture content in liquid composites in a cylindrical cavity at a single frequency, 138–148, 2006, J. M. Catalá-Civera A.J. Canós, F.L. Peñaranda-Foix, E. de los Reyes-Davó, Figure 14.7.)

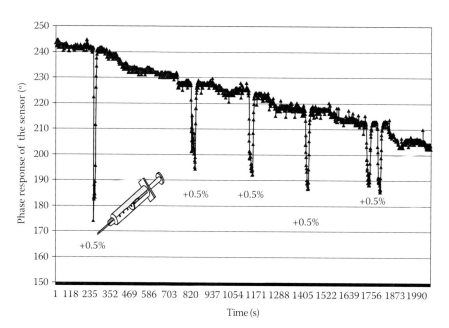

FIGURE 14.11   Phase response of the microwave sensor. The water content added in each step is displayed with percentage in volume. (Reproduced with kind permission from Springer Science + Business Media: *Advances in Microwave and Radio Frequency Processing*, Microwave non-destructive evaluation of moisture content in liquid composites in a cylindrical cavity at a single frequency, 138–148, 2006, J. M. Catalá-Civera, A.J. Canós, F.L. Peñaranda-Foix, E. de los Reyes-Davó, Figure 14.8.)

percentage in volume. As shown, the phase response of the sensor is very sensible to water content. Since water is directly injected into the pipe, a high concentration of water goes through the sensor during each of the adding processes, and therefore, sharp and momentary changes on the response can be seen. After the water injection, the phase takes its original value and then, as polyol is mixed with water in the tank, the response follows the degradation degree. It is possible to identify variations lower than 0.25% of moisture.

The use of phase monitoring at a single frequency in the resonator during the degradation process was found as a simple, cheap, robust and accurate mechanism of control, showing the unique capabilities of microwaves to follow changes in material properties.

REFERENCES

1. B. Clarke, Dielectric measurements and other measurements relevant to RF and microwave processing, *RF Industrial Processing Club Meeting*, National Physical Laboratory, U.K., 2006.
2. E. Nyfors, Industrial microwave sensors. A review. *Subsurf. Sens. Tech. Appl.*, 1(1), 2000.
3. B. Clarke, Ed., *A Guide to Characterization of Dielectric Materials at RF and Microwave Frequencies*, *NPL Guide*, National Physical Laboratory, London, U.K., 2003.
4. J. Krupka, Frequency domain complex permittivity measurements at microwave frequencies, *Meas. Sci. Technol.*, 17, R55–R70, 2006.
5. A.P. Gregory and R.N. Clarke, Dielectric metrology with coaxial sensors, *Meas. Sci. Technol.*, 18, 1372–1386, 2007.
6. A. Raj, W.S. Holmes, and S.R. Judah, Wide bandwidth measurement of complex permittivity of materials using coplanar lines, *IEEE Trans. Instrum. Meas.*, 50(4), 2001.
7. J.C. Anderson, *Dielectrics*, Science Paperbacks, Modern Electrical Studies, Chapman and Hall, London, U.K., 1964.
8. B. García-Baños, J.M. Catalá-Civera, A.J. Canós et al., Design rules for the optimization of the sensitivity of open-ended coaxial microwave sensors for monitoring changes in dielectric materials, *Meas. Sci. Technol.*, 16, 1186–1192, 2005.
9. A.P. Gregory and B. Clarke, Tables of the complex permittivity of dielectric reference liquids at frequencies up to 5 GHz, NPL Report MAT 23, ISSN 1754-2979, March 2009.
10. Rodhe and Schwarz, Measurement of dielectric material properties, Application Note, Application Center Asia/Pacific, RAC0607-0019, CY Kuek 07, 2006.
11. F.L. Penaranda-Foix and J.M. Catala-Civera, Circuital analysis of cylindrical structures applied to the electromagnetic resolution of resonant cavities, *Passive Microwave Components and Antennas*, Vitaliy Zhurbenko (Ed.), ISBN: 978-953-307-083-4, IN TECH, 2010.
12. J. Krupka, K. Derzakowski, B. Riddle et al., A dielectric resonator for measurements of complex permittivity of low loss dielectric materials as a function of temperature, *Meas. Sci. Technol.*, 9, 1751, 1998.
13. D.V. Blackham, An improved technique for permittivity measurements using a coaxial probe, *IEEE Trans. Instrum. Meas.*, 46(5), 1084–1092, October 1997.
14. D. Polo, P. Plaza-Gonzalez, B. García-Baños et al., Design of a low cost reflectometer coefficient system at microwave frequencies, *X Int. Conf. Microw. RF Heat.*, 235, 2005.
15. G. Chin Hock, C.K. Chakrabarty, M.H. Badjian et al., Super-heterodyne interferometer for non-contact dielectric measurements on millimeter wave material, *IEEE Int. RF Microw. Conf. Proc. RFM 2008*, Malaysia, 7–10, 2008.

16. B. García-Baños, J.M. Catalá-Civera, F.L. Peñaranda-Foix et al., Microwave sensor system for continuous monitoring of adhesive curing processes, *Meas. Sci. Technol.*, 23, 035101, 2012.

17. L.F. Chen, C.K. Ong, C.P. Neo et al., *Microwave Electronics, measurement and materials characterization*, John Wiley and Sons, West Sussex, England, 2004.

18. A.J. Canós, J.M. Catalá-Civera, F.L. Peñaranda-Foix et al., A novel technique for deembedding the unloaded resonance frequency from measurements of microwave cavities, *IEEE Trans. Microw. Theory Tech.*, 54(8), 3407–3416, 2006.

19. D. Xu, L. Liu, and Z. Jiang, Measurement of the dielectric properties of biological substances using an improved open-ended coaxial line resonator method, *IEEE Trans. Microw. Theory Tech.*, 35(12), 1987.

20. D. Kajfez, *Q Factor*, Vector Fields, Oxford, MS, 1994.

21. B. García-Baños, A.J. Canós, F.L. Peñaranda-Foix et al., Noninvasive monitoring of polymer curing reactions by dielectrometry, *IEEE Sens. J.*, 11(1), 62–70, January 2011.

22. B. García-Baños, R. Perez-Paez, J.M. Catala-Civera et al., Dielectric characterization of water in oil emulsions flowing through a pipe, *Proceedings of the 12th International Conference on Microwave and High Frequency Heating-AMPERE-*, Karlsruhe, Germany, 2009.

23. G. Oertel, *Polyurethane Handbook*, 2nd edn., Hanser, Munich, 1994.

24. M.C. Hegg, A. Ogale, A. Mescher et al., Remote monitoring of resin transfer molding processes by distributed dielectric sensors *J. Compos. Mater.*, 39(17), 1519–1538, 2005.

25. R. Casalini, S. Corezzi, A. Livi et al., Dielectric parameters to monitor the crosslink of epoxy resins, *J. Appl. Polym. Sci.*, 65(1), 17–25, 1997.

26. J.M. Catala-Civera, F. Peñaranda-Foix, B. Garcia-Baños et al., Microwave sensor system for monitoring the curing of polymer thermosets, *The 8th International Conference on Electromagnetic Wave Interaction with Water and Moist Substances (ISEMA)*, Espoo, Finland, 2009.

27. M.J. Lodeiro and D.R. Mulligan, *Cure monitoring techniques for polymer composites, adhesives and coatings*, NPL, Measurement Good Practice Guide, 75, February 2005.

28. B. García-Baños, F.L. Peñaranda-Foix, P.J. Plaza-Gonzalez et al., Microwave sensor for continuous monitoring of adhesives curing, *Proceedings of the 13th International Conference on Microwave and High Frequency Heating -AMPERE-*, Toulouse, France, 2011.

15 Microwave Reflectometry for Sensing Applications in the Agrofood Industry

Andrea Cataldo, Egidio De Benedetto, and Giuseppe Cannazza

CONTENTS

15.1 INTRODUCTION

This chapter describes some of the most promising applications of microwave reflectometry (MR) for monitoring and sensing purposes in the agrofood industry. The interest towards MR originates from the fact that this technique can satisfy several contrasting requirements, such as low implementation cost, real-time response, possibility of remote control, reliability and adequate measurement accuracy. Typically, in MR measurements, an electromagnetic (EM) signal is propagated into the system under test (SUT): the analysis of the reflected signal along with a specific data processing is used to retrieve the desired information on the SUT [1]. Applications of MR are numerous and cover a wide range of fields; in fact, thanks to the versatility of the approach, this technique has proven useful for several applications, such as

- Localization of faults along cables [2–4]
- Geotechnical engineering for assessing distributed pressure profiles [5–10]
- Measurement of liquid levels (also in stratified liquids) [11–13]

- Monitoring applications in civil engineering for sensing crack/strain in reinforced concrete structures [14], for fault location on concrete anchors [15], for monitoring cement hydration [16], etc.
- Water-leak detection in underground metal pipes [17]

A comprehensive overview of the most important and promising applications of MR can be found in [18]. On the other hand, this chapter focuses on three specific applications in which MR is employed for monitoring purposes in the agrofood industry. In the following paragraphs, first, the basic theoretical principles behind MR are recalled and the adopted measurement strategies are discussed. Finally, some interesting test cases related to MR-based monitoring of agrofoods are presented; in particular, the following applications are considered:

1. Moisture measurement of granular agrofood materials [19]
2. Quality control of vegetable oils [20,21]
3. Monitoring of dehydration process of fruit and vegetables [22]

Measurement techniques: An MR measurement system generally consists of two main elements: (1) the unit for generating and receiving the EM signal and (2) the measurement probe (or sensing element).

MR measurements can be performed either in time domain (time-domain reflectometry – TDR) or in frequency domain (frequency domain reflectometry – FDR): the choice of using either approach depends on the intended application. In general, TDR instruments are less expensive and often also portable; as a result, TDR is often considered strategically suitable for *in situ* application (e.g. in a production line). On the other hand, employing FDR instruments may often ensure a better measurement accuracy. An optimal trade-off of instrument portability/low cost and measurement accuracy can be achieved through a combined TD/FD approach (i.e. by performing TDR measurements and extrapolating, through suitable processing techniques, the corresponding FD information) [23].

In TDR measurements, a step-like EM signal is propagated, along a probe, through the SUT. The output of TDR measurements is a reflectogram, that is, the reflection coefficient $\rho(t)$ along the sections of the probe is displayed as a function of time (or as a function of the travelled electric distance). The behaviour of $\rho(t)$ is strictly associated with the impedance variations along the electrical path travelled by the EM signal. In turn, the electric distance travelled by the EM signal, L_{app}, is related to the dielectric characteristics of the SUT:

$$L_{app} = \sqrt{\varepsilon_{app}}\, L = \frac{ct_t}{2} \qquad (15.1)$$

where
$\quad$ L is the length of the used probe
$\quad$ $c \cong 3 \cdot 10^8$ m/s is the velocity of light in free space
$\quad$ t_t is the travel time
$\quad$ ε_{app} is the apparent relative dielectric permittivity of the material in which the probe is inserted [24]

The quantity ε_{app} depends on the relative complex dielectric permittivity of the considered material [25]; for low-loss, low-dispersive materials, ε_{app} is approximately equal to the real part of the dielectric permittivity (ε_r') and can be considered constant with frequency [23].

The analysis of TDR waveform directly leads to the evaluation of L_{app}, whose value is used in the subsequent data processing for retrieving the desired information.

In FDR measurements, generally, the quantity to be measured is the reflection scattering parameter, $S_{11}(f)$. Traditionally, measurements performed directly in FD rely on vector network analyzers (VNAs): in this case, the excitation stimulus is a sinusoidal signal whose frequency is swept over the desired range of analysis. One of the advantages of FD measurements is that there are well-established error correction models that, through appropriate calibration procedures [such as the short–open–load (SOL) calibration], can be used to reduce the influence of systematic errors [26]. The FDR approach is often used for the dielectric spectroscopy of materials and for impedance characterization of electronic devices and components.

Finally, the adoption of the combined TDR/FDR approach allows taking advantage of the benefits of both the approaches. In particular, estimating the $S_{11}(f)$ from TDR measurements can help disclose useful information that is masked in time domain (e.g. multiple dielectric relaxation) [27]. This strategy is regarded as a powerful tool for guaranteeing simultaneously low cost and portability of the instruments and measurement accuracy [28–30]. However, there are some crucial aspects that need to be considered in order to perform an accurate TD/FD transformation, thus reducing errors on the assessment of the spectral response of the system and in the extraction of $S_{11}(f)$ [31].

Sensing element: The sensing element (or probe) is responsible for the interaction of the stimulus signal with the SUT, and it is the ultimate factor that influences the accuracy of results. Since MR senses the changes in impedance, it is extremely important to employ a probe with a well-known impedance profile; in this way, it is easier to discriminate and interpret the impedance variations due to the SUT.

Several probe configurations can be used: multifilar, coaxial, planar, etc. In general, the design of coaxial probes is quite simple, and for this configuration, the impedance profile (in the transverse electromagnetic (TEM) propagation mode) can be determined from the transmission line (TL) theory [27,32]:

$$Z(f) = \frac{60}{\sqrt{\varepsilon_r^*(f)}} \cdot \ln(b/a) \tag{15.2}$$

where

$Z(f)$ is the impedance of the probe filled with the considered material

$\varepsilon_r^*(f)$ is the complex relative dielectric permittivity of the material filling the probe

b is the inner diameter of the outer conductor

a is the outer diameter of the inner conductor

Coaxial probes are widely used for monitoring and diagnostics on liquids [20,22].

On the other hand, for soil monitoring and for granular material monitoring in general, the multi-rod configuration is a widely used solution. In fact, this configuration allows an easier insertion of the probe into the granular material. In particular, three-rod probes have become a widespread solution since their electric behaviour resembles that of coaxial cells [33,34]. Furthermore, when a noninvasive approach must be preserved, it is possible to use surface probes [35] or antennas [36]. A numerical analysis of some probe configurations can be found in [37].

Dielectric characterization and quality control of materials: The dielectric characteristics of materials are intrinsic features that can be associated to other characteristics of the considered SUT (also to nonelectric characteristics); therefore, dielectric spectroscopy has been employed as a non-destructive diagnostic method for a wide range of materials. The combination of TDR measurement with a suitable TL modelling of the measurement cell can provide accurate results on the dielectric properties of materials. With regard, for example, to the adoption of the TDR/FDR for the accurate evaluation of the Cole–Cole parameters, the typical procedure would include three major steps:

- First, the TDR waveforms are acquired and processed in order to obtain the corresponding $S_{11}(f)$.
- Second, the measurement cell (i.e. probe and material under test) is accurately modelled as a TL, in which the dielectric characteristics of the considered liquid are parameterized through the Cole–Cole formula [38]. The modelled scattering parameter, $S_{11,\text{MOD}}(f, \varepsilon_r^*(f))$, is evaluated.
- Finally, the Cole–Cole parameters of the SUT are evaluated by minimizing the deviations between (1) the measured reflection scattering parameter, $S_{11}(f)$, obtained from the TD/FD approach and (2) the modelled reflection scattering parameter obtained from the implemented TL model of the probe plus SUT, $S_{11,\text{MOD}}(f)$ [27].

As a matter of fact, dielectric properties of foods are a useful tool for controlling the quality of food products [39–41]. Also, knowing the dielectric characteristics of food products is important for allowing the optimization of several industrial processing, for example, in terms of power absorption during microwave drying process [42] or in terms of effectiveness of radio-frequency heating for disinfestation of fruit [43], for controlling the salting process of pork meat [44] and so on.

15.2 MOISTURE-CONTENT MEASUREMENT OF GRANULAR AGROFOOD MATERIALS

In the agrofood industry, monitoring water content of granular materials (such as coffee, flour) is extremely important; in fact, water content of these materials is strictly associated to their quality status: a high amount of water may lead to the growth of microorganisms and bacteria and could compromise the storage of the material.

A wide variety of methods can be used for estimating water content of porous media, ranging from destructive (gravimetric) to non-destructive methods (gamma radiation probe, neutron probe, etc.). The gravimetric sampling, for example, is a highly accurate method; nevertheless, since it requires medium samples to be removed from the medium mass, it is extremely time-consuming and difficult to implement on a process line. As for non-destructive methods, such as the neutron scattering method [45] and the gamma ray attenuation method [46], they are accurate as well; nevertheless, their higher costs and caution to avoid possible health hazards limit their adoption [19].

On the other hand, the adoption of TDR can be an excellent candidate for monitoring water content of agrofood materials. TDR is a highly flexible technique; it can guarantee low measurement uncertainty and continuous measurements. Additionally, measurements have only a small dependence on temperature. Furthermore, TDR allows the possibility of controlling several signals through multiplexing.

Basically, in TDR, the value of the volumetric water content, θ, is inferred from dielectric permittivity measurements. In fact, the relative permittivity of water (approximately 80) is considerably higher than the typical relative permittivity of agrofood materials; therefore, the presence of water increases the overall permittivity of the 'moistened' material [47]. In TDR-based method, moisture content is determined from measurements of ε_{app}, which can be easily evaluated from the TDR waveform, by applying Equation 15.1; the value of ε_{app} is then substituted in a mathematical model, which gives the value of moisture as a function of ε_{app}. In the literature, there are a number of models that can be used; however, most of them are material-specific and none has a universal validity [48]. A broad classification of the models that describe the relationship between $\varepsilon_{app} - \theta$ can be made by distinguishing into empirical and/or partly deterministic approaches. The former category simply fits mathematical expressions (i.e. polynomial whose coefficients must be determined for each specific material) to measured data; the latter considers dielectric mixing models, which take into account the dielectric characteristic and the volume fractions of each constituent (e.g. solid, water and air) of the considered material.

15.2.1 EMPIRICAL MODELS

One of the most widely used empirical models is described by Topp's equation, which is mostly used for soils [49,50]:

$$\theta = 4.3 \times 10^{-6} \varepsilon_{app}^3 - 5.5 \times 10^{-4} \varepsilon_{app}^2 + 2.92 \times 10^{-2} \varepsilon_{app} - 5.3 \times 10^{-2} \qquad (15.3)$$

In this equation, the coefficients are already determined, and the expression is considered generally valid for materials with low content of organic materials. However, when dealing with more complex system, specific calibration curves must be individuated for each considered material. These models have some shortcomings (such as the limited possibility of extrapolation outside the moisture range of the original set of experiments [48]); nevertheless, they are routinely used.

As aforementioned, the empirical approach simply fits mathematical expressions to measured data; polynomial regression equations are typically used to individuate the relationship for different types of materials. In this way, the TDR-measured ε_{app} can be related to θ; these curves are called calibration curves. Once the calibration curve (for the specific material) is individuated, to determine the unknown moisture level, it is enough to measure the dielectric constant and to retrieve the moisture level from the calibration curve [33].

Typically, the $\varepsilon_{app} - \theta$ calibration curve can be assessed as follows. Samples of the considered material are moistened at prefixed values of moisture (θ), and the corresponding dielectric constant (ε_{app}) is measured through the TDR method. The (ε_{app}, θ) points are fitted through a non-linear regression method [33].

15.2.2   DIELECTRIC MIXING MODELS

A more accurate $\theta - \varepsilon_{app}$ relationship can be obtained through dielectric mixing models, which take into account the volume fraction and the dielectric constant of each single constituent that is present in the material under test (MUT), the porosity, the effect of the geometrical arrangement of the medium components, etc. Therefore, the dielectric mixing models are suitable for the evaluation of the $\varepsilon_{app} - \theta$ relationship for nonhomogeneous granular materials, with low dry bulk density, with large amount of bound water or with relatively large permittivity of the solid phase. Despite the extensive scientific efforts that have been devoted to obtain data on the dielectric properties of agrofoods, the development of specific dielectric models for each considered material, relating the dielectric properties of agrofood materials to the corresponding moisture content, is still an open issue [19].

Partly deterministic and probabilistic models: Several dielectric mixing models have been proposed in literature. In the present case study, two families of models are considered: partly deterministic models and probabilistic models.

With regard to the first category, two main models can be considered: (1) the so-called α-model, based on the semiempirical equations proposed by Birchak et al. [51], and (2) the Maxwell-De Loor (MD) model, a theoretical model proposed by De Loor and based on Maxwell's equation [52]. In these two models, the water fraction is considered as a single phase. To overcome this simplification and to take into account the effect of the water portion close to the medium surface (which is water whose permittivity value is lower than that of the free water), Dobson et al. [53] extended the two aforementioned models, thus deriving a four-component system that suitably considers the effects of free water, bound water, solid phase and air.

Finally, the theoretical model of Ansoult [54] belongs to a particular family of the partly deterministic models, the so-called probabilistic model. This model is based on the random propagation of the pulse in a porous medium that is represented as an array of capacitors. In this approach, the granular material is schematized as a set of three capacitors, each associated to the dielectric constants (ε_{ij}) of the different sample components; eventually, a computational algorithm is used to evaluate the overall equivalent dielectric constant [19].

15.2.3   EXPERIMENTAL COMPARISON OF DIFFERENT MODELS FOR MOISTURE-CONTENT MEASUREMENTS ON GRANULAR MATERIALS

In this section, a comparative approach between different empirical and deterministic models for the TDR-based estimation of water content is described. For each considered material, the functional relations between dielectric properties and the corresponding moisture levels are also assessed.

The following models are considered: (i) a 3rd-degree polynomial with four 'adjusted' parameters obtained by calibration procedures on the materials under test, (ii) the polynomial equation of Topp with its original parameters, (iii) the partly deterministic simplified models based on the theory of mixtures of three components, (iv) the partly deterministic simplified models based on the theory of mixtures of four components, (v) the MD model and (vi) the probabilistic model of Ansoult, based on the random propagation of the EM pulse in the soil.

The aforementioned models were tested on three feedstuff materials: corn, corn flour and bran. TDR measurements were performed through the HyperLabs 1500 (which is a portable TDR unit); the used probe was a 30 cm long three-wire metallic probe (Campbell Scientific CS610).

Materials were progressively moistened at prefixed (known) levels of moisture in the ranges reported in Table 15.1. For each moisture level, the corresponding ε_{app} was measured. As a result, for each material, the corresponding empirical calibration curve $\varepsilon_{app} - \theta$ was obtained.

The $\varepsilon_{app} - \theta$ curves obtained for all the considered models were derived by superimposing the TDR-measured ε_{app} values, and the different models were compared with the empirical calibration curve. Results are shown in Figure 15.1. Repeated measurements showed that the maximum percentage relative standard uncertainty in ε_{app} is less than 8%.

The root mean square error (RMSE) between the actual values of moisture and the values obtained from the models was taken as a figure of merit of the different $\varepsilon_{app} - \theta$ relations. The RMSE values are reported in Table 15.2. It can be seen that the empirical Topp's equation is the least accurate model; conversely, the best fitting is observed for the partly deterministic models. This is attributable to the fact that the effect of porosity and of the additional physical components is explicitly taken into account. More specifically, the MD model predicts the $\varepsilon_{app} - \theta$ behaviour for bran and corn flour better than it does for the other materials; this is due to the fact that bran and corn flour contain the highest percentage of hygroscopic water. On the other hand, for corn samples, α-model succeeds better than the other models in characterizing the $\varepsilon_{app} - \theta$ relation [19].

TABLE 15.1

Moisture-Content Range of the Considered Agrofood Materials

Material	Moisture Levels Range
	(cm³/cm³)
Corn	0–0.25
Corn flour	0–0.22
Bran	0–0.21

Source: Cataldo, L. et al., *Comput. Standards Interfaces*, 32(3), 86, 2010. Copyright © Elsevier 2009.

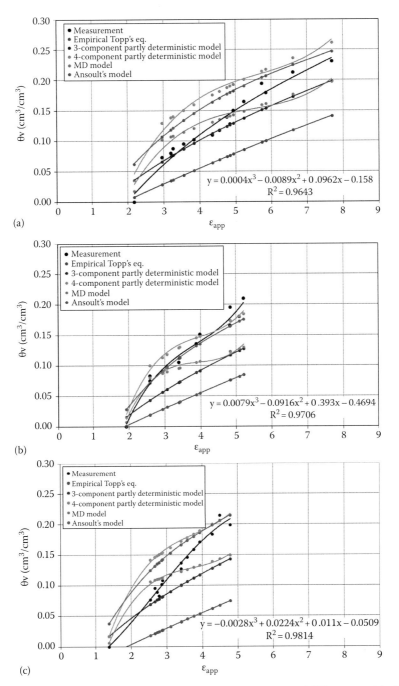

FIGURE 15.1 Comparison among experimental measurements and different empirical and partly deterministic models for (a) corn flour, (b) corn and (c) bran. Interpolation functions are also reported. (From Cataldo, L. et al., *Comput. Standards Interfaces*, 32(3), 86, 2010. Copyright © Elsevier 2009.)

TABLE 15.2
RMSE Values between the Measured and the Modelled Values of Water Content, for Feedstuff Materials

Model	Corn	Corn Flour	Bran
Topp's equation	0.08	0.07	0.09
α-Model (3 components)	0.05	0.02	0.04
α-Model (4 components)	0.02	0.05	0.04
MD model	0.04	0.02	0.03
Ansoult's model	0.02	0.03	0.03

Source: Cataldo, L. et al., *Comput. Standards Interfaces,* 32(3), 86, 2010. Copyright © Elsevier 2009.

15.3 QUALITY CONTROL OF VEGETABLE OILS

Typically, the *in situ* evaluation of the Cole–Cole parameters for industrial quality control purposes is not an easy task, especially when reference data are missing. This is the case, for example, with vegetable oils, for whose dielectric characteristics only scarce reference data are available [55,56]. Additionally, since quality control of vegetable oils is becoming more stringent (especially for avoiding adulteration), the estimation of these dielectric parameters may be used as an indicator for certifying the product quality.

Although highly sophisticated methods for the analysis of edible oils are available, many of them are not easily applicable for routine or continuous monitoring (e.g. gas chromatography [57] and liquid chromatography [58]), whereas others (e.g. Fourier transform infrared (FTIR) [59], Raman spectroscopy [60] and nuclear magnetic resonance (NMR) [61]) are rather complex and highly expensive. On the other hand, other simpler methodologies that are commonly adopted for routine analysis (such as those based on chemical titrations) do not provide much information. As a matter of fact, all of aforementioned methods are affected by some limitations: in particular, they cannot be performed on the process line and they are very laborious.

The following experimental procedure describes a method for evaluating the Cole–Cole parameters of several types of edible oils. The Cole–Cole parameters are different for each material and they define its spectral signature. The five parameters are the static dielectric permittivity (ε_s), the permittivity at infinite frequency (ε_∞), the relaxation frequency (f_r), the static electrical conductivity (σ) and a parameter that describes the spread in relaxation frequency (β) [38]. Starting from traditional TDR measurements, the evaluation of the Cole–Cole parameters was carried out following the TL-based procedure described in Section 15.1. This is done in view of possible practical applications, which may be useful, for example, in the online monitoring of the characteristics of oils throughout a production process.

Transmission line modelling of the measurement cell: Figure 15.2 shows the picture and the schematic diagram of the used probe. The TL model of the measurement cell

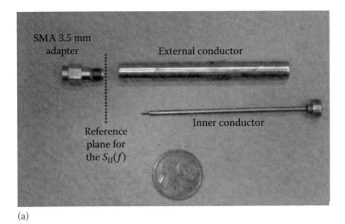

(a)

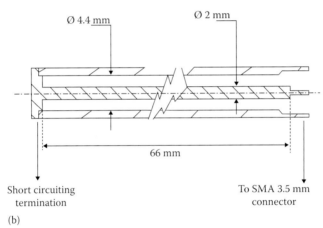

(b)

FIGURE 15.2 Picture of the used coaxial probe. (a) Schematic configuration of the used coaxial probe (b). (From Cataldo, E. et al., *Measurement*, 43(8), 1031, 2010. Copyright © Elsevier 2010.)

was implemented in the AWR Microwave Office® simulator. The basic TL model includes a length of short-circuited coaxial probe, with the physical dimensions of the actual probe, filled with a material characterized through its relative permittivity and loss tangent. These parameters, in turn, vary with frequency according to the Cole–Cole model, described by four parameters (ε_s, ε_∞, f_r, β).

The implemented TL model includes not only the Cole–Cole parameters of the liquid under test as minimization variables but also some lumped circuit elements that account for additional parasitic effects that cannot be compensated for through traditional SOL calibration procedures (the values and the nature of the lumped circuit elements were assessed through preliminary TDR measurements on well-referenced materials: these measurements allowed the optimization of the TL model).

Description of the procedure and discussion of results: The probe, filled with the liquid under test, was connected to the TDR instrument, and the corresponding waveform was acquired. For each TDR measurement, also the waveforms corresponding to the SOL calibration standards were acquired. The collected TD data were processed through the fast Fourier transform (FFT)-based algorithm described in [31], thus extracting the corresponding calibrated $S_{11}(f)$ [20].

Finally, the Cole–Cole parameters are extracted by minimizing (over the Cole–Cole parameters) the difference between the FFT-extracted $S_{11}(f)$ and the modelled $S_{11,MOD}(f)$. However, to obtain a more accurate evaluation of the Cole–Cole parameters, it is advisable to reduce the number of variables in the final minimization routine. To this purpose, the values of the static permittivity were assessed through an alternative method, and the minimization routine was performed over the remaining three Cole–Cole parameters (i.e. ε_∞, f_r and β).

The evaluation of the ε_s of the oils was performed through capacitive measurements carried out with an inductance, capacitance and resistance (LCR) meter. LCR measurements were performed using the same coaxial probe but removing the short circuit at the distal end. The probe (filled with the liquid under test) was considered as a cylindrical capacitor whose capacitance ideally depends on the liquid filling the space between the cylindrical plates. The values of the static permittivity were considered as known (fixed) values in the subsequent minimization routines carried out on FFT TDR data, employing the procedure described in Section 15.1: as expected, since the successive minimizations were performed only on three Cole–Cole parameters, the accuracy of the final results of the minimizations is enhanced (electrical conductivity is negligible for edible oils) [20].

Table 15.3 summarizes the Cole–Cole parameters and the corresponding standard deviations for 10 different types of vegetable oils.

The results obtained for the Cole–Cole parameters through the proposed combined procedure are in good agreement with the scarce data available in the literature. Therefore, the proposed technique can be useful for quality monitoring of oils.

TABLE 15.3

Averaged Cole–Cole Parameters for 10 Types of Vegetable Oil (at 20.0°C). Standard Deviation Values Are Also Reported

Type of Oil	ε_s	$\sigma_{\varepsilon s}$	ε_∞	$\sigma_{\varepsilon\infty}$	f_r (MHz)	σ_{fr} (MHz)	β	σ_β
Olive (ac. = 0.3%)	3.08	0.01	2.39	0.01	315	11	0.33	0.01
Olive (ac. = 1.2%)	3.14	0.01	2.38	0.01	288	9	0.36	0.02
Olive (ac. = 1.6%)	3.19	0.03	2.36	0.01	259	4	0.40	0.01
Olive (ac. = 4.0%)	3.19	0.03	2.34	0.02	249	5	0.42	0.02
Peanut	3.05	0.01	2.40	0.01	334	5	0.28	0.01
Sunflower	3.12	0.04	2.40	0.01	292	5	0.31	0.01
Corn	3.11	0.02	2.41	0.01	309	6	0.29	0.02
Castor	4.69	0.01	2.56	0.01	122	2	0.42	0.01
Various seeds	3.10	0.02	2.43	0.01	371	5	0.27	0.01
Soybean	3.09	0.01	2.41	0.02	390	7	0.30	0.02

Source: Cataldo, E. et al., *Measurement*, 43(8), 1031, 2010. Copyright © IEEE 2009.

Results show that the relaxation frequency appears to be the Cole–Cole parameter that differs the most among the considered oils. On this basis, a reliable, real-time, and *in situ* quality control procedure may rely on relaxation frequency as the key parameter for identifying different kinds of oil. To this purpose, results may be collected in a database (together with the associated confidence intervals) and used as reference values for future 'offline' measurements on unknown oils. More advances and additional details on the assessment of the procedure can be found in [21].

15.4 MONITORING OF THE DEHYDRATION PROCESS OF FRUIT AND VEGETABLES

Another industrial application of MR relates to the monitoring of the osmotic dehydration (OD) process of fruit and vegetables. OD is a process that is employed for the partial removal of water from fruit and vegetables, thus obtaining a significant increase of their shelf life. In fact, by reducing water content, the growth of microorganisms is inhibited, and the rate of degradation reactions is reduced.

In the OD process, the product is immersed in a hypertonic solution, thus promoting two countercurrent flows: (i) water motion from biological tissues to hypertonic solution and (ii) osmotic agents flow towards the product. OD has been extensively used for pre-dehydration of pineapples [62], mushrooms [63], melons [64], bananas [65] and many other perishable goods in general. Nevertheless, as reported in [66], the adoption of OD in industrial production processes is hindered by problem with the overall management of the concentrated solutions. In fact, some practical aspects make managing and controlling the osmotic solution the bottleneck of the OD process. In particular, there are still some major concerns regarding the appropriate individuation of the loss in dewatering capacity and the possibility of reusing the spent solution, so as to make the process more economically advantageous [67].

On such bases, problems related to the OD control and osmotic solution management are still actual, and currently, no methodologies can successfully and rapidly represent the solution to the aforementioned needs. As a matter of fact, the osmotic pressure (π) might represent the primary parameter whose estimation would lead to the identification of the quality status of the osmotic solution. However, many food-industry processes typically involve osmotic solutions with high value of π (up to 250–300 bar), which, in practice, seriously limit the possibility of realizing specific devices for measurements. Additionally, the high uncertainties and the long times needed for measurement are other drawbacks that make the direct measurement of π extremely difficult.

On the other hand, water activity (a_w) is a reliable indicator of the status of the osmotic solution, since it is strictly related to the dewatering capacity of the solution. In fact, the OD proceeds until the osmotic solution and the food product have reached the same a_w value [68]. Nevertheless, although this parameter can be a suitable indicator in the OD management, not only the instrumentation for measuring it is quite costly, but also measurements usually have to be performed in laboratory (i.e. samples of the solutions should be taken and subsequently analysed offline), thus limiting the possibility of continuous and *in situ* monitoring.

To overcome this limitation, an alternative procedure, based on MR, for identifying the status of the industrial osmotic solutions for control and management purposes has been developed. In particular, it was demonstrated that there is a specific linear relationship between characteristic parameters of the solution (i.e. a_w, degree Brix and, consequently, π) and the corresponding measured reflection coefficient. As a direct practical consequence, once the functional relationship is individuated for the specific OD process, it can be used for the subsequent online monitoring in the process line or for the optimization of the involved process parameters. The clear advantage of resorting to MR is that measurements can be performed *in situ* and, most importantly, can provide results in real time, thus allowing to promptly intervene on the process [22].

The proposed methodology was adopted also in a real test case involving the OD of tomatoes, which is a typical application of food industry. In this case, the most appropriate reflectometric parameter was found to be the so-called steady-state reflection coefficient, ρ_∞, which is the value of the reflection coefficient at long times (i.e. when multiple reflections caused by the signal travelling back and forth along the probe have died out). The quantity ρ_∞ is associated to the corresponding variation of a_w and of degree Brix. The experimental results confirm the suitability of the proposed method as a powerful tool for controlling the status of the osmotic solution, thus paving the way for a prompt management of the hypertonic solution in OD processes in the field of food industry.

Sliced tomatoes were immersed in the solution (mass ratio 5:1) of glucose syrup, and the dehydration process was observed for 24 h. TDR measurements were performed at eight prefixed time intervals; results are reported in Figure 15.3. The overall range of variation of the reflection coefficient (from −0.91 to −0.96), corresponding to different dehydration intervals, is rather narrow, even though some well-distinguishable differences can be clearly detected. Referring to Figure 15.3, the long-distance traces

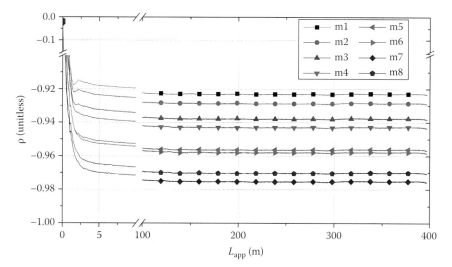

FIGURE 15.3 Long distance TDR measurements performed during the 24 h long cycle on the solution of the process line. (From Cataldo, G. et al., *J. Food Eng.*, 105(1), 186, 2011. Copyright © Elsevier 2011.)

present a slightly increased sensitivity range, anticipating a better performance of the method around the steady-state conditions. In fact, due to the elevate presence of polar species, such as the high concentration of NaCl, the resultant dielectric response is strongly attenuated; hence, significant variations of the reflected signal can be located prevalently around the static condition. In other words, the solution at the initial time has a high electrical conductivity (approximately 3080 µS/cm); as the dehydration proceeds, the electrical conductivity value increases up to approximately 5600 µS/cm, due to the agents' flow between tomatoes and solutions. However, thanks to the broadband frequency sensitivity of the TDR method, also little variation in static electrical conductivity of osmotic solutions can be detected.

To individuate the relation between the reference values and the measured data, in Figures 15.4 and 15.5, the long-distance reflection coefficients are plotted against a_w and degree Brix (both evaluated through dedicated methods), respectively, and a linear fitting is imposed. A good linear proportionality is observed since the values of the adjusted R-square indicate the correctness of the linearity assumption for the $\rho_\infty - a_w$ and for the $\rho_\infty - °Bx$ relations. As a result, differently from the sugar-based solutions previously considered (which presented a linear trend at higher-frequency measured data), for this kind of industrial solutions and for the intended control purposes, the most suitable measurement parameter is ρ_∞. Therefore, the adoption of the specific calibration curves ($\rho_\infty - a_w$ or $\rho_\infty - °Bx$) is the key for controlling the status of high-conductivity osmotic solutions.

On a side note, it is worth mentioning that the osmotic pressure values can be predicted through appropriate models, starting from the inferred values of a_w. This implies that, should there be the specific requirement to calibrate TDR measurements against

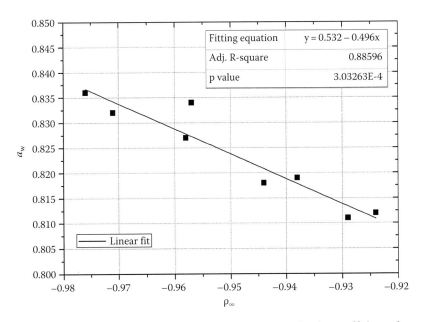

FIGURE 15.4 Behaviour of a_w as a function of the steady-state reflection coefficient value, ρ_∞. (From Cataldo, G. et al., *J. Food Eng.*, 105(1), 186, 2011. Copyright © Elsevier 2011.)

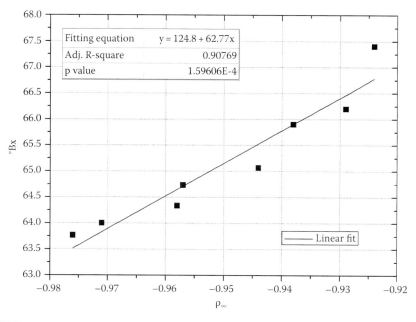

FIGURE 15.5 Behaviour of °Bx as a function of the steady-state reflection coefficient value, ρ_∞. (From Cataldo, G. et al., *J. Food Eng.*, 105(1), 186, 2011. Copyright © Elsevier 2011.)

the corresponding π values, the proposed method would still be successfully applicable even without the difficult direct measurement of the osmotic pressure (given that the independent measurement of a_w and the consideration of a robust model can be suitably considered).

15.5 CONCLUSIONS

In this chapter, a brief overview of current achievements of the use of MR for monitoring and diagnostic purposes was given. The different approaches (TDR, FDR and TDR/FDR) were addressed, and the pros and cons were described. Additionally, the different strategies for enhancing the final measurement accuracy were presented. Finally, the major applications (and the obtained results) of MR in the field of agrofood monitoring were discussed, thus providing a global picture of the open issues and of the advances of this measurement technique. Results demonstrate that MR is a powerful technique that can be effectively employed in a number of practical applications; in particular, the high versatility and other intrinsic characteristics (such as real-time response and possibility of remote control) have fostered the interest towards the adoption of MR for monitoring purposes. An additional major strength of MR lies in the high adaptability of the technique also to pre-existing conditions (i.e. MR-based monitoring systems can be easily implemented even on already-operating production line). Finally, also the implementation costs of MR-based solutions are highly competitive (especially when resorting to TDR instruments), thus making such systems extremely appealing not only for large-scale applications but also for small-scale monitoring purposes.

REFERENCES

1. A. M. O. Mohamed, *Principles and Applications of Time Domain Electrometry in Geoenvironmental Engineering*. Taylor & Francis Group, Great Britain, 2006.
2. K. O'Connor and C. H. Dowding, *Geomeasurements by Pulsing TDR Cables and Probes*. CRC Press, Boca Raton, FL, January 1999.
3. D. E. Dodds, M. Shafique, and B. Celaya, TDR and FDR identification of bad splices in telephone cables, in *Proceedings of IEEE Canadian Conference on Electrical and Computer Engineering*, 838–841, May 2006.
4. S. Wu, C. Furse, and C. Lo, Noncontact probes for wire fault location with reflectometry, *IEEE Sensors Journal*, 6(6), 1716–1721, 2006.
5. A. Scheuermann and C. Huebner, On the feasibility of pressure profile measurements with time-domain reflectometry, *IEEE Transactions on Instrumentation and Measurement*, 58(2), 467–474, 2009.
6. A. Scheuermann, C. Hubner, H. Wienbroer, D. Rebstock, and G. Huber, Fast time domain reflectometry (TDR) measurement approach for investigating the liquefaction of soils, *Measurement Science and Technology*, 21, 025104, 2010.
7. A. Corsini, A. Pasuto, M. Soldati, and A. Zannoni, Field monitoring of the Corvara landslide (Dolomites, Italy) and its relevance for hazard assessment, *Geomorphology*, 66(1–4), 149–165, 2005.
8. W. F. Kane, T. J. Beck, and J. Hughes, Applications of time domain reflectometry to landslide and slope monitoring, in *Proceedings of the 2nd International Symposium and Workshop on Time Domain Reflectometry for Innovative Geotechnical Applications*, pp. 305–314, 2001.
9. A. M. O. Mohamed and R. A. Said, Detection of organic pollutants in sandy soils via TDR and eigendecomposition, *Journal of Contaminant Hydrology*, 76, 235–249, 2005.
10. A. McClanahan, S. Kharkovsky, A. R. Maxon, R. Zoughi, and D. D. Palmer, Depth evaluation of shallow surface cracks in metals using rectangular waveguides at millimeter-wave frequencies, *IEEE Transactions on Instrumentation and Measurement*, 59(6), 1693–1704, 2010.
11. A. Cataldo, L. Catarinucci, L. Tarricone, F. Attivissimo, and E. Piuzzi, A combined TD-FD method for enhanced reflectometry measurements in liquid quality monitoring, *IEEE Transactions on Instrumentation and Measurement*, 58(10), 3534–3543, 2009.
12. A. Cataldo, L. Tarricone, F. Attivissimo, and A. Trotta, Simultaneous measurement of dielectric properties and levels of liquids using a TDR method, *Measurement*, 41(3), 307–319, 2008.
13. C. P. Nemarich, Time domain reflectometry liquid levels sensors, *IEEE Instrumentation & Measurement Magazine*, 4(4), 40–44, 2001.
14. S. Sun, D. J. Pommerenke, J. L. Drewniak, G. Chen, L. Xue, M. A. Brower, and M. Y. Koledintseva, A novel TDR-based coaxial cable sensor for crack/strain sensing in reinforced concrete structures, *IEEE Transactions on Instrumentation and Measurement*, 58(8), 2714–2725, 2009.
15. C. Furse, P. Smith, and M. Diamond, Feasibility of reflectometry for nondestructive evaluation of prestressed concrete anchors, *IEEE Sensors Journal*, 9(11), 1322–1329, 2009.
16. N. E. Hager, III and R. C. Domszy, Monitoring of cement hydration by broadband time-domain-reflectometry dielectric spectroscopy, *Journal of Applied Physics*, 96(9), 5117–5128, 2004.
17. A. Cataldo, G. Cannazza, E. De Benedetto, and N. Giaquinto, A new method for detecting leaks in underground water pipelines, *IEEE Sensors Journal*, 12(6), 1660–1667, 2012.
18. A. Cataldo, E. De Benedetto, and G. Cannazza, *Broadband Reflectometry for Enhanced Diagnostics and Monitoring Applications*. Springer, Berlin, Germany, 2011.

19. A. Cataldo, L. Tarricone, M. Vallone, G. Cannazza, and M. Cipressa, TDR moisture measurements in granular materials: from the siliceous sand test-case to the applications for agro-food industrial monitoring, *Computer Standards and Interfaces*, 32(3), 86–95, 2010.

20. A. Cataldo, E. Piuzzi, G. Cannazza, and E. De Benedetto, Dielectric spectroscopy of liquids through a combined approach: evaluation of the metrological performance and feasibility study on vegetable oils, *IEEE Sensors Journal*, 9(10), 1226–1233, 2009.

21. A. Cataldo, E. Piuzzi, G. Cannazza, and E. De Benedetto, Classification and adulteration control of vegetable oils based on microwave reflectometry analysis, *Journal of Food Engineering*, 112(4), 338–345, 2012.

22. A. Cataldo, G. Cannazza, E. De Benedetto, C. Severini, and A. Derossi, An alternative method for the industrial monitoring of osmotic solution during dehydration of fruit and vegetables: a test-case for tomatoes, *Journal of Food Engineering*, 105(1), 186–192, 2011.

23. A. Cataldo and E. De Benedetto, Broadband reflectometry for diagnostics and monitoring applications, *IEEE Sensors Journal*, 11(2), 451–459, 2011.

24. *Time Domain Reflectometry Theory*, Application Note 1304–2, Palo Alto, CA, May 2006.

25. D. A. Robinson, S. B. Jones, J. M. Wraith, D. Or, and S. P. Friedman, A review of advances in dielectric and electrical conductivity measurement in soils using time domain reflectometry, *Vadose Zone Journal*, 2, 444–475, 2003.

26. *Applying Error Correction to Network Analyzer Measurements*, Application Note 1287-3, Santa Clara, CA, 2002.

27. R. Friel and D. Or, Frequency analysis of time-domain reflectometry (TDR) with application to dielectric spectroscopy of soil constituents, *Geophysics*, 64(3), 707–718, 1999.

28. T. J. Heimovaara, W. Bouten, and J. M. Verstraten, Frequency domain analysis of time domain reflectometry waveforms: 2. a four-component complex dielectric mixing model for soils, *Water Resources Research*, 30, 201–209, 1994.

29. T. J. Heimovaara, Frequency domain analysis of time domain reflectometry waveforms: 1. measurement of the complex dielectric permittivity of soils, *Water Resources Research*, 30(2), 189–199, 1994.

30. S. B. Jones and D. Or, Frequency domain analysis for extending time domain reflectometry water content measurement in highly saline soils, *Soil Science Society of America Journal*, 68, 1568–1577, 2004.

31. A. Cataldo, L. Catarinucci, L. Tarricone, F. Attivissimo, and A. Trotta, A frequency-domain method for extending TDR performance in quality determination of fluids, *Measurement Science and Technology*, 18(3), 675–688, 2007.

32. T. J. Heimovaara, J. A. Huisman, J. A. Vrugt, and W. Bouten, Obtaining the spatial distribution of water content along a TDR probe using the SCEM-UA bayesian inverse modeling scheme, *Vadose Zone Journal*, 3, 1128–1145, 2004.

33. A. Cataldo, G. Cannazza, E. De Benedetto, L. Tarricone, and M. Cipressa, Metrological assessment of TDR performance for moisture evaluation in granular materials, *Measurement*, 42(2), 254–263, 2009.

34. S. J. Zegelin, I. White, and D. R. Jenkins, Improved field probes for soil water content and electrical conductivity measurement using time domain reflectometry, *Water Resources Research*, 25, 2367–2376, 1989.

35. J. S. Selker, L. Graff, and T. Steenhuis, Noninvasive time domain reflectometry moisture measurement probe, *Soil Science Society of America Journal*, 57, 934–936, 1993.

36. A. Cataldo, G. Monti, E. De Benedetto, G. Cannazza, and L. Tarricone, A noninvasive resonance-based method for moisture content evaluation through microstrip antennas, *IEEE Transactions on Instrumentation and Measurement*, 58(5), 1420–1426, 2009.

37. P. A. Ferré, J. H. Knight, D. L. Rudolph, and R. G. Kachanoski, A numerically based analysis of the sensitivity of conventional and alternative time domain reflectometry probes, *Water Resources Research*, 36(9), 2461–2468, 2000.

38. K. S. Cole and R. H. Cole, Dispersion and absorption in dielectrics: I. alternating current characteristics, *Journal of Chemical Physics*, 9(4), 341–351, 1941.

39. N. Miura, S. Yagihara, and S. Mashimo, Microwave dielectric properties of solid and liquid foods investigated by time-domain reflectometry, *Journal of Food Science*, 68(4), 1396–1403, 2003.

40. S. O. Nelson, W. Guo, S. Trabelsi, and S. J. Kays, Dielectric spectroscopy of watermelons for quality sensing, *Measurement Science and Technology*, 18, 1887–1892, 1999.

41. S. O. Nelson and S. Trabelsi, Dielectric spectroscopy of wheat from 10 MHz to 1.8 GHz, *Measurement Science and Technology*, 17, 2294–2298, 2006.

42. V. Changrue, V. Orsat, G. Raghavan, and D. Lyew, Effect of osmotic dehydration on the dielectric properties of carrots and strawberries, *Journal of Food Engineering*, 88, 280–286, 2008.

43. S. L. Birla, S. Wang, J. Tang, and G. Tiwari, Characterization of radio frequency heating of fresh fruits influenced by dielectric properties, *Journal of Food Engineering*, 89, 390–398, 2008.

44. M. Castro-Giráldez and P. J. Fito, Application of microwaves dielectric spectroscopy for controlling pork meat (Longissimus dorsi) salting process, *Journal of Food Engineering*, 97, 484–490, 2010.

45. W. Gardner and D. Kirkham, Determination of soil moisture by neutron scattering, *Soil Science*, 73, 391–401, 1951.

46. R. Reginato and C. van Bavel, Soil water measurement with gamma attenuation, *Soil Science Society of America Proceedings*, 28, 721–724, 1964.

47. A. Cataldo, M. Vallone, L. Tarricone, G. Cannazza, and M. Cipressa, TDR moisture estimation for granular materials: an application in agro-food industrial monitoring, *IEEE Transactions on Instrumentation and Measurement*, 58(8), 2597–2605, 2009.

48. R. Cerny, Time-domain reflectometry method and its application for measuring moisture content in porous materials: a review, *Measurement*, 42(3), 329–336, 2009.

49. K. Noborio, Measurement of soil water content and electrical conductivity by time domain reflectometry: a review, *Computers and Electronics in Agriculture*, 31(3), 213–237, 2001.

50. G. C. Topp, J. L. Davis, and A. P. Annan, Electromagnetic determination of soil water content: measurements in coaxial transmission lines, *Water Resources Research*, 16, 574–582, 1980.

51. J. R. Birchak, D. G. Gardner, J. E. Hipp, and J. M. Victor, High dielectric constant microwave probes for sensing soil moisture, *Proceedings of the IEEE*, 62, 93–98, 1974.

52. G. P. De Loor, Dielectric properties of heterogeneous mixtures, PhD dissertation, University of Leiden, Leiden, the Netherlands, 1956.

53. M. C. Dobson and M. T. Hallikainen, Microwave dielectric behavior of wet soil-part II: Dielectric mixing models, *IEEE Transactions on Geoscience and Remote Sensing*, 23(1), 35–46, 1985.

54. M. Ansoult, L. W. De Backer, and M. Declercq, Statistical relationship between apparent dielectric constant and water content in porous media, *Soil Science Society of America Journal*, 49, 47–50, 1985.

55. H. Lizhi, K. Toyoda, and I. Ihara, Dielectric properties of edible oils and fatty acids as a function of frequency, temperature, moisture and composition, *Journal of Food Engineering*, 88, 151–158, 2008.

56. T. S. Ramu, On the high frequency dielectric behavior of castor oil, *IEEE Transactions on Electrical Insulation*, 14, 136–141, 1979.

57. G. Morchio, A. DiBello, C. Mariani, and E. Fedeli, Detection of refined oils in virgin olive oil, *Rivista Italiana Sostanze Grasse*, 66, 251–257, 1989.
58. A. H. El-Hamdy and N. K. El-Fizga, Detection of olive oil adulteration by measuring its authenticity factor using reversed-phase high-performance liquid chromatography, *Journal of Chromatography*, 708, 351–355, 1995.
59. A. Tay, R. K. Singh, S. S. Krishnan, and J. P. Gore, Authentication of olive oil adulterated with vegetable oils using fourier transform infrared spectroscopy, *Lebensmittel Wissenschaft und-Technologie*, 35, 99–103, 2002.
60. F. Guimet, J. Ferré, and R. Boque, Rapid detection of olive-pomace oil adulteration in extra virgin olive oils from the protected denomination of origin Siurana using excitation-emission fluorescence spectroscopy and three-way methods of analysis, *Analytica Chimica Acta*, 544, 143–152, 2005.
61. R. Sacchi, F. Adeo, and L. Polillo, 1H and 13C NMR of virgin olive oil. An overview, *Magnetic Resonance in Chemistry*, 35, 133–145, 1997.
62. G. E. Lombard, J. C. Oliveira, P. Fito, and A. Andres, Osmotic dehydration of pineapple as a pre-treatment for further drying, *Journal of Food Engineering*, 85(2), 277–284, 2008.
63. E. Torringa, E. Esveld, R. van den Berg, and P. Bartels, Osmotic dehydration as a pre-treatment before combined microwave-hot-air drying of mushrooms, *Journal of Food Engineering*, 49, 185–191, 2001.
64. S. Rodrigues and F. A. N. Fernandes, Dehydration of melons in a ternary system followed by air-drying, *Journal of Food Engineering*, 80, 678–687, 2007.
65. F. A. N. Ferndandes, S. Rodrigues, O. C. P. Gaspareto, and E. L. Oliveira, Optimization of osmotic dehydration of bananas followed by air-drying, *Journal of Food Engineering*, 77, 188–193, 2006.
66. M. Dalla Rosa and F. Giroux, Osmotic treatments (OT) and problems related to the solution management, *Journal of Food Engineering*, 49, 223–236, 2001.
67. J. Warczok, M. Ferrando, F. Lopez, A. Pihlajamaki, and C. Guell, Reconcentration of spent solutions from osmotic dehydration using direct osmosis in two configurations, *Journal of Food Engineering*, 80, 317–326, 2007.
68. G. V. Barbosa-Canovas and H. Vega-Mercado, *Dehydration of Food*. Chapman & Hall, New York, 1996.

Part IV

Point-of-Care and Biosensors

16 Applications of Sensor Systems in Prognostics and Health Management

Shunfeng Cheng and Michael Pecht

CONTENTS

Prognostics and health management (PHM) is an emerging discipline consisting of technologies and methods to assess the reliability of a product in its actual life cycle conditions to determine the advent of failure and mitigate system risk [1]. PHM generally combines sensing and interpretation of environmental, operational and performance-related parameters to assess the health of a product and predict its remaining useful life [1–3]. The benefits of conducting PHM for a product include (1) providing advance warning of failures; (2) minimizing unscheduled maintenance, extending maintenance cycles and maintaining effectiveness through timely repair actions; (3) reducing the life cycle cost of equipment by decreasing inspection costs, downtime and inventory; and (4) improving qualification and assisting in the design and logistical support of fielded and future systems [1,2].

Traditionally, PHM has been implemented using a data-driven approach, a model-based approach or a fusion approach that fuses the data-driven and model-based approaches [1,2]. Parameter monitoring and the analysis of acquired data using prognostic models are fundamental steps for these PHM methods. In data-driven approaches, *in situ* monitoring of environmental and operational parameters of the product is required, and then the complex relationships and trends in the data can be captured by data-driven methods without the need for specific failure models. Model-based approaches are based on an understanding of the physical processes and interrelationships among the different components or subsystems of a product.

The physical processes and interaction information are monitored by various parameters. For example, physics-of-failure (PoF)-based prognostic methods, which are one category of model-based approaches, integrate *in situ* monitored parameter data from sensor systems with models that identify the deviation or degradation of a product from an expected normal condition and predict of the future state of reliability [1,2]. A fusion prognostic approach integrates the data-driven and model-based prognostic approaches, and so parameter monitoring is required too.

PHM has several requirements for parameter monitoring. (1) PHM requires monitoring multiple parameters. It is possible to monitor thousands of parameters in the entire life cycle of the product to provide health status information. PHM may monitor operational and environmental parameters, as well as performance conditions of the product, including temperature, vibration, shock, pressure, acoustic levels, strain, stress, voltage, current, humidity levels, contaminant concentration, usage frequency, usage severity, usage time, power and heat dissipation. In each case, a variety of monitoring features, such as magnitude, variation, peak level and rate of change, may be required in order to obtain the characteristics of the parameters. Figure 16.1 and Table 16.1 show an example of PHM for an automobile to show the complexity of a PHM application [1]. (2) PHM also requires monitoring parameters during all stages of the product life cycle, including manufacturing, shipment, storage, handling and operation. Failures may be accumulated and occur in all these stages. (3) PHM should have minimum adverse influences on the performance and reliability of the monitored product and should have relatively low cost. This requires that all additional parts, including sensor systems, are selected carefully to minimize the adverse effects on the monitored host products [1].

Parameter monitoring relies heavily on sensor systems to provide long-term, accurate, *in situ* information to PHM for anomaly detection, fault isolation and rapid failure prediction. In this chapter, the method to identify the parameters to monitor for PHM will be presented first. The considerations of how to select appropriate sensor systems for a PHM application, including the required performance, electrical and physical attributes, reliability and cost of the sensor systems, will be discussed. The features of current commercially available sensor systems for PHM and emerging trends in the

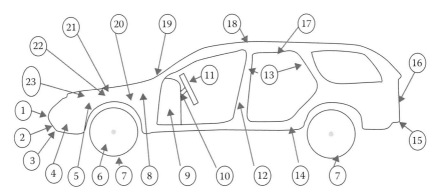

FIGURE 16.1 Example of PHM application for an automobile. (From Cheng, S. et al., *Sensors*, 10, 5774, 2010.)

TABLE 16.1
Sensors in Figure 16.1

Number (Location)	Monitoring and Sensing	Number (Location)	Monitoring and Sensing
1	Collision avoidance, night vision, front crash detection and forward obstacle sensor	13	Side airbag deployment sensor
2	Vehicle distance sensor	14	Angular acceleration (suspension)
3	Road condition sensor	15	Emissions sensor
4	Side obstacle sensor	16	Back-up collision, rear vision camera, rear radar, rear obstacle sensor
5	Oil/fuel pressure and flow monitoring	17	Temperature and humidity sensor and comfort control
6	Speed sensors	18	Rollover detection
7	Tyre pressure monitoring	19	Rain sensor and wiper control
8	Fire detection sensor	20	Power train control module
9	Driver monitoring sensor	21	Throttle position monitoring and control
10	Steering angle sensor and stability control	22	Battery monitoring
11	Yaw and acceleration sensors for airbag deployment	23	Ignition and engine control monitoring
12	Side crash detection		

Source: Cheng, S. et al., *Sensors*, 10, 5774, 2010.

technologies of sensor systems for PHM will also be presented. Several case studies will be introduced to demonstrate the applications of sensor systems for PHM.

16.1 PARAMETER IDENTIFICATION

In general, the parameters to be monitored can contain any available variables, including operational and environmental loads, as well as performance parameters. However, some of the available variables may not be necessary for PHM. For a specific PHM application, the parameters to be monitored can be identified based on their relationship to functions that are crucial for the safety and reliability of the product, are likely to be implicated in catastrophic failures, are essential for mission completeness or result in long downtimes. Selection is also based on knowledge of the critical parameters established by qualification testing and by past experience and field failure data from similar products [1].

A systematic method, failure modes, mechanisms and effects analysis (FMMEA), can be used to identify the parameters to be monitored. FMMEA is a methodology used to identify the critical failure mechanisms for all potential failure modes of a product under expected operational and environmental conditions,

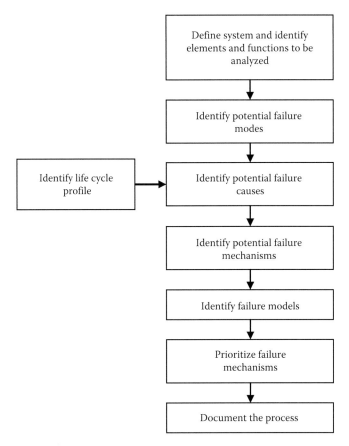

FIGURE 16.2 Flowchart of the FMMEA process. (From Cheng, S. et al., *Sensors*, 10, 5774, 2010.)

as shown in Figure 16.2 [1–5]. The output of the FMMEA process includes a list of critical failure modes and mechanisms that enable us to identify the parameters to monitor and determine the locations to place the sensor systems. FMMEA uses the life cycle profile (LCP) of a product along with the design information to identify the critical failure mechanisms affecting that product. An LCP is a forecast of the events and their associated environmental and usage conditions, and the rates of change of those conditions, that a product may experience from manufacture to end of life. Examples of these conditions include temperature, humidity, pressure, vibration, shock, chemical environments, radiation, contaminants, current, voltage and power.

The product is divided into lower-level subassemblies for investigation based on various categories such as functions, structures and modules. These subassemblies are the potential sites of failures. For each possible failure site, possible failure modes are analysed. A failure mode is the manner in which a failure is observed by methods such as visual inspection, electrical measurement or other tests and measurements [4–6]. For each failure mode, the potential failure causes should be identified.

A failure cause is the specific process, design and/or environmental condition that initiated the failure and whose removal will eliminate the failure. Possible failure causes are identified by investigation of the conditions of the life cycle of the product, including manufacturing/assembly, test, storage, transportation and handling, operation and maintenance [1,4].

Failure mechanisms are the processes by which specific combinations of physical, electrical, chemical and mechanical stresses induce failures. The potential failure mechanisms are identified for each failure mode and site based on causes, loads and design (geometry and material). FMMEA prioritizes the failure mechanisms based on their occurrence and severity. Identification of failure mechanisms enables determination of the major environmental and operational stresses (parameters) and where the sensor systems should be placed.

Cheng et al. used FMMEA to determine parameters that need to be monitored for PHM of resettable circuit protection devices [3,7] and multilayer ceramic capacitors [1]. Using a resettable circuit protection device as an example here, an understanding of the structure and the operational principle is necessary to define the system and identify the elements and functions to be analysed. Shown in Figure 16.3, a polymer positive temperature coefficient (PPTC) resettable circuit protection device consists of lead, foil, conductive polymer composite (carbon black particles and polymer) and outside coating. The resistance of the device increases when the temperature increases. At normal temperatures, the polymer is in a semi-crystalline state, in which the polymer contains both crystalline and amorphous regions. The conductive particles in the semi-crystalline polymer form many conductive paths, which allow current to flow through the PPTC device without interruption. However, if the temperature rises to the crystalline-to-amorphous (C $\rightarrow$ A) phase transition temperature of the polymer, some of the crystalline regions in the polymer will become amorphous and the conductive paths break. This results in a large nonlinear increase in the resistance of the PPTC device in a narrow temperature range. When the temperature decreases, the

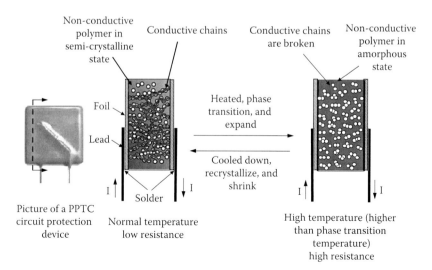

FIGURE 16.3 Structure of a radial through-hole PPTC resettable circuit protection device.

polymer recrystallizes and the resistance restores to the normal range. Based on this information, the subsystem or component needed to be analysed includes the conductive polymer composite, the interconnection between the foil and polymer composite, the solder between the foil and lead and the outside coating. These are the potential failure sites. Other FMMEA results, including potential failure modes, causes and mechanisms analyses, are shown in Table 16.2 [3,6]. Based on the FMMEA analysis, the parameters to be monitored are shown in Table 16.3.

TABLE 16.2
FMMEA Results of Circuit Protection Devices Shown in Figure 16.3

Potential Failure Sites	Potential Failure Modes	Potential Failure Causes	Potential Failure Mechanisms
Conductive polymer composite	Abnormal trip behaviours (e.g. trip at normal current, no trip at fault current)	Increase in the heat dissipation resistance, the cracks or gaps at the interconnections, the changes in the polymer properties and the changes in the distribution of the CB particles	Degradation of the polymer, aggregation of CB particles
Interconnection between foil and polymer composite	Shift in resistance and surface temperature	CTE mismatch, gaps, manufacturing defects	Fatigue
Solder between lead and foil	Shift in resistance	Cracks	Fatigue
Outside coating	Cracks, separation with foil, shift in surface temperature	CTE mismatch, deformation, manufacturing defects	Fatigue

TABLE 16.3
Parameters to be Monitored for Resettable Circuit Protection Device

Parameters		Can It Be Monitored *In Situ*?
Trip Time		Yes
Resistance	Resistance after reset	Yes
	Resistance during trip	Yes (by measure current and voltage of the device during trip)
Surface temperature		Yes
Current	Current through the devices at normal condition	Yes
	Trickle current	Yes
	Actual hold current	No
Voltage across the device		Yes

16.2   REQUIREMENTS FOR SENSOR SYSTEMS FOR PHM

After identifying the parameters to be monitored for PHM of a product, appropriate sensor systems should be selected. Many sensor systems can be found to monitor the parameters; however, not all of them can be used for PHM. PHM has its requirements for sensor systems, including the performance needs, physical attributes, electrical features (e.g. power, memory, data processing and data transmission), reliability and cost of the sensor systems [8]. There is no sensor system that fits all PHM applications, so each application needs to determine its particular needs to find the best sensor system for it. Table 16.4 is a summary of these requirements of sensor systems for PHM applications. Based on these requirements, an integrated sensor system, as shown in Figure 16.4 [1,2,9], with multiple sensing abilities, miniature size, light weight, low power consumption, long-range and high-rate data transmission, large onboard memory, fast onboard data processing, low cost and high reliability is advantageous to PHM applications [1,2].

The Centre for Advanced Life Cycle Engineering (CALCE) developed sensor system selection guidance from commercial sensor systems for PHM applications [2]. The selection of a sensor system for PHM applications requires analysis of the application, identification of the parameters to be monitored and all the requirements for the sensor system and prioritization of these requirements based on the specific application. Then, sensor system candidates should be identified and evaluated based on these requirements. Finally, some trade-offs, such as a trade-off between costs and sensor system performance, must be made to select proper sensor systems [2].

16.3   FEATURES OF CURRENT COMMERCIALLY AVAILABLE SENSOR SYSTEMS FOR PHM APPLICATIONS

In order to determine the commercial availability of sensor systems that can be used in PHM for electronic products and systems, a survey was conducted by CALCE recently on 33 sensor systems from 23 manufacturers. The survey only included commercially available sensor systems having features desirable for PHM. The concerned characteristics of sensor systems included parameters to be sensed, power supply and power management capability, onboard memory and memory management functions, data transmission methods, availability of onboard data processing, size, weight and cost. The data were collected from the manufacturer's website, product datasheets, e-mails and evaluations of demo products. Appendix A of [2] shows the information for these 33 sensor systems.

Based on the results of the survey, features of current commercially available sensor systems were identified. Currently, many sensor systems can measure multiple parameters (e.g. temperature, humidity, vibration and pressure) using built-in sensors or add-on external sensors. Many sensor systems have onboard power supply and onboard power management capabilities. Some sensor systems have diverse onboard data storage capacities and embedded signal processing algorithms that enable data compression or simplification prior to data transfer. Current sensor

TABLE 16.4

Summary of Requirements of Sensor Systems for PHM Applications

Items	Requirements and Explanation
Multiparameter monitoring	PHM requires integration of many different parameters to assess the health status, detect and isolate the faults and predict the remaining life of a product. Individual sensor systems that can monitor multiple parameters, such as temperature, humidity, vibration and pressure; simplify PHM; and reduce the cost. Sensor systems may contain multiple sensing elements and support multiple external sensing elements or a combination of these.
Sensor system performance	The requirements for parameter features should be translated into the requirements of sensor system performance, such as measurement range, dynamic range, accuracy, sensitivity, repeatability, resolution, frequency, response, hysteresis, linearity, response time, stabilization time and sampling rate.
Physical characteristics of sensor systems	The physical characteristics of a sensor system include its size, weight, shape, packaging and mounting of sensors into the host. In some PHM applications, the size of the sensor may become one of the most significant selection criteria due to limitations on the available space for attaching the sensor or the inaccessibility of locations to be sensed. The weight of the sensor should also be considered in certain PHM applications, such as for vibration and shock measurements using accelerometers, since the added mass can change the system response. Users should also consider the shape (round, rectangular or flat) of the sensor system; the packaging materials, such as metal or plastic; and the method for attaching or mounting the sensor, for example, glue, adhesive tape, magnets, fixtures or screws (bolts).
Power and power management	Power consumption is an essential characteristic of a sensor system, as it determines how long the sensor system can function independently. PHM can use non-battery-powered or battery-powered sensor systems, depending on the specific application. However, power management is preferred in sensor system selection for PHM to optimize the power consumption of the sensor system in order to extend its operating time.
Onboard memory and memory management	Onboard memory is the memory contained within a sensor system. It can be used to store collected data as well as information pertaining to the sensor system (e.g. sensor identity, battery status). Onboard memory allows for high data sampling and save rates. It also enables onboard data analysis. Sensor systems for PHM should have memory management functions, which allow a user to configure, allocate, monitor and optimize the utilization of memory. Examples of common memory management functions include programmable sampling modes and sampling rates and memory status management.
Onboard data processing	Onboard processing can significantly reduce the number of data points and, thus, free up memory for more data storage. This, in turn, reduces the volume of data that must be transmitted to a base station or computer and, hence, results in lower power consumption by data transmission. Onboard processing can be set to provide real-time updates for taking immediate action, such as powering off the equipment to avoid accidents or catastrophic failures. It can also be set to provide prognostic horizons to conduct future repair and maintenance activities. However, the abilities of the onboard processor are limited by power supply and onboard memory.

TABLE 16.4 (continued)
Summary of Requirements of Sensor Systems for PHM Applications

Items	Requirements and Explanation
Data transmission	In general, the methods for data transmission are either wired or wireless. Wired data transmission can offer high-speed transmission, but this transmission is limited by the need for transmission wires, and the cost is increased by wires. Wireless transmission refers to the transmission of data over a distance without the use of a hard-wired connection. A lot of wireless technologies can be used for wireless data transmission of sensor systems, for example, RFID, Bluetooth, Wi-Fi (IEEE 802.11), UWB, Certified WUSB, WiMAX (Worldwide Interoperability for Microwave Access and IEEE 802.16) and ZigBee (IEEE 802.15.4). When selecting the wireless technology to use for a particular application, the user should consider the range and rate of communication, power consumption, ease of implementation and data security.
Cost	Selection of the proper sensor system for a given PHM application must include an evaluation of the cost. The cost evaluation should address the total cost of ownership including the purchase, installation, maintenance and replacement of sensor systems. In fact, the initial purchase cost of a product can be less than 20% of the product's total lifetime cost [10].
Reliability	The reliability of a sensor system requires the ability of the sensor system to perform a necessary function under stated conditions for a stated period. One strategy to improve the reliability of sensor systems is to use multiple sensors (redundancy) to monitor the same product or system. By using redundancies, the risk of losing data due to sensor system failure is reduced, but the cost increases. Some technologies, such as sensor validation, can also improve the reliability of sensor systems.

Sources: Cheng, S. et al., *Sensors*, 10, 5774, 2010; Pecht, M., *Prognostics and Health Management of Electronics*, Wiley-Interscience, New York, 2008.

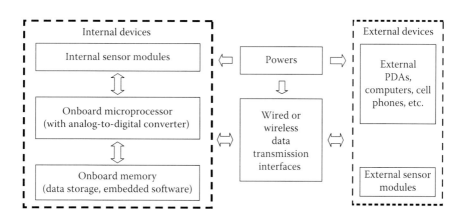

FIGURE 16.4 Integrated sensor system for PHM. (From Cheng, S. et al., *Sensors*, 10, 5774, 2010; Pecht, M., *Prognostics and Health Management of Electronics*, Wiley-Interscience, New York, 2008.)

systems use various wireless technologies including radio-frequency identification (RFID), Bluetooth, Wi-Fi, ZigBee, ultra-wideband (UWB) and Wireless USB. Current commercially available sensor systems can perform multiple functions using their own power management, data storage, signal processing and wireless data transmission [1,2].

Current sensor systems have several limitations that constrain the applications for PHM. The first are size and weight. PHM for electronics systems with a high density of components requires small, lightweight sensor systems that can be placed on the circuit board and that minimize the weight increase and potential adverse effects on the reliability of the monitored electrical system. Secondly, onboard power is one of the main limitations of the current commercially available sensor systems, especially for wireless data transmission sensor systems. The main onboard power is a battery, which needs to be replaced or recharged when it is used up. Higher-capacity batteries with small sizes and lightweight or battery-free power are needed for sensor systems to operate longer in PHM applications. Thirdly, the onboard data processing ability should also be improved. In the survey, only a few sensor systems had onboard data processing ability, and they only had very simple functions such as data reduction. Onboard data processing provides timely information about the health of the system and can reduce the cost of the entire PHM application. Onboard data processing requires high-speed processors, large-capacity memory, more power and processing algorithms. The processing algorithms can be developed based on the specific PHM application [1].

16.4   CASES OF SENSOR SYSTEMS IN PHM APPLICATIONS

A specific PHM application may not use all the methods and requirements discussed earlier to determine the best sensor systems. For example, if the parameters are easy to identify or already identified, FMMEA is unnecessary to identify the parameters to be monitored. A U.S. military project named 'SWORDS' required monitoring the temperature, humidity and vibration of the 'SWORDS' unmanned controlled vehicle to report its health status [9]. Thus, an RF-based wireless ePrognostic sensor system for PHM applications, shown in Figure 16.5 [9], was selected for this application. This sensor system includes multiple sensor elements which can monitor temperature, humidity, motion, shock and vibration. It integrates onboard processors, memories, data communication electronics and a high-capacity wafer-thin battery into laminated layers. The sensor system is flexible and similar in size to a typical credit card. The sensor system incorporates prognostic software that can analyse the collected data *in situ* and provide timely prognostic information. Gu et al. used a sensor module to monitor the usage conditions of a laptop [11]. Similar to the ePrognostic sensor tag, the sensor module is the size of a credit card. It was inserted into the computer express card slot to collect temperature, humidity and vibration loads. This sensor module had its own memory and battery so that it could record the data independently of the computer system's status (on/off), which made it possible to collect load conditions during shipping, storage and field transport [11].

Gu et al. conducted health monitoring and prognostics of electronic components mounted on a printed circuit board (PCB) under vibration loading [12,13]. In this project, the parameters to be monitored were determined by a virtual qualification

(a)

(b)

FIGURE 16.5 (**See colour insert.**) ePrognostic sensor system in a PHM application (a) U.S. military 'SWORDS' unmanned ground vehicle and (b) placement of sensor tags. (Red arrows point to the ePrognostic sensor tags.)

tool based on FMMEA. According to the analysis results, the strain, vibration and shock signals should be monitored. Strain gauges were used to monitor the strain of the PCB. Strain gauge data were collected by a NI SCXI1314 data acquisition card incorporated with the LabVIEW program. The accelerometer was mounted in the middle of the PCB and the data were collected by Shock and Vibration Environment Record (SAVER). SAVER could be coded for time-triggered and signal-triggered functions. In the time-triggered mode, data were recorded at a fixed time interval, for example, 1000 data per second. In the signal-triggered mode, the data were recorded only when the signal met a threshold. Gu et al. converted the life cycle loads collected by the accelerometer and strain gauges to the electronic component interconnects' stress values, which were then used in a vibration failure fatigue model for damage assessment. The sensor systems and the prediction model monitored and predicted the health status of a group of electronic components mounted on the PCB.

In a project for anomaly detection and failure prognostics of resettable circuit protection devices, Cheng et al. firstly used FMMEA to identify the parameters to be monitored, shown in Table 16.3, in which the current through the device, the voltage across the device and the surface temperature of the device were monitored *in situ* [3,6]. Resistance during the trip (the process of the device changing from a low resistance to high resistance status) was calculated by the Ohm theory using the monitored current and voltage. Resistance after tripping was measured by a data logger using a four-wire connection. Trip time (the time needed for tripping the device) was calculated by the difference between the time when the fault current occurred and the time when the current decreased to the hold current. This was measured by a current metre or sensors. Figure 16.6 shows the sensor systems used in the experiment. This project used two sensor systems. The first was the four-channel programmable power supply, Agilent 6705, which monitored the current, voltage and trip time. The second was the multiple function data logger, Agilent 34970A, which monitored the voltage, temperature (by thermocouples) and resistance. A computer controlled the data logger using a LabVIEW program and controlled the four-channel power supply by a VEE program. Data were transmitted to the computer through wires. Anomaly detection and failure prognostics algorithms were run in the computer to provide the health status of the resettable circuit protection devices. The sensor systems shown in Figure 16.6 are only for the experiment; however, alternatives can be found to measure the voltage, current, temperature, resistance and trip time of the PPTC resettable devices in actual PHM applications. These sensor systems and the required data collection, anomaly detection and failure prediction software can be built into the host system that is using the PPTC resettable circuit protection devices [3,7].

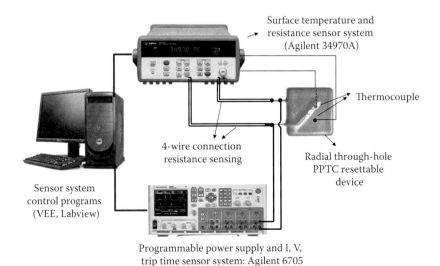

FIGURE 16.6 Sensor systems setup for PPTC circuit protection device monitoring.

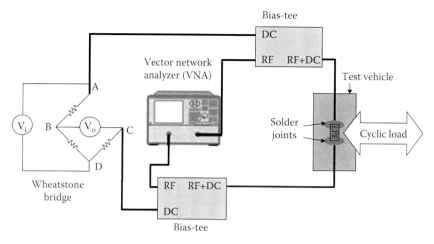

FIGURE 16.7 TDR sensing circuit for interconnection health monitoring.

Kwon et al. used time-domain reflectometry (TDR) as a nondestructive sensing method for electronic interconnect failure mechanisms, including solder joint cracking and solder pad cratering [14]. The sensing circuit is shown in Figure 16.7. The whole sensing system included a low-pass filter that was soldered in the circuit, two bias tees for the simultaneous monitoring of the TDR reflection coefficient and DC resistance, a Wheatstone bridge for DC resistance measurement and a vector network analyser for TDR reflection coefficient measurement. The test results consistently demonstrated that the TDR reflection coefficient gradually decreased as the solder pad separated from the circuit board, whereas it increased during solder joint cracking. Traditional test methods based on electrical resistance monitoring cannot distinguish between failure mechanisms and do not detect degradation until an open circuit has been created. In contrast, the TDR reflection coefficient can be used as a sensing method for the determination of interconnect failure mechanisms, as well as for early detection of the degradation associated with those mechanisms [14].

In PHM applications, some sensor systems can directly identify the faults and failure mechanisms of a product [1]. For example, resistance measurement can be used to isolate the open or short locations in a simple circuit, and corrosion sensors can use electrochemical impedance spectroscopy (EIS) to monitor the corrosion of structures directly [15]. For complex electrical packages, some techniques or sensor systems can be used to isolate the failure sites. For example, the scanning superconductive quantum interference device (SQUID) microscopy can be used to detect shorts in microprocessors, 3D x-ray radiography/tomography can image various levels of interconnections, and scanning acoustic microscopy can detect interfacial delaminations and defects in packages [16]. Other sensor systems using electromagnetic nondestructive testing technologies [17], ultrasonic guided wave technologies [18] or optical technologies [19] can detect cracks inside a product [1].

16.5   EMERGING TRENDS IN SENSOR TECHNOLOGY FOR PHM

With the development of electronics manufacturing technologies and new materials, electronic products continue to decrease in size. PHM for these products requires that sensor technology be headed towards extreme miniaturization, battery-free power or ultra-low power consumption and intelligent wireless networks [1]. As micro-electromechanical systems (MEMS) or nano-electromechanical systems (NEMS) and smart material technologies mature, MEMS sensors or nanosensors will integrate the sensing element, amplification, analogy-to-digital converter and memory cells into one microchip [1].

Although the power supply for sensor systems is still a challenge, the development of new materials and energy technologies provides more chances to generate effective power supply solutions. For example, development of ultra-low power consumption electronics and energy-harvesting technologies will enable battery-free sensor systems in the future, especially for use in embedded, remote and other inaccessible monitoring conditions [1].

Ultra-low power electronics will enable future sensor systems to consume much less power. For example, in May 2010, Intel released its new generation of Atom ultra-low power processors based on the 45 nm technology. 'Collectively these new chips deliver significantly lower power including >50× reduction in idle power, >20× reduction in audio power, and 2–3× reductions across browsing and video scenarios – all at the platform level when compared to Intel's previous-generation product. These power savings translate into >10 days of standby, up to 2 days of audio playback and 4–5 h of browsing and video battery life' [20].

Energy harvesting is a process to extract energy from the environment or from a surrounding system and convert it into usable electrical energy. For example, current research on energy-harvesting sources includes converting sunlight, thermal gradient, human motion, body heat, wind, vibration, radio power and magnetic coupling into electrical energy [21–25]. Some large-scale energy-harvesting schemes, such as wind turbines and solar cells, have made the transition from research to commercial products. For sensor systems, small-scale energy harvesting, including implanted medical sensors and sensors on aerospace structures, is being developed and used [1]. Some physical effects, such as electromagnetic, piezoelectric, electrostatic and thermoelectric effects, are usually utilized to convert energy [1]. For example, the mechanical vibration inside a device or ambient mechanical vibration can be converted into electrical energy by piezoelectric materials or electromagnetic induction [1].

Another trend in sensor systems is wireless sensor networks, which integrate wireless technologies and sensor network structures. Wireless transmission is heading towards a long transmission distance, high transmission rate, high security and low power consumption. Sensor networks consist of multiple sensor nodes that are capable of communicating with each other and collaborating on the same sensing goal [1]. This allows data from multiple sensors to be combined or fused to obtain inferences that may not be possible from a single sensor. Furthermore, sensor nodes in the network will have built-in diagnostic and prognostic capabilities, which will make the entire sensor network more functional [1]. The integration of all of the aforementioned technology accelerates the development of wireless intelligent sensor networks [1].

16.6 SUMMARY

In this chapter, the requirements of sensor systems for PHM applications were discussed. PHM requires monitoring the parameters of the product in its life cycle to assess the health status of the product, identify abnormal conditions and predict the remaining life of the product [1]. As a device for *in situ* monitoring of the actual life cycle of a product, sensor systems should be integrated with multiple sensing abilities; low power consumption; low-cost, long-range and high-rate data transmission (wireless or wired); large onboard memory capacity; fast onboard data processing abilities; miniature weight and size; and high reliability [1].

Results of a survey of the current commercially available sensor systems showed that many sensor systems have some features suitable for PHM applications, such as multiple sensing capabilities, onboard power and power management ability, onboard memory and management ability, wireless data transmission and onboard data processing ability [1,2]. However, the survey also identified several unmet needs of sensor systems for PHM applications, including the size and weight of the sensor systems and the limitations of onboard power supply, memory capacity and onboard data processing ability. Several technologies, such as MEMS or NEMS, energy-harvesting techniques, ultra-low power consumption circuits and intelligent wireless sensor network techniques, are emerging to provide more powerful sensor systems in the future [1].

In a specific PHM application, the parameters to be monitored should be identified first. A failure mechanism–based method, FMMEA, can be used to help analyse the potential failure mechanisms and determine the parameters and the location to place the sensor systems. The requirements of PHM applications for sensor systems regarding sensor performance, electrical and physical attributes, reliability, cost and availability must also be understood. Some trade-offs must be made in selecting the proper sensor systems as well [1,2].

REFERENCES

1. S. Cheng, M. Azarian, and M. Pecht, Sensor system for prognostics and health management, *Sensors*, 10, 5774–5797, 2010.
2. M. Pecht, *Prognostics and Health Management of Electronics*. Wiley-Interscience, New York, 2008.
3. S. Cheng, K. Tom, and M. Pecht, Failure precursors for polymer resettable fuses, *IEEE Transactions on Devices and Materials Reliability*, 10(3), 374–380, 2010.
4. S. Ganesan, V. Eveloy, D. Das, and M. Pecht, Identification and utilization of failure mechanisms to enhance FMEA and FMECA, *Proceedings of the IEEE Workshop on Accelerated Stress Testing and Reliability (ASTR)*, Austin, TX, October 3–5, 2005.
5. IEEE Standard 1413.1-2002. IEEE Guide for Selecting and Using Reliability Predictions Based on IEEE 1413. IEEE Standard, 2003.
6. M. Pecht and A. Dasgupta, Physics-of-failure: An approach to reliable product development, *Journal of the Institute of Environmental Sciences,* 38, 30–34, 1995.
7. S. Cheng, K. Tom, and M. Pecht, Anomaly detection of polymer resettable circuit protection devices, Accepted by *IEEE Transactions on Devices and Materials Reliability*, accepted in September 2011.
8. B. Tuchband, S. Cheng, and M. Pecht, Technology assessment of sensor systems for prognostics and health monitoring, *IMAPS on Military, Aerospace, Space and Homeland Security: Packaging Issues and Applications (MASH)*, May 2007.

9. S. Cheng, K. Tom, L. Thomas, and M. Pecht, A wireless sensor system for prognostics and health management, *IEEE Sensors Journal*, 10(4), 856–862, 2010.

10. Space Age Control. Sensor total cost of ownership. Online White paper. Available at: http://www.spaceagecontrol.com/s054a.htm, accessed on July 22, 2009.

11. J. Gu, N. Vichare, E. Tinsley, and M. Pecht, Computer usage monitoring for design and reliability tests, *IEEE Transactions on Components and Packaging Technologies*, 32(3), 550–556, September 2009.

12. J. Gu, D. Barker, and M. Pecht, Health monitoring and prognostics of electronics subject to vibration load conditions, *IEEE Sensors Journal*, 9(11), 1479–1485, November 2009.

13. J. Gu, D. Barker, and M. Pecht, Prognostics of electronics under vibration using acceleration sensors, *Proceeding for 62nd Meeting of the Society for Machinery Failure Prevention Technology (MFPT)*, pp. 253–263, Virginia Beach, VA, May 2008.

14. D. Kwon, M. H. Azarian, and M. Pecht, Non-destructive sensing of interconnect failure mechanisms using time domain reflectometry, *IEEE Sensors Journal*, 11(5), 1236–1241, May 2011.

15. G. Davis, C. Dacres, and L. Krebs, *In-situ* corrosion sensor for coating testing and screening, *Materials Performance*, 39(2), 46, 2000.

16. M. Pacheco, Z. Wang, L. Skoglund, Y. Liu, A. Medina, A. Raman, R. Dias, D. Goyal, and S. Ramanathan, Advanced fault isolation and failure analysis techniques for future package technologies, *Intel Technology Journal*, 9(4), 337–352, 2005.

17. A. Sophian, G. Tian, and S. Zairi, Pulsed magnetic flux leakage techniques for crack detection and characterization, *Sensors and Actuators A: Physical*, 125(2), 186–191, 2006.

18. J. Van Velsor, S. Owens, and J. Rose, Guided waves for nondestructive testing of pipes, US. Patent: US 2009/0150094 A1. Published Date: June 11, 2009.

19. K. Wan and C. Leung, Fiber optic sensor for the monitoring of mixed mode cracks in structures, *Sensors and Actuators A: Physical*, 135(2), 370–380, 2007.

20. Intel Corporation, New Atom™ Processor Platform Using Significantly Lower Power Readies Intel for Smartphone, Tablet Push, available at: http://files.shareholder.com/downloads/INTC/1727524358x0x371749/e011b4d1–9bb7–4b0c-98ac-7000aa2053ee/INTC_News_2010_5_5_Technology_Leadership.pdf, accessed on March 9, 2012.

21. D. N. Fry, D. E. Holcomb, J. K. Munro, L. C. Oakes, and M. J. Maston, Compact portable electric power sources, Oak Ridge National Laboratory Report, ORNL/TM-13360, 1997.

22. P. Glynne-Jones and M. White, Self-powered systems: a review of energy sources, *Sensor Review*, 21, 91–97, 2001.

23. J. Roundy, *Energy Scavenging for Wireless Sensor Nodes with a Focus on Vibration to Electricity Conversion,* PhD dissertation, Department of Mechanical Engineering, University of California, Berkeley, CA, 2003.

24. A. Qiwai, P. Thomas, C. Kellogg, and J. Baucom, Energy harvesting concepts for small electric unmanned systems, *Proceedings of SPIE*, 5387, 84–95, 2004.

25. A. Paradiso and T. Starner, Energy scavenging for mobile and wireless electronics, *IEEE Pervasive Computing*, 4, 18–27, 2005.

17 Rigid Body Motion Capturing by Means of Wearable Inertial and Magnetic MEMS Sensor Assembly

Towards the Reconstitution of the Posture of Free-Ranging Animals in Bio-Logging

Hassen Fourati, Noureddine Manamanni,
Lissan Afilal and Yves Handrich

CONTENTS

17.1   INTRODUCTION

The rigid body attitude and orientation estimation problems are highly motivated by various applications. For example, in rehabilitation and biomedical engineering (Zhou et al. 2006), the attitude is used in stroke rehabilitation exercises to record patients' movements in order to provide adequate feedback for the therapist. In human motion tracking and biomechanics (O'Donovan et al. 2007), the attitude serves as a tool for physicians to perform long-term monitoring of the patients and to study human movements during everyday activities. Moreover, attitude estimation is extensively used in tracking handheld microsurgical instruments (Ang et al. 2004). In aerial and marine vehicles (Mahony et al. 2008), the attitude is used to achieve a stable controller.

Recently, the problem of attitude and orientation tracking has been treated in Bio-logging. The latter stands in the intersection of animal behaviour and bioengineering and aims at obtaining new information from the natural world and providing new insights into the hidden lives of animal species (Ropert-Coudert et al. 2009; Rutz and Hays 2009). Bio-logging generally involves a free-ranging animal-attached electronic device (also called bio-logger) that records aspects of the animal's biology (behaviour, movement, physiology) (Bost et al. 2007; Halsey et al. 2007) and its environment. Thirty years ago, several tagging technologies such as satellite tracking (the Argos system) (Le Boeuf et al. 2000) and time-depth recorders (TDRs) (Kooyman 2004) have been used to provide a basic knowledge on the function of free-ranging organisms. The recent advances in electronic miniaturization, sensors and digital information processing provided researchers studying animal's biology with a high level of detail and across the full range of ecological scales.

Many marine and terrestrial animals are studied during their daily activities. The posture and orientation tracking of these free-ranging animals represents one of the recent biology aspects studied in Bio-logging. Indeed, some scientific researches started to focus on this topic using low-cost sensors based on microelectromechanical system (MEMS) technology as a 3-axis accelerometer and a 3-axis magnetometer. The obvious advantage of this new approach is the gain access to the 3D space, which is the key to good understanding of the diving strategies observed in the afore-mentioned predators (Elkaim et al. 2006). The main question to answer is how it is possible to extract the gravity components of the body animal (Johnson and Tyack 2003; Watanabe et al. 2005; Wilson et al. 2008). This information is exploited later to deduce the corresponding attitude and consequently the dynamic body acceleration (DBA).

In this chapter, we propose the addition of 3-axis gyroscope measurements to the sensors already used (a 3-axis accelerometer and a 3-axis magnetometer) in Bio-logging. The use of a gyroscope with an accelerometer and a magnetometer, mounted in triad configuration, in Bio-logging has not been considered yet, to the author's knowledge. In our opinion, it can improve the estimation precision of the attitude especially during dynamic situation of the animal motion (Mahony et al. 2008; Fourati et al. 2009, 2011a). The main idea of the algorithm is to use a complementary filter coupled with a Levenberg–Marquardt algorithm (LMA) to process the measurements from a 3-axis gyroscope, a 3-axis magnetometer and a 3-axis accelerometer. The proposed approach combines a strap-down system, based on the time integral of the angular velocity, with the LMA that uses the Earth's magnetic field and the gravity vector to compensate the attitude predicted by the gyroscope. It is important to note that the resulting structure is complementary: high bandwidth rate gyro measurements are combined with low bandwidth vector observations (gravity and Earth's magnetic field) to provide an accurate attitude estimate. Thanks to the knowledge of the estimated attitude, it is now possible to reconstitute the DBA of an animal in order to evaluate its daily diary (Wilson et al. 2008) (sleeping, walking/flying, running and hunting) and provide important insights into some of the stresses faced by free-ranging animals, especially the king penguin and badger. Based on the values of the DBA, the problem of 3D position estimation in the case of pedestrian locomotion can be addressed in future works in Bio-logging to reconstruct the trajectory of an animal.

This chapter is organized as follows: Section 17.2 describes the problem statement and motivation for motion estimation in Bio-logging. Section 17.3 details the attitude parameterization and the sensor measurement models used in this work. Section 17.4 details the structure of the proposed complementary filter for attitude estimation. Section 17.5 is devoted to experimental results and comparisons to illustrate the effectiveness of the proposed algorithm. Finally, Section 17.6 summarizes the main conclusions of the chapter.

17.2   MOTIVATION AND PROBLEM FORMULATION

Recent technological advances have revolutionized the approach of the animals in their environment and have enabled researchers in biology and ecophysiology to leave their laboratories to study these adaptations on the animal models living freely in their natural environment. Bio-logging has been introduced as the science that studies the behaviour, physiology, ecology and environment properties of free-living animals (bioclimatic, global change, etc.) that are often beyond the border of our visibility or experience. Bio-logging has found its origin in the marine environment (Kooyman 2004) and has diversified into the study of flying and terrestrial species. This scientific area refers often to the study of free-ranging animals in their natural environment through miniaturized electronic devices, called bio-loggers (Naito 2004), usually attached to their bodies. These systems measure and record biological parameters or physico-chemical properties related to the individual and/or its environment using various types of sensors (luminosity, pressure, velocity, etc.). The loggers provide time tracking of physical and biological parameters over periods ranging from several hours to several months or sometimes a year and at sampling

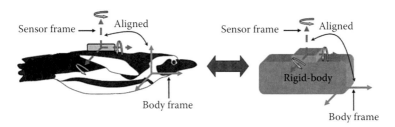

FIGURE 17.1   Schematic diagram of how an inertial measurement unit is attached to a penguin.

rates ranging from minutes to several times per second. The king penguin and badger are the major biological models studied in Strasbourg University thanks to the Biologging technology. Biologists are recently interested in reconstructing the motion of these animals (3D attitude and position) under several acceleration profiles, to be able to study their behaviour during long periods.

In this chapter, we are interested in proposing a robust alternative approach to estimate the attitude or orientation of a rigid body (Fourati 2010), which represents an animal's body, to be applied later in the case of penguin (see Figure 17.1). To achieve this goal, we use a wearable inertial and magnetic MEMS sensor assembly based on an inertial measurement unit (IMU) composed of a 3-axis accelerometer, a 3-axis magnetometer and a 3-axis gyroscope. Furthermore, the estimated attitude is used to calculate three components of DBA of an animal, which provides biologists with important information about the energy budgets of free-living animals. This work will serve in future to address the problem of 3D position estimation in the case of animal pedestrian locomotion, based on attitude and DBA estimations.

17.3   MATERIALS AND METHODS

17.3.1   RIGID BODY ATTITUDE AND COORDINATE SYSTEMS

A rigid body is considered as a solid formed from a finite set of material points with deformable volume (Goldstein 1980). Generally, the rigid body attitude represents the direction of its principal axes relative to a reference coordinate system, and its dynamics expresses the change of object orientation. In the navigation field, the attitude estimation problem requires the transformation of measured and computed quantities between various frames. The rigid body attitude is based on measurements gained from sensors attached to this latter. Indeed, inertial sensors (accelerometer, gyroscope, etc.) are attached to the body platform and provide inertial measurements expressed relative to the instrument axes. In most systems, the instrument axes are nominally aligned with the body-platform axes. Since the measurements are performed in the body frame, we describe in Figure 17.2 the orientation of the body-fixed frame $B(X_B, Y_B, Z_B)$ with respect to the Earth-fixed frame $N(X_N, Y_N, Z_N)$, which is tangent to the Earth's surface (local tangent plane, LTP). This local coordinate is particularly useful to express the attitude of a moving rigid body on the surface of the Earth (Grewal et al. 2001). The X_N-axis points true north. The Z_N-axis points towards

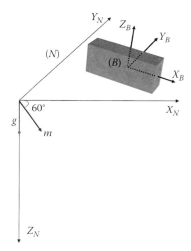

FIGURE 17.2   The coordinate system (B) of a rigid body represented in the Earth-fixed frame (N).

the interior of the Earth, perpendicular to the reference ellipsoid. The Y_N-axis completes the right-handed coordinate system, pointing east (NED: north, east, down).

17.3.2   MATHEMATICAL MODEL OF ATTITUDE REPRESENTATION

In this chapter, the quaternion algebra is used to describe the rigid body attitude. The unit quaternion, denoted by q, is expressed as

$$q = q_0 + q_{vect} = q_0 1 + q_1 i + q_2 j + q_3 k \in H \tag{17.1}$$

where
$q_{vect} = q_1 i + q_2 j + q_3 k$ represents the imaginary vector
q_0 is the scalar element

H can be written as

$$H = \left\{ q / q^T q = 1, \; q = \begin{bmatrix} q_0 & q_{vect}^T \end{bmatrix}^T, \; q_0 \in \mathfrak{R}, \; q_{vect} = [q_1 \quad q_2 \quad q_3]^T \in \mathfrak{R}^{3 \times 1} \right\} \tag{17.2}$$

The rotation matrix in terms of quaternion can be written as

$$M_N^B(q) = \begin{bmatrix} 2\left(q_0^2 + q_1^2\right) - 1 & 2(q_1 q_2 + q_0 q_3) & 2(q_1 q_3 - q_0 q_2) \\ 2(q_1 q_2 - q_0 q_3) & 2\left(q_0^2 + q_2^2\right) - 1 & 2(q_0 q_1 + q_2 q_3) \\ 2(q_0 q_2 + q_1 q_3) & 2(q_2 q_3 - q_0 q_1) & 2\left(q_0^2 + q_3^2\right) - 1 \end{bmatrix} \tag{17.3}$$

We invite the reader to refer to suggest Kuipers (1999) for more details about quaternion algebra.

17.3.3   3-Axis Inertial/Magnetic Sensor Package Measurement Models

The sensor configuration consists of a 3-axis accelerometer, a 3-axis magnetometer and a 3-axis gyroscope containing MEMS technologies. A detailed study of these sensors is given in Beeby et al. (2004).

17.3.3.1   3-Axis Accelerometer

An accelerometer measures the acceleration of the object that it supports. In our case, three accelerometers are mounted in orthogonal triad in a rigid body, such that their sensitive axes coincide with the principal axes of inertia of the moving body. The output of a 3-axis accelerometer in the body-fixed frame (B) is given by the following measurement vector (Guerrero-Castellanos 2008):

$$f = M_N^B(q)(a + G) + \delta_f \qquad (17.4)$$

where $G = \begin{bmatrix} 0 & 0 & g \end{bmatrix}^T$ and $a = \begin{bmatrix} a_x & a_y & a_z \end{bmatrix}^T$ represent, respectively, the gravity vector and the DBA of the rigid body, given in the Earth-fixed frame (N). $\delta_f \in \mathfrak{R}^3$ is a noise vector assumed to be independent, white and Gaussian. $M_N^B(q)$ is the rotation matrix defined in (17.3), reflecting the transition between the frames (N) and (B).

17.3.3.2   3-Axis Magnetometer

A magnetometer is a device for measuring the direction and intensity of a magnetic field, especially the Earth's magnetic field. The output of a 3-axis magnetometer in the body-fixed frame (B) is given by the following measurement vector (Guerrero-Castellanos 2008):

$$h = M_N^B(q)m + \delta_h \qquad (17.5)$$

where m is the magnetic field expressed in the Earth-fixed frame (N) by

$$m = \begin{bmatrix} m_x & 0 & m_z \end{bmatrix}^T = \begin{bmatrix} \|m\|\cos(I) & 0 & \|m\|\sin(I) \end{bmatrix}^T \qquad (17.6)$$

δ_h is a white Gaussian noise and $M_N^B(q)$ is expressed in (17.3). The parameters of the theoretical model of the geomagnetic field m closest to reality can be deduced from Astrosurf (2012).

17.3.3.3   3-Axis Gyroscope

A gyroscope is an inertial sensor that measures the angular velocity of reference attached to the sensor compared to an absolute reference frame along one or more axes (Titterton and Weston 2004). The output of a 3-axis gyroscope in the body-fixed frame (B) is given by the measurement vector (Guerrero-Castellanos 2008):

$$\omega_G = \omega + b + \delta_G \tag{17.7}$$

where

$\omega \in \mathfrak{R}^3$ is the real angular velocity
$b \in \mathfrak{R}^3$ is a slowly time-varying function (Beeby et al. 2004) also called bias
δ_G is a white Gaussian noise.

17.4 COMPLEMENTARY FILTER FOR ATTITUDE ESTIMATION

In this chapter, the objective is to design an attitude estimation algorithm based on inertial and magnetic MEMS sensors. The application in mind is related to a free-ranging animal case in Bio-logging (Fourati et al. 2011b). By considering the rigid body kinematic model, a complementary filter is proposed in order to take advantage from the good short-term precision given by rate gyro integration and the reliable long-term accuracy provided by accelerometer and magnetometer measurements. This leads to better attitude estimates (Mahony et al. 2008). It is important to note that the resulting approach structure is complementary: high bandwidth rate gyro measurements are combined with low bandwidth vector observations to provide an accurate attitude estimate (Brown and Hwang 1997).

17.4.1 RIGID BODY KINEMATIC MOTION EQUATION

The rigid body motion can be described by the attitude kinematic differential equation (Shuster 1993), which represents the time rate of attitude variation, expressed in a quaternion term q, as a result of the rigid body angular rates measured by the gyroscope:

$$\dot{q} = \frac{1}{2} \begin{bmatrix} -q_{vect}^T \\ I_{3\times3} q_0 + \left[q_{vect}^\times \right] \end{bmatrix} \omega_G \tag{17.8}$$

where $q = \begin{bmatrix} q_0 & q_{vect}^T \end{bmatrix}^T$ is the unit quaternion that denotes the mathematical representation of the rigid body attitude between two frames: body-fixed frame (B) and Earth-fixed frame (N). Note that $q_{vect} = [q_1 \quad q_2 \quad q_3]^T$ represents the vector part of q. It is customary to use quaternion instead of Euler angles since they provide a global parameterization of the body orientation and are well suited for calculations and computer simulations.

ω_G represents the angular velocity vector expressed in (B) and $I_{3\times3}$ is the identity matrix of dimension 3.

$\left[q_{vect}^{\times}\right]$ represents the standard vector cross product (the skew-symmetric matrix), which is defined as

$$\left[q_{vect}^{\times}\right] = \begin{bmatrix} q_1 \\ q_2 \\ q_3 \end{bmatrix}^{\times} = \begin{bmatrix} 0 & -q_3 & q_2 \\ q_3 & 0 & -q_1 \\ -q_2 & q_1 & 0 \end{bmatrix} \tag{17.9}$$

17.4.2 DESIGN OF STATE MODEL

Let us consider the following system model (S_1) composed of (17.8) with the output y that represents the linear measurement model. The output $y \in \mathfrak{R}^6$ of this system is built by stacking the accelerometer and magnetometer measurements:

$$(S_1): \begin{cases} \begin{bmatrix} \dot{q}_0 \\ \dot{q}_1 \\ \dot{q}_2 \\ \dot{q}_3 \end{bmatrix} = \frac{1}{2} \begin{bmatrix} -q_{vect}^T \\ I_{3\times3}q_0 + \left[q_{vect}^{\times}\right] \end{bmatrix} \omega_G = \frac{1}{2} \begin{bmatrix} -q_1\omega_{Gx} - q_2\omega_{Gy} - q_3\omega_{Gz} \\ q_0\omega_{Gx} - q_3\omega_{Gy} + q_2\omega_{Gz} \\ q_3\omega_{Gx} + q_0\omega_{Gy} - q_1\omega_{Gz} \\ q_1\omega_{Gy} - q_2\omega_{Gx} + q_0\omega_{Gz} \end{bmatrix} \\ y = \begin{bmatrix} f_x & f_y & f_z & h_x & h_y & h_z \end{bmatrix}^T \end{cases} \tag{17.10}$$

By considering the rigid body kinematic equation and the linear measurement model y, the proposed system (S_1) can take advantage of the good short-term precision given by the rate gyro integration and the reliable long-term accuracy provided by accelerometer and magnetometer measurement fusion (Brown and Hwang 1997; Fourati et al. 2010), which leads to improve the quaternion estimation.

17.4.3 ATTITUDE COMPLEMENTARY FILTER DESIGN

The aim of this approach is to ensure a compromise between the accuracy provided by short-term integration of the gyroscope data and the long-term measurement precision obtained by the accelerometer and the magnetometer. To compensate for the drifts on the estimated quaternion that are observed during the integration of the differential equation (17.8), a correction term T is introduced in this equation based on a quaternion product $\otimes$. We propose the following complementary filter:

$$(F): \begin{bmatrix} \dot{\hat{q}}_0 \\ \dot{\hat{q}}_1 \\ \dot{\hat{q}}_2 \\ \dot{\hat{q}}_3 \end{bmatrix} = \frac{1}{2} \begin{bmatrix} -\hat{q}_1\omega_x - \hat{q}_2\omega_y - \hat{q}_3\omega_z \\ \hat{q}_0\omega_x - \hat{q}_3\omega_y + \hat{q}_2\omega_z \\ \hat{q}_3\omega_x + \hat{q}_0\omega_y - \hat{q}_1\omega_z \\ \hat{q}_1\omega_y - \hat{q}_2\omega_x + \hat{q}_0\omega_z \end{bmatrix} \otimes T \tag{17.11}$$

where $\hat{q} = \begin{bmatrix} \hat{q}_0 & \hat{q}_1 & \hat{q}_2 & \hat{q}_3 \end{bmatrix}^T \in \mathfrak{R}^4$ represents the estimated quaternion. The correction term T is calculated from a fusion approach of accelerometer and magnetometer data. The quaternion product introduced in (17.11) allows us to merge the magnetic and inertial measurements.

Let us present the method for calculating the correction term T. We consider the modelling error $\delta(\hat{q}) = (y - \hat{y})$. The estimated output is given by $\hat{y}$:

$$\hat{y} = \begin{bmatrix} \hat{f}_x & \hat{f}_y & \hat{f}_z & \hat{h}_x & \hat{h}_y & \hat{h}_z \end{bmatrix}^T \tag{17.12}$$

Measurements of the estimated accelerations $\hat{f}_x, \hat{f}_y$ and $\hat{f}_z$ can be calculated by assuming that the DBA a is low ($\|a\|_2 \ll \|G\|_2$) (Fourati et al. 2010). Thus, we obtain

$$\hat{f} = \begin{bmatrix} 0 & \hat{f}_x & \hat{f}_y & \hat{f}_z \end{bmatrix}^T = \hat{q}^{-1} \otimes G_q \otimes \hat{q} \tag{17.13}$$

where $G_q = \begin{bmatrix} 0 & 0 & 0 & 9.8 \end{bmatrix}^T$: quaternion representation of the gravity vector $G = \begin{bmatrix} 0 & 0 & 9.81 \end{bmatrix}^T$.

Measurements of the estimated Earth's magnetic field $\hat{h}_x$, $\hat{h}_y$ and $\hat{h}_z$ can be calculated as

$$\hat{h} = \begin{bmatrix} 0 & \hat{h}_x & \hat{h}_y & \hat{h}_z \end{bmatrix}^T = \hat{q}^{-1} \otimes m_q \otimes \hat{q} \tag{17.14}$$

where $m_q = \begin{bmatrix} 0 & m_x & 0 & m_z \end{bmatrix}^T$: quaternion representation of the Earth's magnetic field $m = \begin{bmatrix} m_x & 0 & m_z \end{bmatrix}^T$.

The minimization of the modelling error $\delta(\hat{q})$ is performed using a regression method that minimizes the scalar squared error criterion function $\xi(\hat{q})$ related to $\delta(\hat{q})$:

$$\xi(\hat{q}) = \delta(\hat{q})^T \delta(\hat{q}) \tag{17.15}$$

In this chapter, the LMA (Marquardt 1963) is used to minimize the non-linear function $\xi(\hat{q})$. This choice reflects the robustness demonstrated by this algorithm compared to other methods such as Gauss–Newton or gradient (Dennis and Schnabel 1983).

The unique solution to this problem can be written in the following form (Deutschmann et al. 1992):

$$\eta(\hat{q}) = K\delta(\hat{q}) \tag{17.16}$$

where $K = k[X^T X + \lambda I_{3\times3}]^{-1} X^T$ is the gain of the filter used to minimize the error $\delta(\hat{q})$. $X \in \mathfrak{R}^{6\times3}$ is the Jacobian matrix defined by

$$X = -2\begin{bmatrix} [f^\times] & [h^\times] \end{bmatrix}^T = -2\begin{bmatrix} 0 & -f_z & f_y & 0 & -h_z & h_y \\ f_z & 0 & -f_x & h_z & 0 & -h_x \\ -f_y & f_x & 0 & -h_y & h_x & 0 \end{bmatrix}^T \tag{17.17}$$

The constant λ is chosen to ensure the non-singularity of the minimization problem. The constant k determines the crossover frequency of the latter. It is used to tune the balance between measurement noise suppression and response time of the filter. Generally, it combines low bandwidth accelerometer/magnetometer readings with high bandwidth gyroscope measurements. Notice that the complementary filter has a better convergence when k is chosen somewhere between 0.1 and 1 (Mahony et al. 2008). $\eta(\hat{q})$ represents a part of the correction term T. To achieve the quaternion product in (17.11), the term T must be of dimension 4. So T is constructed as follows:

$$
T = \begin{bmatrix} 1 & 0 & 0 & 0 & 0 & 0 & 0 \\ 0 & & & & & & \\ 0 & & & K & & & \\ 0 & & & & & & \end{bmatrix} \begin{bmatrix} 1 \\ \delta(\hat{q}) \end{bmatrix} \tag{17.18}
$$

The scalar part of quaternion error is chosen to 1 to force the error quaternion to represent small angles of rotation (Deutschmann et al. 1992). Finally, the complementary filter can be written as follows:

$$
(F): \begin{bmatrix} \dot{\hat{q}}_0 \\ \dot{\hat{q}}_1 \\ \dot{\hat{q}}_2 \\ \dot{\hat{q}}_3 \end{bmatrix} = \frac{1}{2} \begin{bmatrix} -\left(\hat{q}_1\omega_x + \hat{q}_2\omega_y + \hat{q}_3\omega_z\right) \\ \left(\hat{q}_0\omega_x - \hat{q}_3\omega_y + \hat{q}_2\omega_z\right) \\ \left(\hat{q}_3\omega_x + \hat{q}_0\omega_y - \hat{q}_1\omega_z\right) \\ \left(\hat{q}_1\omega_y - \hat{q}_2\omega_x + \hat{q}_0\omega_z\right) \end{bmatrix} \otimes \begin{bmatrix} 1 & 0 & 0 & 0 & 0 & 0 & 0 \\ 0 & & & & & & \\ 0 & & & K & & & \\ 0 & & & & & & \end{bmatrix} \begin{bmatrix} 1 \\ \delta(\hat{q}) \end{bmatrix}
$$

$$\tag{17.19}$$

17.5   EXPERIMENTAL VALIDATION

17.5.1   Experimental Tool for Attitude Estimation: Inertial Measurement Unit MTi-G

In order to evaluate the efficiency of the proposed complementary filter in real-world applications, an experimental setup was developed, resorting to an inertial and magnetic sensor assembly. The goal is to obtain an estimation of the quaternion that represents the orientation of a rigid body and to investigate its accuracy under various conditions. For the experiments, the *MTi-G* from Xsens Motion Technologies (Xsens Technologies 2012) was employed. This MEMS device is a miniature, lightweight, 3D calibrated digital output sensor (3D acceleration from an accelerometer, 3D angular rate from a gyroscope and 3D magnetic field data from a magnetometer), a GPS-enhanced attitude and heading reference system with built-in bias, sensitivity and temperature compensation. The *MTi-G* outputs data at a rate of 100 Hz and records them on a computer (see Figure 17.3). In addition, this device is designed to track the body 3D attitude output in quaternion representation using an embedded Extended Kalman filter algorithm. The calibration procedure to obtain the gain, offsets and non-orthogonality of the sensors was performed by the manufacturer of the sensor module.

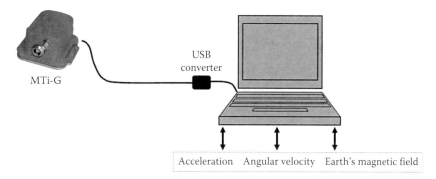

FIGURE 17.3   Inertial measurement unit *MTi-G*.

It is important to note that the *MTi-G* device serves as tool for the evaluation of the complementary filter efficiency and cannot be suitable for use in the Bio-logging field due to its dependence on an energy source as well as its heavy weight. In the following set of experiments, the calibrated data from the *MTi-G* are used as input to the complementary filter.

17.5.2   EVALUATION TEST AND ESTIMATION ATTITUDE ANALYSIS IN FREE MOVEMENT OF AN ANIMAL

In this set of experiments, the accuracy of the complementary filter is evaluated during the free motion of a domestic animal (a dog). The *MTi-G* is attached to the back of the animal with its *xyz*-axes aligned with those of the animal. The path followed by the animal in a football stadium is captured as shown in Figure 17.4.

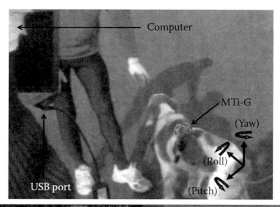

FIGURE 17.4   The *MTi-G* attached to the back of a dog – description of the dog motion.

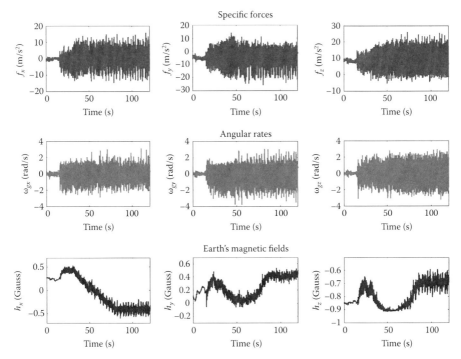

FIGURE 17.5   Inertial and magnetic measurements recorded from the *MTi-G*.

Inertial/magnetic measurements and attitude (in quaternion representation) are recorded using the *MTi-G* during the motion of the dog (see Figure 17.5) and transmitted to a computer via a USB port. Based on the measurements recorded by the accelerometer, we note that the animal motion consists of two acceleration profiles, one corresponding to the low frequencies of motion (during walk) and the other, rather, to the high frequencies (during trot and canter). The acceleration profile varies between $[-15, 15\ \mathrm{m/s^2}]$ for f_x and f_y and $[-5, +25\ \mathrm{m/s^2}]$ for f_z. The increase in the acceleration level between the natural gaits is due to the DBA a of the dog that is more important during the trot and the canter.

The recorded inertial and magnetic measurements from the *MTi-G* are used to estimate the attitude, using the proposed complementary filter. The calculated attitude from the *MTi-G* is considered as reference to the dog's motion. Figure 17.6 plots the evolution of the difference between the calculated quaternion using the *MTi-G* and the one estimated by the proposed approach. Although some parts of the motion are with high dynamics, we can remark that the errors on quaternion's components don't exceed 0.03 on q_0, q_1, q_2 and 0.05 on q_3. To provide more clarity to the reader, we also represent the attitude estimation results of the same movement using the Euler angles (roll, pitch and yaw). Figure 17.7 shows the evolution of the difference between the Euler angles estimated by the complementary filter and the *MTI-G*.

It is clear that this mismatch between the estimated attitude by our approach and the *MTi-G* is small. Then, one can conclude about the performance of the complementary filter in attitude estimation of the animal body even in dynamic situations.

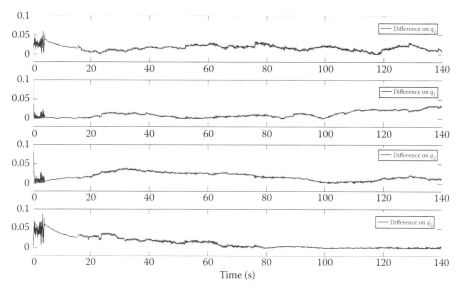

FIGURE 17.6 Differences between quaternion's estimates provided by the complementary filter and the *MTi-G* during the motion of the dog.

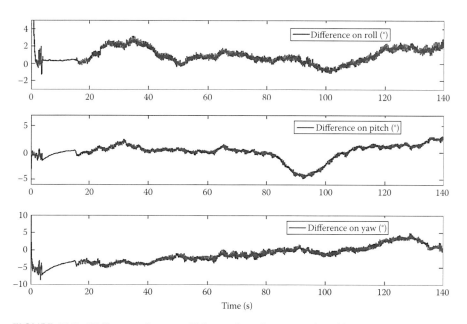

FIGURE 17.7 Differences between Euler angle estimates produced by the complementary filter and the *MTi-G* during the motion of the dog.

Although our approach didn't exploit GPS data as done in *MTi-G*, it is able to recon-struct the orientation of the dog given by the *MTi-G* with a small error.

17.5.3   PERFORMANCE COMPARISON WITH PREVIOUS BIO-LOGGING WORKS

We propose in this section a comparative study between the performance of the atti-tude estimation obtained from three methods: the complementary filter and two other approaches that have been proposed in Bio-logging that we called method_1 (Wilson et al. 2008) and method_2 (Watanabe et al. 2005). Both approaches use only a combina-tion of triaxial accelerometer and magnetometer and provide an attitude estimation in Euler angle representation. The purpose of this comparison is to analyse the performance of the complementary filter and to prove if it is possible to make an improvement of the attitude estimation in Bio-logging and show the interest to add gyroscope in such an application. This comparison is performed in the case of the experimental test on the dog, presented earlier, and we used the measurements recorded by the *MTi-G*. To compare the three methods, the estimated quaternion from the complementary filter is converted to Euler angles using the formulas presented in Phillips et al. (2001). The estimation results obtained separately from the three approaches (method_1, method_2 and the comple-mentary filter) are compared with those provided by the internal algorithm of the *MTi-G*.

17.5.3.1   Attitude Estimation

The results of this comparison, illustrated in Figure 17.8, show the errors obtained from the difference between the estimates of Euler angles calculated by the *MTi-G* and those provided by the three methods. The smallest difference was obtained with the complementary filter. This error does not exceed 5° on the three Euler angles even in high-frequency movements of the animal, where the DBA is important. The estimation errors obtained by method_1 and method_2 are around 10° for roll and pitch angles and 20° for the yaw angle. These large errors are mainly due to the approximations established in these two methods, since the accelerometer does not extract the attitude during the dynamic situations of movement. These high-frequency dynamics are present during the motion of the dog.

Performance analysis of each method can also be established using the root mean square difference (*RMSD*). This criterion quantifies the difference between the Euler angles calculated by the *MTi-G*, considered as reference, and those estimated by each method. The *RMSD* was calculated as

$$RMSD_{sliding}(k) = \sqrt{\frac{\sum_{i=k}^{n+k}\left(x_i - \hat{x}_i\right)^2}{N}} \qquad (17.20)$$

where
  x_i is the Euler angle measured by the *MTi-G* algorithm
  $\hat{x}_i$ is the Euler angle estimated by the chosen method (complementary filter, method_1 or method_2)
  N is the time interval ($T = 2$)

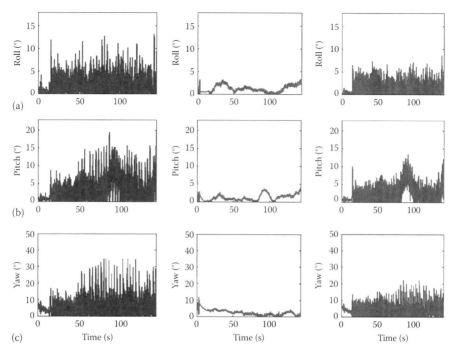

FIGURE 17.8 Estimation errors of Euler angles during the motion of the dog – (a) difference between *MTi-G* and method_1, (b) difference between *MTi-G* and complementary filter and (c) difference between *MTi-G* and method_2.

TABLE 17.1

Average of the $RMSD_{Sliding}$ Corresponding to Euler Angles for Each Method during the Experiment on the Dog

Methods	Complementary Filter	Method_1	Method_2
Average of the $RMSD_{Sliding}$ (ROLL)	0.934	1.6144	1.1846
Average of the $RMSD_{Sliding}$ (PITCH)	0.8609	2.4962	1.8019
Average of the $RMSD_{Sliding}$ (YAW)	5.0426	19.1813	12.6655

An average of $RMSD_{sliding}$ on the Euler angles for each method is subsequently established in Table 17.1. Note that the $RMSD_{sliding}$ values relating to the three Euler angles are obtained also with the complementary filter. This highlights the improvements we were able to make at the attitude estimation, compared to the two methods developed in Bio-logging.

17.5.3.2 Dynamic Body Acceleration Estimation

In this subsection, we are interested in the calculation of the DBA of the animal during its movement. This acceleration relates solely to the movement of the animal's body. To calculate the DBA, we used the attitude estimation $\hat{q}$ obtained from the complementary filter during the movement of the dog. The following equation is used:

$$\hat{a} = inv\left(M_N^B(\hat{q})\right)f - G \qquad (17.21)$$

where the rotation matrix $M_N^B(\hat{q})$ is expressed in (17.3), $G \in \mathfrak{R}^3$ is the gravity vector and $f \in \mathfrak{R}^3$ represents the measurements of the accelerometer.

We calculate later the norm of the acceleration using the following equation:

$$\|\hat{a}\|_2 = \sqrt{\hat{a}_x^2 + \hat{a}_y^2 + \hat{a}_z^2} \qquad (17.22)$$

Similarly, we calculated the attitude by method_1 and method_2. The attitude values obtained from each method are used to calculate the DBA of the animal using (17.21). Finally, the norm of the acceleration is calculated using (17.22). We report in Figure 17.9 the results of this comparison by establishing the difference between the norm of acceleration obtained from the *MTi-G* and the one provided by each method (complementary filter, method_1 and method_2). The smallest difference is obtained with the complementary filter. Indeed, the errors of the complementary filter do not exceed 0.7 m/s², but they reach 3 m/s² for method_1 and 2 m/s² for method_2. These results demonstrate the improvements made by the proposed approach in calculating the DBA of an animal. We recall that a more precise calculation of the DBA will allow biologists to better assess the energy expenditure of an animal.

Similarly, we used the $RMSD_{sliding}$ given in (17.20) to measure the difference between the norm of DBA calculated by the *MTi-G* (reference) and the one estimated by each method.

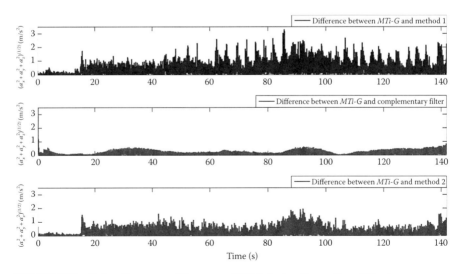

FIGURE 17.9   Estimation error of the norm of DBA during the motion of the dog.

TABLE 17.2

Average of the $RMSD_{Sliding}$ Corresponding to the Norm of DBA for Each Method during the Experiment on the Dog

Methods	Complementary Filter	Method_1	Method_2
Average of the $RMSD_{Sliding}$	0.1168	0.3351	0.1929

We used for that the following notations

x_i is the norm of DBA calculated by the *MTi-G*

$\hat{x}_i$ is the norm of DBA estimated by the chosen method (complementary filter, method_1 or method_2)

Table 17.2 shows the averages of the $RMSD_{sliding}$ corresponding to the norm of DBA for each method. Note that we obtained the smallest value of this average with the complementary filter. We conclude that this criterion reflects the filter's ability to provide a more accurate calculation of DBA.

17.6 CONCLUSION

This chapter presents the design and experimental results of a quaternion-based complementary filter for animal body motion tracking using inertial/magnetic sensor modules containing orthogonally mounted triads of accelerometers, angular rate sensors and magnetometers. The complementary filter was designed in order to be able to produce highly accurate orientation estimates without resorting to GPS data. The filter design makes use of a simple kinematic motion equation to describe the system model. The filter design is further simplified by preprocessing accelerometer and magnetometer data using the LMA. The modelling error produced by the LMA is provided as input to the filter along with angular rate data. Some experiments are carried out on free motion of an animal through sensor measurements provided by an IMU. From the experiments designed to validate filter performance, this approach was shown to work well. Future works will focus on designing a low-cost, lightweight and embedded prototype for this application.

REFERENCES

Ang, W. T., Khosla, P. K., and Riviere, C. N. 2004. Kalman filtering for real-time orientation tracking of handheld microsurgical instrument. *IEEE/RSJ International Conference on Intelligent Robots and Systems*, Sendai, Japan, pp. 2574–2580.

Astrosurf. September 2012. Available: http://www.astrosurf.com

Beeby, S., Ensell, G., Kraft, M., and White, N. 2004. *MEMS Mechanical Sensors*. Boston, MA: Artech House Publishers.

Bost, C. A., Handrich, Y., Butler, P. J., Fahlman, A., Halsey, L. G., Woakes, A. J., and Ropert-Coudert, Y. 2007. Change in dive profiles as an indicator of feeding success in king and Adélie penguins. *Deep-Sea Research II* 54(3–4): 248–255.

Brown, R. G. and Hwang, P. Y. C. 1997. *Introduction to Random Signal and Applied Kalman Filtering*. 3rd edn. New York: John Wiley.

Dennis, Jr. J. E. and Schnabel, R. B. 1983. *Numerical Methods for Unconstrained Optimization and Nonlinear Equations*. Englewood, NJ: Prentice Hall.

Deutschmann, J., Bar-Itzhack, I., and Galal, K. 1992. Quaternion normalization in spacecraft attitude determination. *AIAA Astrodynamics Conference*, Washington, DC, pp. 27–37.

Elkaim, G. H., Decker, E. B., Oliver, G., and Wright, B. 2006. Marine Mammal Marker (MAMMARK) dead reckoning sensor for In-Situ environmental monitoring. *IEEE Position, Location and Navigation Symposium*, Monterey, CA, April 2006, pp. 976–987.

Fourati, H. 2010. Contributions à l'estimation d'attitude chez l'animal ou l'homme par fusion de données inertielles et magnétiques: de la reconstitution de la posture vers la navigation à l'estime: une application au Bio-logging. PhD dissertation, Strasbourg University, Strasbourg, France.

Fourati, H., Manamanni, N., Afilal, L., and Handrich, Y. 2009. A rigid body attitude estimation for Bio-logging application: A quaternion-based nonlinear filter approach. *IEEE/RSJ International Conference on Intelligent Robots and Systems (IROS)*, St. Louis, MO, pp. 558–563.

Fourati, H., Manamanni, N., Afilal, L., and Handrich, Y. 2011a. A nonlinear filtering approach for the attitude and dynamic body acceleration estimation based on inertial and magnetic sensors: Bio-logging application. *IEEE Sensors Journal* 11(1): 233–244.

Fourati, H., Manamanni, N., Afilal, L., and Handrich, Y. 2011b. Posture and body acceleration tracking by inertial and magnetic sensing: Application in behavioural analysis of free-ranging animals. *Biomedical Signal Processing and Control (BSPC)*, 6(1): 94–104.

Fourati, H., Manamanni, N., Benjemaa, A., Afilal, L., and Handrich, Y. 2010. A quaternion-based complementary sliding mode observer for attitude estimation: Application in free-ranging animal motions. *IEEE Conference on Decision and Control (CDC)*, Atlanta, GA, pp. 5056–5061.

Goldstein, H. 1980. *Classical Mechanics*. 2nd edn. Reading, MA: Addison-Wesley.

Grewal, M. S., Weill, L. R., and Andrews, A. P. 2001. *Global Positioning Systems, Inertial Navigation, and Integration*. Hoboken, NJ: John Wiley & Sons, Inc.

Guerrero-Castellanos, J. F. 2008. Estimation de l'attitude et commande borné en attitude d'un corps rigide: Application à un hélicoptère à quatre rotors. Ph.D dissertation, Joseph Fourier University, Grenoble, France.

Halsey, L. G., Handrich, Y., Fahlman, A., Schmidt, A., Bost, C. A., Holder, R. L., Woakes, A. J., and Butler, P. J. 2007. Fine-scale analyses of diving energetics in king penguins Aptenodytes patagonicus: how behaviour affect costs of a foraging dive. *Marine Ecology Progress Series* 344: 299–309.

Johnson, M. P. and Tyack, P. L. 2003. A digital acoustic recording tag for measuring the response of wild marine mammals to sound. *IEEE Journal of Oceanic Engineering* 28(1): 3–12.

Kooyman, G. L. 2004. Genesis and evolution of bio-logging devices: 1963–2002. *Memoirs of the National Institute of Polar Research* 58: 148–154.

Kuipers, J. B. 1999. *Quaternion and Rotation Sequences*. Princeton, NJ: Princeton University Press.

Le Boeuf, B. J., Crocker, D. E., Costa, D. P., Blackwell, S. P., Webb, P. M., and Houser, D. S. 2000. Foraging ecology of northern elephant seals. *Ecological Monographs* 70(3): 353–382.

Mahony, R., Hamel, T., and Pflimlin, J. M. 2008. Nonlinear complementary filters on the special orthogonal group. *IEEE Transactions on Automatic Control* 53(5): 1203–1217.

Marquardt, D. W. 1963. An algorithm for the least-squares estimation of nonlinear parameters. *SIAM Journal of Applied Mathematics* 11(2): 431–441.

Naito, Y. 2004. New steps in bio-logging science. *Memoirs of National Institute of Polar Research* 58: 50–57.

O'Donovan, K. J., Kamnik, R., O'Keeffe, D. T., and Lyons, G. M. 2007. A inertial and magnetic sensor based technique for joint angle measurement. *Journal of Biomechanics* 40(12): 2604–2611.

Phillips, W., Hailey, C., and Gebert, G. 2001. A review of attitude representations used for aircraft kinematics. *AIAA Journal of Aircraft* 38(4): 718–737.

Ropert-Coudert, Y., Beaulieu, M., Hanuise, N., and Kato, A. 2009. Diving into the world of biologging. *Endangered Species Research* 10: 21–27.

Rutz, C. and Hays, G. C. 2009. New frontiers in biologging science. *Biology Letters* 5(3): 289–291.

Shuster, M. D. 1993. A survey of attitude representations. *Journal of the Astronautical Science* 41(4): 493–517.

Titterton, D. H. and Weston, J. L. 2004. *Strapdown Inertial Navigation Technology*. 2nd edn. Stevenage, U.K.: The institution of Electrical Engineers.

Watanabe, S., Izawa, M., Kato, A., Ropert-Coudert, Y., and Naito, Y. 2005. A new technique for monitoring the detailed behaviour of terrestrial animals: A case study with the domestic cat. *Applied Animal Behaviour Science* 94(1): 117–131.

Wilson, R., Shepard, E. L. C., and Liebsch, N. 2008. Prying into the intimate details of animal lives: use of a daily diary on animals. *Endangered Species Research* 4: 123–137.

Xsens Technologies. June 2011. Available: http://www.xsens.com

Zhou, H., Hu, H., Harris, N. D., and Hammerton, J. 2006. Applications of wearable inertial sensors in estimation of upper limb movements. *Biomedical Signal Processing and Control* 1(1): 22–32.

18 Point-of-Care Testing Platform with Nanogap-Embedded Field-Effect Transistors

Maesoon Im and Yang-Kyu Choi

CONTENTS

18.1 INTRODUCTION

The global *in vitro* diagnostics industry is predicted to grow rapidly in the coming years to reach a value of approximately US$ 52 billion by the end of 2013, with a 5% annual growth rate from 2011 to 2013 [1]. Among the various *in vitro* diagnostic constituents, point-of-care testing (POCT) constitutes the most lucrative share of the global market.

Because human life expectancy is higher than in the past, there is accordingly increased interest in healthcare as a means of enjoying a longer life in good health. Along with the development of medical technologies, consumers are paying more attention to personal health monitoring systems that use biosensors to provide accurate, easy and inexpensive medical checkups and diagnostics to identify many diseases in their early stages. Medical treatment costs can be significantly reduced if diseases are diagnosed as early as possible. To reduce the temporal or spatial limitations in such medical checkups, healthcare should be easily available at low cost. Thus, ubiquitous POCT devices can benefit people all over the world who have limited access to medical resources.

Blood glucose testers may be the most well-known POCT device. In addition, POCT systems can be found in a broad range of tests for electrolytes, glucose, cholesterol, cardiac or coagulation markers and others. The spectrum of POCT is likely to be widened as healthcare becomes a major concern with rising standards of living. In particular, the possible occurrence of a large-scale human pandemic such as bird flu or swine flu makes POCT devices more attractive, as individual medical checkups can be performed in private places rather than in public hospitals or clinics. Such at-home testing may help restrict the spread of highly contagious diseases by minimizing infectious contact.

Low cost is essential for POCT devices. In particular, costly POCT systems are incommensurate with the needs of end users in developing countries. For low cost, POCT systems can take advantage of the advanced semiconductor business. The mass production of transistor-based sensors and electric interfaces can provide compact and inexpensive platform without complex transducers. In this chapter, we describe the detection of various biomolecules with field-effect transistor (FET)-based sensor arrays with general-purpose readout platforms.

18.1.1 Biosensor Techniques for Point-of-Care Testing Applications

Numerous researchers have demonstrated various detection methods for biosensor applications. Fluorescent imaging [2] is a popular and conventional method for biologists and other scientists in sensing applications. Several biomolecules have been successfully detected using the physical or chemical properties of the materials such as mechanical resonance [3] or deflection [4], electrochemical currents [5] and conductance [6–9]. Arlett et al. summarized the mechanical biosensor field in a recent review in *Nature Nanotechnology* [10]. However, in terms of the integration of transducers, electrical detection methods are superior to all other detection methods, as no additional steps or components are required to generate a signal. For example, fluorescent imaging [2] requires a fluorescent labelling process and an excitation light source, and some mechanical detection methods require ultrahigh-vacuum equipment [11]. Because reducing the size of the overall system is important to maximize its portability, the necessity of these additional parts can present a sizeable drawback.

Biosensors should meet the following four criteria for a promising POCT system. (1) Label-free: Biosensors should not require a labelling process for the detection of target molecules because most labelling processes are costly and time consuming. (2) Selectivity: To avoid false positives and false negatives, the receptor should have

TABLE 18.1
Comparison of Biosensor Types: Fluorescent Imaging [2],
Microcantilever Deflection Sensing [4], Si-Nanowire Conductance
Sensing [9] and FET-Based Sensing [11]

Sensing Method	Fluorescent	Microcantilever	Si-Nanowire	FET
Labelling	Label necessary	*Label-free*	*Label-free*	*Label-free*
Sensitivity	*Very high*	*Very high*	*Very high*	*High*
Integration of transducer	Difficult	Difficult	Difficult	*Easy*

good selectivity for the analyte. Selectivity is the most critical criterion for the reliable operation of a POCT system. (3) Sensitivity: Biosensors should have high sensitivity to diagnose diseases. However, ultra sensitivity is not required because, upon positive diagnosis by the POCT system, the analyte should be accurately re-analysed to determine the exact phase of the disease with cutting-edge technologies in a central laboratory. (4) Integrability: For inexpensive POCT systems, the biosensor should be easily integrated with user-interface electronics to minimize the system size and production cost.

Table 18.1 compares a few representative sensing methods. In particular, FET biosensors [11] are attractive because they can be mass produced using well-established complementary metal–oxide–semiconductor (CMOS) fabrication processes and can easily be coupled with integrated circuits using other CMOS devices for readout, signal processing and other required functions. Because of the aforementioned advantages, FET-based biosensors have been continuously reported [12–14] for the label-free detection of target molecules.

18.1.2 LABEL-FREE DETECTION BASED ON FIELD-EFFECT TRANSISTORS

Since its first reported development in the 1970s, the ion-sensitive FET (ISFET) has been widely studied for use in biosensors for the detection of various biomolecules [15]. Compared with traditional FET devices, the ISFET uses a reference electrode placed in an electrolyte rather than a conventional gate structure. As a pH sensor, the ISFET returns the modulation of drain-to-source currents as a function of the pH level of the solution contacting the sensing surface and the potential of a reference electrode. In addition to pH sensor applications, ISFETs in various designs have been utilized in diverse areas. Schöning and Poghossian published a comprehensive review of the biological applications of ISFETs [11]. Additionally, semiconducting nanowires have been actively studied for the detection of various biomolecules [9,16,17]. The research to date on semiconducting nanowires for biomolecular sensing has been reviewed by several authors [18,19].

ISFETs have some drawbacks including drift [20,21] and hysteresis [22]. Because in biological applications ISFETs are in contact with electrolyte solutions with high ion concentrations, the ISFETs sometimes show unstable behaviour. Although the causes of drift and hysteresis are not fully understood, leakage currents and temperature fluctuations have been identified as sources of instability [21]. Moreover, the electrical signals of ISFETs and semiconducting nanowires have a strong dependency on the Debye length [23], which is a strong function of the ionic strength of the electrolyte [11,24]. This phenomenon, known as Debye screening, is undesirable in POCT applications because it is difficult to accurately control the ionic concentration of the electrolyte and the sample solution outside of the laboratory.

As an alternative to ISFETs and semiconducting nanowires, a dielectric-modulated FET (DMFET) [25] was proposed by Im et al. In the first version of the DMFET [25], the gate dielectric layer was partially etched to create nanogaps underneath both sides of the gate electrode. The working principle of the DMFET is quite simple: by sacrificially etching the gate dielectric, the threshold voltage (V_{Th}) of the DMFET is increased because of its reduced dielectric constant. When biomolecules are immobilized in the nanogaps, they increase the dielectric constant in the nanogaps, which results in a shift in the V_{Th} of the DMFET. In addition to high sensitivity and selectivity [25], the DMFET has advantages over ISFETs or semiconducting nanowire devices because the DMFET can detect biomolecules that have weak or neutral charges. Additionally, DMFETs display a large signal change [25], which is preferable in POCT devices [26].

A few DMFET designs have been tested for the detection of biomolecules [14,25,27]. In addition to experimental verifications, the V_{Th} shift in DMFET has also been analytically modelled [28]. However, to avoid incorrect diagnoses with a POCT device, the V_{Th} changes in multiple DMFETs should be collectively analysed. Moreover, the readout platform can enhance the throughput of data analysis for DMFETs in an array form.

This chapter introduces a new POCT platform for the detection of avian influenza (AI) with a nanogap-embedded separated double-gate FET (DGFET) array (a new version of DMFET, hereafter referred to as 'nanogap-DGFET') and a general-purpose readout platform. The nanogap-DGFET was fabricated using the standard microfabrication technologies used for CMOS transistors. The contents of this chapter include work previously reported by the authors [14].

18.2   NANOGAP-EMBEDDED FIELD-EFFECT TRANSISTOR BIOSENSOR

18.2.1   Nanogap-DGFET

As previously mentioned, several types of nanogap-embedded FETs have been reported, although these devices can be further improved [25,27]. To enhance the mechanical and fluidic properties of such sensors, a double-gate structure was proposed, as shown in Figure 18.1. In the newly designed nanogap-DGFET, nanogaps are created on both sides of a nanowire channel in the transistor.

In the previous nanogap FETs [25,27], the possibility of stiction (permanent collapse of the nanogap caused by the surface tension of the sample liquid) was a major concern that is not easily addressed. Liquid can remain in the nanogaps after the etching of a sacrificial layer or the immobilization of biomolecules, and the nanogaps are prone to collapse because of the surface tension of the entrained liquid during drying.

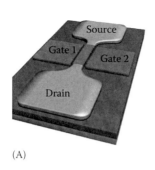
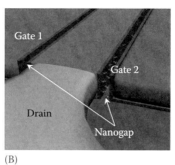

(A)   (B)

FIGURE 18.1 (A) Schematic diagram of a nanogap-DGFET. (B) Magnified view of nanogaps near the drain side. (Reprinted with permission from Im M., Ahn J.-H., Han J.-W., Park T. J., Lee S. Y., and Choi, Y.-K., Development of a point-of-care testing platform with a nanogap-embedded separated double-gate field effect transistor array and its readout system for detection of Avian influenza, *IEEE Sensors J.*, 11, 351–360. Copyright 2011 IEEE.)

Despite the surface tension of the liquid, the nanogap-DGFET is not vulnerable to stiction problems because the forces on both sides of the nanogap are balanced.

The vertical nanogaps in the nanogap-DGFET have an advantage over planar nanogaps because they allow the verification of the introduction of biomolecules into the nanogap. As described in the following, gold nanoparticles were used to visualize the introduction of a sample liquid and biomolecules into the nanogaps without destroying the device.

The use of nanogap-DGFETs has several advantages over conventional FET-based biosensor devices [11–13]. Generally, threshold voltage changes, which constitute a representative sensing metric that defines a sensing window in FET-based biosensors, are in the range of tens of millivolts for the detection of biomolecules. However, nanogap FETs, including nanogap-DGFETs, show larger threshold voltage changes during the detection of biomolecules, which is a desirable characteristic in POCT devices [25,27]. It is also worth noting that the nanogap-DGFET is not affected by the Debye length [23], which is difficult to control, especially for a layperson, in samples such as blood serum, urine or saliva, which are among the primary analytes in POCT applications. However, nanogap-DGFET devices are insensitive to the Debye length because the measurement occurs in a dry state, as explained in the next section. This Debye-screening-free feature arises from the working principle of the nanogap-DGFET, which is based on permittivity changes.

Recently, Ahn et al. reported the enhancement of detection sensitivity using independent bias control in a double-gate structure [29]. This control method is another advantage of the nanogap-DGFET because it can be used to improve a sensor's limit of detection.

18.2.2 Fabrication of a Nanogap-DGFET Array

To avoid cross-contamination issues in POCT applications, the nanogap-DGFET was fabricated in the form of a disposable sensor cartridge, as in previous studies [30–32]. Additionally, the sensor cartridge was designed with 36 nanogap-DGFETs to collect

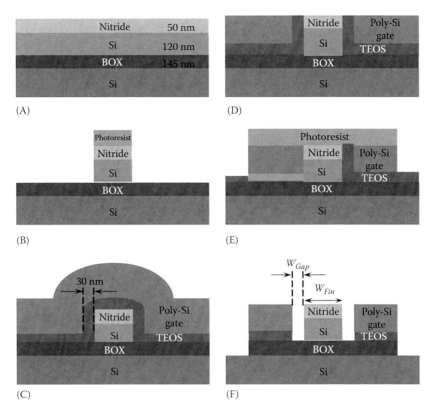

FIGURE 18.2 **(See colour insert.)** Fabrication processes for the nanogap-DGFET: (A) silicon nitride deposition (50 nm) on SOI wafer, (B) active layer lithography, photoresist trimming and dry etching, (C) TEOS oxide (30 nm) and n+ poly-Si (300 nm) deposition, (D) CMP with a stopping layer of silicon nitride, (E) gate poly-Si patterning and source/drain implantation and (F) nanogap formation by TEOS oxide removal. (Reprinted with permission from Im M., Ahn J.-H., Han J.-W., Park T. J., Lee S. Y., and Choi Y.-K., Development of a point-of-care testing platform with a nanogap-embedded separated double-gate field effect transistor array and its readout system for detection of Avian influenza, *IEEE Sensors J.*, 11, 351–360. Copyright 2011 IEEE.)

a statistically meaningful set of data. The readout platform for the nanogap-DGFET array is described later in this chapter. This section describes the fabrication of the nanogap-DGFET, which is fully compatible with conventional CMOS technology.

Figure 18.2 schematically illustrates the entire fabrication process, starting with a boron-doped (9–18 Ω·cm) 8 in. <100> *p*-type silicon-on-insulator (SOI) wafer. The wafer initially had a 120 nm thick silicon active layer on a 145 nm thick buried oxide layer. As shown in Figure 18.2, the thickness of the silicon active layer determines the depth (or height) of the final nanogaps. A nitride layer 50 nm thick was first deposited using low-pressure chemical vapour deposition (LPCVD). This silicon nitride layer acts as an etching stopper in the subsequent chemical–mechanical polishing (CMP) of the polycrystalline silicon (poly-Si) layer. For the source/drain and nanowire channels in the nanogap-DGFET, a photoresist was patterned by KrF optical lithography (at a wavelength of 193 nm). The patterned photoresist was further trimmed in a partial ashing process to achieve a pattern finer than the drawn size of the photomask. After the etching of the active silicon layer, the photoresist was completely removed by an additional ashing process. A layer of tetraethylorthosilicate (TEOS) oxide 30 nm thick was subsequently deposited as a sacrificial layer for the nanogaps. TEOS is a desirable sacrificial layer material because it has a good conformal deposition and a fast etch rate in hydrofluoric acid (HF). The width of the nanogaps is determined by the thickness of the deposited TEOS oxide because of its conformal deposition. As a gate electrode material, an *in situ* phosphorous-doped ($>10^{21}$ cm^{-2}) poly-Si layer was sequentially deposited. The CMP process was employed to planarize the protruding poly-Si gate over the nanowire channel. In the CMP step, the TEOS oxide layer was partially exposed to air, as illustrated in Figure 18.2. The poly-Si gate layer was patterned by lithography and dry etching to form the gate electrode in the designated locations. The source/drain region was doped by arsenic implantation (5×10^{15} cm^{-2}, 30 keV) using the pre-existing photoresist patterns for the poly-Si gate. After the removal of the photoresist, the dopants were activated by a rapid thermal annealing at 1,000°C for 10 s.

Thirty-six nanogap-DGFET devices with the same dimensions as a nanowire (fin), with a width (W_{Fin}) and a gate length (L_G) as depicted in Figure 18.3, were deployed in a 6×6 array on a single die with dimensions of 3×3 mm^2. The devices were simultaneously measured with a custom-designed readout circuitry, as described in the following section. A 500 nm thick silicon dioxide (SiO$_2$) layer was deposited by plasma-enhanced chemical vapour deposition (PECVD) for electrical isolation between the nanogap-DGFETs and the subsequent aluminium interconnection layer. The PECVD SiO$_2$ layer was patterned by dry etching with a photoresist masking layer to open electrical contact holes between the source/drain/gate and the upper aluminium interconnection lines. The sensing region, with dimensions of 10×10 μm^2,

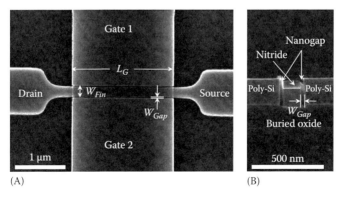

(A)   (B)

FIGURE 18.3 SEM images of the fabricated device: (A) a top view of the nanogap-DGFET and (B) a cross-sectional view of the etched nanogaps. (Reprinted with permission from Im M., Ahn J.-H., Han J.-W., Park T. J., Lee S. Y., and Choi, Y.-K., Development of a point-of-care testing platform with a nanogap-embedded separated double-gate field effect transistor array and its readout system for detection of Avian influenza, *IEEE Sensors J.*, 11, 351–360. Copyright 2011 IEEE.)

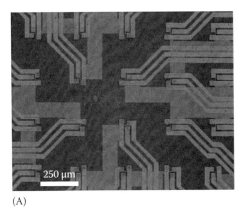

(A)   (B)

FIGURE 18.4   (A) Optical microscope image of the fabricated sensor array after metallization. (B) A photograph of the fabricated nanogap-DGFET array chip. (Reprinted with permission from Im M., Ahn J.-H., Han J.-W., Park T. J., Lee S. Y., and Choi, Y.-K., Development of
a point-of-care testing platform with a nanogap-embedded separated double-gate field effect
transistor array and its readout system for detection of Avian influenza, *IEEE Sensors J.*, 11,
351–360. Copyright 2011 IEEE.)

was also formed on top of the nanowire channel in the nanogap-DGFETs by etching
the PECVD SiO_2 layer with a dilute HF solution (10:1). In the fabrication process for
these nanogap-DGFETs, the uniformity of the nanogap dimensions can be improved
by complete removal of the sacrificial TEOS oxide when a sacrificial layer partially
remains in the planar nanogap device [25].

Figure 18.3A shows a fabricated nanogap-DGFET device after the wet etching of
the TEOS oxide. The device shown here has a W_{Fin} of 190 nm and an L_G of 2 μm. To
verify the complete etching of the TEOS oxide inside the nanogap, a cross-sectional
scanning electron microscope (SEM) image of the etched nanogap was obtained,
as shown in Figure 18.3B. An optical microscope image of the fabricated nanogap-
DGFET array with aluminium interconnection lines is also shown in Figure 18.4A.
After the metallization step, the fabricated wafer was diced while protected by the
positive photoresist on the front side. Figure 18.4B shows a nanogap-DGFET array
chip after the removal of the protective photoresist. To minimize the number of pads,
every nine nanogap-DGFETs have a common poly-Si source, which is indicated in
yellow in Figure 18.4A. For each nanogap-DGFET in the array, three bonding pads
were assigned: two bonding pads for the two gates (i.e. each double gate) and the
other bonding pad for the drain. Consequently, 112 bonding pads were placed in
an area of 75 × 75 μm^2 with a pitch of 100 μm on the array chip, as illustrated in
Figure 18.4B.

18.2.3   VERIFICATION OF NANOGAP WETTING

There is some uncertainty about the flow of a liquid sample into such small nanogaps.
We previously performed numerical calculations and simulations to theoretically
confirm the wetting of the nanogaps [33]. In addition to the 3D simulation, the

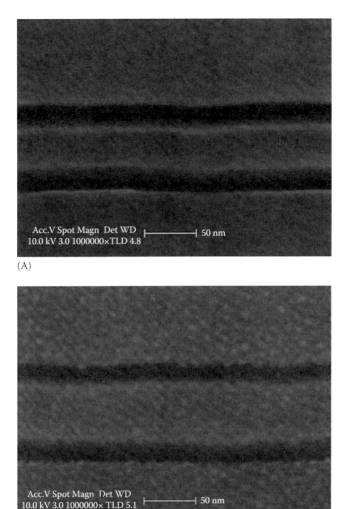

FIGURE 18.5 SEM images of a nanogap-DGFET with an 80 nm wide silicon nanowire channel (A) before and (B) after gold nanoparticle binding.

nanogap wetting was experimentally confirmed with gold nanoparticles immobilized inside the nanogaps, as shown in Figure 18.5. The nanogap-DGFET device was immersed in an H_2SO_4:H_2O_2 (1:1) solution for 10 min to introduce hydroxyl groups (−OH) onto the nanogap surface. A self-assembled monolayer (SAM) was formed on the hydroxylated silicon surface by dipping it in a 1% aminopropyltriethoxysilane (APTES) solution in ethanol for 30 min. After the ethanol rinse, the sample was baked on a hot plate at 120°C for 10 min. Finally, gold nanoparticles in suspension were immobilized on the silicon surface by immersion for 9 h and 30 min. In Figure 18.5, gold nanoparticles with a diameter of 5 nm are visible on the silicon surface and on the sidewalls of the nanogaps. In addition to the 3D simulation [33], these nanoparticles verify the wetting of the nanogaps.

18.2.4 ELECTROSTATIC ANALYSIS OF NANOWIRE CHANNEL

As shown in the SEM images of the fabricated nanogap-DGFET, stiction did not occur during the drying of the etchant after the etching of the sacrificial layer in the fabrication process, which indicates that the stiction caused by capillary forces should be negligible in the repetitive nanogap wetting and drying typically encountered in biological experiments [34]. However, we decided to investigate the possibility of operational stiction arising from electrostatic forces between the gate electrode and the nanowire channel when gate bias voltages are applied to a nanogap-DGFET device based on previous reports of electrostatic stiction and pull-in states in doubly clamped nanotube devices with dimensions similar to our nanogap-DGFETs [35,36]. To analyse the electrostatic behaviour of the nanowire channels in the nanogap-DGFET, 3D simulations were conducted using multiphysics simulation software (CFD-ACE+). In the worst case, the nanowire channel may be suspended, as shown in Figure 18.6, if the etching of the sacrificial TEOS oxide is performed for long enough to etch the buried oxide layer underneath the silicon nanowire channel. Figure 18.7 shows the displacement of the nanowire channel by electrostatic forces in the topside gate at different bias voltages. In these simulations, a single gate was biased with 5, 10, 30 or 50 V, and another gate was grounded. The electrostatic deflection of the nanowire channel was estimated to be less than 1 nm up to a gate bias of 10 V. This result indicates that the nanogap-DGFET devices should not be disrupted by electrostatic forces during operation. Moreover, the nanowire channel should be stable, as it is tightly clamped to the buried oxide layer underneath it, which makes the electrostatic movement of the nanowire channel negligible.

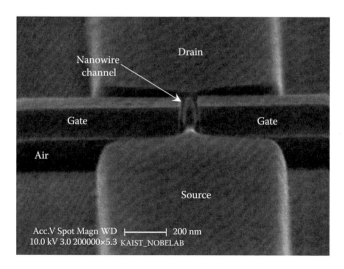

FIGURE 18.6 Suspended nanowire channel after over-etching of a buried oxide layer.

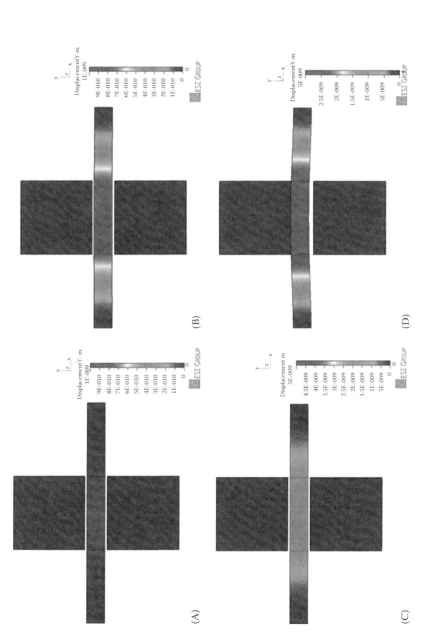

FIGURE 18.7 **(See colour insert.)** Simulation results for electrostatic movements of the nanowire channel: (A) $V_G = 5$ V, (B) $V_G = 10$ V, (C) $V_G = 30$ V, and (D) $V_G = 50$ V.

18.3   IMPLEMENTATION OF THE POINT-OF-CARE TESTING PLATFORM

18.3.1   READOUT CIRCUIT DESIGN FOR THE NANOGAP-DGFET ARRAY

To collect a statistically meaningful number of data points simultaneously, the nanogap-DGFET array was fabricated with 36 devices in a single cartridge. For POCT applications, the nanogap-DGFET array requires supporting electronics for end-user interfacing. For reliable operation of the microelectronics, the readout circuitry was designed on a separate chip so that it could be protected from liquid contact during biomolecular detection processes.

In the nanogap-DGFET array, biomolecules can be detected by monitoring threshold voltage (V_{Th}) changes in the transistors. To analyse a V_{Th} shift before and after the immobilization of biomolecules in the nanogaps, electrical current flowing between the drain and the source (I_{DS}) should be measured as a function of gate bias voltage (V_{GS}). This function defines the transfer characteristics of a transistor, and a V_{Th} change can be further analysed by comparing gate bias voltages that evoke certain amounts of current.

The readout circuitry was designed for the concurrent readout of drain-to-source current signals in the nanogap-DGFET arrays, as represented schematically in Figure 18.8 [14]. The main purpose of this circuit is to convert a current signal into a voltage signal using a current mirror and a charge integration capacitor. The input stage has a negative feedback amplifier to stably maintain the drain–source voltage of the nanogap-DGFET. This stabilization feedback loop is helpful in excluding the channel-length modulation arising from increased drain bias in the nanogap-DGFETs.

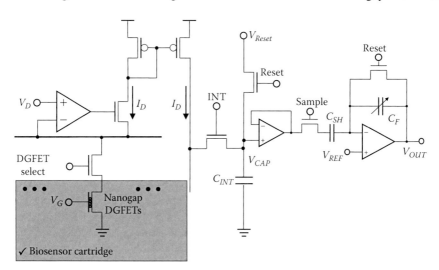

FIGURE 18.8   Schematic representation of the readout circuit for a nanogap-DGFET array. (Reprinted with permission from Im M., Ahn J.-H., Han J.-W., Park T. J., Lee S. Y., and Choi, Y.-K., Development of a point-of-care testing platform with a nanogap-embedded separated double-gate field effect transistor array and its readout system for detection of Avian influenza, *IEEE Sensors J.*, 11, 351–360. Copyright 2011 IEEE.)

While the stable drain bias is maintained, the I_{DS} is duplicated by a current mirror that delivers charge to an integration capacitor (C_{INT}). After a controllable charge integration time, the voltage of the integration capacitor is buffered by a voltage follower and is sampled by a correlated double sampling (CDS) unit, which is a widely used technique for noise reduction [37]. In this readout operation, the amplifier is the most important component in terms of noise susceptibility, and it should be designed to increase the signal-to-noise ratio in the analogue front end and minimize power consumption. A data acquisition card (PCI-6259, National Instruments) with 16-bit analogue-digital-converters (ADCs) was used to collect the final voltage outputs (V_{OUT}) using a program written in LabVIEW (National Instruments).

The LabVIEW program plots the transfer curves (I_{DS}–V_{GS}) from the calculated I_{DS} of the nanogap-DGFET after reading out the V_{OUT} based on the following equation:

$$I_{DS} = \frac{C_{INT} \times (V_{Reset} - V_{Sample})}{T_{INT}} = (V_{OUT} - V_{REF}) \times \frac{C_F}{C_{SH}} \times \frac{C_{INT}}{T_{INT}} \qquad (18.1)$$

where

T_{INT} is the integration time of the current flowing into the integration capacitor (C_{INT})
V_{Reset} is the reset voltage of C_{INT}
V_{Sample} is the sampled voltage of the C_{INT} after the integration time
V_{REF} is the reference voltage of the charge amplifier

The capacitances of the integration capacitor (C_{INT}) and the sampling capacitor (C_{SH}) are both 100 pF. The value of the feedback capacitor (C_F) is chosen in the range of 5–100 pF for different amplifications according to the input current level.

Figure 18.9 shows simulation results [38] with an I_{DS} of 1 μA, a T_{INT} of 10 μs, a C_F of 10 pF, a V_{Reset} of 1 V and a V_{REF} of 2 V. After the integration period, V_{CAP} changes by 100 mV; the corresponding V_{OUT} change is 1.001 V after the sampling. This simulation thus demonstrated the successful operation of the current integration and the CDS unit.

The readout circuitry was fabricated using a Samsung 0.35 μm CMOS process (one poly-Si and four metal layers). The IC Design Education Center (IDEC) at KAIST supported the fabrication with a multiproject wafer (MPW) program. The readout circuitry chip has dimensions of 4.35 × 4.35 mm²; the core, with dimensions of 1.5 × 1.5 mm², is composed of current mirrors in eight columns, current integrators in eight rows, one CDS unit and sensor and capacitor decoders. The sensor decoders enable random access by the nanogap-DGFETs, and the capacitor decoders select the appropriate capacitor size for the charge amplifier. In future studies, more sophisticated function blocks that include ADCs could be monolithically integrated to further miniaturize the system.

18.3.2 Implementation of the Measurement System

Conventionally, the electrical performance of semiconductor devices including transistors is measured with semiconductor parameter analyzers. This measurement requires a tedious and time-consuming process of probing the contact pads.

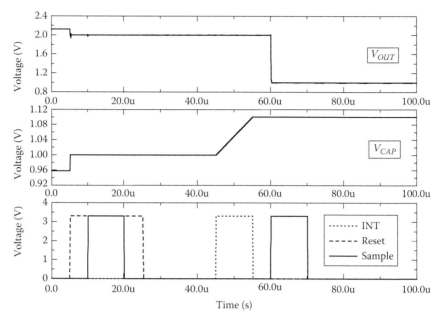

FIGURE 18.9   SPICE simulation results of readout circuitry. (Reprinted with permission from Im M. and Choi, Y.-K. Low noise current readout circuit for point-of-care testing application, in *Korean Conference Semiconductor*, Daejeon, Republic of Korea, pp. 450–451. Copyright 2009 IEEE.)

For the simultaneous measurement of the multiple nanogap-DGFET devices, a dedicated measurement system was implemented with the previously described readout circuitry. The fabricated nanogap-DGFET array was connected to a single printed circuit board (PCB) with gold wire bonding, as shown in Figure 18.10A. Black epoxy was applied to protect the bonding wires from physical damage. Because the nanogap-DGFETs were located at the centre of the chip, we were able to avoid the blocking of the sensing area with epoxy resin, as illustrated in Figure 18.10B. Figure 18.10C shows a flat, flexible cable that was used for the electrical connection between the PCB for the nanogap-DGFET array and the PCB for the readout circuitry.

Figure 18.11 shows the entire platform, which includes a nanogap-DGFET array chip on a PCB. A block diagram showing the signal paths is presented in Figure 18.12. In addition to a readout circuitry chip, controlling and biasing blocks were placed on a main PCB. To generate synchronized switching signals and decoder input signals, a Verilog program was coded into an electro-programmable logic device (EPM7128AETC100-7, Altera). A 16-bit ADC of a National Instruments data acquisition card (PCI-6259) was used to acquire the output voltage signal (V_{OUT}) from the readout circuitry in digital form. These functions for the control signals and ADC can be integrated into the readout circuit chip in future implementations to generate a more compact system. The LabVIEW program was used to coordinate the functions of the Altera chip and the data acquisition card. For pseudo-real-time monitoring,

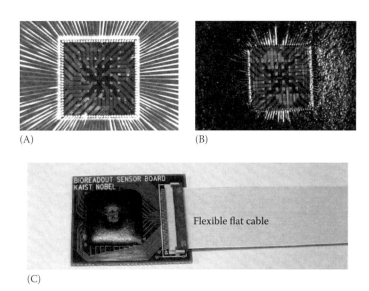

(A) (B)

(C)

FIGURE 18.10 Photographs of the fabricated device: (A) after wire bonding, (B) after epoxy moulding of the bonding wires and (C) after mounting on a PCB as a biosensor cartridge. (Reprinted with permission from Im M., Ahn J.-H., Han J.-W., Park T. J., Lee S. Y., and Choi, Y.-K., Development of a point-of-care testing platform with a nanogap-embedded separated double-gate field effect transistor array and its readout system for detection of Avian influenza, *IEEE Sensors J.*, 11, 351–360. Copyright 2011 IEEE.)

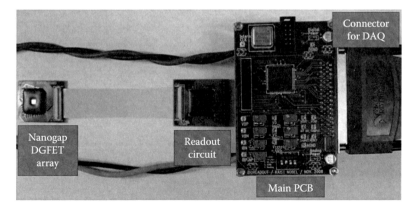

FIGURE 18.11 Photograph of the whole measurement setup with three pieces of PCB. (Reprinted with permission from Im M., Ahn J.-H., Han J.-W., Park T. J., Lee S. Y., and Choi, Y.-K., Development of a point-of-care testing platform with a nanogap-embedded separated double-gate field effect transistor array and its readout system for detection of Avian influenza, *IEEE Sensors J.*, 11, 351–360. Copyright 2011 IEEE.)

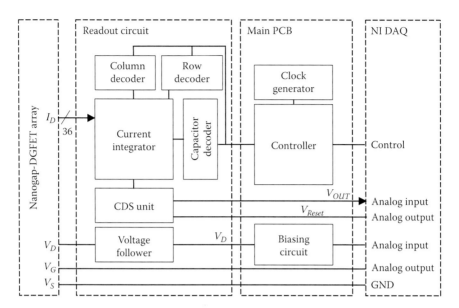

FIGURE 18.12 Block diagram of the implemented system for the readout of a nanogap-DGFET array. (Reprinted with permission from Im M., Ahn J.-H., Han J.-W., Park T. J., Lee S. Y., and Choi Y.-K., Development of a point-of-care testing platform with a nanogap-embedded separated double-gate field effect transistor array and its readout system for detection of Avian influenza, *IEEE Sensors J.*, 11, 351–360. Copyright 2011 IEEE.)

the threshold voltage (V_{Th}) changes of the nanogap-DGFET array are plotted and compared when the user presses a button in the LabVIEW program.

18.4   DETECTION OF AVIAN INFLUENZA

18.4.1   Detection Principle

The name of the first nanogap-embedded transistor biosensor, dielectric-modulated FET (DMFET), simply indicates the detection principle for biomolecules. Im et al. first reported the use of the DMFET and experimental results for streptavidin–biotin detection [25]. After this pioneering study, various nanogap-embedded transistor structures have been studied and used for the detection of diverse biomolecules such as prostate cancer markers [39], AI viruses [27], C-reactive protein [40] and DNA [41]. In all of the previous studies, the biomolecular detection principle is the change in threshold voltage (V_{Th}) that originates from the altered dielectric constant in the nanogap because of the presence of the biomolecules [25], assuming that any charge effects from the biomolecules are negligible. When the gap between the semiconducting channel and the gate electrode is filled with biomolecules, the drain-to-source current (I_{DS}) flowing in the semiconducting channel (i.e. the n-type channel in the fabricated nanogap-DGFET in this work) is modulated by the V_{Th} shift, which mainly arises from the change in the gate capacitance (C_G).

Generally, the V_{Th} of a FET is expressed as follows [42]:

$$V_{Th} = V_{FB} + 2\left|\phi_p\right| + \frac{\left|Q_{dep}\right|}{C_G} \tag{18.2}$$

where

V_{FB} is the flat band voltage
$2|\phi_p|$ is the surface potential
$|Q_{dep}|$ is the depletion layer charge
C_G is the gate capacitance

When the gate dielectric layer is removed by etching the sacrificial layer to form nanogaps, the dielectric constant of the empty nanogap region becomes unity ($\varepsilon_{air} = 1$). If the nanogap is filled with biomolecules with dielectric constants greater than unity ($\varepsilon_{biomolecule} > 1$), the C_G increases because it is proportional to the dielectric constant. Accordingly, the V_{Th} becomes lower as more biomolecules enter the nanogap and are immobilized on the nanogap surfaces. A more detailed analysis of the V_{Th} shift caused by the immobilized materials is presented in Choi et al. [28] and the supporting information of Gu et al. [27], although the structures analysed therein are different from the nanogap-DGFET described in this chapter. If the charge effect arising from the biomolecules cannot be neglected, one type of channel, i.e. either an n-channel or a p-channel in a CMOS, can be preferentially utilized according to its charge polarity. A detailed analysis with a consideration of the charge and dielectric constant effects is provided elsewhere [41]. A noteworthy feature of the double-gate structure is that it can be used to enhance the detection sensitivity by independent bias voltage control on two separate gate electrodes [29]. This ability demonstrates the feasibility of an operational implementation of nanogap-DGFETs for sensitivity improvement.

18.4.2 IMMOBILIZATION OF AVIAN INFLUENZA

To demonstrate the successful operation of the readout platform implemented with the fabricated nanogap-DGFET array, AI antigen (AIa) and antibody were successively detected by tracing changes in threshold voltages. The detection results are discussed in the following section. Here, we address the immobilization method for these biomolecules.

Generally, for biosensor applications, the functionalization of the sensing surface is a major concern. The sensing surface must be properly functionalized to immobilize a target analyte. Suitable functionalization should also prevent nonspecific binding and thereby reduce the possibility of false positives. The following experiments with the AIa and antibody on the fabricated nanogap-DGFET array did not require complicated surface modification processes. With the aid of silica-binding proteins (SBPs) [27], the AIa and antibody were selectively bound inside the nanogap, as illustrated in Figure 18.13. These polypeptides demonstrate high affinity for inorganic materials [43,44]. In other studies, genetically engineered proteins were fused

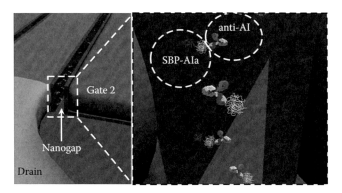

FIGURE 18.13   Schematic diagrams of immobilized SBP-AIa and anti-AI inside a nanogap. (Reprinted with permission from Im M., Ahn J.-H., Han J.-W., Park T. J., Lee S. Y., and Choi Y.-K., Development of a point-of-care testing platform with a nanogap-embedded separated double-gate field effect transistor array and its readout system for detection of Avian influenza, *IEEE Sensors J.*, 11, 351–360. Copyright 2011 IEEE.)

with various disease markers such as those for severe acute respiratory syndrome (SARS) [45], hepatitis B virus [46] and AI [27].

To use SBPs for AI detection, an SBP was fused with AIa on one end; the other end of the SBP has a strong affinity for silica (SiO_2) surfaces. The detailed protein preparation procedures were described in a previous study [27]. When the solution of SBP fused with AIa (SBP-AIa) was applied to the nanogap-DGFET devices, the solution was introduced into the nanogap, as described elsewhere [33]. With a 100 µL droplet of 10 mM phosphate-buffered saline (PBS, pH 7.4) containing 25 µg/mL SBP-AIa, a reaction time of 1 h was used at room temperature. The nanogap-DGFET array was then rinsed with running deionized water and dried with a mild N_2 stream. AI antibody (anti-AI) at a concentration of 10 µg/mL in a 100 µL PBS droplet was selectively bound on the SBP-AIa immobilized in the nanogap because of the specificity between the SBP-AIa and the anti-AI [27]. The additional reaction was allowed to proceed for 1 h at room temperature. Finally, the nanogap-DGFET array was washed with deionized water and dried with an N_2 stream.

18.4.3   Statistical Analysis of Threshold Voltage Shifts in a Nanogap-DGFET Array

For POCT applications, it is essential to avoid the use of bulky instruments for analysis. Thus, rather than using the parameter analyzer that is typically used for FET device analysis, the fabricated nanogap-DGFET array was analysed using the customized readout platform described in the previous section. The successful operation of the readout circuit had been confirmed in testing with MOSFETs prior to the readout of the nanogap-DGFETs [14]. The AIa and antibody had been detected with a nanogap-DGFET array with a fin width (W_{Fin}) of 190 nm, a length (L_G) of 2 µm and a nanogap width (W_{Gap}) of 27 nm. In the nanogap-DGFET array, the drain-to-source (I_{DS}) versus gate-to-source voltage (V_{GS}) graph (the transfer characteristic) was suitably altered, and Figure 18.14 shows the curves of the

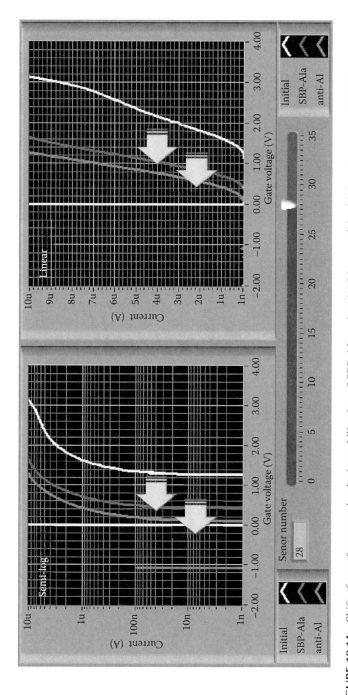

FIGURE 18.14 Shift of transfer curves by the immobilization of SBP-Ala and anti-AI in one of the fabricated nanogap-DGFET arrays. From right to left: transfer curves for the bare nanogap (right-most curve), after immobilization of SBP-Ala (middle curve) and after the immobilization of anti-AI (left-most curve), respectively. (Reprinted with permission from Im M., Ahn J.-H., Han J.-H., Han J.-W., Park T. J., Lee S. Y., and Choi Y.-K., Development of a point-of-care testing platform with a nanogap-embedded separated double-gate field effect transistor array and its readout system for detection of Avian influenza, *IEEE Sensors J.*, 11, 351–360. Copyright 2011 IEEE.)

28th nanogap-DGFET in the 6 × 6 array of devices. The transfer characteristic in the initial state (right-most curve) moved leftwards (middle curve) as the SBP-AIa was immobilized in the nanogap, which represents a negative shift of the threshold voltage (V_{Th}), as expected. After the anti-AI immobilization, the transfer characteristic also moved leftwards (i.e. to a lower V_{Th}) because of the additional binding of biomolecules. Both shifts were caused by the increment of the dielectric constant in the nanogap. The data acquisition and analysis were performed in LabVIEW.

During the measurement, a drain-to-source voltage (V_{DS}) of 1 V was used, and both gates in the nanogap-DGFET were biased together with same voltage (V_{GS}), which was swept from −2 to 4 V in steps of 10 mV. The nanogap-DGFET in Figure 18.14 demonstrated V_{Th} shifts of −0.84 V (from 1.27 to 0.43 V) and −0.32 V (from 0.43 to 0.11 V) with the SBP-AIa and anti-AI, respectively. These reduced V_{Th} changes demonstrate that the SBP-AIa and the anti-AI strengthen the field effect of the applied gate bias because of the increased gate dielectric constant inside the nanogap. The V_{Th} values were extracted by the constant-current method [47] at $I_{DS} = 10$ nA.

Figure 18.15 shows the changes of V_{Th} and I_{DS} collected from the 6 × 6 nanogap-DGFET array. As discussed earlier and shown in Figure 18.15A, the V_{Th} values decreased by 1.25 and 0.35 V, respectively, after the sequential immobilizations of SBP-AIa and anti-AI. Consequently, the I_{DS} values were increased, as shown in Figure 18.15B, because of the reduction in V_{Th} as the biomolecules were bound in the nanogap. The relatively smaller change induced by the anti-AI antibody can be explained by two reasons: the lower anti-AI concentration and a reduced binding affinity because of the previous attachment of biomolecules. We would expect the magnitude of the signal change induced by anti-AI to be improved by the additional optimization of the SBP-AIa concentration and nanogap width.

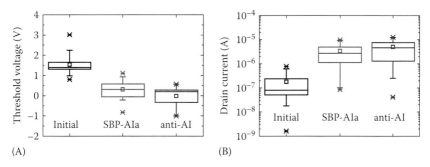

(A)   (B)

FIGURE 18.15   (A) Statistical data set for the initial threshold voltages for the nanogap, after the immobilization of SBP-AIa and after the immobilization of anti-AI. (B) Statistical data set for the drain-to-source currents at $V_{GS} = 1.5$ V in the aforementioned three steps. The average currents are 0.20, 2.95 and 4.16 μA for the bare nanogap, the SBP-AIa-immobilized nanogap and the anti-AI-immobilized nanogap, respectively. (−, max; ×, 99%; □, mean; ×, 1%; −, min; box, 25%–75%). (Reprinted with permission from Im M., Ahn J.-H., Han J.-W., Park T. J., Lee S. Y., and Choi Y.-K., Development of a point-of-care testing platform with a nanogap-embedded separated double-gate field effect transistor array and its readout system for detection of Avian influenza, *IEEE Sensors J.*, 11, 351–360. Copyright 2011 IEEE.)

18.5   CONCLUSIONS AND OUTLOOK

Label-free electrical detection of AIa and antibody was demonstrated with nanogap-DGFETs and a customized readout platform for a POCT application. The readout circuitry enabled high-speed data collection from a large number of nanogap-DGFETs, which were designed as disposable cartridges. The gate dielectric constant was modulated by the AIa and antibody immobilized within the nanogap. Changes in the dielectric constant were reflected in the threshold voltages and the drain currents of the nanogap-DGFETs. These electrical signals were successfully analysed with the aid of the readout platform. The implemented platform is expected to be useful for the label-free detection of various disease markers using genetically engineered SBPs fused with different types of biomarkers. In a nanogap-DGFET array, the multiplexed detection of various biomolecules can be performed by linking various receptors to designated nanogap-DGFETs, thereby expediting the detection of multiple disease markers [48]. Although this system was tested with an AIa and antibody diluted in a phosphate buffer solution, biomarkers will be detected in whole-blood samples in future research.

In this work, the nanogap-DGFET array was fabricated as a separate chip using processes that are fully compatible with conventional semiconductor technology. To statistically analyse the signals from larger numbers of nanogap-DGFETs, the arrays could possibly be monolithically integrated on top of readout circuitry using thin-film transistor (TFT) technologies [49]. The integration of the nanogap-DGFETs with other standardized FETs in the interface circuitry would provide benefits for the readout platform including noise reduction for more accurate detection and simple interconnections for further miniaturization. This research could pave the way for new combinations of commercialized CMOS technology and CMOS-based biosensors to enable mass production at low cost. Samsung has already demonstrated the multiple stacking of single-crystal TFTs for their SRAM products [50] and flash memory [51] as alternatives to pre-existing memory applications. Moreover, a previous study reported that the polysilicon (poly-Si) TFT shows high device performance without a crystallization process when the poly-Si TFT size is comparable to or smaller than the poly-Si grain size [49]. Therefore, with the aid of three-dimensionally stacked integration of poly-Si active layers, an extremely small biosensor, e.g., a dust-mote-sized sensor, may be feasible in the near future.

ACKNOWLEDGEMENTS

This work was partially supported by the National Research Foundation (NRF) of Korea grant funded by the Korean Ministry of Education, Science and Technology (MEST) (No. 2011-0020487). It was also partially supported by the National Research and Development Program (NRDP, 2011-0002182) for the development of biomedical function monitoring biosensors. This work was also sponsored by the Korean MEST. Readout circuit fabrication was partially supported by IDEC. Maesoon Im would like to thank the Brain Korea 21 project, the School of Information Technology, KAIST, for the financial support provided in 2009. The authors would like to thank Dr. Jae-Hyuk Ahn, Dr. Jin-Woo Han, Dr. Tae Jung Park and Prof. Sang Yup Lee for their help in the fabrication and analysis of nanogap-DGFET devices.

REFERENCES

1. Online: Available: http://www.researchandmarkets.com/reports/604967/global_in_vitro_diagnostic_market_analysis
2. S. W. Oh, J. D. Moon, H. J. Lim, S. Y. Park, T. Kim, J. B. Park, M. H. Han, M. Snyder, and E. Y. Choi, Calixarene derivative as a tool for highly sensitive detection and oriented immobilization of proteins in a microarray format through noncovalent molecular interaction, *FASEB Journal,* 19, 1335–1337, 2005.
3. C. C. Y. T. Yang, X. L. Feng, K. L. Ekinci, and M. L. Roukes, Zeptogram-scale nanomechanical mass sensing, *Nano Letters,* 6, 583–586, 2006.
4. J. Fritz, M. K. Baller, H. P. Lang, H. Rothuizen, P. Vettiger, E. Meyer, H.-J. Güntherodt, C. Gerber, and J. K. Gimzewski, Translating biomolecular recognition into nanomechanics, *Science,* 288, 316–318, 2000.
5. T. G. Drummond, M. G. Hill, and J. K. Barton, Electrochemical DNA sensors, *Nature Biotechnology,* 21, 1192–1199, 2003.
6. X. D. Cui, A. Primak, X. Zarate, J. Tomfohr, O. F. Sankey, A. L. Moore, T. A. Moore, D. Gust, G. Harris, and S. M. Lindsay, Reproducible measurement of single-molecule conductivity, *Science,* 294, 571–574, 2001.
7. G. S. McCarty, Monitoring the addition of molecular species to electrodes utilizing inherent electronic properties, *Journal of Applied Physics,* 99, 064701.1–7, 2006.
8. M. A. Reed, C. Zhou, C. J. Muller, T. P. Burgin, and J. M. Tour, Conductance of a molecular junction, *Science,* 278, 252–254, 1997.
9. F. Patolsky, B. P. Timko, G. Zheng, and C. M. Lieber, Nanowire-based nanoelectronic devices in the life sciences, *MRS Bulletin,* 32, 142–149, 2007.
10. J. L. Arlett, E. B. Myers, and M. L. Roukes, Comparative advantages of mechanical biosensors, *Nature Nanotechnology,* 6, 203–215, 2011.
11. M. J. Schöning and A. Poghossian, Recent advances in biologically sensitive field-effect transistors (BioFETs), *Analyst,* 127, 1137–1151, 2002.
12. D.-S. Kim, J.-E. Park, J.-K. Shin, P. K. Kim, G. Lim, and S. Shoji, An extended gate FET-based biosensor integrated with a Si microfluidic channel for detection of protein complexes, *Sensors and Actuators B: Chemical,* 117, 488–494, 2006.
13. T. Sakata and Y. Miyahara, Direct transduction of allele-specific primer extension into electrical signal using genetic field effect transistor, *Biosensors and Bioelectronics,* 22, 1311–1316, 2007.
14. M. Im, J.-H. Ahn, J.-W. Han, T. J. Park, S. Y. Lee, and Y.-K. Choi, Development of a point-of-care testing platform with a nanogap-embedded separated double-gate field effect transistor array and its readout system for detection of Avian influenza, *IEEE Sensors Journal,* 11, 351–360, 2011.
15. P. Bergveld, Thirty years of ISFETOLOGY What happened in the past 30 years and what may happen in the next 30 years, *Sensors and Actuators B: Chemical,* 88, 1–20, 2003.
16. E. Stern, J. F. Klemic, D. A. Routenberg, P. N. Wyrembak, D. B. Turner-Evans, A. D. Hamilton, D. A. LaVan, T. M. Fahmy, and M. A. Reed, Label-free immunodetection with CMOS-compatible semiconducting nanowires, *Nature,* 445, 519–522, 2007.
17. E. Stern, A. Vacic, N. K. Rajan, J. M. Criscione, J. Park, B. R. Ilic, D. J. Mooney, M. A. Reed, and T. M. Fahmy, Label-free biomarker detection form whole blood, *Nature Nanotechnology,* 5, 138–142, 2009.
18. E. Stern, A. Vacic, and M. A. Reed, Semiconducting nanowire field-effect transistor biomolecular sensors, *IEEE Transactions on Electron Devices,* 55, 3119–3130, 2008.
19. K.-I. Chen, B.-R. Li, and Y.-T. Chen, Silicon nanowire field-effect transistor-based biosensors for biomedical diagnosis and cellular recording investigation, *Nano Today,* 6, 131–154, 2011.

20. J.-C. Chou and C.-N. Hsiao, Drift behavior of ISFETs with a-Si:H-SiO$_2$ gate insulator, *Materials Chemistry and Physics,* 63, 270–273, 2000.
21. C. G. Jakobson, M. Feinsod, and Y. Nemirovsky, Low frequency noise and drift in ion sensitive field effect transistors, *Sensors and Actuators B: Chemical,* 68, 134–139, 2000.
22. J. C. Chou and C. N. Hsiao, The hysteresis and drift effect of hydrogenated amorphous silicon for ISFET, *Sensors and Actuators B: Chemical,* 66, 181–183, 2000.
23. W. M. Siu and R. S. C. Cobbold, Basic properties of the electrolyte-SiO$_2$-Si system: Physical and theoretical aspects, *IEEE Transactions on Electron Devices,* ED-26, 1805–1815, 1979.
24. E. Stern, R. Wagner, F. J. Sigworth, R. Breaker, T. M. Fahmy, and M. A. Reed, Importance of the debye screening length on nanowire field effect transistor sensors, *Nano Letters,* 7, 3405–3409, 2007.
25. H. Im, X.-J. Huang, B. Gu, and Y.-K. Choi, A dielectric-modulated field-effect transistor for biosensing, *Nature Nanotechnology,* 2, 430–434, 2007.
26. M. J. Tierney, J. A. Tamada, R. O. Potts, R. C. Eastman, K. Pitzer, N. R. Ackerman, and S. J. Fermi, The GlucoWatch biographer: A frequent automatic and noninvasive glucose monitor, *Annals of Medicine,* 32, 632–641, 2000.
27. B. Gu, T. J. Park, J.-H. Ahn, X.-J. Huang, S. Y. Lee, and Y.-K. Choi, Nanogap field-effect transistor biosensors for electrical detection of Avian influenza, S*mall,* DOI: 10.1002/smll.200900450, 2009.
28. J.-M. Choi, J.-W. Han, S.-J. Choi, and Y.-K. Choi, Analytical modeling of a nanogap-embedded FET for application as a biosensor, *IEEE Transactions on Electron Devices,* 57, 3477–3484, 2010.
29. J.-H. Ahn, S.-J. Choi, J.-W. Han, T. J. Park, S. Y. Lee, and Y.-K. Choi, Double-gate nanowire field effect transistor for a biosensor, *Nano Letters,* 10, 2934–2938, 2010.
30. I. R. Lauks, Microfabricated biosensors and microanalytical systems for blood analysis, *Accounts of Chemical Research,* 31, 317–324, 1998.
31. J. W. Choi, A. Puntambekar, C. C. Hong, C. Gao, X. Zhu, R. Trichur, J. Han, S. Chilukuru, M. Dutta, and S. Murugesan, A disposable plastic biochip cartridge with on-chip power sources for blood analysis, in *Proceedings of IEEE Conference Micro Electro Mechanical System,* Kyoto, Japan, 2003, pp. 447–450.
32. B. D. DeBusschere and G. T. A. Kovacs, Portable cell-based biosensor system using integrated CMOS cell-cartridges, *Biosensors Bioelectrons,* 16, 543–556, September 2001.
33. Maesoon. Im and Y.-K. Choi, Numerical analysis and simulation of fluidics in nanogap embedded separated double-gate field effect transistor for biosensor, in *New Perspectives in Biosensors Technology and Applications,* P. A. Serra, Ed., Intech, ISBN: 978-953-307-448-1, 2011, pp. 229–244, Available: http://www.intechopen.com/books/new-perspectives-in-biosensors-technology-and-applications/numericalanalysis-and-simulation-of-fluidics-in-nanogap-embedded-separated-double-gate-field-effect.
34. N. Tas, T. Sonnenberg, H. Jansen, R. Legtenberg, and M. Elwenspoek, Stiction in surface micromachining, *Journal of Micromechanics and Microengineering,* 6, 385–397, 1996.
35. N. Pugno, C. H. Ke, and H. D. Espinosa, Analysis of doubly clamped nanotube devices in the finite deformation regime, *Journal of Applied Mechanics,* 72, 445–449, 2005.
36. G. R. H. Sadeghian and E. Malekpour, Pull-in phenomenon investigation of nanoelectromechanical systems, *Journal of Physics: Conference Series,* 34, 1123–1126, 2006.
37. C. C. Enz and G. C. Temes, Circuit techniques for reducing the effects of op-amp imperfections: Autozeroing, correlated double sampling, and chopper stabilization, *Proceedings of the IEEE,* 84, 1584–1614, 1996.
38. M. Im and Y.-K. Choi, Low noise current readout circuit for point-of-care testing application, in *Korean Conference Semiconductor,* Daejeon, Republic of Korea, 2009, pp. 450–451.

39. J.-H. Ahn, M. Im, and Y.-K. Choi, Label-free electrical detection of PSA by a nanogap field effect transistor, in *Twelfth International Conference on Miniaturized Systems for Chemistry and Life Sciences*, San Diego, CA, 2008, pp. 979–981.

40. J.-H. Ahn, J.-Y. Kim, M. Im, J.-W. Han, and Y.-K. Choi, A nanogap-embedded nanowire field effect transistor for sensor applications: Immunosensor and humidity sensor, in *14th International Conference on Miniaturized Systems for Chemistry and Life Sciences*, Groningen, the Netherlands, 2010, pp. 1301–1303.

41. C.-H. Kim, C. Jung, K.-B. Lee, H. G. Park, and Y.-k. Choi, Label-free DNA detection with a nanogap embedded complementary metal oxide semiconductor, *Nanotechnology*, 22, 135502, 2011.

42. R. S. Muller and T. I. Kamins, *Device Electronics for Integrated Circuits*, 3rd edn. New York: Wiley, 2003.

43. T. J. Park, S. Y. Lee, S. J. Lee, J. P. Park, K. S. Yang, K.-B. Lee, S. Ko et al., Protein nanopatterns and biosensors using gold binding polypeptide as a fusion partner, *Analytical Chemistry*, 78, 7197–7205, 2006.

44. K. Taniguchi, K. Nomura, Y. Hata, T. Nishimura, Y. Asami, and A. Kuroda, The Si-tag for immobilizing proteins on a silica surface, *Biotechnology and Bioengineering*, 96, 1023–1029, April 15, 2007.

45. T. J. Park, M. S. Hyun, H. J. Lee, S. Y. Lee, and S. Ko, A self-assembled fusion protein-based surface plasmon resonance biosensor for rapid diagnosis of severe acute respiratory syndrome, *Talanta*, 79, 295–301, 2009.

46. S. Zheng, D.-K. Kim, T. J. Park, S. J. Lee, and S. Y. Lee, Label-free optical diagnosis of hepatitis B virus with genetically engineered fusion proteins, *Talanta*, 82, 803–809, 2010.

47. X. Zhou, K. Y. Lim, and W. Qian, Threshold voltage definition and extraction for deep-submicron MOSFETs, *Solid State Electronics*, 45, 507–510, March 2001.

48. G. Zheng, F. Patolsky, Y. Cui, W. U. Wang, and C. M. Lieber, Multiplexed electrical detection of cancer markers with nanowire sensor arrays, *Nature Biotechnology*, 23, 1294–1301, October 2005.

49. M. Im, J.-W. Han, H. Lee, L.-E. Yu, S. Kim, C.-H. Kim, S. C. Jeon et al., Multiple-gate CMOS thin-film transistor with polysilicon nanowire, *IEEE Electron Device Letters*, 29, 102–105, January 2008.

50. S.-M. Jung, H. Lim, W. Cho, H. Cho, C. Yeo, Y. Kang, D. B. J. Na et al., Highly area efficient and cost effective double stacked S^3 (stacked single-crystal Si) peripheral CMOS SSTFT and SRAM cell technology for 512 M bit density SRAM, San Francisco, CA, 2004, pp. 265–268.

51. S.-M. Jung, J. Jang, W. Cho, H. Cho, J. Jeong, Y. Chang, J. Kim et al., Three dimensionally stacked NAND flash memory technology using stacking single crystal Si layers on ILD and TANOS structure for beyond 30 nm node, in *IEDM*, San Francisco, CA, 2006, pp. 37–40.

19 Bionanotechnological Advances in Neural Recording and Stimulation

Alper Bozkurt and George C. McConnell

CONTENTS

19.1 INTRODUCTION

Our understanding of how neurons code and communicate information and how they function in networks is bound by the limitations of the physical connections that we make with the neural tissue. The current set of tools and techniques commercially available to interface with neural tissue do not efficiently sense or induce neural activity and suffer from poor accessibility to most of the neural system. Bionanotechnological tools hold the promise to improve these interfaces beyond what is commercially available. This chapter surveys and highlights some of the many achievements in this field along with future challenges that must be faced in order realize the full potential of bionanotechnological approaches in neural recording and stimulation.

Most of the developments in neural stimulation and recording have focused on the application of the latest advances in microfabrication of silicon and other electronic materials to the classical intra- and extracellular electrophysiology techniques (Najafi 1997, Banks 1998, Stieglitz and Meyer 1998, Wise 2005). Here, we focus on the methods benefiting from the latest developments in nanoscale technologies and tools such as nanoparticles, nanowires, carbon nanotubes (CNT), quantum dots and silicon nanoscale transistors. Since these materials are on the scale of subcellular components such as ion channels, lipid bilayer membrane and signalling proteins, direct and selective control over these electrophysiologically relevant, neuroanatomical parts is possible by design of the cell-to-synthetic material interface. The magnetic, optical and electrical properties of these materials also provide a new playground to improve our understanding of how neurons function, based on their recorded activity, and how we can manipulate them, by stimulation. Engineering better solutions for neural recording and stimulation each have their own unique challenges. Recording of neural activity has been done to understand how neurons, individually and in networks, function both normally and pathologically. Therefore, neural recording aims to unlock the mysteries of the native neural system without altering its function. The stimulation of neural tissue with embedded technologies, on the other hand, provides opportunities to modulate the firing rates and/or patterns of neurons and thereby treat several neural disorders and pathologies. For each of the nanotechnologies discussed in this chapter, we focused more on either the stimulation side or the recording side of the neural interface based on the emphasis of the literature. Any discussion on the application of nano-engineered materials must include potential toxicity concerns; therefore, relevant toxicity studies are presented.

19.2 CARBON NANOTUBES

CNT, sheets of graphene rolled into a tubular structure, have unique chemical, mechanical, electrical and thermal properties that have generated high interest for use in biomedical applications. The exceptional strength and high conductivity of CNT, in addition to a chemical structure that can be modified using functional groups, make them well suited for neural engineering applications. In this section, emphasis will be placed on progress in the application of CNT for use in neural recording and stimulating.

CNT are classified as single-walled (SWCNT), double-walled (DWCNT) or multiwalled (MWCNT). In the case of SWCNT, with a single graphene sheet, the thickness of the diameter of a single tube is ~1 nm and the length of a tube on the other hand can range 12 orders of magnitude larger. DWCNT and MWCNT contain two or more concentric nanotubes (NTs), respectively.

Widespread clinical use of implanted electrodes for neural prosthetics will require a predictably high percentage of electrodes with good-quality neural signals for several years after the implant. Following the initial implantation of electrodes, a breakdown of the blood–brain barrier occurs and an inflammatory response ensues, which are both persistent and neurodegenerative (Biran 2005, McConnell et al. 2009). One approach to reduce the initial injury and persistent inflammatory response is to reduce the dimensions of the recording electrode. The inflammatory response to 50 μm diameter stainless steel microwires is less than that for 200 μm (Thelin et al. 2011). By extension, the ideal chronic electrode would be as small as possible while maintaining the structural integrity necessary to penetrate the brain tissue with an electrical impedance low enough to achieve a high signal-to-noise (SNR) ratio for single-unit neural recording. Initial work in this area suggests that carbon may be more suitable than noble metals for chronic electrodes. The electrical impedance properties of reducing carbon nanofibre (CNF) electrodes into the nanometre range suggest that CNF is better suited for biosensing applications than noble metals at the nanoscale (Siddiqui et al. 2010).

Instead of replacing noble metal electrodes entirely with CNTs, another approach to improve the long-term performance of chronic neural recordings is to coat noble metals with CNTs in order to decrease electrical impedance and increase charge transfer. Platinum (Pt) electrodes electrochemically co-deposited with polypyrrole (PPy) and SWCNTs displayed both of these characteristics. In addition, increased neuronal density and decreased astrogliosis were also observed in the vicinity of the electrodes (Lu et al. 2010). CNT coatings are also being investigated as release vehicles of anti-inflammatory drug. Luo et al. developed a system of drug release, which can be controlled by electrical stimulation (Luo et al. 2011). The advantage of this strategy is that the drug can be stored until a time when it might most effectively reduce the inflammatory response. The system uses the inner cavity of MWCNTs as nanoreservoirs to load the anti-inflammatory drug dexamethasone, with the open ends of the MWCNTs sealed with PPy films.

In addition to the use of CNT-coated electrodes for enhancement of neural recordings, CNT-coated electrodes have shown enhancement of neural stimulation in acute experiments (Keefer et al. 2008). This is due to the decrease in electrode impedance and increase in charge transfer, with more than double the maximum charge density of CNT-coated electrodes compared to iridium oxide electrodes of similar size (Phely-Bobin 2006).

One area ripe for application of nanotechnology is deep brain stimulation (DBS). DBS is an effective neurosurgical treatment that uses electrical stimulation of the brain to treat the motor symptoms of Parkinson's disease, essential tremor and dystonia. Technical improvements such as a reduction in electrode size (~1 mm diameter) and the ability for closed-loop stimulation using feedback from neuronal activity and dopamine levels may further advance this technology (Li and Andrews 2007). NASA Ames is developing a CNF array to address these limitations capable of precise, multisite electrical stimulation and recording along with monitoring of dopamine levels in the basal ganglia (Li and Andrews 2007). In addition to improvements to the electrode design, nanotechnology may 1 day lead to miniaturization of other hardware including the cable connections between the stimulation leads and implantable pulse generator (Albanese and Romito 2011).

Biocompatibility is of paramount importance for neural interfaces. The toxicity of NTs is linked to the specific way in which they interface to the neural tissue. Direct dispersion of NTs in liquids was found to have toxic effects on cells (Shvedova 2003, Shevdova 2011, Lam et al. 2004, Shi Kam 2004,

Warheit et al. 2004, Cui et al. 2005, Sato et al. 2005), while solid phase coatings were found to be well tolerated (Mattson et al. 2000, Gheith et al. 2005, Lovat et al. 2005, Wang et al. 2006). *In vitro* studies show that a CNT substrate does not interfere with neuron cell adhesion, neurite extension or synapse formation (Gheith et al. 2005, Malarkey and Parpura 2007). Increased electrical connectivity on CNT mats has also been observed, possibly caused by increased conductivity between neurons (Lovat et al. 2005). However, only a narrow range of conductivity promotes neurite outgrowth (Malarkey et al. 2009), which could be explained by an inhibitory effect of CNT on endocytosis (Malarkey et al. 2008). Other studies have shown that CNT substrates alter the growth cone of the neurites in terms of its branching and density (Webster et al. 2004, Hu et al. 2005, Liopo et al. 2006) and increase the firing rate of neurons (Lovat et al. 2005), suggesting that CNT alter normal function. While these *in vitro* results look promising, future *in vivo* studies assessing toxicity are necessary.

19.3   CONDUCTIVE POLYMER NANOTUBES AND NANOFIBRES

In addition to CNT coatings, conductive polymer (CP) NTs and nanofibres (NFs) have been used to coat microelectrodes in order to lower electrical impedance and improve charge transfer. One approach first spins dexamethasone-loaded biodegradable CPNF onto an uncoated microelectrode (Abidian et al. 2009). An alginate hydrogel coating is then added and poly(3,4-ethylenedioxythiophene) (PEDOT) is electrochemically polymerized on the electrode sites.

PPy and PEDOT are two types of CP that have been extensively studied for coating neural electrodes to improve the performance of long-term neural recordings. CPNTs possess clear advantages over CP films electrically, structurally and in its interaction with tissue. CPNTs of PPy and PEDOT have lower impedance and higher charge-capacity density (Abidian et al. 2010), as well as a larger apparent diffusion coefficient than the corresponding film (Abidian and Martin 2008). These electrical advantages are thought to be due to the high effective surface area filled with supporting electrolyte due to the open nanotubular structure, which facilitates ion diffusion and exchange with the electrolyte. Structurally, PPy and PEDOT NT have stronger adherence to silicon dioxide surfaces compared to their film counterparts (Abidian et al. 2010). Biologically, dorsal root ganglia cultured on PPy and PEDOT NT have longer neurites compared to PPy and PEDOT film (Abidian et al. 2010). While these *in vitro* results look promising, functional studies *in vivo* have revealed that strategies in addition to improving the electrical properties of the electrode are necessary. Chronic recordings from PEDOT NT-coated microelectrodes in rats showed an initial increase in the average number of quality units, most likely due to a decreased noise floor (Abidian and Martin 2009). However, a decrease in SNR was observed over time in both coated and uncoated electrodes, similar to that reported using PEDOT film coating (Ludwig et al. 2006). One such approach is to use CPNTs for stimulation-controlled drug release of anti-inflammatory drug. In this way, anti-inflammatory drug could be delivered at crucial time points during the inflammatory response when the drug's effect might be most effective (Abidian et al. 2006).

19.4   POLYMER NANOFIBRES AND ULTRATHIN FILMS

Bionanotechnologically structured scaffolds enabling the tailoring of dimension, topography and biochemical processes are being explored to enhance cell survival and nerve regeneration.

Restoration of motor control to amputees requires a long-term peripheral nerve interface to bridge the gap between the remaining nerve and external electronics (Dhillon 2004). Regeneration across long gaps requires formation of a fibrin cable for Schwann cells to cross (Belkas et al. 2005). One strategy to promote Schwann cell migration, and thereby direct axonal repair, is to provide topographical cues to stimulate endogenous nerve repair mechanisms (Kim et al. 2008a, Clements 2009). Aligned NFs promoted Schwann cell migration coincident with regeneration of axons across long gaps, including formation of neuromuscular junctions and functional recovery. Random NFs, on the other hand, failed to induce regeneration (Figure 19.1) (Kim et al. 2008b).

He et al. reported a novel application of layer-by-layer assembly to build nanoscale coatings of laminin on silicon electrodes. *In vitro*, the coated electrodes showed significantly enhanced adhesion and differentiation of chick cortical neurons compared to bare silicon dioxide controls (Figure 19.2) (He and Bellamkonda 2005). Further *in vivo* testing showed a decreased astrocytic and reactive microglia response 4 weeks post-implant (He et al. 2006).

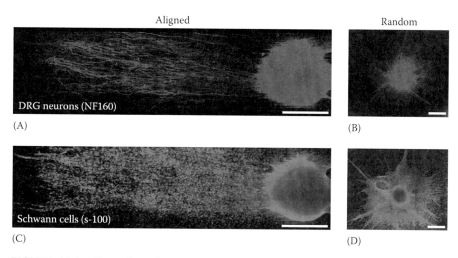

FIGURE 19.1   (See colour insert.) Aligned NFs direct axonal growth of dorsal root ganglion (DRG) neurons in the orientation of fibre film *in vitro*. (A) Aligned NF film showing the preferred direction of axonal outgrowth in the direction of the NFs. (B) Random NF film showed no preferred direction in axonal growth. (C,D) Schwann cell migration was similarly preferentially directed by (C) aligned NFs but not (D) random NFs. NF films were double immunostained for NF160, a marker for axons, and S-100, a marker for Schwann cells. All scale bars = 500 μm. (Adapted from *Biomaterials*, 29, Kim, Y.-T., Haftel, V.K. Kumar, S. Bellamkonda, R.V., The role of aligned polymer fiber-based constructs in the bridging of long peripheral nerve gaps, 3117–3127, Copyright 2008a, with permission from Elsevier.)

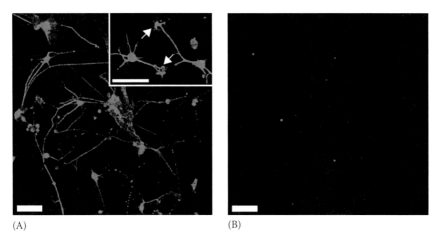

(A)   (B)

FIGURE 19.2 **(See colour insert.)** Nanofilm laminin coating promotes neuronal adhesion and outgrowth *in vitro.* (A) Chick cortical cultures growing on a silicon dioxide surface with a laminin coating (~11 nm thick) consisting of alternating bilayers of laminin and polyethyleneimine. Inset: Higher magnification with arrows showing growth cones. (B) Bare silicon dioxide control was not conducive to neuronal attachment. Cells were visualized with anti-α-tubulin immunohistochemistry. All scale bars = 50 μm. (Adapted from *Biomaterials, 26,* He, W. and Bellamkonda, R. V., Nanoscale neuro-integrative coatings for neural implants, 2983–2990, Copyright 2005, with permission from Elsevier.)

19.5 FUNCTIONALIZED NANOPARTICLES

Organic polymer-based nanoparticles have been used to deliver drugs to support neurite outgrowth or modulate inflammatory response, thanks to their higher surface-to-volume ratio (Eavarone et al. 2000, Saul et al. 2003, 2006, Kim and Martin 2006). Their optical and magnetic properties have let them perform the task of a contrast agent to image brain activity or neural pathologies given their ability to penetrate the blood–brain barrier (Kreuter 2001, West and Halas 2003, Xie et al. 2005, Ruan et al. 2006, Warnement et al. 2007). The surface modification of inorganic nanoscale particles, in particular, has provided new opportunities to monitor neural activity and has recently been utilized to excite the neural tissue. A wide range of inorganic materials have been used to synthesize magnetic, metallic and semiconductor nanoparticles for neural applications (Na et al. 2007, Kim et al. 2008a, Jun et al. 2008). Magnetic nanoparticles are typically composed for oxides of irons and alloys of iron and gold (Na et al. 2007, Liu et al. 2008). The typical application of these nanoparticles in neurology has been targeted for heat generation in the presence of AC fields to induce hyperthermia and kill brain tumours (Jordan and Maier-Hauff 2007, Jun et al. 2008).

The semiconductor nanoparticles, on the other hand, can be used in the form of quantum dots and rods with very broad absorption and narrow excitation spectra. Optically, excitation of nanoparticles causes a dipole moment due to the electron trapping at the nanoparticle surface where the charge separation occurs between the electron and hole in the exciton (Jackson 1999). When integrated to the cell membrane, optical excitation of nanoparticles exhibits a temporary electrical dipole moment with theoretically sufficient strength to deform the protein structure of the neighbouring voltage-gated ion channels and to let sodium ions in as a result of electrostatic interactions (Atkins 1994, Malvindi et al. 2008). The dipole moment may also block the ion transport through the ion channels in the vicinity, and this phenomenon was tested with calcium sulfide (CdS) nanoparticles (Winter et al. 2001). The distance needs to be smaller than 2 nm to be able to initiate an action potential or to silence it (Winter et al. 2005).

Optical excitation of asymmetric core/shell cadmium selenide (CdSe)/CdS nano-rods (Carbone et al. 2007) embedded into the tentacle neurons of freshwater coelen-terate *Hydra vulgaris* caused a tentacle writhing behaviour at a whole animal level (Malvindi et al. 2008).

Another light-stimulated nerve-signalling device was obtained with quantum-confined mercury telluride (HgTe) nanoparticles forming photovoltaic thin films. Cultured neuroblastoma and glioma cell line over these thin films were stimulated with a 532 nm laser. This photostimulus resulted in photoelectric current generation as light quanta absorbed in quantum-confined nanoparticle thin films, and cell depolarization was observed during the direct current (DC) phase of the photostimulation, thanks to the resistive coupling between the cell and nanoparticles (Cassagneau et al. 1998, Helland et al. 2007, Pappas et al. 2007). In another study, hippocampal neurons were cultured on plasmonic nanoparticle templates and action potentials were recorded opti-cally through localized surface plasmon resonance (Zhang et al. 2009).

The inner surface of pulled glass microtip electrodes was also coated with light-activatable semiconductor (lead selenide (PbSe)) nanoparticle thin film. These pho-toelectrodes were able to produce neural firing patterns in rat hippocampus slices and olfactory bulb. The focal illumination with near-infrared light at 830 nm wavelength depolarized and activated the neocortical and olfactory bulb neurons in the intact brain. Neural stimulation could be modulated in time and intensity. These electrodes eliminate the direct contact between neurons and PbSe and thus the related neural toxicity potential (Zhao et al. 2009).

Optical properties of semiconductor nanoparticles offer alternative spectroscopic tools with superior temporal resolution and SNR ratio with respect to the tradi-tional voltage-sensitive dyes to observe the dynamic behaviour of neural membrane potentials. Combined with surface functionalization, the unique optical property of a quantum dot/rod has been used to label and track neurons (Pathak et al. 2006, Maysinger et al. 2007) as well as the individual neurotransmitter receptors embed-ded in the neural cell membrane (Dahan et al. 2003). These semiconductor nanopar-ticles are generally made from CdSe or cadmium telluride (CdTe).

Transduction of magnetic energy into biological signal can be achieved by cou-pling synthesized magnetic nanoparticles with proteins. Then an externally applied magnetic field is transduced into heating, force/pull, torque and aggregation. The application of the magnetic field offers almost unlimited penetration depth through the body and makes parallel stimulation of freely moving animals possible. Remote neural stimulation by nanoparticle heating was achieved by application of 40 MHz oscillation of a 1.3 Gauss magnetic field inducing Neel relaxation on the 2D arrays of supermagnetic nanoparticles and delivering a local (tens of nm) heating limited to the cell membrane with constant temperature inside the cell. The temperature change was verified and quantified through fluorescence measurements. This significantly

raised the membrane temperature from 30°C to 43°C and changed the conforma-
tional state of ion channels within 1–5 s. Resulting depolarization of the membrane
and the elicited action potential were imaged with voltage-sensitive dye. This stimulation was demonstrated to induce thermal avoidance behaviour of *Caenorhabditis
elegans* triggered by the noxious temperature where the forward motion was halted
and retraction behaviour was observed. It also induced action potentials on hippocampal neurons transfected with TRPVR1 and eCFP and coated with nanoparticles
(Figure 19.3) (Huang et al. 2010).

Fe₃O₄ core supermagnetic nanoparticles were also coated with a thermoresponsive
hydrogel matrix shell with incorporated payload of a neurotoxic agent (NIPA-M).
These, then, were injected intravascularly to dogs and captured magnetically around
the cardiac tissue. The application of magnetic field contracted the hydrogel and

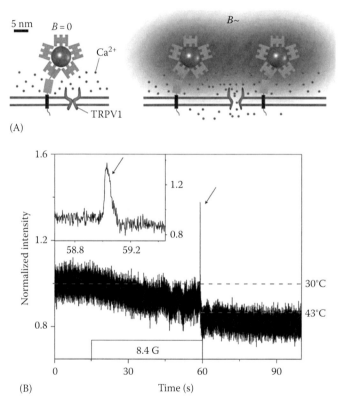

(A)

(B)

FIGURE 19.3 (A) Local heating induced opening of TRPV1 with applied RF magnetic
field to membrane-anchored streptavidin–DYlight549-coated supermagnetic nanoparticles.
(B) Action potentials induced in nanoparticle attached cultured rat hippocampal neurons as
local temperature was raised from 30°C to 43°C with application of RF magnetic field. The
arrow shows the action potential and the inset shows a magnified view of the action potential.
(Adapted with permission from Macmillan Publishers Ltd. *Nat. Nanotechnol.*, Huang, H.,
Delikanli, S., Zeng, H., Ferkey, D.M., and Pralle, A., Remote control of ion channels and
neurons through magnetic-field heating of nanoparticles, 5, 602–606, Copyright 2010.)

released the payload. The neurotoxin denervated the ganglionated plexi in the heart and avoided atrial fibrillation (Yu et al. 2010).

The toxicity of nanoparticles depends on the material type of the core and the presence of any coating around it, the size of the nanoparticle and the aggregation site. Semiconductor nanoparticles that incorporate cadmium, such as CdS, CdTe and CdSe, were found to decrease cell viability (Chan et al. 2006) and cause changes in cell morphology (Lovric et al. 2005a) even after very short-term exposure. However, it was tolerated when conjugated with the nerve growth factor (Vu et al. 2005). These results can be extended to other heavy metals.

The use of nonspecifically bound single nanoparticles rather than thin films imposes a challenge due to endocytosis. It has been shown with rat neonatal cortical cells cultured over CdS and CdTe quantum dot thin films that the lack of film stability caused degradation in 3–5 days in primary neuron cultures (Winter et al. 2005).

The main cause of toxicity was speculated to be generation of reactive oxygen species affecting neuron's metabolic activity and altering the cell membrane permeation, eventually causing cell death (Lovric et al. 2005b). Some bare nanoparticles were found to cause genetic alteration (Vu et al. 2005, Chan et al. 2006, Choi et al. 2007, Choi 2008), which was avoided with silane or poly(ethylene glycol) coating. Other coating materials causing efficient reduction in toxicity are bovine serum albumin (Derfus et al. 2004, Lovric et al. 2005a) and complexes of silane and phosphonate (Kirchner et al. 2005, Zhang et al. 2006). The reduction of nanoparticle size increased the toxicity due to the increased surface-to-volume ratio and easier access to the nucleus (Shiohara et al. 2004, Kirchner et al. 2005, Lovric et al. 2005b). Metal nanoparticles such as Au, Fe_3O_4 and Fe_2O_3 were found to be less toxic with respect to semiconductor nanoparticles (Ding et al. 2005, Gurr et al. 2005, Sayes et al. 2005, Lin et al. 2006, Buzea et al. 2007, Fischer and Chan 2007, Bastus et al. 2008, Clift et al. 2008, Diaz et al. 2008).

19.6 FIELD-EFFECT TRANSISTORS AND SILICON NANOWIRES

Silicon becomes an improved substrate for culturing neural cells when coated with a thin layer of thermally grown silicon dioxide. The thin layer of oxide suppresses the transfer of electrons and thus avoids the concomitant electrochemical processes that lead to a corrosion of silicon, which may damage the cell, though cell adhesion may be an issue. The latest developments in the well-established semiconductor technology allow for nanoscale transistors in direct contact with cells. Nanoscale field-effect transistors (FETs) may set the basis for massive parallel recording and stimulation with very-large-scale integration (VLSI) of several devices as test systems for pharmaceutical development, prosthetic devices and neural network engineering.

The requirement of a tight and stable coupling between the gate oxide and the neural cell limits the use of FETs as sensors and stimulators for *in vivo* applications; therefore, most of the FET-based systems were used to understand the basic neural function and the signal processing in neural networks *in vitro* (Fromherz 2003). Traditionally, nanoscale FETs with exposed gate structures have been capacitively coupled directly to neurons *in vitro*. The gate oxides of these structures are exposed to the electrolyte and a tight seal is required with neural cells. The region between neural cell and the gate oxide functions as a bidirectional interface, allowing recording and stimulation of neural activity. The neural activity leads to an ionic current through the membrane, which gives rise to extracellular potential in this region. The associated electrical field across the gate oxide is then probed by an FET (Vassanelli and Fromherz 1998). FETs were able to distinguish capacitive, ohmic and voltage-gated currents dominating the region between cell and gate oxide during neural activity of cultured mammalian neurons (Fromherz et al. 1991, 1993, Jenker and Fromherz 1997, Offenhäusser et al. 1997, Weis and Fromherz 1997, Schätzthauer and Fromherz 1998). The transistor operation enables amplification right on the recording side and lowers the noise substantially (Yoon and Tokumitsu 2000). The integration of neural recording electrodes with FETs has allowed for higher resolution, higher SNR ratio and longer-term recordings with respect to the clamped electrodes without interfering with cellular activity *in vitro*. In one of the earlier examples, leach neurons were capacitively coupled to the gate of an FET, where the change in the extracellular potential was detected by changes in current flow through the source drain channel (Fromherz et al. 1991).

High-density FETs have potential to integrate neural dynamics and digital electronics. To achieve this, the cell–silicon junction should be well understood with accompanying models (Massobrio et al. 2007). FET-based stimulation requires a capacitive current through the gate oxide to induce an extracellular potential above the threshold potential that gives rise to an ionic current with the opening of voltage-gated ion channels. A point contact model was developed to simulate intra- and extracellular voltage and to understand the effect of voltage-step stimulation on membrane leakage currents, capacitive currents and voltage-gated ion currents. This model was validated with stimulation of voltage- and current-clamped leech and snail neurons. The model suggests that a successful stimulation requires a small distance, high specific resistance, large area of contact with cell and high area specific gate oxide capacitance, thereby imposing a challenge to nanoscale miniaturization efforts (Fromherz and Stett 1995).

In another study, two synaptically connected snail neurons were both capacitively coupled to two FET devices. The first FET was used to stimulate the first neuron, which excited the second neuron through its synaptic connection, and the resulting extracellular changes due to induced action potential were detected at the source drain current of the second FET (Zeck and Fromherz 2001). FETs were also used to establish neural memory by enhancing excitatory chemical synapses. For this, two FETs were used in a positive feedback configuration, one to record the postsynaptic potential and the second to apply capacitively coupled stimulation pulses to the pre-synaptic cell (Kaul et al. 2004). These were some of the earliest demonstrations of external manipulation of natural neural networks at the micron and submicron level. Several similar studies were performed on networks of other invertebrate (Fromherz et al. 1991, 1995, Zeck and Fromherz 2001, Bonifazi and Fromherz 2002, Merz and Fromherz 2005) and vertebrate neural cells *in vitro* (Figure 19.4) (Offenhausser et al. 1997a,(NSL), b(BB), Voelker and Fromherz 2005). VLSI of FETs has been possible, thanks to advances in microelectronics and microchip fabrication. This allowed for fabrication of planar FET-based electrodes with more than 128×128 active recording sites in 1 mm^2 (Lambacher et al. 2004). These electrodes were used

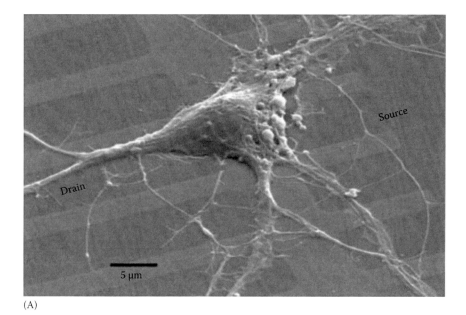

(A)

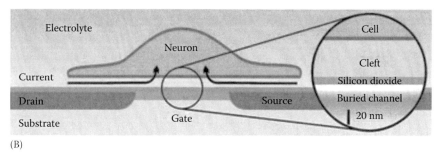

(B)

FIGURE 19.4   (See colour insert.) (A) Scanning electron microscopy (SEM) (colourized) of rat hippocampal neuron on a buried channel FET. (B) Schematic cross section of the top set-up. (Adapted from Voelker, M. and Fromherz, P: Signal transmission from individual mammalian nerve cell to field-effect transistor. *Small.* 2005. 1. 206–210. Copyright Wiley-VCH Verlag GmbH & Co. KGaA. Reproduced with permission.)

to image electrical field potentials of large neural networks where brain slices were cultured over the recording sites (Hutzler et al. 2006). The SNR ratio in FET-based recording applications has suffered from the inherent noise caused by electron tunnelling between silicon and traps in the gate oxide, especially at lower frequencies. This was mitigated by burying the channel a few nanometres below the neuron–gate oxide interface to allow for recording very low extracellular potentials of interfaced neurons (Voelker 2006).

Towards *in vivo* application, FETs were incorporated into traditional Utah and Michigan electrodes for neuroprosthetic applications. The metal electrode sites at the tip of the shanks were connected to the FET amplifiers via metal interconnects for on-site preamplification to improve the SNR ratio (Wise 2005). FET-based signal conditioners, analogue-to-digital converters and radio-frequency (RF) transmitters were also combined with the amplifiers for wireless recording and stimulation applications.

Nanowires with controllable structural and electronic properties were used in FET configurations to offer superior (several orders of magnitude) sensitivity with respect to traditional FETs (Schoning and Poghossian 2002, Soldatkin et al. 2003, Sant et al. 2004). With active areas of two orders of magnitude smaller than planar FETs (less than 100 nm), nanowire FETs offer sensitivities on the order of 300 µV − 3 mV and submillisecond temporal resolution. In these, silicon semiconductor nanowires were aligned on the oxide surface deposited over silicon substrate and connected source and drain to form an FET channel (Cui et al. 2001, Patolsky et al. 2004, 2006, Zheng et al. 2005, Stern et al. 2007). The coating of nanowires with cell adhesive and inert factors helped mediate tight junctions at desired locations and avoided unwanted junctions. Thanks to these coatings, neurites were extended in desired directions (Patolsky et al. 2006). The curvature provided by the FET topology decreased the cell-to-electrode distance with respect to the perfectly flat surfaces and therefore made the junction with the neural membrane even tighter. The small size of the sensing parts made localized recordings from extremely localized small regions (less than axonal and dendritic diameters) possible, where multiple electrodes can be defined with a pitch size of 10 nm to 10 µm.

The neurons cultured *in vitro* over nanowire FETs formed 'artificial synapses', which detected action potentials where the extracellular potential modulated the conductance of the nanowire channel. Negative potentials caused increased conductance on a p-type nanowire due to carrier accumulation, and positive potentials caused depletion and hence decreased the conductance (Figure 19.4) (Voelker and Fromherz 2005). Moreover, these artificial synapses were used to apply excitatory stimulation pulses through capacitive coupling (Steidl et al. 2006). Independently addressable nanowire FETs were distributed along the dendrite and axon, where simultaneous propagation of the signal across both regions was observed (Patolsky et al. 2006). The temporally resolved recordings on these spatially separated nanowires demonstrated (1) no change in peak amplitude and sharp peak shape during propagation along the axon and (2) reduced amplitude and broadening along the dendrites. Spike propagation rate was also quantified to be higher in axons compared to dendrites and similar values obtained with conventional patch clamp-based methods (Stuart and Sakmann 1994, Nowak and Bullier 1998). By applying hyperpolarizing pulses to nanowires, generation of action potentials was inhibited and propagation was blocked (Casey and Blick 1969, Bhadra and Kilgore 2004).

The multiplexed mapping with 2D nanowire FET arrays has a great potential to reveal the functional connectivity in natural neural networks. Small active area (0.06 µm²) silicon-nanowire FETs were interfaced to acute brain slices, providing spatial resolution better than 30 µm (Qing et al. 2010).

Cortical neurons with 500 µm long axon were grown on 50 addressable 20 nm nanowire elements to quantify axon/dendrite propagation simultaneously on a single cell (Figure 19.5). Hybrid neuron/nanowire FET devices were also demonstrated to realize complementary electronics and NOR logic gates (Patolsky et al. 2006).

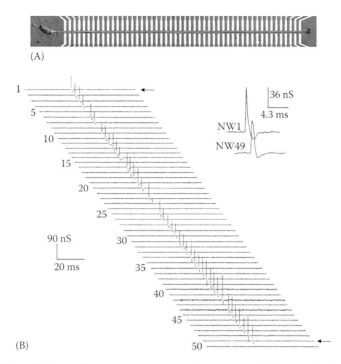

FIGURE 19.5   (A) Optical image of an axon crossing an array of 50 nanowire channels with 10 μm pitch. (B) Recorded waveforms from the 50 nanowire array. (Adapted with permission from Patolsky, F. et al., *Science,* 313, 1100, 2006.)

Pentacene-based organic FETs were also proposed for *in vivo* implantable applications with flexible electronics. Individually addressable active switching matrices were fabricated on polyimide substrate and encapsulated in Parylene C. Acting as a voltage-controlled current source, these devices drove drain source current through neural electrodes for stimulation rather than the capacitive stimulation used in silicon-based devices where a tight sealing is required between the device and the neuron. Flexible substrate also mitigated the strain mismatch between the stiffer silicon-based FET and the soft neural tissue. These devices were successfully tested on sciatic nerve on *Xenopus laevis* frog (Feili et al. 2008). As another example, silicon-nanowire FET arrays fabricated on flexible polymeric substrates simultaneously recorded voltage from the different parts of embryonic chick heart (Timko et al. 2009). Silicon-nanowire transistors were also successfully coated with continuous lipid bilayer and peptide pores to function like neurons where voltage and chemically gated ion transport were achieved through the membrane pores for ionic to electronic signal transduction (Misra et al. 2009).

Nanowires were also used to interface neurons without FET configuration. Vertical silicon nanowires were penetrated into dissociated cultures of rat cortical neurons to be used as scalable intracellular electrodes (Figure 19.6) (Robinson et al. 2012). A degenerately doped silicon core combined with a sputter deposited titanium/gold tip established electrical connectivity, while silicon dioxide shell

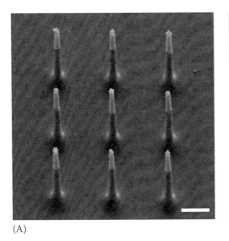
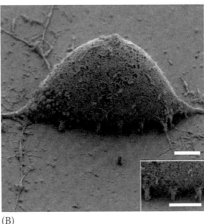

(A)   (B)

FIGURE 19.6 **(See colour insert.)** (A) SEM (colourized) image of an electrode array composed of nine silicon nanowires with metal-coated tips (grey) and insulating silicon dioxide (blue). Scale bar is 1 μm. (B) SEM image of a rat cortical cell on top of the pad shown on the left where silicon nanowires are interfacing with the cellular membrane (inset). Both scale bars are 2.5 μm. (Adapted with permission from Macmillan Publishers Ltd. *Nat. Nanotechnol.*, Robinson, J. T., Jorgolli, M., Shalek, A. K., Yoon, M.-H., Gertner, R. S., and Park, H. Vertical nanowire electrode arrays as a scalable platform for intracellular interfacing to neuronal circuits, 7, 180–184, Copyright 2012.)

encapsulation prevented current leakage from forming tight seals with the cell membrane. Penetration of nanowires was achieved by permeabilizing the membrane with a short voltage pulse and penetration was verified with the change in membrane potential. A current clamp was built with these electrodes to perform high-fidelity intracellular stimulation. Multiple neighbouring neurons were also penetrated with similar nanowire electrodes and magnitude, sign and latency of inhibitory, and excitatory postsynaptic potentials were recorded with stimulation of presynaptic neural cells. Stimulation and recording of multiple neurons in parallel helped map multiple individual synaptic connections.

Nanowires have been used to establish nanoscale electrodes to interface neural tissue. The challenge in this case has been the amplification of electrode impedance with miniaturization. It is also not a trivial task to address each electrode individually in such a massive array. Probes made out of nanowires were introduced to the microvasculature of the brain through a catheter to record surrounding neural signals without interfering with blood flow (Llinas et al. 2005). Vertically aligned nanowires were also grown on traditional implantable neural electrodes to provide almost ten times the surface area of planar electrodes, significantly reducing the impedance to provide superior electrochemical properties (Yoon et al. 2008).

Coating electrically active sites of platinum electrodes with platinum nanowires provided 600 times effective surface area enlargement, significantly lowering the impedance, reducing the cut-off frequency by 3 orders of magnitude, improving the charge storage capacity by a factor of 1000 and widening the water window by 0.25 V. No sign of toxicity was observed in *in vitro* cultures (Jin et al. 2011).

Epitaxial gallium phosphide nanowires grown vertically from a gallium phosphide surface (Seifert et al. 2004) were used to form monodisperse arrays to measure cellular forces *in vitro*. Displacement of nanowire tips was used to estimate the cellular forces during neural growth with a spatial resolution of 1 µm. An average of 15 pN was measured on the growth cones of dissociated cells from mouse dorsal root ganglia (Hallstrom et al. 2010). No sign of acute or chronic toxicity was observed on these cells (Hallstrom et al. 2007).

The capability of mechanical properties of nanowires to measure brain micromotion and mechanical stress was also suggested through finite element model analysis of nanowire electrodes embedded in the tissue (Zhu et al. 2011).

Atomic force microscopy was used to characterize with nanometre resolution the ultrafine morphological responses of neuroblastoma cells to neural stimulation (Shenai et al. 2004).

19.7 NANOBIOPHOTONIC METHODS (OPTOGENETICS AND NANO-NEUROSURGERY)

To understand cellular basis of neural computation, neural circuitry needs to be driven with appropriate stimuli. Although arrays of stimulation electrodes have been the most common method to achieve this (Buzsaki 2004), the concerns related to tissue damage caused by electrochemical side reactions have been a limiting factor (Stefan et al. 2001, Merrill et al. 2005). Optical structures of nanoscale biomolecules have been used as a novel way to non-invasively and wirelessly stimulate the neurons with precise temporal and spatial resolution. To replace the traditional current-injection methods and to achieve transduction of optical signal-to-neural response, three nanoscale photostimulation techniques were introduced: activation of a biochemically caged neurotransmitter (glutamate)-containing compound (Callaway and Katz 1993), chemical rendering of ion channels to respond to light (Banghart et al. 2004) and genetic incorporation of light-sensitive proteins into light-insensitive cells (Zhang et al. 2006). Called as 'optogenetics', the final method uses specific genetic sequences (opsins) to enable the modulation of ion channels in the presence of light. The convergence of nanotechnology with optogenetics may result in startling and challenging advances over the coming decades (Chamber and Kramer 2008). Optogenetics can be combined with calcium imaging to enable all-optical, high-speed and genetically targeted stimulation and recording of neural activity (Olek et al. 2004). This tool offers promising opportunities to model physiological neural disorders (Gradinaru et al. 2007, Airan et al. 2007a,b).

In optogenetic studies, light-gated ion channels shift the membrane potential and induce stimulation, silencing and modulating neural activity. Channelrhodopsin-2 is a nonspecific cation (H^+, Na^+, K^+, Ca^{+2}) channel activated with blue light at ~480 nm to induce photostimulation (Boyden et al. 2005). Halorhodopsin is a chloride pump activated with yellow light at ~580 nm to induce photoinhibition (Han and Boyden 2007, Zhang et al. 2007). Together, multicolour optical activation (depolarization) and silencing (hyperpolarization) of neural activity were made possible in millisecond ranges (Chow et al. 2010).

In another example, photoactive alpha subunits of the G protein were introduced through genetic targeting and were used as an on–off switch for neural activity (Zemelman et al. 2002). This was demonstrated to induce certain behaviours such as wing beating in Drosophila.

Surgical inhibition of neural activity could be achieved by femtosecond laser acting like a pair of scissors (Steinmeyer et al. 2010). Axotomy of *C. elegans* was performed by this method after which most of the severed axons regenerated within 24 h and grew new branches to reach the target (Yanik et al. 2004). A high-throughput surgery and screening system was developed to test the effects of several neuroregenerative drugs and chemicals on zebrafish (Pardo-martin et al. 2010, Samara et al. 2010).

Selective lesions on cortical neurons were induced by spatial localization of multiphoton excitation (Sacconi 2007). Single dendrites were dissected with submicron precision without causing any visible damage to the surrounding dendrites and axons (Sacconi 2012). The very short duration of the pulse focused photons in a very small volume with very high precision, avoiding damage to the surrounding tissue.

19.8   CONCLUSION

Understanding the neural system and manipulating it to treat neural diseases require tools capable of interfacing with neural systems at the molecular level. There is an urgent need to develop nanoscale strategies that improve our capability to record and stimulate neural tissue: in the short term, to enhance the efficiency of the interface and in the long term, to revolutionize the current methods to be able to treat several neural disorders. In the short term, nanoscale coating of current neural electrodes with NT and wires would provide topological and electrochemical augmentation for enhanced cell adhesion and viability in addition to superior charge injection capacity for safer and more energy-efficient stimulation. Though demonstrated only *in vitro*, FET-based electrode sites with nanoscale features would increase the sensitivity and capacitive coupling with neurons. The limitations imposed by the requirement of small gaps between the neuron and the device can be mitigated with coating or growth of nanostructures on these devices. Combined with controllable structural surface and electrical properties, the simultaneous detection and modulation of action potentials *in vitro* with high spatial and temporal resolution makes nanowire FETs a very promising tool to probe neural function. Its compatibility with flexible polymers improves the nanowire FET's biocompatibility, neural viability and adhesion and may extend its use beyond cultured neural networks and brain slices to *in vivo* applications including neuroprosthetics. Nanowire FETs were used to detect, stimulate, inhibit or reversibly block signal propagation along the specific pathways, while simultaneously mapping signal flow through the network. This provides a promising tool to understand spike propagation mechanisms, to investigate synaptic processing in neural networks and to build hybrid circuits for bioelectronic information processing. In the long term, optical and magnetic properties of nanoparticles and engineered biomolecules will enable remote wireless stimulation and recording of neural activity. Although nanoparticles have been approved by the FDA and some preliminary toxicity studies exist supporting the aforementioned bionanotechnological tools, long-term studies of toxicity in animal models are needed before these nanotechnologies can be used clinically.

REFERENCES

Abidian, M. R., Corey, J. M., Kipke, D. R., Martin, D. C. 2010. Conducting-polymer nanotubes improve electrical properties, mechanical adhesion, neural attachment, and neurite outgrowth of neural electrodes. *Small* 6: 421–429.

Abidian, M. R., Kim, D.-H., Martin, D. C. 2006. Conducting-polymer nanotubes for controlled drug release. *Advanced Materials* 18: 405–409.

Abidian, M. R., Ludwig, K. A., Marzullo, T. C., Martin, D. C., Kipke, D. R. 2009. Interfacing conducting polymer nanotubes with the central nervous system: chronic neural recording using poly(3,4-ethylenedioxythiophene) nanotubes. *Advanced Materials* 21: 3764–3770.

Abidian, M. R., Martin, D. C. 2008. Experimental and theoretical characterization of implantable neural microelectrodes modified with conducting polymer nanotubes. *Biomaterials* 29: 1273–1283.

Abidian, M. R., Martin, D. C. 2009. Multifunctional nanobiomaterials for neural interfaces. *Advanced Functional Materials* 19: 573–585.

Airan, R. D., Hu, E. S., Vijaykumar, R., Roy, M., Meltzer, L. A., Deisseroth, K. 2007. Integration of light-controlled neuronal firing and fast circuit imaging. *Current Opinion in Neurobiology* 17: 587–592.

Airan, R. D., Meltzer, L. A., Roy, M., Gong, Y., Chen, H., Deisseroth, K. 2007. High-speed imaging reveals neurophysiological links to behavior in an animal model of depression. *Science* 317: 819–823.

Albanese, A., Romito, L. 2011. Deep brain stimulation for Parkinson's disease: where do we stand? *Frontiers in Neurology* 2: 33.

Atkins, P. 1994. *Physical Chemistry,* 5th edn., A8. New York: W. H. Freeman and Co.

Banghart, M., Borges, K., Isacoff, E., Trauner, D., Kramer, R. H. 2004. Light-activated ion channels for remote control of neuronal firing. *Nature Neuroscience* 7: 1381–1386.

Banks, D. 1998. Neurotechnology. *IEEE Engineering Science Education* 7: 135–144.

Bastus, N. G., Casals, E., Vazquez-Campos, S., Puntes, V. 2008. Reactivity of engineered inorganic nanoparticles and carbon nanostructures in biological media. *Nanotoxicology* 2: 99–112.

Belkas, J. S., Munro, C. A., Shoichet, M. S., Midha, R. 2005. Peripheral nerve regeneration through a synthetic hydrogel nerve tube. *Restorative Neurology and Neuroscience* 23: 19–29.

Bhadra, N., Kilgore, K. L. 2004. Direct current electrical conduction block of peripheral nerve. *IEEE Transactions on Neural Systems and Rehabilitation Engineering* 12: 313–324.

Biran, R., Martin, D. C., Tresco, P. A. 2005. Neuronal cell loss accompanies the brain tissue response to chronically implanted silicon microelectrode arrays. *Experimental Neurology* 195(1): 115–126.

Bonifazi, P., Fromherz, P. 2002. Silicon chip for electronic communication between nerve cells by non-invasive interfacing and analog-digital processing. *Advanced Materials* 14: 1190–1193.

Boyden, E. S., Zhang, F., Bamberg, E., Nagel, G., Deisseroth, K. 2005. Millisecond-timescale, genetically targeted optical control of neural activity. *Nature Neuroscience* 8: 1263–1268.

Buzea, C., Pacheco, I. I., Robbie, K. 2007. Nanomaterials and nanoparticles: sources and toxicity. *Biointerphases* 2: MR17.

Buzsáki, G. 2004. Large-scale recording of neuronal ensembles. *Nature Neuroscience* 7: 446–451.

Callaway, E. M., Katz, L. C. 1993. Photostimulation using caged glutamate reveals functional circuitry in living brain slices. *Proceedings of the National Academy of Sciences of the United States of America* 90: 7661–7665.

Carbone, L., Nobile, C., De Giorgi, M. et al. 2007. Synthesis and micrometer-scale assembly of colloidal CdSe/CdS nanorods prepared by a seeded growth approach. *Nano Letters* 7: 2942–2950.

Casey, K. L., Blick, M. 1969. Observations on anodal polarization of cutaneous nerve. *Brain Research* 13: 155–167.

Cassagneau, T., Mallouk, T. E., Fendler, J. H. 1998. Layer-by-layer assembly of thin film zener diodes from conducting polymers and CdSe nanoparticles. *Journal of the American Chemical Society* 120: 7848–7859.

Chambers, J. J., Kramer, R. H. 2008. Light-activated ion channels for remote control of neural activity. *Methods in Cell Biology* 90: 217–232.

Chan, W.-H., Shiao, N.-H., Lu, P.-Z. 2006. CdSe quantum dots induce apoptosis in human neuroblastoma cells via mitochondrial-dependent pathways and inhibition of survival signals. *Toxicology Letters* 167: 191–200.

Choi, A. O., Brown, S. E., Szyf, M., Maysinger, D. 2008. Quantum dot-induced epigenetic and genotoxic changes in human breast cancer cells. *Journal of Molecular Medicine* 86(3): 291–302.

Choi, A. O., Cho, S. J., Desbarats, J., Lovrić, J., Maysinger, D. 2007. Quantum dot-induced cell death involves Fas upregulation and lipid peroxidation in human neuroblastoma cells. *Journal of Nanobiotechnology* 5: 1–13.

Chow, B. Y., Han, X., Dobry, A. S. et al. 2010. High-performance genetically targetable optical neural silencing by light-driven proton pumps. *Nature* 463: 98–102.

Clements, I. P., Kim, Y., English, A. W., Lu, X., Chung, A., Bellamkonda, R. V. 2009. Thin-film enhanced nerve guidance channels for peripheral nerve repair. *Biomaterials* 30(23–24): 3834–3846.

Clift, M. J. D., Rothen-Rutishauser, B., Brown, D. M. et al. 2008. The impact of different nanoparticle surface chemistry and size on uptake and toxicity in a murine macrophage cell line. *Toxicology and Applied Pharmacology* 232: 418–427.

Cui, D., Tian, F., Ozkan, C. S., Wang, M., Gao, H. 2005. Effect of single wall carbon nanotubes on human HEK293 cells. *Toxicology Letters* 155: 73–85.

Cui, Y., Wei, Q., Park, H., Lieber, C. M. 2001. Nanowire nanosensors for highly sensitive and selective detection of biological and chemical species. *Science* 293: 1289–1292.

Dahan, M., Lévi, S., Luccardini, C., Rostaing, P., Riveau, B., Triller, A. 2003. Diffusion dynamics of glycine receptors revealed by single-quantum dot tracking. *Science* 302: 442–445.

Derfus, A. M., Chan, W. C. W., Bhatia, S. N. 2004. Probing the cytotoxicity of semiconductor quantum dots. *Nano Letters* 4: 11–18.

Dhillon, G. S., Lawrence, S. M., Hutchinson, D. T., Horch, K. W. 2004. Residual function in peripheral nerve stumps of amputees: implications for neural control of artificial limbs. *The Journal of Hand Surgery* 29(4): 605–615.

Díaz, B., Sánchez-Espinel, C., Arruebo, M. et al. 2008. Assessing methods for blood cell cytotoxic responses to inorganic nanoparticles and nanoparticle aggregates. *Small* 4: 2025–2034.

Ding, L., Stilwell, J., Zhang, T. et al. 2005. Molecular characterization of the cytotoxic mechanism of multiwall carbon nanotubes and nano-onions on human skin fibroblast. *Nano Letters* 5: 2448–2464.

Eavarone, D. A., Yu, X., Bellamkonda, R. V. 2000. Targeted drug delivery to C6 glioma by transferrin-coupled liposomes. *Journal of Biomedical Materials Research* 51: 10–14.

Feili, D., Schuettler, M., Stieglitz, T. 2008. Matrix-addressable, active electrode arrays for neural stimulation using organic semiconductors-cytotoxicity and pilot experiments *in vivo*. *Journal of Neural Engineering* 5: 68–74.

Fischer, H. C., Chan, W. C. W. 2007. Nanotoxicity: the growing need for *in vivo* study. *Current Opinion in Biotechnology* 18: 565–571.

Fromherz, P. 2003. Neuroelectronic interfacing: semiconductor chips with ion channels, nerve cells, and brain. In *Nanoelectronics and Information Technology*, ed. R. Waser, pp. 781–810. Berlin, Germany: Wiley-VCH.

Fromherz, P., Muller, C. O., Weis, R. 1993. Neuron transistor: electrical transfer function measured by the patch-clamp technique. *Physical Review Letters* 71: 4079–4083.

Fromherz, P., Offenhäusser, A., Vetter, T., Weis, J. 1991. A neuron-silicon junction: a Retzius cell of the leech on an insulated-gate field-effect transistor. *Science* 252: 1290–1293.

Fromherz, P., Stett, A. 1995. Silicon-neuron junction: capacitive stimulation of an individual neuron on a silicon chip. *Physical Review Letters* 75: 1670–1674.

Gheith, M. K., Sinani, V. A., Wicksted, J. P., Matts, R. L., Kotov, N. A. 2005. Single-walled carbon nanotube polyelectrolyte multilayers and freestanding films as a biocompatible platform for neuroprosthetic implants. *Advanced Materials* 17: 2663–2670.

Gradinaru, V., Thompson, K. R., Zhang, F. et al. 2007. Targeting and readout strategies for fast optical neural control *in vitro* and *in vivo*. *The Journal of Neuroscience* 27: 14231–14238.

Gurr, J.-R., Wang, A. S. S., Chen, C.-H., Jan, K.-Y. 2005. Ultrafine titanium dioxide particles in the absence of photoactivation can induce oxidative damage to human bronchial epithelial cells. *Toxicology* 213: 66–73.

Hällström, W., Lexholm, M., Suyatin, D. B. et al. 2010. Fifteen-piconewton force detection from neural growth cones using nanowire arrays. *Nano Letters* 10: 782–787.

Hällström, W., Mårtensson, T., Prinz, C. et al. 2007. Gallium phosphide nanowires as a substrate for cultured neurons. *Nano Letters* 7: 2960–2965.

Han, X., Boyden, E. S. 2007. Multiple-color optical activation, silencing, and desynchronization of neural activity, with single-spike temporal resolution. *PloS One* 2: e299.

He, W., Bellamkonda, R. V. 2005. Nanoscale neuro-integrative coatings for neural implants. *Biomaterials* 26: 2983–2990.

He, W., McConnell, G. C., Bellamkonda, R. V. 2006. Nanoscale laminin coating modulates cortical scarring response around implanted silicon microelectrode arrays. *Journal of Neural Engineering* 3: 316–326.

Helland, A., Wick, P., Koehler, A., Schmid, K., Som, C. 2007. Reviewing the environmental and human health knowledge base of carbon nanotubes. *Environmental Health Perspectives* 115: 1125–1131.

Hu, H., Ni, Y., Mandal, S.K. et al. 2005. Polyethyleneimine functionalized single-walled carbon nanotubes as a substrate for neuronal growth. *The Journal of Physical Chemistry B* 109: 4285–4289.

Huang, H., Delikanli, S., Zeng, H., Ferkey, D.M., Pralle, A. 2010. Remote control of ion channels and neurons through magnetic-field heating of nanoparticles. *Nature Nanotechnology* 5: 602–606.

Hutzler, M., Lambacher, A., Eversmann, B., Jenkner, M., Thewes, R., Fromherz, P. 2006. High-resolution multitransistor array recording of electrical field potentials in cultured brain slices. *Journal of Neurophysiology* 96: 1638–1645.

Jackson, J. D. 1999. *Classical Electrodynamics,* 3rd edn., pp. 150–190. New York: Wiley.

Jenkner, M., Fromherz, P. 1997. Bistability of membrane conductance in cell adhesion observed in a neuron transistor. *Physical Review Letters* 79: 4705–4708.

Jin, Y.-H., Daubinger, P., Fiebich, B. L., Stieglitz, T. 2011. A novel platinum nanowire-coated neural electrode and its electrochemical and biological characterization. *Proceedings of IEEE MEMS Conference,* Boston, MA, pp. 1003–1006.

Jordan, A., Maier-Hauff, K. 2007. Magnetic nanoparticles for intracranial thermotherapy. *Journal of Nanoscience and Nanotechnology* 7: 4604–4606.

Jun, Y.-W., Seo, J.-W., Cheon, J. 2008. Nanoscaling laws of magnetic nanoparticles and their applicabilities in biomedical sciences. *Accounts of Chemical Research* 41: 179–189.

Kaul, R., Syed, N., Fromherz, P. 2004. Neuron-semiconductor chip with chemical synapse between identified neurons. *Physical Review Letters* 92: 1–4.

Keefer, E. W., Botterman, B. R., Romero, M. I., Rossi, A. F., Gross, G. W. 2008. Carbon nanotube coating improves neuronal recordings. *Nature Nanotechnology* 3: 434–439.

Kim, D.-H., Martin, D. C. 2006. Sustained release of dexamethasone from hydrophilic matrices using PLGA nanoparticles for neural drug delivery. *Biomaterials* 27: 3031–3037.

Kim, Y.-T., Haftel, V. K., Kumar, S., Bellamkonda, R. V. 2008a. The role of aligned polymer fiber-based constructs in the bridging of long peripheral nerve gaps. *Biomaterials* 29: 3117–3127.

Kim, J., Lee, J. E., Lee, S. H. et al. 2008b. Designed fabrication of a multifunctional polymer nanomedical platform for simultaneous cancer-targeted imaging and magnetically guided drug delivery. *Advanced Materials* 20: 478–483.

Kirchner, C., Liedl, T., Kudera, S. et al. 2005. Cytotoxicity of colloidal CdSe and CdSe/ZnS nanoparticles. *Nano Letters* 5: 331–338.

Kreuter, J. 2001. Nanoparticulate systems for brain delivery of drugs. *Advanced Drug Delivery Reviews* 47: 65–81.

Lam, C.-W., James, J. T., McCluskey, R., Hunter, R. L. 2004. Pulmonary toxicity of single-wall carbon nanotubes in mice 7 and 90 days after intratracheal instillation. *Toxicological Sciences* 77: 126–134.

Lambacher, A., Jenkner, M., Merz, M. et al. 2004. Electrical imaging of neuronal activity by multi-transistor-array (MTA) recording at 7.8 micrometer resolution. *Applied Physics A* 79: 1607–1611.

Li, J., Andrews, R. J. 2007. Trimodal nanoelectrode array for precise deep brain stimulation: prospects of a new technology based on carbon nanofiber arrays. In: *Operative Neuromodulation*, eds. D. E. Sakas, and B. Simpson, pp. 537–545. Berlin, Germany: Springer.

Lin, W., Huang, Y.-W., Zhou, X.-D., Ma, Y. 2006. *in vitro* toxicity of silica nanoparticles in human lung cancer cells. *Toxicology and Applied Pharmacology* 217: 252–259.

Liopo, A. V., Stewart, M. P., Hudson, J., Tour, J. M., Pappas, T. C. 2006. Biocompatibility of native and functionalized single-walled carbon nanotubes for neuronal interface. *Journal of Nanoscience and Nanotechnology* 6: 1365–1374.

Liu, H. L., Wu, J. H., Min, J. H., Kim, Y. K. 2008. Synthesis of monosized magnetic-optical AuFe alloy nanoparticles. *Journal of Applied Physics* 103: 07D529.

Llinás, R. R., Walton, K. D., Nakao, M., Hunter, I., Anquetil, P. A. 2005. Neuro-vascular central nervous recording/stimulating system: using nanotechnology probes. *Journal of Nanoparticle Research* 7: 111–127.

Lovat, V., Pantarotto, D., Lagostena, L. et al. 2005. Carbon nanotube substrates boost neuronal electrical signaling. *Nano Letters* 5: 1107–1110.

Lovrić, J., Bazzi, H. S., Cuie, Y., Fortin, G. R. A., Winnik, F. M., Maysinger, D. 2005. Differences in subcellular distribution and toxicity of green and red emitting CdTe quantum dots. *Journal of Molecular Medicine* 83: 377–385.

Lovrić, J., Cho, S. J., Winnik, F. M., Maysinger, D. 2005. Unmodified cadmium telluride quantum dots induce reactive oxygen species formation leading to multiple organelle damage and cell death. *Chemistry and Biology* 12: 1227–1234.

Lu, Y., Li, T., Zhao, X. et al. 2010. Electrodeposited polypyrrole/carbon nanotubes composite films electrodes for neural interfaces. *Biomaterials* 31: 5169–5181.

Ludwig, K. A., Uram, J. D., Yang, J., Martin, D. C., Kipke, D. R. 2006. Chronic neural recordings using silicon microelectrode arrays electrochemically deposited with a poly(3,4-ethylenedioxythiophene) (PEDOT) film. *Journal of Neural Engineering* 3: 59–70.

Luo, X., Matranga, C., Tan, S., Alba, N., Cui, X. T. 2011. Carbon nanotube nanoreservoir for controlled release of anti-inflammatory dexamethasone. *Biomaterials* 32: 6316–6323.

Malarkey, E. B., Fisher, K. A., Bekyarova, E., Liu, W., Haddon, R. C., Parpura, V. 2009. Conductive single-walled carbon nanotube substrates modulate neuronal growth. *Nano Letters* 9: 264–268.

Malarkey, E. B., Parpura, V. 2007. Application of carbon nanotubes in neurobiology. *Neurodegenerative Diseases* 4: 292–299.

Malarkey, E. B., Reyes, R. C., Zhao, B., Haddon, R. C., Parpura, V. 2008. Water soluble single-walled carbon nanotubes inhibit stimulated endocytosis in neurons. *Nano Letters* 8: 3538–3542.

Malvindi, M. A., Carbone, L., Quarta, A., Tino, A., Manna, L., Pellegrino, T., Tortiglione, C. 2008. Rod-shaped nanocrystals elicit neuronal activity *in vivo*. *Small* 4: 1747–1755.

Massobrio, G., Massobrio, P., Martinoia, S. 2007. Modeling and simulation of silicon neuron-to-ISFET junction. *Journal of Computational Electronics* 6: 431–437.

Mattson, M. P., Haddon, R. C., Rao, A. M. 2000. Molecular functionalization of carbon nanotubes and use as substrates for neuronal growth. *Journal of Molecular Neuroscience* 14: 175–182.

Maysinger, D., Behrendt, M., Lalancette-Hébert, M., Kriz, J. 2007. Real-time imaging of astrocyte response to quantum dots: *in vivo* screening model system for biocompatibility of nanoparticles. *Nano Letters* 7: 2513–2520.

McConnell, G. C., Rees, H. D., Levey, A. I., Gutekunst, C.-A., Gross, R. E., Bellamkonda, R. V. 2009. Implanted neural electrodes cause chronic, local inflammation that is correlated with local neurodegeneration. *Journal of Neural Engineering* 6: 056003.

Merrill, D. R., Stefan, I. C., Scherson, D. A., Mortimer, J. T. 2005. Electrochemistry of gold in aqueous sulfuric acid solutions under neural stimulation conditions. *Journal of the Electrochemical Society* 152: E212.

Merz, M., Fromherz, P. 2005. Silicon chip interfaced with a geometrically defined net of snail neurons. *Advanced Functional Materials* 15: 739–744.

Misra, N., Martinez, J. A., Huang, S. C. J., Wang, Y., Stroeve, P., Grigoropoulos, C. P. 2009. Bioelectronic silicon nanowire devices using functional membrane proteins. *Proceedings of the National Academy of Sciences of the United States of America* 106: 13780–13784.

Na, H. B., Lee, J. H., An, K. et al. 2007. Development of a T1 contrast agent for magnetic resonance imaging using MnO nanoparticles. *Angewandte Chemie (International ed. in English)* 46: 5397–5401.

Najafi, K. 1997. Handbook of microlithography, micromachining, and microfabrication. *SPIE-International Society for Optical Engine,* Chicago: 517.

Nowak, L. G., Bullier, J. 1998. Axons, but not cell bodies, are activated by electrical stimulation in cortical gray matter. *Experimental Brain Research* 118: 477–488.

Offenhäusser, A., Sprössler, C., Matsuzawa, M., Knoll, W. 1997a. Electrophysiological development of embryonic hippocampal neurons from the rat grown on synthetic thin films. *Neuroscience Letters* 223: 9–12.

Offenhausser, A., Sprossler, C., Matsuzawa, M., Knoll, W. 1997b. Field-effect transistor array for monitoring electrical activity from mammalian neurons in culture. *Biosensors and Bioelectronics* 12: 819–826.

Olek, M., Ostrander, J., Jurga, S., Mo, H. 2004. Layer-by-layer assembled composites from multiwall carbon nanotubes with different morphologies. *Nano Letters* 4: 1889–1895.

Pappas, T. C., Wickramanyake, W. M. S., Jan, E., Motamedi, M., Brodwick, M., Kotov, N. A. 2007. Nanoscale engineering of a cellular interface with semiconductor nanoparticle films for photoelectric stimulation of neurons. *Nano Letters* 7: 513–519.

Pardo-Martin, C., Chang, T.-Y., Koo, B. K., Gilleland, C. L., Wasserman, S. C., Yanik, M. F. 2010. High-throughput *in vivo* vertebrate screening. *Nature Methods* 7: 634–636.

Pathak, S., Cao, E., Davidson, M. C., Jin, S., Silva, G. A. 2006. Quantum dot applications to neuroscience: new tools for probing neurons and glia. *The Journal of Neuroscience* 26: 1893–1895.

Patolsky, F., Timko, B. P., Yu, G., Fang, Y., Greytak, A. B., Zheng, G., Lieber, C. M. 2006. Detection, stimulation, and inhibition of neuronal signals with high-density nanowire transistor arrays. *Science* 313: 1100–1104.

Patolsky, F., Zheng, G., Hayden, O., Lakadamyali, M., Zhuang, X., Lieber, C. M. 2004. Electrical detection of single viruses. *Proceedings of the National Academy of Sciences of the United States of America* 101: 14017–14022.

Phely-Bobin, T. S., Tiano, T., Farrell, B., Fooksa, R., Robblee, L., Edell, D. J., Czerw, R. 2006. Carbon nanotube based electrodes for neuroprosthetic applications. *MRS Proceedings* 926: 1–6.

Qing, Q., Pal, S. K., Tian, B. et al. 2010. Nanowire transistor arrays for mapping neural circuits in acute brain slices. *Proceedings of the National Academy of Sciences of the United States of America* 107: 1882–1887.

Robinson, J. T., Jorgolli, M., Shalek, A. K., Yoon, M.-H., Gertner, R. S., Park, H. 2012. Vertical nanowire electrode arrays as a scalable platform for intracellular interfacing to neuronal circuits. *Nature Nanotechnology* 7: 180–184.

Ruan, G., Agarwal, A., Smith, A. M., Gao, S., Nie, S. 2006. Quantum dots as fluorescent labels for molecular and cellular imaging. *Reviews in Fluorescence* 3: 181–193.

Sacconi, L., O'Connor, R. P., Jasaitis, A., Masi, A., Buffelli, M., Pavone, F. S. 2007. In vivo multiphoton nanosurgery on cortical neurons. *Journal of Biomedical Optics* 12(5): 050502.

Samara, C., Rohde, C. B., Gilleland, C. L., Norton, S., Haggarty, S. J., Yanik, M. F. 2010. Large-scale *in vivo* femtosecond laser neurosurgery screen reveals small-molecule enhancer of regeneration. *Proceedings of the National Academy of Sciences of the United States of America* 107: 18342–18347.

Sant, W., Pourciel-Gouzy, M. L., Launay, J. et al. 2004. Development of a creatinine-sensitive sensor for medical analysis. *Sensors and Actuators B: Chemical* 103: 260–264.

Sato, Y., Yokoyama, A., Shibata, K., et al. 2005. Influence of length on cytotoxicity of multi-walled carbon nanotubes against human acute monocytic leukemia cell line THP-1 *in vitro* and subcutaneous tissue of rats *in vivo*. *Molecular Biosystems* 1: 176–182.

Saul, J. M., Annapragada, A. V., Bellamkonda, R. V. 2006. A dual-ligand approach for enhancing targeting selectivity of therapeutic nanocarriers. *Journal of Controlled Release* 114: 277–287.

Saul, J. M., Annapragada, A., Natarajan, J. V., Bellamkonda, R. V. 2003. Controlled targeting of liposomal doxorubicin via the folate receptor *in vitro*. *Journal of Controlled Release* 92: 49–67.

Sayes, C. M., Gobin, A. M., Ausman, K. D., Mendez, J., West, J. L., Colvin, V. L., 2005. Nano-C60 cytotoxicity is due to lipid peroxidation. *Biomaterials* 26: 7587–7595.

Schätzthauer, R., Fromherz, P. 1998. Neuron-silicon junction with voltage-gated ionic currents. *The European Journal of Neuroscience* 10: 1956–1962.

Schoening, M. J., Poghossian, A. 2002. Recent advances in biologically sensitive field-effect transistors (BioFETs). *Analyst* 127: 1137–1151.

Seifert, W., Borgström, M., Deppert, K. et al. 2004. Growth of one-dimensional nanostructures in MOVPE. *Journal of Crystal Growth* 272: 211–220.

Shenai, M. B., Putchakayala, K. G., Hessler, J. A., Orr, B. G., Banaszak Holl, M. M., Baker, J. R. 2004. A novel MEA/AFM platform for measurement of real-time, nanometric morphological alterations of electrically stimulated neuroblastoma cells. *IEEE Transactions on Nanobioscience* 3: 111–117.

Shi Kam, N. W., Jessop, T. C., Wender, P. A., Dai, H. 2004. Nanotube molecular transporters: internalization of carbon nanotube-protein conjugates into mammalian cells. *Journal of the American Chemical Society* 126(22): 6850–6851.

Shiohara, A., Hoshino, A., Hanaki, K., Suzuki, K., Yamamoto, K. 2004. On the cyto-toxicity caused by quantum dots. *Microbiology and Immunology* 48: 669–675.

Shvedova, A., Castranova, V., Kisin, E., Schwegler-Berry, D., Murray, A., Gandelsman, V., Baron, P. 2003. Exposure to carbon nanotube material: assessment of nanotube cytotoxicity using human keratinocyte cells. *Journal of Toxicology and Environmental Health Part A* 66(20): 1909–1926.

Siddiqui, S., Arumugam, P. U., Chen, H., Li, J., Meyyappan, M. 2010. Characterization of carbon nanofiber electrode arrays using electrochemical impedance spectroscopy: effect of scaling down electrode size. *ACS Nano* 4: 955–961.

Soldatkin, A. P., Montoriol, J., Sant, W., Martelet, C., Jaffrezic-Renault, N. 2003. A novel urea sensitive biosensor with extended dynamic range based on recombinant urease and ISFETs. *Biosensors and Bioelectronics* 19: 131–135.

Stefan, I. C., Tolmachev, Y. V., Nagy, Z. et al. 2001. Theoretical analysis of the pulse-clamp method as applied to neural stimulating electrodes. *Journal of the Electrochemical Society* 148: E73.

Steidl, E.-M., Neveu, E., Bertrand, D., Buisson, B. 2006. The adult rat hippocampal slice revisited with multi-electrode arrays. *Brain Research* 1096: 70–84.

Steinmeyer, J. D., Gilleland, C. L., Pardo-Martin, C. et al. 2010. Construction of a femtosecond laser microsurgery system. *Nature Protocols* 5: 395–407.

Stern, E., Klemic, J. F., Routenberg, D. A. et al. 2007. Label-free immunodetection with CMOS-compatible semiconducting nanowires. *Nature* 445: 519–522.

Stieglitz, T., Meyer, J. U. 1998. *Microsystem Technology in Chemistry and Life Science,* eds. A. Manz, and H. Becker, pp. 131–162. Berlin, Germany: Springer.

Stuart, G. J., Sakmann, B. 1994. Active propagation of somatic action potentials into neocortical pyramidal cell dendrites. *Nature* 367: 69–72.

Thelin, J., Jörntell, H., Psouni, E. et al. 2011. Implant size and fixation mode strongly influence tissue reactions in the CNS. *PloS One* 6: e16267.

Timko, B. P., Cohen-Karni, T., Yu, G., Qing, Q., Tian, B., Lieber, C. M. 2009. Electrical recording from hearts with flexible nanowire device arrays. *Nano Letters* 9: 914–918.

Vassanelli, S., Fromherz, P. 1998. Rapid communication transistor records of excitable neurons from rat brain. *Applied Physics A* 66: 459–463.

Voelker, M., Fromherz, P. 2005. Signal transmission from individual mammalian nerve cell to field-effect transistor. *Small* 1(2): 206–210.

Vu, T. Q., Maddipati, R., Blute, T. A., Nehilla, B. J., Nusblat, L., Desai, T. A. 2005. Peptide-conjugated quantum dots activate neuronal receptors and initiate downstream signaling of neurite growth. *Nano Letters* 5: 603–607.

Wang, K., Fishman, H. A., Dai, H., Harris, J. S. 2006. Neural stimulation with a carbon nanotube microelectrode array. *Nano Letters* 6: 2043–2048.

Warheit, D. B., Laurence, B. R., Reed, K. L., Roach, D. H., Reynolds, G. A. M., Webb, T. R. 2004. Comparative pulmonary toxicity assessment of single-wall carbon nanotubes in rats. *Toxicological Sciences* 77: 117–125.

Warnement, M. R., Tomlinson, I. D., Rosenthal, S. J. 2007. Fluorescent imaging applications of quantum dot probes. *Current Nanoscience* 3: 274–284.

Webster, T. J., Waid, M. C., McKenzie, J. L., Price, R. L., Ejiofor, J. U. 2004. Nanobiotechnology: carbon nanofibres as improved neural and orthopaedic implants. *Nanotechnology* 15: 48–54.

Weis, R., Fromherz, P. 1997. Frequency dependent signal transfer in neuron transistors. *Physical Review* E 55: 877–889.

West, J. L., Halas, N. J. 2003. Engineered nanomaterials for biophotonics applications: improving sensing, imaging, and therapeutics. *Annual Review of Biomedical Engineering* 5: 285–292.

Winter, J. O., Gomez, N., Korgel, B. A., Schmidt, C. E., Engineering, B. 2005. Quantum dots for electrical stimulation of neural cells. In *Nanobiophotonics and Biomedical Applications II*, eds. A. N. Cartwright, and M. Osinki, 5705: pp. 235–246. Proceedings of SPIE.

Winter, B. J. O., Liu, T. Y., Korgel, B. A., Schmidt, C. E. 2001. Recognition molecule directed interfacing between semiconductor quantum dots and nerve cells. *Advanced Materials* 13: 1673–1677.

Wise, K. D. 2005. Silicon microsystems for neuroscience and neural prostheses. *IEEE Engineering in Medicine and Biology* 24: 22–29.

Xie, Y., Ye, L., Zhang, X. et al. 2005. Transport of nerve growth factor encapsulated into liposomes across the blood-brain barrier: *in vitro* and *in vivo* studies. *Journal of Controlled Release* 105: 106–119.

Yanik, M. F., Cinar, H., Cinar, H. N., Chisholm, A. D., Jin, Y., Ben-Yakar, A. 2004. Functional regeneration after laser axotomy. *Nature* 432: 822.

Yoon, H., Deshpande, D. C., Ramachandran, V., Varadan, V. K. 2008. Aligned nanowire growth using lithography-assisted bonding of a polycarbonate template for neural probe electrodes. *Nanotechnology* 19: 025304.

Yoon, S.-M., Tokumitsu, E. 2000. Ferroelectric neuron integrated circuits using $S_2Bi_2Ta_2O_9$-Gate FET's and CMOS Schmitt-Trigger oscillators. *IEEE Transactions on Electron Devices* 47: 1630–1635.

Yu, L., Scherlag, B. J., Dormer, K., Nguyen, K. T., Pope, C., Fung, K.-M., Po, S. S. 2010. Autonomic denervation with magnetic nanoparticles. *Circulation* 122: 2653–2659.

Zeck, G., Fromherz, P. 2001. Noninvasive neuroelectronic interfacing with synaptically connected snail neurons immobilized on a semiconductor chip. *Proceedings of the National Academy of Sciences of the United States of America* 98: 10457–10462.

Zemelman, B. V., Lee, G., Miesenbok, M. 2002. Selective photostimulation of genetically charged neurons. *Neuron* 33: 15–22.

Zhang, J., Atay, T., Nurmikko, A. V. 2009. Optical detection of brain cell activity using plasmonic gold nanoparticles. *Nano Letters* 9: 519–524.

Zhang, T., Stilwell, J. L., Gerion, D. et al. 2006. Cellular effect of high doses of silica-coated quantum dot profiled with high throughput gene expression analysis and high content cellomics measurements. *Nano Letters* 6: 800–808.

Zhang, F., Wang, L.-P., Boyden, E. S., Deisseroth, K. 2006. Channelrhodopsin-2 and optical control of excitable cells. *Nature Methods* 3: 785–792.

Zhang, F., Wang, L.-P., Brauner, M. et al. 2007. Multimodal fast optical interrogation of neural circuitry. *Nature* 446: 633–639.

Zhao, Y., Larimer, P., Pressler, R. T., Strowbridge, B. W., Burda, C. 2009. Wireless activation of neurons in brain slices using nanostructured semiconductor photoelectrodes. *Angewandte Chemie (International ed. in English)* 48: 2407–2410.

Zheng, G., Patolsky, F., Cui, Y., Wang, W. U., Lieber, C. M. 2005. Multiplexed electrical detection of cancer markers with nanowire sensor arrays. *Nature Biotechnology* 23: 1294–1301.

Zhu, R., Huang, G. L., Yoon, H., Smith, C. S., Varadan, V. K. 2011. Biomechanical strain analysis at the interface of brain and nanowire electrodes on a neural probe. *Journal of Nanotechnology in Engineering and Medicine* 2: 031001.

Part V

Ultra-Low-Power
Biomedical Systems

20 Advances in Ultra-Low-Power Miniaturized Applications for Health Care and Sports
The Sensium™ Platform

Miguel Hernandez-Silveira, Su-Shin Ang and Alison Burdett

CONTENTS

20.1 INTRODUCTION

Recent wearable ambulatory monitoring technologies consist of wireless body-worn devices capable of monitoring vital signs (e.g. respiration rate (RR), heart rate (HR)) and physical activity (PA) levels that allow patients to freely ambulate, exercise and perform activities of daily living either within the hospital environment, their home or their community. Unfortunately, the majority of existing prototypes and commercially available products tend to be designed and manufactured with multiple off-the-shelf components, resulting in costly bulky devices prone to excessive power consumption.

Owing to the technological advances in application-specific integrated circuits (ASICs), it is now possible to develop custom biomedical systems-on-chip (SoCs), intended for production of low-cost and miniaturized wireless wearable monitoring devices that can be easily attached to the body and operate for reasonable periods of time from a single button-cell battery. Toumaz has been a pioneer in the development of ultralow-power SoCs – such as the Sensium™, an ASIC specially designed for monitoring physiological parameters when interfaced with appropriate body-worn sensors. This microchip, combined with appropriate software, has opened new avenues for the development of products intended for both ambulatory clinical use (i.e. inpatient or outpatient) and sports applications. Thus, Sensium™ enables real-time monitoring and smart interpretation of multiple vital signs, leading to potential reduction of costs of health care by simply connecting individuals wirelessly to hospital networks in affordable and unobtrusive manner.

In this chapter, we provide the reader with an overview of the Sensium™ technology health-care platform, with an emphasis on the development, evaluation, optimization and implementation of embedded biomedical algorithms suitable for body-worn devices. Furthermore, we share our experience of the engineering process leading to the ultimate incorporation of these algorithms into our microchips, as well as discuss trade-offs faced when implementing the software within limited hardware resources, without compromising the performance in terms of accuracy and reliability of the information processed and conveyed by our body-worn sensor nodes. In summary, this chapter includes a simple explanation of our methodology towards implementation of ultralow-power health-care platforms, ranging from the microchips to the software components forming part of the systems.

20.2   SENSIUM™ MICROCHIP

The Sensium™ SoC (TZ1031) is a microchip capable of enabling ubiquitous medical vital-sign monitoring and interpretation when interfaced with appropriate sensors and electrodes and represents the state of the art in terms of health-care functionality and ultralow power consumption. As shown in Figure 20.1, this ASIC comprises three main stages: the sensor interface, the wireless transceiver and the digital baseband [1].

The sensor interface contains essential circuitry (including bias current/voltage, amplification and filtering) for measuring both electrocardiograms (ECG) and respiratory activity by means of impedance pneumography (IP), using a single-lead configuration (i.e. two electrodes), and body temperature using an external thermistor. The outputs of these circuits (i.e. amplified and filtered biosignals with frequency components ranging from DC to 250 Hz) are fed into a 12-bit sigma-delta analogue-to-digital converter (ADC). This comprises a third-order switched op-amp architecture with a 64 times oversampling ratio. In addition to the sensor interface, the TZ1031 possesses an SPI port, which provides a means of interfacing external micro-engineered inertial sensors (e.g. a triaxial accelerometer or gyroscope) with the IC.

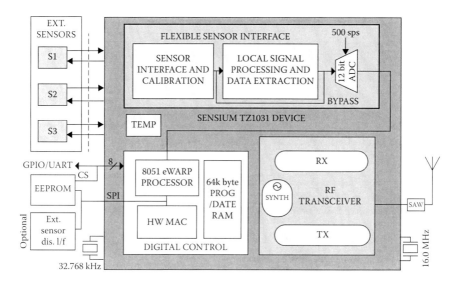

FIGURE 20.1   The Sensium™ TZ1031 SoC. (Courtesy of ISSCC.)

The wireless transceiver was custom designed to enable 100 kbps transmission at a 1 V supply with peak currents below 3 mA. A two-stage zero-IF structure based on a sliding IF technique was included in the receive path, allowing lower current consumption (compared to a single-stage direct-conversion architecture) while maintaining adequate filtering and noise profiling. The PA stage was designed to deliver −10 dBm into a matched antenna load, within a typical range of 10 m indoors [1].

The digital section comprises an 8-bit microcontroller core (8051), peripherals, memory, timers and medium access controller (MAC). The MAC protocol block was designed to ensure ultralow-power operation while maintaining secure/robust performance and control of the RF section (e.g. listen-before-transmit (LBT) algorithm to avoid collisions), link establishment, data transfer and sleep management. Since these key functions are implemented in custom hardware, the embedded microcontroller is available to run the application code containing the sensor processing algorithms.

This enables the possibility of fitting all the circuitry in adequate form factors and enclosures (discussed in the following section), providing means for continuous unobtrusive monitoring of PA, body temperature and heart and respiratory rhythms/conditions.

20.3   SENSIUM™ WIRELESS BODY NODE

Wireless body-sensor networks (WBSNs) comprise sensor nodes placed in one or several parts of the body. These nodes are usually designed to monitor physiological and/or biomechanical information and must be capable of gathering, storing and often processing data locally before transmission to a fixed or mobile base station.

The Sensium™ wireless body-sensor node is available in two versions: a disposable version mainly intended for clinical (in-hospital) health care and a reuseable version intended for outpatient monitoring and sports applications.

20.3.1 DISPOSABLE BODY-WORN NODE

The disposable body-worn node (the 'digital plaster') is designed for one-time usage, and it is the preferred choice if cost and hygiene are important considerations. In the general ward of a hospital, this node can be used to monitor noncritical patients as part of an early warning system. The node is hermetic and made of clinical graded dressing materials. Furthermore, this disposable node not only is waterproof but also ergonomic and fitted with medical foams to isolate the circuits from the body.

The disposable wireless body node comprises a Sensium™ TZ1031 chip and associated external circuitry together on a PCB housed inside a plaster-like casing (Figure 20.2a). Flexible PCB tracks are extended to single-snap connectors situated at each extreme of the plaster. These connectors are used to affix the plaster to the subject by means of clinical ECG pre-gelled electrodes, usually placed on the thoracic region as shown in Figure 20.2b.

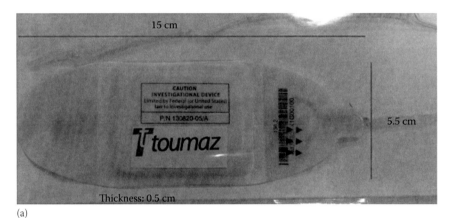

(a)

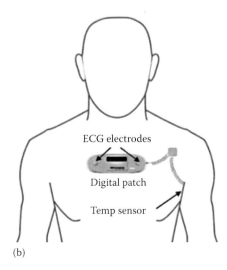

(b)

FIGURE 20.2 (a) Wireless body-worn device. (b) Location. (From Shoon-Shiong, P. et al., *Electron. Lett. Semiconductors Personalized Med.*, 47(26), S26, 2011.)

The wireless and unobtrusive nature of the digital plaster allows the subject to ambulate freely without the hindrance of wires and heavy equipment. This digital plaster uses a 1.5 V button battery. The lifespan of the plaster is of the order of seven days, sufficient for a typical patient stay in the hospital general ward. ECG, IP and the temperature are sequentially collected in cyclical fashion every 2 min. At each interval, ECG and respiratory signals are recorded as segments of 30 and 60 s, respectively, whereas only 1 s of temperature is acquired during each cycle. This functionality allows the acquisition circuitries to be switched off periodically in order to reduce energy consumption, as well as reduce the amount of data that needs to be stored within the node. The latter is important as this digital plaster has a limited storage capacity of 32 KB. For these reasons, ECG and respiratory signals are processed locally to obtain average values of HR and RR. Thus, the plaster not only is capable of storing up to 3 h of HR, RR and temperature values when in roaming mode (i.e. out of the range of the base station) but also minimizes the radio transmission payload.

20.3.2   REUSABLE BODY-WORN NODE

The form factor of the reuseable node is similar to the disposable one. The main differences are in terms of material, as well as how firmly the node may be attached to the body. More specifically, the reuseable node is necessarily made of clinical grade rubber material, to allow disinfection and reuse. In addition, the battery of this non-disposable device can be either replaced or recharged. These features are important in cases where long-term monitoring is required, that is, outpatient health care. Another field of application of this reusable technology lies in the sport arena. In this context, the sensor node needs to be firmly attached to the subject to reduce the impact of motion artefacts on the acquired signals. Therefore, mechanisms ensuring extra adhesion were considered in our design. Moreover, we included triaxial inertial sensors in this version (i.e. accelerometer and gyroscope) besides the ECG and respiratory sensing capabilities present in the disposable version.

This node can be configured to stream either raw data from sensors or processed values from the bio-algorithms continuously and simultaneously. This feature is important in sports monitoring where abrupt bursts of high-intensity activities and rapid fluctuations in vital signs such as HR and RR are common. Continuous monitoring is even more important for outpatient applications and elderly monitoring at home or when ambulating in the community. In these scenarios, information provided by inertial sensors is of paramount importance. Furthermore, information from these sensors can be combined with physiological information to provide accurate estimates of calories expenditure due to PA and fitness, useful for monitoring obese, sedentary and diabetic patients [3].

In summary, the need for continuous monitoring in both sports and outpatient scenarios supersedes the demands for energy efficiency, particularly in cases of individual and team sports, where the typical duration of the events is relatively short (e.g. 90 min in football matches and no more than 12 h in a marathon), and hence, long-lasting battery charge is not crucial.

20.4   SENSIUM™ PLATFORMS

The ambulatory monitoring application domain can be broadly categorized into two groups. In the first group, the subjects of interest ambulate within a well-defined zone and the zone of operation is relatively small. Both patients within the hospital boundaries and sport team players on court are examples. In the second group, it is difficult to define the zone of ambulation and this zone is usually large. Examples include cross-country cyclists or marathon runners. For this reason, two types of Sensium™ platforms have been designed. The first type of platform is static, and the zone of operation is populated by static base stations in order to ensure that a persistent connection is maintained between each node and a bridge at all times. This type of platform is targeted at applications with well-defined zones of operation. The second type of platform is a fully portable bridge that can either be in the form of a separate accessory carried by or attached to each subject or a small dongle that can be plugged into a portable table or mobile phone. This portable platform was envisioned for applications with ill-defined or large zones of ambulation. In each case, both the disposable or reuseable versions of the node may be used, depending on the needs of the application.

20.4.1   HEALTH-CARE PLATFORM

In most hospitals, the vital signs of normal patients within the general wards are monitored intermittently, typically at 4 h intervals. In certain cases, a normal patient can deteriorate rapidly within this period, even before the medical staff notices their condition. Reports from the National Institute for Health and Clinical Excellence published in 2005 indicated that significant illness or even death occurs, which might otherwise have been avoided had the deterioration been discovered early. This provides the motivation for the development of early warning systems to rapidly identify these potential deteriorations.

Figure 20.3 illustrates the concept of our current health-care platform. A digital plaster is attached to each person requiring inpatient care, and the patient is free to

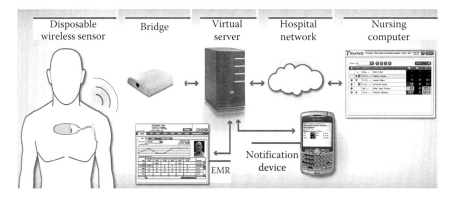

FIGURE 20.3   The Sensium™ health-care platform. (From Shoon-Shiong, P. et al., *Electron. Lett. Semiconductors Personalized Med.*, 47(26), S26, 2011.)

move within the hospital boundaries. A wireless link is maintained between the plaster and the base station, to allow the transmission of vital signs at regular intervals. Each base station is in turn connected to a virtual server, which not only provides mass storage for patient information but also sends notifications to members of the medical staff when the vital signs change beyond predetermined thresholds. Hence, this allows patients to be tracked and monitored in an unobtrusive manner either while they are staying in their rooms or when they are moving within the hospital boundaries. To ensure a comprehensive coverage, base stations are located in several locations throughout the hospital, such as corridors, bedrooms and x-ray units.

In addition, this platform can also be implemented for follow-up, home outpatient monitoring. In this context, data can be safely stored in a remote data server, so that medium and long-term trends in the vital signs can be monitored by a separate terminal that has secure access to such cloud. This feature is very useful for monitoring patients that have been discharged from the hospital with long-term chronic conditions, such as sleep apnoea [4], chronic fatigue syndrome/myalgic encephalomyelitis [5] and post-stroke [6]. Moreover, other types of patients can also benefit from the option of accessing his or her data through a web application on his or her personal computer. It is believed that this option would be suitable for physically capable patients (e.g. diabetics, obese and sedentary) who need to maintain an exercise regime and are amenable to the idea of joining a social network, comprising members suffering from similar conditions. In this manner, these patients can exchange information in a convenient way and remain motivated in their exercise regime by sharing their achievements and even competing with each other.

When the patient wanders out of the range of one base station, the digital plaster seamlessly disconnects with that base station and establishes a connection with the nearest neighbouring base station. This roaming capability extends the range of ambulation, without disrupting the monitoring process. On the other hand, if the patient moves to a point outside the zone of connection, the vital signs are stored locally on the plaster (up to 3 h of storage) and retransmitted to the nearest hot spot when a connection is re-established. This situation cannot be maintained indefinitely due to the limited memory resources that are available on the platform. In fact, this particular system was designed to work within well-defined areas of ambulation such as hospitals or home buildings. Therefore, the use of this architecture could also be extrapolated to some sports applications, for example, team games taking place in a stadium in which base stations can be distributed accordingly within the boundaries.

20.4.2 Full Portable Concept

To overcome the constraints imposed by the static nature of the health-care platform, an alternative system based on the mobile phone network has been developed. This system is illustrated in Figure 20.4. Instead of a fixed base station, a wireless link is maintained between a lightweight and small portable base station and the digital patches, which are located on different parts of the body. This can be implemented either as a software application on a mobile phone or as a portable small and lightweight hardware component with cellular mobile capabilities. Where it is possible to establish a link between the base station and the GSM/3G network, the vital signs

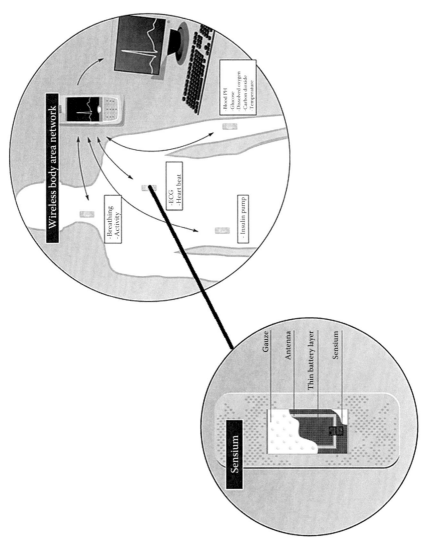

FIGURE 20.4   The Sensium™ mobile application platform. (Courtesy of ISSCC.)

are immediately sent to a remote server for storage. These data could be relayed to a PC for display over the internet, so that the vital signs of athletes could be monitored in real time for the duration of the competition. In cases where patients are undergoing outpatient treatment, these data could also be sent to hospitals, clinics or other third-party institutions so that objective measures may be used in the treatment process or as a warning system for patients who are at risk of suffering relapse of their conditions (epilepsy or post-myocardial infarction). When the bridge is out of range of GSM base station, the data are buffered within its storage memory. At present, 'smartphones' with 16 gigabytes of storage capacity or more are readily available at a reasonable cost. Therefore, depending on the number and type of vital signs that are being monitored, several days of data may potentially be stored on the mobile phone alone. In addition, the processing resources on the mobile phone may be leveraged to obtain signals with higher signal-to-noise ratios and subsequently more accurate information.

Consequently, the Sensium™ mobile application platform is useful for monitoring physiological and biomechanical information in a number of application domains, including sports and disease management. The flexibility of this platform is particularly suited to sports activities that cover a wide area, such as marathons and competitive cycling, as well as patients who are suffering from chronic conditions (sometimes recurrent and life threatening, such as certain cardiac arrhythmias) but who require periodic surveillance when at home or elsewhere in their community.

20.5 BIOMEDICAL ALGORITHMS

20.5.1 R&D Process

The research and development (R&D) framework of biomedical algorithms has to be flexible and comprehensive in terms of the exploration and selection stages. Nonetheless, it also has to be rigorous in the implementation and testing procedures of the final algorithms, in order to ensure reliability and accuracy of the software in its ultimate operating environment.

Figure 20.5 describes the methodology involved in the research and development process. First, an extensive exploration of the current state of the art should be carried out in order to determine the initial set of potential algorithms that could be implemented to perform a task with little or no adaptation/optimization. Alternatively, new techniques could also be designed. New approaches may involve the creation of a new algorithm or a combination of existing methods that would result in an enhanced version. In any case, the candidate set of algorithms should then be pruned by evaluating them based on the accuracy and computational complexity. A preliminary dataset may be necessary for this purpose, and a reference technique can be used to determine the relative reliability and performance of each candidate algorithm. For example, if the algorithm is intended for evaluation of normal and abnormal ECG rhythms, then the 'development' dataset may be obtained from well-established sources such as the MIT/BIH arrhythmia database [7]. In this case, the reference values to confirm the reliability and performance of the algorithm can be obtained from the annotation files of this database. Likewise, when

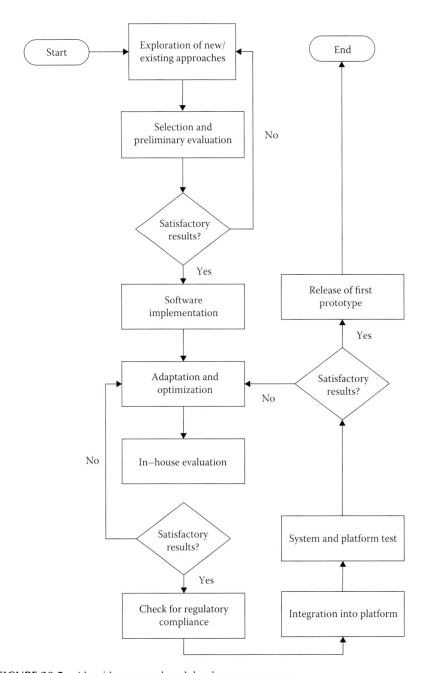

FIGURE 20.5 Algorithm research and development process.

developing either an HR or an RR calculation algorithm, synthetic data obtained from patient simulators could be initially used. Here, the reference values are known 'a priori' as indicated in the simulator screen or labels. Once these algorithms passed this preliminary test, a more elaborate test including data collection from healthy volunteers can be performed within our biomedical laboratory or with the collaboration of an academic institution (provided ethical approval is granted in both cases). Well-established methods (e.g. qualified nurse measuring the pulse at the wrist) and third-party clinical equipment (e.g. medical vital-sign monitors) can perhaps be used as valid references in this particular situation. An important aspect to consider in any case is that the resultant dataset must be statistically acceptable in terms of sample size. In addition, synthetic signals must also be as similar as possible to real signals, since aspects such as their morphology and the sensitivity of each algorithm to noise (e.g. originating from motion) are important factors in the evaluation. Algorithms can be initially implemented and evaluated using mathematical software such as MATLAB® and other statistical tools.

Although some of the existing algorithms are perhaps extremely accurate, they might not be suitable for implementation on embedded processors. Crude estimates of computational complexity can be obtained by calculating the required number of potential multiplications and accumulative (MAC) operations that could take place inside the arithmetic logic unit of the Sensium™, as well as the potential amount of storage required for variables (code space requirements are usually evaluated in the implementation phase). The feasibility of these algorithms can then be established by comparing these estimates with the available resources on the target platform. Further exploration may be required if none of the candidate algorithms proved to be reliable.

Once an algorithm has been chosen, it is implemented using a programming language suitable for the target platform (e.g. ANSI or embedded C). At this stage, the application programming interface (API) functions should be defined. This step will ease the integration of the algorithm into the platform, as well as provide a common interface at the time of porting the algorithm across different platforms. For example, the algorithm may be implemented as a binary targeted at the eWARP 8051 processor contained in the TZ1031, a dynamic-linked library (DLL) that can be used on a PC platform or even as a '.MEX' executable [8] that can be invoked for simulation purposes in the MATLAB environment. In these cases, the algorithm kernel is invoked by API functions that are defined in a common header file.

Modifications may be required to adapt and optimize the selected algorithm for the target platform. For example, numeric quantities such as coefficients contained in digital filters and other mathematical expressions/models may be strongly linked to the type of sensors used for the acquisition of the physiological or biomechanical signals that stem from the body-worn node. Therefore, algorithms need to be calibrated with actual experimental data in order to obtain adequate values for these parameters. For example, meta-heuristic optimization techniques such as simulated annealing [9] may be used to obtain the model parameters, depending on the nature of the data and the problem. In simple terms, simulated annealing operates on the assumption that the optimum solution lies relatively close to the 'seed' (initial estimate of the model parameters). Therefore, by starting with a good seed, it is highly probable that the algorithm will arrive at a near-optimal solution that is 'good enough', provided that the assumption is true and the solution space has a limited number of local minima.

In addition, the algorithm should be optimized in terms of its memory usage and computational speed. In the context of the 8051 processor contained inside the Sensium™ chip, this will imply conversion of standard (ANSI) C code into embedded C. For example, a binary variable is often defined as an 8-bit 'char'. This implies that the most significant 7 bits are redundant, so an eightfold memory saving can be achieved by storing 8 binary variables in a single byte rather than storing them separately. This memory saving is particularly substantial when the memory compaction operation is applied to large arrays. In addition, the Sensium™ contains a memory hierarchy, with a large memory bank that has a long access time and a fast register bank that has a much lower capacity. Therefore, by locating heavily accessed variables in registers rather than the main memory, the execution speed of the algorithm can be reduced. Further reductions in execution time may be achieved by making use of fixed-point arithmetic instead of floating point. These time-savings also increase with the number of arithmetic operations involved. Furthermore, divisions could also be replaced by shift operations, provided that the divisor may be expressed as an integer power of two, so as to reduce execution time.

Once the algorithm has been adapted and optimized, it should be exposed to further evaluation. Thus, the algorithm should be retested with a new and more comprehensive dataset, so as to determine its reliability and limitations in an objective manner. For example, when the HR calculation algorithm implemented on the Sensium™ was first tested, ECG data were collected from healthy individuals to compile the first evaluation dataset. In contrast, the data used for the second evaluation were collected from patients suffering from a range of arrhythmias and other conditions. Consequently, the ECG signals showed nontypical morphologies including exceptionally low QRS peak amplitudes and varying R-to-R (R–R) intervals (at rest). Using this comprehensive dataset, it was then possible to determine those scenarios where the algorithm would produce reliable estimates or failed to do so.

Rigorous evaluation techniques are necessary to determine if an algorithm is fit for purpose. It is essential to select the appropriate technique based on the nature of the algorithm as well as the objectives of the evaluation. For instance, an algorithm might be designed to accept or reject results (e.g. HR) based on certain features (e.g. the amount of noise in the signal, the amount of variation in the peak-to-peak intervals) of the input signal. Frequently, it is not possible to accept or reject signals with complete accuracy, based on these features. Figure 20.6 illustrates this problem: the probability density functions (pdfs) of a feature for accepted and rejected results are shown. The overlap between these pdfs represents 'an area of fuzziness' where an accurate distinction between reliable and unreliable results cannot be made based on this feature alone. Indeed, by situating the threshold for the feature in between the pdfs, the classification error rate will be proportional to this 'area of fuzziness'. More specifically, classification events can be categorized into four different types: true positives ((TP) the algorithm correctly accepts a result), true negatives ((TN) the algorithm correctly rejects a result), false positives ((FP) the algorithm incorrectly

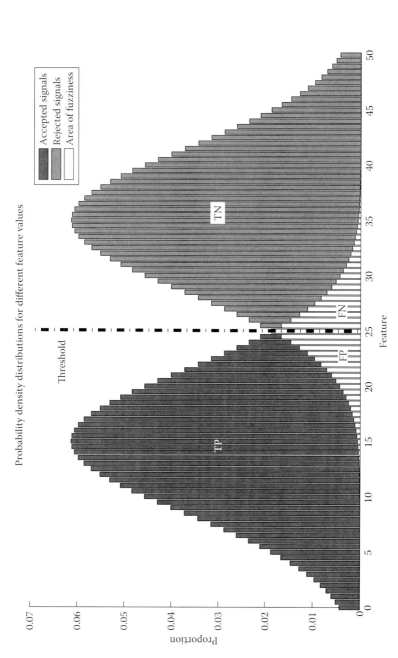

FIGURE 20.6 Probability density distributions of a feature for accepted and rejected signals; classification events including TP, TN, FP and FN are depicted.

accepts a result) and false negatives ((FN) the algorithm incorrectly rejects a result). By reducing the threshold value, the number of FPs can be reduced at the expense of the number of FNs and vice versa. Clearly, it is important to consider all four types of events when determining the classification accuracy of the algorithm.

The four different types of classification events can be combined to give a more holistic picture of classification accuracy. The positive predictivity (P+) and sensitivity (Se) indicators can be used for this purpose. Sensitivity refers to the ability of the algorithm to accept the correct results, whereas positive predictivity refers to the ability of the algorithm to discriminate between correct and incorrect results. These indicators are expressed in Equations 20.1 and 20.2 [10]:

$$Se = \frac{TP}{TP + FN} \times 100 \qquad (20.1)$$

$$P+ = \frac{TP}{TP + FP} \times 100 \qquad (20.2)$$

Another type of evaluation involves the statistical comparison between the results that are generated concurrently by the algorithm and the reference device. In this case, the null hypothesis is as follows: the mean values of the results generated by algorithm and the reference device are equal. Before this hypothesis can be tested, it is necessary to check if each dataset is normally distributed. Several types of normality tests (e.g. Kolmogorov–Smirnoff) may be used for this purpose. During these tests, the normality of the dataset is determined by the 'goodness of fit' between the dataset and standard Gaussian curves. Assuming that the dataset is normally distributed, the Student's t-test can be applied to the dataset to determine the validity of the null hypothesis. This test yields a 'p' value, where a high 'p' value indicates that the observed difference between the dataset is a random occurrence (both techniques are interchangeable), whereas a low 'p' value indicates the differences between the datasets are statistically significant. Specifically, the differences between the datasets are statistically significant with a confidence level of 95% if $p < 0.05$, and the null hypothesis is rejected. On the other hand, the Wilcoxon signed-rank test can be used for paired datasets that are non-Gaussian. In simple words, differences between each pair of results (generated at the same instance by the algorithm and reference device) are weighted (ranked) according to their absolute value before being aggregated for the calculation of the 'p' value. Also, paired results that are identical are discarded. Similar to the t-test, the null hypothesis may be accepted or rejected by considering the 'p' value.

In many cases, statistically significant differences between the reference device and the algorithm may not be clinically relevant. For example, small errors in the HR calculation are often unnoticeable to doctors and unlikely to affect their diagnosis. Nevertheless, it is important to quantify this error in a meaningful manner. A typical way of comparing two normally distributed and paired datasets is by calculating the Pearson's correlation coefficient (r). High absolute values of r imply that both datasets are linearly dependent. The Spearman's rank correlation coefficient may be used to quantify the extent of dependence for datasets that are non-Gaussian. It is important to note that the correlation coefficient has to be interpreted carefully while one is trying to ascertain the similarity between two datasets. Indeed, two techniques that are in agreement should have a high correlation coefficient. On the other hand, an 'r' does not automatically mean that there is an agreement between the two techniques. Rather, it simply means that both datasets are linearly dependent and that the data collection process is well controlled. Consequently, correlation analysis has to be augmented with other indicators before one can conclude if there is an agreement between two techniques.

One technique that can be used to determine if there is agreement between two techniques is the use of the Bland–Altman plot [11]. An example of such a plot is shown in Figure 20.8. For each pair of results, (X_i, Y_i), the corresponding location on the Bland–Altman plot is $((X_i + Y_i)/2, X_i - Y_i)$. If there is an agreement between both techniques, the resulting plot should be tightly clustered around the horizontal axis. In Figure 20.7, the mean error between both techniques is non-zero and positive. Consequently, the bias (horizontal line in the middle of the figure) is situated above the axis.

The confidence interval may be used to characterize the extent of variation between the two techniques; the 95% confidence interval is shown in (20.3). However, as the sample size (n) becomes large, the confidence interval becomes too narrow to reflect the variation in the dataset. For this reason, the 95% prediction interval (PI) is used

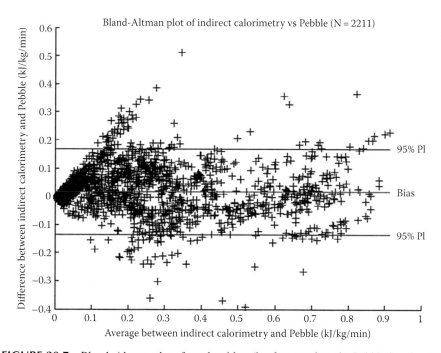

FIGURE 20.7 Bland–Altman plot of an algorithm (implemented on the Pebble Sensium™ device) versus a reference technique (indirect calorimetry).

instead. This measure is expressed in Equation 20.4 and is similar to the confidence interval with the exception of the sample size. These bounds are also shown using the two horizontal lines (bounding the data points) in Figure 20.7:

$$\text{Confidence interval} = \left[X' - 1.96\frac{s}{\sqrt{n}},\, X' + 1.96\frac{s}{\sqrt{n}} \right] \qquad (20.3)$$

where
   X' is the sample mean
   s is the standard deviation
   n is the number of samples

$$\text{Prediction interval} = \left[X' - 1.96s,\, X' + 1.96s \right] \qquad (20.4)$$

The statistical techniques that are used to evaluate two datasets have been discussed earlier. There are cases where it is necessary to compare three or more datasets. For instance, one might like to compare several HR algorithms and a reference device. In such a situation, the analysis of variance (ANOVA) may be used to determine if the differences between the mean values of these datasets are a random occurrence. In addition to this null hypothesis test, the non-parametric F-test can be used to quantify differences between these datasets based on the inter-variance between datasets, with respect to intra-variance within each dataset.

After passing these rigorous evaluations, the next step was to incorporate the whole dataset (or part of it) as testing vectors in an automated unit test, as it provides software developers with tools to retest the code during the code refactoring and system integration processes. Our experience suggests the implementation of this 'developer's testing unit' in standard C, so that it can be ported to most platforms (e.g. PC and embedded software). If the performance of the algorithm falls short of expectations, further adaptation or optimization would be necessary. Another important step to follow after passing in-house evaluation is code refactoring and verification, in order to ensure it complies with industrial standards, such as Misra C [12] (standard used in safety critical industries).

After attaining full compliance, the algorithm is ready for integration into the target platform. As a result, its modules should work seamlessly with other components of the platform, including the data acquisition sensors and the RF. At this stage, it is important to design rigorous test cases to ensure reliability and adequate performance. As a first test, it is important to ensure that the correct vital signs have been transmitted to the bridge. For instance, the patch could be connected to a patient simulator that is generating ECG signals, with known HR values. The HR that is transmitted to the bridge should then be compared with the reference value so as to determine if the system was operating correctly. In addition, it is pertinent to ascertain the range of the radio in realistic scenarios, so that the rate of false alarms, as a result of broken connections, may be reduced. In addition, usability tests at the system level and focus groups are also essential to determine possible problems when the product is deployed in the field. Thereby, based on both feedback from users and the perceptions of the multidisciplinary team involved in the focus group, further adaptations of the algorithms would perhaps be needed. For example, patients might feel that the device is too cumbersome and uncomfortable. Therefore, the form factor of the device would need to be reconsidered/redesigned not only to reduce its size but also to enhance its ergonomics. Once the results from the system and platform-level tests are satisfactory, it is then the right time to release the first prototype.

An important aspect to bear in mind is that the core algorithms were written in standard C, so they can be implemented in different platforms. However, additional 'wrapping' code may need to be produced in order to adapt the interface (e.g. input/output arguments) across different applications. For instance, in a PC application running under Microsoft Windows, an algorithm may be integrated as DLL. In this case, wrappers must be written to satisfy the syntax and structure of the selected programming environment (e.g. C#, C++, Visual Basic, Delphi) in which the user interface and other components of the platform are implemented.

20.5.2   CHALLENGES AND TRADE-OFFS

Several challenges were found during the development and evaluation processes of each algorithm. Due to the ambulatory nature of our technology, the most common problem is the corruption of the acquired signals by motion artefacts. When analysing data from research and clinical trials, we noted that this issue was more notorious in respiration signals resulting from IP. Furthermore, some of these IP signals were clinically irrelevant, as they were recorded when the patient was talking, coughing and swallowing. There were also situations when a body node was faulty due to manufacturing defects or when the electrodes are faulty or not applied properly. In all these cases, digital signal-processing techniques can be useful for cleaning the signals and/or for identification of spurious events. However, these techniques must be chosen carefully, bearing in mind that they could work reliably under constraints imposed by the platform. This topic is further elaborated in subsequent sections.

20.5.3   HEART RATE CALCULATION ALGORITHM

HR is probably the most widely used physiological parameters in the medical and sports monitoring arena. Its calculation requires a reliable algorithm capable of detecting heartbeats from both clean and noisy ECG signals. Furthermore, the robustness and reliability requirements for this algorithm are more demanding in ambulatory applications – that is, it has to be energy efficient, be robust against motion artefacts and other sources of noise and work accurately within the memory and processing constraints of the body-sensor node. Therefore, an extensive review was carried out on existing algorithms during the design process. Among the set of candidate algorithms that were considered, the algorithm that was found to meet these requirements was the one that was designed by Hamilton and Tompkins [13].

Thus, the HR calculation algorithm that was deployed on the Sensium™ platform was derived from [8], and it comprises two stages (Figure 20.8). The preprocessing stage was responsible for emphasizing the QRS component of each beat, attenuating the P and T waves as well as noise. This stage facilitated accurate peak detection in

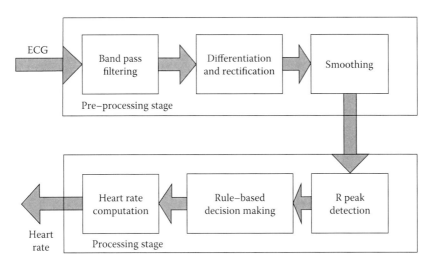

FIGURE 20.8   HR computation algorithm.

the subsequent processing stage. A further set of rules is applied in the selection of the actual QRS peaks, to allow the identification of authentic R peaks. Finally, the HR value is computed as a function of these peak locations.

In order to achieve its goals, the preprocessing stage first filters the ECG signal to eliminate unwanted components outside the predominant frequency band of the QRS complexes (8–16 Hz). Subsequently, the signal is differentiated and rectified to emphasize high-frequency regions of the signal where the QRS complexes were likely to reside. In addition, smoothing was applied to reduce the number of local peaks that were not R waves.

After the preprocessing stage, peak detection takes place throughout the resulting signal. A set of rules (based on the physiological characteristics of R peaks seen in the ECG) is then applied to determine if each of the obtained peaks corresponds to a genuine R peak. For example, since the maximum HR is no greater than 300 bpm, the interval between each pair of R peaks should be at least 200 ms. In addition, the QRS complexes contain higher-frequency components than the P and T waveforms, and the maximum gradients of consecutive QRS complexes tend to be similar among patients. Thus, peaks from T and QRS waveforms can be distinguished based on their maximum gradients. Moreover, the HR of normal human subjects tends to decrease slowly. Therefore, to further account for genuine changes in the R peak amplitude, the amplitude thresholds for R peak identification were adaptively adjusted when no new R peak (above the current R peak amplitude threshold) was discovered within 1.5 R–R intervals (this interval was computed as a function of previous R–R intervals).

As reported by Hamilton and Tompkins [8], this algorithm demonstrated to be effective in predicting HR with continuous ECG signals. In other words, it was found to have sensitivity and positive predictive values of greater than 99% when tested with the MIT/BIH database [14]. However, these testing signals were acquired from patients resting in bed and with more than two electrodes positioned in standard

ECG leads configuration. In contrast, the digital patch is affixed to the chest in a non-standard position using pre-gelled ECG electrodes inter-separated by a distance of approximately 10 cm (Figure 20.3). As a result, ECG signals acquired using this configuration will differ from those measured following standard positions. Based on our experience, in many cases ECG signals from the patch are characterized by more pronounced T waves with QRS smaller in amplitude. Consequently, larger number of false detections can occur, misleading the calculation of the HR. Another important consideration is that the original algorithm was developed for continuous and prolonged ECG signals. Conversely, in the case of our health-care platform, the body-sensor node collects intermittent epochs of ECG (30 s of duration) in a cyclical fashion (every 2 or 10 min), aiming to reduce power consumption. For this reason, the algorithm suffered a number of modifications in order to adapt and improve its performance prior to its implementation into our ambulatory systems. Last but not the least, the ambulatory nature of the acquired data requires a distinct approach capable of discriminating clean ECG epochs from those severely corrupted by motion artefacts, faulty hardware, bad electrode contact and other sources of interference.

A preliminary dataset was created from 10 healthy volunteers (who were resting, sitting and cycling) for the purpose of evaluating the algorithm and tuning the parameters of the algorithm for optimum performance. This dataset was collected using our digital plaster and a reference device (Philips IntelliVue MP30 patient monitor).

The algorithm was first adapted to work with intermittent ECG segments – the major difference between the adapted and original algorithm is the manner in which the HR was reported. Specifically, the original algorithm was designed to report an instantaneous HR for every QRS peak that was detected, whereas the adapted algorithm reported one HR for each segment of the ECG signal; this HR was computed as a function of the R–R intervals that were detected in the epoch. The median operator was used within this function so as to reduce the influence of outliers (caused by spurious peaks) on the computed HR. A second adaptation was implemented to accommodate for the alterations seen in the ECG as a result of placing the electrodes in the non-standard position. In the peak detection algorithm, the QRS peak threshold was dynamically adjusted using previous QRS and noise peak amplitudes. A fixed gain factor, alpha, was used to determine the rate of adjustment and the final threshold value. This gain factor is readjusted to account for more pronounced T waves within the ECG signal using the preliminary dataset. For this purpose, an error metric was formulated as the sum of absolute difference (SAD) between the HR values simultaneously collected from a valid reference device and the Toumaz HR calculation algorithm. Subsequently, a software simulation was carried out where the HR calculation algorithm was applied to the dataset iteratively. At each iteration, the gain factor was changed and the corresponding error was calculated. The results of the simulation are shown in Figure 20.9, where different values of *alpha* and their corresponding errors are shown. From this analysis, the desirable value for *alpha* was selected where the SAD is minimized and employed in the adapted version of the algorithm.

A third adaptation was the inclusion of a confidence indicator module, whose function is to verify the validity of the computed HR. In simple words, if the calculated HR results from a corrupted ECG segment, then the result must be discarded

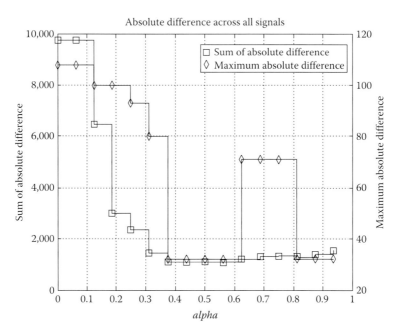

FIGURE 20.9 Plot of the sum of absolute differences and the maximum absolute differences for different values of *alpha*.

and labelled as invalid. Several versions of the confidence indicator were considered in the design process. Specifically, it was found that reliable HRs can be computed from ECG signals with relatively stable R–R intervals and peak amplitudes, whereas these parameters tended to be less stable for corrupted signals. Therefore, the amounts of variation in the R–R intervals and the peak amplitudes are considered as features for the detection of corrupted signals. In addition, compared with reliable signals, the number of non-QRS peaks is generally higher for corrupted signals due to noise peaks. Therefore, the proportion of non-QRS peaks is considered as another viable feature.

After a desirable set of features have been identified, thresholds have to be determined for each feature. For example, if the proportion of non-QRS peaks with respect to the total number of discovered peaks exceeded $T\%$, the ECG segment is deemed to be corrupted. Otherwise, the ECG segment is considered a reliable signal. The determination of thresholds is a delicate process, involving trade-offs between the number of false positives (FPs) and false negatives (FNs). In this context, an FP is defined as a signal that is accepted as a valid signal when it should be rejected. On the other hand, an FN is defined as a signal that is rejected as an invalid signal when it should be accepted. These trade-offs are detailed in the 'R&D Process' section and probability distribution functions shown in Figure 20.6. The algorithm can be made more conservative (e.g. by reducing the threshold T) or more liberal in accepting signals. While a liberal algorithm is undesirable because of high frequency of erroneous HRs that are produced, an overly conservative algorithm is unacceptable if it results in a high false alarm rate (the patient is falsely deemed to be at risk due

TABLE 20.1
Different Testing Datasets for the HR Calculation Algorithm

Source	Total Number of Epochs	Description
Rigel 333	146	Patient simulator episodes involving normal ECG morphologies but with normal and abnormal rhythms ranging from 30 to 240 bpm
SimMan®	101	Patient simulator episodes involving different ECG morphologies and normal and abnormal rhythms, common in the clinical environment
Normal Subject	20	Episodes recorded from one normal subject following a protocol for testing a number of 'lead off' conditions
Patients	1101	Episodes collected from patients during a clinical trial, involving 44 individuals with high BMI, pulmonary oedema, cardiac arrhythmias and low-voltage QRS complexes resulting from ventricular hypertrophy

to the large number of invalid signals produced by the algorithm). The final version of the confidence indicator module provided the best compromise between accuracy and the number of false alarms.

After the adaptation process, the algorithm was tested with a relatively large dataset, comprising of synthetic data (from patient simulators) and ECG signals that were collected from healthy volunteers and patients while resting. The size and description of each dataset are shown in Table 20.1. Comparisons were made between the results from the Toumaz HR algorithm and the valid reference. Before performing an in-depth statistical analysis, the data were tested for normality using the Kolmogorov–Smirnov test, part of the commercial statistical analysis software used in this investigation. The results demonstrated that the data were not normally distributed. Therefore, a non-parametric statistical test (Wilcoxon signed-rank test, two tailed, $p = 0.05$) was chosen to evaluate the difference between both the HR calculation algorithm and the HR algorithm of the reference device.

Using the dataset that was collected from the Rigel 333 patient simulator, the HR calculation algorithm attained an accuracy of 100%. However, it rejected some signals, but the sensitivity and positive predictivity of the algorithm were still large at 94.5% and 100%, respectively. From the SimMan® dataset, it was found that all the RRs produced by the algorithm were within ±2 bpm of the reference values. Also, the algorithm was unable to work for a number of arrhythmias, including atrial fibrillation (AF) and ventricular tachycardia (VT). In total, the algorithm was able to calculate accurate HRs in 95% out of 98% of the total number of signals of the dataset. A thorough statistical analysis (detailed in the preceding section: R&D Process) is performed on patients data and a high level of agreement was found between the digital plaster and the patient monitor.

The HR calculation algorithm has a number of limitations. Firstly, it was found that the algorithm cannot provide valid HRs for ECG signals with extremely low

QRS peak amplitudes (less than 8 ADC counts). Secondly, the confidence indicator is ineffective for certain arrhythmias such as AF and VT where there are substantial variations in the R–R intervals. As part of future work, the HR calculation algorithm can be improved by preprocessing the signal using more advanced techniques, as well as making use of a more sophisticated confidence indicator. For instance, adaptive filters may be used to achieve a higher signal-to-noise ratio if a second ECG channel can be made available as a reference signal. A better confidence indicator could be designed by identifying better features. In addition, sophisticated classifiers can be used to more accurately identify valid and invalid signals. They include intelligent machine learning techniques such as artificial neural networks and support vector machines (SVMs) as classifiers.

20.5.4 Respiratory Rate Calculation via Impedance Pneumography

Deterioration of the normal pattern of respiration is often referred as the primary cause of critical illness and the most common reason for admission of patients in intensive care [15]. In addition, alterations of the RR are also indicators of the physical strains imposed on the human body during strenuous PA or exercise. There are direct and indirect methods to measure RR (please refer to [16] for detailed information. Of these, IP not only is a method approved by the FDA but also offers a practical alternative involving the use of low-cost disposable sensors. Another advantage is that respiratory activity can be sensed with the same pair of electrodes used to acquire ECG. This technique has been successfully used in different monitoring applications for apnoea in infants [17] and adults [18] and for other sleep disorders in patients with cardiovascular diseases [19]. Thus, we opted for the on-chip implementation of the IP method as part of the Sensium™ family.

The working principle of IP is as follows: variations of the air volume of the lungs resulting from inhalation and exhalation while breathing induce small and periodic changes in the body impedance (between 0.1 and 3 Ω) that can be detected as voltage fluctuations across the thoracic region when using appropriate instrumentation. The latter involves the injection of a low-intensity ($\leq$100 µA) high-frequency (i.e. from 20 to 100 kHz) AC current through the chest wall using conventional pre-gelled surface (skin) electrodes. Thus, the typical pattern of a clean waveform resulting from IP respiratory activity is quasiperiodic by nature and resembles a sine waveform. Unfortunately, this is not always the case as the IP technique is very sensitive to motion artefacts, affecting the quality of the IP respiration signals. In simple terms, perturbation of the equilibrium of ionic charges around the electrode–electrolyte interface occurs as a result of mechanical strains imposed by movement on the electrodes. Consequently, such a perturbation results in unwanted alterations of the voltaic potential of the electrodes, adding errors to the dynamic potential difference signal measured/recorded across the thoracic region [20]. As mentioned earlier in this chapter, some of the signals generated by this method (in the absence of motion artefacts) are not clinically relevant. For example, human voice results from vibration of the vocal chords induced by air outflow. Thus, the rhythmic respiration pattern is disrupted when the person is talking. Likewise, food or drink intake interrupts the natural pattern of breathing. Therefore, resultant IP signals are not clinically relevant in these cases and hence must be discarded. All these factors together with the current platform limitations (i.e. intermittent respiration segments of 60 s each, lack of additional channel for reference signals from other sensors and electrodes and limited memory and processing capabilities in the body-sensor node) introduced defiant challenges in the respiration algorithm design – that is, must be simple but without compromising the reliability and robustness of the RR estimations. Thereby, the respiratory algorithm designed here does not involve the use of expensive computational DSP techniques to preprocess the signal. Similar to other approaches [9], it relies on a systematic set of rules to determine the validity of the signal instead. Then, the algorithm either returns an RR value when the signal corresponds to genuine respiration activity or an 'invalid data' message when the signal is distorted or clinically irrelevant.

The flow chart of the respiratory algorithm is shown in Figure 20.10. One-minute segments of respiratory signals are first passed through the motion detection and discriminator module, which, based on the slope and amplitude characteristics of the respiratory waveform, is capable of rejecting those signals severely deteriorated and fragmented by motion. If accepted, the signals are subsequently fed to a conditioning stage that first performs data centring and gain adjustment. The latter is necessary not only because the respiration event detector (explained later) requires positive and negative values but also because input offset errors are variable between digital plasters.

Thereafter, the signals are filtered to cancel the in-band noise produced by dynamic changes in the impedance of the heart as it contracts and expands (these signal distortions are commonly known as 'heart bumps'). It has been shown in past that a self-tunable analogue hardware biquad filter is effective in removing 'heart bumps' [21]. Therefore, we implemented a digital software version of this type of filter that uses preceding HR information to adjust its cut-off frequency in accordance with the frequency of the present heart contaminants. Once this artefact is removed, the signal is further processed by first-order recursive low- and high-pass filters, whose coefficients were calculated for a sampling rate of 25 Hz and cut-off frequencies of 2 and 0.1 Hz, respectively. Thus, other sources of contamination outside the respiration signal bandwidth (e.g. muscle and mains noise) are successfully removed. The coefficients of the biquad transfer function vary according to the input HR. Since typical HRs vary within a range of 30–300 bpm (0.5–5 Hz), multiple pole-zero analysis was performed (in steps of 0.5 Hz) across this range of input frequencies so as to ascertain the stability of the filters. Since the remaining filters have transfer functions with constant coefficients, a single pole-zero analysis is required to determine if the filters are stable.

After preprocessing the respiratory signal, the algorithm searches for remainder motion artefacts, only identifiable by their non-periodic behaviour. For such purposes, the algorithm first invokes the respiration events detection module and then searches for periodicity irregularities in inhalation-to-inhalation (I–I) and exhalation-to-exhalation (E–E) intervals.

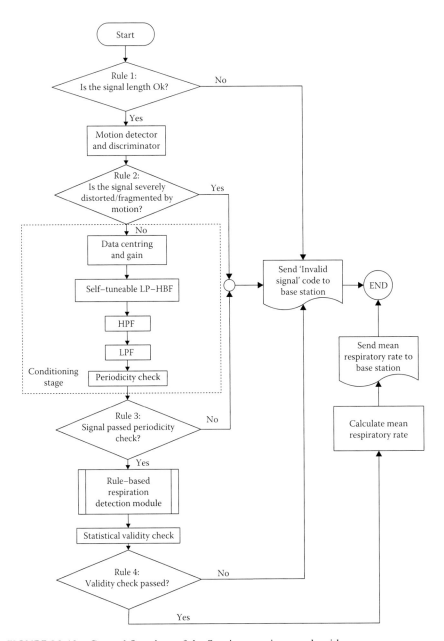

FIGURE 20.10   General flowchart of the Sensium respiratory algorithm.

Prior to the final validity check (i.e. rule 4), the algorithm runs once again the rule-based respiration events detector. This module consists of a three-point sliding window peak/trough detector and a routine that identifies the respiration events (i.e. inhalation peaks and exhalation troughs) based on a number of rules and thresholds associated with typical amplitude and time domain features of the IP respiration signals (Figure 20.11).

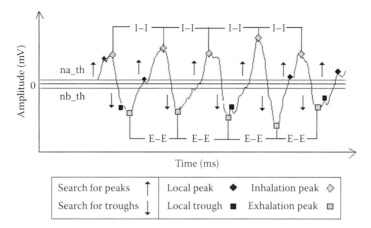

FIGURE 20.11 Detection of respiration events. The algorithm searches for peaks when the sliding window moves upwards above the noise margin (na_th). In contrast, the algorithm searches for troughs when the sliding window moves downwards to the minimum values of the signal below the noise margin (nb_th). I–I and E–E intervals comprise the distances between maximum peak values and between minimum trough values obtained when the searching direction changes as the sliding window passes through the zero-crossing points.

Subsequently, the statistical validity check first evaluates if the number of respiration peaks and troughs detected in the previous stage is the minimum necessary for calculation of the breath rate per minute (BrPM), and then it verifies whether or not certain metrics of statistical dispersion of I–I and E–E exceed a preset percent value. Finally, if the signal passed the last validity check, the algorithm would return the mean RR from the stored I–I and E–E intervals found in 1 min of data. Otherwise, the algorithm would return a code, which would be sent to the base station to indicate that the signal was invalid.

Similar to the HR case, the respiration algorithm was evaluated in terms of its ability to distinguish between corrupted and unreliable signals, as well as the accuracy of the accepted signals with respect to a reference device (Philips IntelliVue MP30 patient monitor). Similar to the HR, a test dataset is created from signals that were generated using a patient simulator (Rigel 333), artificial signals that were created by software as well as signals that were acquired from 10 healthy volunteers (500 test signals in epochs of 1 min were acquired from these subjects).

The results of the assessment first revealed that the algorithm successfully rejected those severely distorted, avoiding false-positive results. However, the approach was shown to be conservative by rejecting some traces from which the rate could have been extracted with accuracy if they were accepted. When the algorithm was applied to signals that were generated by the patient simulator, as well as artificially generated signals with white noise that were superimposed on it, the algorithm was consistently able to produce RRs that were within ±2 BrPM of the reference values, between 4 and 120 BrPM. A statistical analysis revealed that there were statistically significant differences between the results that were generated by the reference device and the algorithm ($p < 0.05$). On the other hand, the 95% confidence interval was found to be [0.01 0.80] BrPM, indicating that this difference is small.

In summary, the algorithm was found to be accurate in distinguishing between corrupted and unreliable signals, as well as estimating the RR. On the other hand, it was particularly sensitive to motion artefacts, leading to a high signal rejection rate. The limited amount of processing and memory resources hamper the use of more advanced signal-processing techniques to de-noise the signal. However, an alternative algorithm that is more robust against motion artefacts may be useful in certain cases. This algorithm will be discussed in the following section.

20.5.5   ECG-Derived Respiration

As discussed earlier, respiratory (IP) signals are easily distorted by noise, leading to a high rate of signal rejection. This motivated researchers to explore alternative techniques relying on the ECG signal. The first approach consists of reconstructing breathing patterns from normal variations of the heartbeat rhythm induced by respiration – that is, respiratory sinus arrhythmia (RSA) [22]. More specifically, in the autonomous nervous system, the parasympathetic branch regulates the firing rate of the sinoatrial node (SA) by means of the vagus nerve. Specifically, during an exhalation event, there is increased parasympathetic nervous system input to the heart via the vagus nerve, leading to a decrease in the firing rate of the SA and a subsequent reduction in instantaneous HR. Conversely, during an inhalation event, the vagus nerve remains unstimulated, leading to a relative increase in the instantaneous HR.

The second alternative is based on the fact that when the subject is breathing, the electrodes move relative to the heart, and thus, the electrical impedance of the thoracic cavity changes in a rhythmic fashion. This results in a persistent amplitude modulation of the QRS complexes [23] observed in the ECG. Thereby, when considering both approaches, it is possible to construct two waveforms from the ECG signal – that is, one built from periodic changes as induced by respiration in the HR (Figure 20.12d) and the other from fluctuations in QRS peak amplitude (Figure 20.12c). These waveforms are known as ECG-derived respiratory (EDR) waveforms.

Based on previous studies [24], it was possible to design a simple algorithm to construct and interpret EDR waveforms. Figure 20.13 shows the sub-modules within the EDR algorithm. The cubic spline interpolation module was used to recover the smooth, interpolated waveforms directly from the QRS peaks and the instantaneous HR values. The respiratory algorithm, which was described in the preceding section, was used to compute the RRs from these waveforms. Subsequently, a variety of confidence metrics may be used in the selection of RRs. One simple metric that may be used is the mean absolute difference (MAD) in the set of peak-to-peak intervals that were produced by the respiratory algorithms. A lower MAD is indicative of lower variance in the peak-to-peak intervals and therefore a more reliable RR. The MAD provides a crude indicator for the reliability of the waveform. However, it fails to capture other aspects of the signal such as its amplitude and spectral content. A more informed decision about the signal may be made by considering more complex

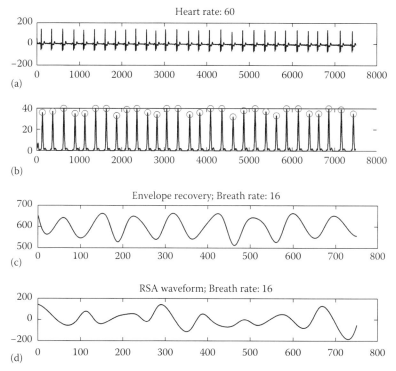

FIGURE 20.12 (a) Original ECG signal. (b) Filtered ECG data. (c) EDR waveforms generated from QRS amplitude modulation. (d) RSA.

metrics, such as the complexity of the waveform [25]. While such techniques are potentially more accurate, they require more computational resources.

The assessment of this algorithm involved several steps. First, the dataset used for verification was collected in two separate experiments. In both, the subjects underwent a variety of activities including step exercises, cycling on a stationary bicycle, walking and running on a treadmill. These subjects underwent higher-intensity activities in the first experiment, compared to the second experiment. In addition, in the second experiment, the subjects had greater freedom in choosing the frequency at which these activities were carried out. Consequently, two datasets with a wide range of RRs were produced from these experiments.

The reference device (MedGraphics CPX Express) that was used for evaluation purposes was the indirect calorimeter. Essentially, this device is a gas analyser that measures the concentration of oxygen and carbon dioxide in the inspired and expired gases from a volunteer. By detecting changes in air pressure, the device was capable of determining the time instances of each inhalation and exhalation event and calculating the resulting instantaneous RR. For the purposes of analysis, these RRs were aggregated over a period of 60 s into a single RR (using the median operator), so that direct comparisons between the reference results and the Toumaz EDR algorithm can be made.

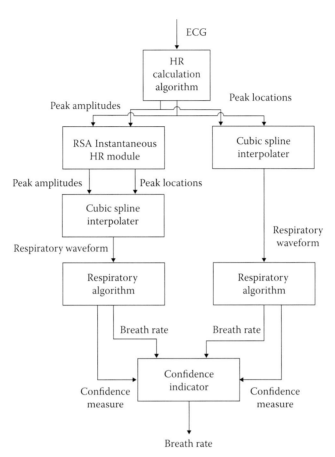

FIGURE 20.13   Top-level diagram of the EDR algorithm.

Consistent with the R&D process, the Kolmogorov–Smirnoff normality test was carried out on both the reference and the EDR algorithm results. These datasets were found to be normally distributed except for the EDR results from experimental dataset one. Therefore, statistically significant differences were found for both datasets ($p < 0.05$). When the datasets were combined, the 95% PI was found to be [−6.85, 11.46] BrPM, and the mean difference was found to be 2.31 BrPM. In addition, the Pearson's correlation coefficient was 0.77 ($p < 0.05$). These findings indicated that while differences between both datasets are statistically significant, these differences are relatively small.

In summary, we found several advantages in using EDR techniques as part of our systems. Firstly, motion artefacts do not drastically affect the ECG ambulatory signals acquired by our wearable devices. Preliminary experiments performed in our laboratory revealed that the technique was robust and reliable when tested with healthy volunteers running on a treadmill at speeds reaching 8 mph. Secondly, since the heart's electrical activity is involuntary by nature, the intrinsic characteristics of the ECG waveforms remain stable and clinically relevant in different situations such as eating, talking and drinking. Thirdly, additional instrumentation for data acquisition (such as in the case of IP) is unnecessary as the EDR algorithm works with single-lead ECG signals. Finally, based on our experience, the high signal-to-noise ratio seen in the ECG indicated that EDR waveform construction can be easily done with minimal computational resources. For these reasons, the EDR technique constitutes an attractive alternative for measuring respiratory frequency in the clinical setting (e.g. studies related to sleep apnoea and Cheyne–Stokes in congestive heart failure) [15] and potentially in the sport arena. However, this approach has some limitations. The RSA phenomenon is very pronounced in individuals who are young and in excellent cardiovascular health, and its effect declines with age [26]. Therefore, the EDR algorithm is perhaps suitable only to determine the RR in certain groups of subjects. In addition, the algorithm is incapable of determining RR if it is more than half of the HR. As each QRS peak represents a sample in the reconstructed respiratory waveform, the effective sampling rate (the number of QRS peaks per second) has to be more than twice the highest frequency component in the respiratory waveform, according to the Nyquist–Shannon sampling theorem.

20.5.6 Physio-Mechanical Algorithms

Until now, we only described algorithms using physiological signals (ECG and IP respiration). The inclusion of the accelerometer outputs expands the set of parameters that can be measured using our body nodes. An example is PA. Indeed, as the lifestyles of individuals become increasingly sedentary, they become susceptible to a wide range of diseases including diabetes and obesity [27]. The physical activity intensity (PAI; measured in calories/kg/min) provides an objective metric by which PA could be measured and for medical professionals to determine if their patients are exercising sufficiently. PA can be measured relatively accurately using the branch equation model [28], which makes use of the HR and the energy present in the tri-axial accelerometer signals (both of these parameters have shown a well-established relationship with oxygen consumption (VO2) and energy expenditure at submaximal workloads). Thus, the method can be used to estimate PA energy consumption during aerobic exercise, and its accuracy is greatly dependent on the type of accelerometers that are used. Nevertheless, we show that the model can be calibrated with comparable accuracy to the original model [29]. Another important aspect is that PAI is also on the type and intensity of activities. Therefore, determining the activity type can potentially increase the accuracy of the model [30]. Different studies showed that it is possible to discriminate between activities from the accelerometer signals and some physiological parameters (e.g. HR) [28]. These approaches usually adopt pattern recognition techniques such as decision trees, neural networks and SVMs [31–33].

Other metrics such as the walking speed and the body tilt can be measured by means of the accelerometer signal. These metrics could be useful as indicators for the assessment of progressive degenerative conditions such as Duchenne's muscular dystrophy [34]. In the latter, sufferers experience increasing muscle loss and increasing difficulty in daily living activities such as walking and maintaining an upright/normal posture when sitting and standing.

20.5.7 ECG COMPRESSION

When the computational requirements of the target application exceed those of the Sensium™ platform, signal compression may be necessary to reduce the bandwidth and energy requirements of transmission. The effective compression ratio is dependent on the signal characteristics. An extensive review has revealed that there exist a large number of studies in the area of ECG data compression [35–39], and a compression rate of up to 120 times has been reported in one case [40]. On the other hand, in selecting the compression algorithm for the target platform, it is essential to balance the energy and computational requirements of the compression algorithm with savings that are derived from reductions in the amount of data that has to be transmitted. Other well-known compression algorithms that are less computationally demanding, with correspondingly lower-compression ratios, include the CORTES [37] and the AZTEC [35] algorithms. These compression techniques exploit redundancies within slowly varying part of the signal while preserving the quality of fast-varying parts of the signal. Note that the power of residual difference (PRD) is a measure of reconstructed signal quality (see Equation 20.5), where x_{org} and x_{rec} refer to the original and reconstructed signals, respectively. The PRD is standard metric that is used as a means of comparison between different compression techniques [add reference]. In addition, the differences in sampling frequencies and precision have to be taken into account when making comparisons between different techniques:

$$PRD = 100 \times \sqrt{\frac{\sum_{i=1}^{N}\left[x_{org}(i) - x_{rec}(i)\right]^2}{\sum_{i=1}^{N} x_{org}^2(i)}} \qquad (20.5)$$

20.6 CONCLUDING REMARKS

This chapter presented a brief description of the Sensium™ platform and its potential use across a wide range of applications, including the sports and health-care domains. The ultralow power consumption, low-cost and wireless characteristics of the body-sensor nodes are important aspects of the system, allowing continuous monitoring in an unobtrusive manner for prolonged periods of time. Nevertheless, these features introduced storage and processing limitations. Therefore, next generations of the Sensium™ microchips will be envisioned to incorporate more advanced hardware features to enhance processing and storage capacity without compromising the energy efficiency of the platform.

The methodology followed to design and develope the embedded algorithms for the Sensium™ was also discussed in this chapter. We hope that researchers and software engineers find this simple procedure useful as a guideline for the creation and preliminary evaluation of reliable biomedical algorithms to be embedded in similar systems.

The ambulatory nature of the Sensium™ technology introduced real challenges in the development of the end-to-end system, particularly in the embedded algorithms.

As discussed earlier, one of the main difficulties found was signal contamination caused by motion. Due to the current simplicity of the body node in terms of hardware and additional reference sensors, it was unfeasible to implement complex DSP noise cancellation techniques in the body-sensor nodes or to transmit segments of raw data for external processing. Nonetheless, we believe that our current rule-based confident indication modules are sufficient to satisfy the current demands of healthcare ambulatory monitoring and some sports applications without compromising the performance and lifespan of the body-sensor nodes. However, these modules are somewhat prone to reject signals from which valid physiological values could be obtained. Future directions will involve the incorporation of machine learning techniques to enhance the quality of discrimination of these confidence indicators. It is also believed that the incorporation of compression algorithms will enable the implementation of external sophisticated and more efficient solutions for processing the data while maintaining the power consumption levels at the minimum.

REFERENCES

1. A. C. Wong, D. McDonagh, G. Karthiresan, O. C. Okundu, O. El-Jamaly, T. C.-K. Chan, P. Paddan, A. J. Burdett. A 1V, Micropower system-on-chip for vital-sign monitoring in wireless body sensor networks. in *IEEE International Solid-State Circuits Conference*, 2008. San Francisco, CA.
2. P. Shoon-Shiong, C. Toumazou, A. Burdett, Ultra-low power semiconductors for wireless vital signs early warning systems. *Electronics Letters on Semiconductors in Personalized Medicine*, 2011. 47(26): S26–S28.
3. D. Andre, D. L. Wolf, Recent advances in free-living physical activity monitoring: A review. *Diabetes, Science, and Technology*, 2007. 1(5): 760–767.
4. J. M. Marin, S. J. Carrizo, E. Vicente, A. G. Agusti, Long-term cardiovascular outcomes in men with obstructive sleep apnoea-hypopnoea with or without treatment with continuous positive airway pressure: An observational study. *The Lancet*, 2005. 365(9464): 1046–1053.
5. K. Fukuda, S. E. Straus, I. Hickie, M. C. Sharpe, J. G. Dobbins, A. Komaroff, The chronic fatigue syndrome: A comprehensive approach to its definition and study. *Annals of Internal Medicine*, 1994. 121: 953–959.
6. Association, A. H., Comprehensive overview of nursing and interdisciplinary care of the acute ischemic stroke patient: A scientific statement from the American Heart Association. *Stroke*, 2009. 40: 2911–2944.
7. G. B. Moody, R. G. Mark, The MIT-BIH Arrhythmia Database on CD-ROM and software for use with it. in *Computers in Cardiology*. 1990. Chicago, IL.
8. Mathworks. *Mex*. 2011 [cited 14 February 2012]; Available from: http://www.mathworks.co.uk/help/techdoc/ref/mex.html
9. D. Bertsimas, J. Tsitsiklis, Simulated annealing. *Statistics*, 1993. 8(1): 10–15.
10. AAMI, Testing and reporting performance results of cardiac rhythm and ST-segment measurement algorithms EC57:1998/(R)2008. Association for Advancement of Medical Instrumentation, Editor 1999. ANSI/AAMI/ISO.
11. D. Altman, J. Bland, Measurement in medicine: The analysis of method comparison studies. *The Statistician*, 1983. 32: 307–317.
12. Association, T.M.I.S.R. MISRA Home Page. 2011 [cited February 14, 2012].
13. J. Pan, W. J. Tompkins, A real-time QRS detection algorithm. *IEEE Transactions on Biomedical Engineering*, 1985. BME-32: 230–236.
14. B.-U. Kohler, C. Hennig, R. Orglmeister, The principles of software QRS detection. *Engineering in Medicine and Biology*, 2002. 21(1): 42–57.

15. S. Kennedy, Detecting changes in the respiratory status of ward patients. *Nursing Standard*, 2007. 21: 42–46.

16. J. M. Van De Water, B. E. Mount, J. R. Barela, R. Schuster, Monitoring the chest with impedance. *Chest*, 1973. 64: 597–603.

17. L. Hamilton, Impedance-measuring respiration monitor for infants. *Annals of Biomedical Engineering*, 1973. 1: 324–332.

18. Y. Yasuda, et al., Modified thoracic impedance plethysmography to monitor sleep apnea syndromes. *Sleep Medicine*, 2005. 6: 215–224.

19. L. Poupard, et al., Use of thoracic impedance sensors to screen for sleep-disordered breathing in patients with cardiovascular disease. *Physiological Measurements*, 2008. 29: 255–267.

20. J. Webster, *Medical Instrumentation: Application and Design* 2009: Wiley India Pvt. Ltd., Boston, MA.

21. A. J. Wilson, C. I. Franks, I. L. Freeston, Methods of filtering the heart-beat artefact from the breathing waveform of infants obtained by impedance pneumography. *Medical and Biological Engineering and Computing*, 1982. 20(3): 293–298.

22. F. Yasuma, J. Hayano, Respiratory sinus arrhythmia: Why does the heartbeat synchronize with respiratory rhythm? *Chest Journal*, 2004. 125: 683–690.

23. G. B. Moody, R. G. Mark, A. Zoccola, S. Mantero, Derivation of respiratory signals from multi-lead ECGs. *Computers in Cardiology*, 1985. 12: 113–116.

24. C. O'Brien, C. Heneghan, A comparison of algorithms for estimation of a respiratory signal from the surface electrocardiogram. *Computers in Biology and Medicine*, 2007. 37: 305–314.

25. M. Aboy, R. Hornero, D. Abasolo, D. Alvarez, Interpretation of the lempel-Ziv complexity measure in the context of biomedical signal analysis. *IEEE Transactions on Biomedical Engineering*, 2006. 53: 2282–2288.

26. W. J. Hrushesky, D. Fader, O. Schmitt, V. Gilbertsen, The respiratory sinus arrhythmia: A measure of cardiac age. *Science*, 1984. 224(4652): 1001–1004.

27. L. L. Marchand, L. R. Wilkens, L. N. Kolonel, J. H. Hankin, L.-C. Lyu, Associations of sedentary lifestyle, obesity, smoking, alcohol use, and diabetes with the risk of colorectal cancer. *IEEE Transactions on Biomedical Engineering*, 1997. 57: 4787–4794.

28. S. Brage, N. Brage, N. Franks, P. W. Ekelund, U. Wong, M. Y. Andersen, L. B. Froberg, K. Wareham, Branched equation model of simultaneous accelerometry and heart rate monitoring improves estimate of directly measured physical activity energy expenditure. *Journal of Applied Physiology*, 2004. 96: 343–251.

29. M.-H. Silveira, S.-S. Ang, T. Mehta, B. Wang, A. Burdett, Implementation and evaluation of a physical activity and energy expenditure algorithm in a Sensium-based body-worn device, in *International Conference on Biomedical Electronics and Devices*, 2012. Algarve, Portugal.

30. V. T. van Hees, U. Ekelund, Novel daily energy expenditure estimation by using objective activity type classification: Where do we go from here? *Journal of Applied Physiology*, 2009. 107: 639–640.

31. S. H. Nawab, S. H. Roy, C. J. De Luca, Functional activity monitoring from wearable sensor data. in *26th Annual International Conference of the IEEE EMBS*. 2004. San Francisco, CA.

32. B. Najafi, K. Aminian, F. Leow, Y. Blanc, Ph. Robert, An ambulatory system for physical activity monitoring in elderly. in *1st Annual International IEEE-EMBS Special Topic Conference on Microtechnologies in Medicine and Biology*. 2000.

33. J. Pärkkä, M. Ermes, P. Korpipää, J. Mäntyjärvi, J. Peltola, I. Korhonen, Activity classification using realistic data from wearable sensors. *IEEE Transactions on Information Technology in Biomedicine*, 2006. 10(1): 119–128.

34. M. H. Brooke, G. M. Fenichel., R. C. Griggs, J. R. Mendell, R. Moxley, J. Florence, W. M. King et al., Duchenne muscular dystrophy: Patterns of clinical progression and effects of supportive therapy. *Neurology*, 1989. 39(4): 475.

35. J. R. Cox, F. M. Nolle, H. A. Fozzard, G. C. Oliver, AZTEC, a pre-processing program for real-time ECG rhythm analysis. *IEEE Transactions on Biomedical Engineering*, 1968. BME-15: 128–129.

36. W. C. Mueller, Arrhythmia detection program for an ambulatory ECG monitor. *Biomedical Science Instrument*, 1978. 14: 81–85.

37. W. J. Thompkins, J. G. Webster, *Design of Microcomputer-Based Medical Instrumentation*. 1981, Englewood Cliffs, NJ; Prentice-Hall.

38. H. Imai, N. Kiraura, Y. Yoshida, An efficient encoding method for electrocardiography using Spline functions. *Systems and Computers in Japan*, 1985. 16: 85–94.

39. U. E. Ruttimann, H. V. Pipberger, Compression of the ECG by prediction or interpolation and entropy encoding. *IEEE Transactions on Biomedical Engineering*, 1979. BME-26: 613–623.

40. W.-S. Chen, L. Hsieh, S.-Y. Yuan, High performance data compression method with pattern matching for biomedical ECG and arterial pulse waveforms. *Computer Methods and Programs in Biomedicine*, 2004. 74: 11–27.

21 Ultra-Low-Power Harvesting Body-Centred Electronics for Future Health Monitoring Devices

Jordi Colomer-Farrarons, Pere Miribel-Catala,
Esteve Juanola-Feliu and Josep Samitier

CONTENTS

21.1 INTRODUCTION

Nowadays, there is increasing interest in the development of sensors capable of monitoring human bodily functions and transmitting the resultant data. Two different approaches to this research are typically followed: external body sensors and implantable devices. In this chapter, we present the state of the art of these two different approaches to bodily sensors. The chapter starts by giving an initial mapping of the different sensors involved in the so-called body sensor networks (BSNs). Then, the first aspect to be addressed in detail concerns the power requirements of the different sensors. An important question that arises is as follows: in what specific ways can energy be supplied to these sensors? Some of them could be powered by external power sources, such as batteries or fuel cells, but in other cases, this option is not viable and other sources must be used. Since today the greenhouse effect is a major concern, the possibility of eliminating external sources ultimately powered by fossil fuels is a key factor. Furthermore, the challenge of removing such elements completely is of great interest in terms of the cost and durability of the systems, though in some cases it is quite complicated to replace them. Special interest is focused particularly on the field of energy harvesting: powering systems from the energy available in their surroundings based on the energy that is generated by human beings and is within us. This in turn generates a new question: How is it possible to extract energy from the body to power electronics and how can sensors be powered from human activity?

Another particular and specific case is presented. If the batteries are removed and powering just relies on the use of an energy-harvesting source, the system could fail. So, if a power management module is incorporated in order to make use of more than just one harvesting source, the system will be more reliable. This concept defines the multi-harvesting approach.

A harvesting module is conceived and then a more advanced stage is envisaged: an implantable subcutaneous event-detecting device.

However, when a prototype is developed in a university environment, how is it then transferred to the real world? Scientists are not isolated in darkrooms, excommunicated from the real world. They must apply the lasted developments for the benefit of society, but the route from the laboratory out into society is quite complicated. The key element here is the technology transfer, and this is particularly important in the fields of energy-harvesting and implantable devices.

These are some of the questions that are addressed in the present work.

21.2   BODY SENSOR NETWORKS: FROM DISCRETE TO INTEGRATED SOLUTIONS FOR PERVASIVE MONITORING

In the last 5 years, interest in the development of the so-called BSNs has increased [1]. The miniaturisation of the main electronic systems involved in such systems, as depicted in Figure 21.1, such as the instrumentation, the signal processing module, communication devices and sensors, has opened up the increasing evolution of this field, where not only single wearable sensor is used. That is, not just wearable solutions are conceived; implantable solutions are introduced in the conception of e-health, as depicted in Figure 21.2, where new trends are introduced into the medical industry.

The technological evolution is defining a new scenario in which it will be possible to monitor patients anywhere and at any time [2] (see Figure 21.3), which is of increasing interest [3]. The placement of a central control node that acts as a master node, with other slave nodes located on or inside the body, monitoring different vital signals, defines a typical wireless BSN (WBSN).

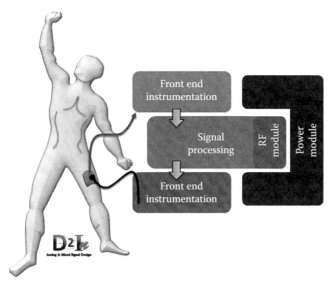

FIGURE 21.1 Miniaturisation of the main involved electronics. (Redrawn from Yang, G.-Z., *Body Sensor Networks*, Springer, New York, 2006.)

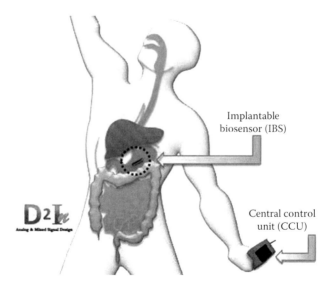

FIGURE 21.2 Concept of an implantable device. (Redrawn from Abouei, J. et al., *IEEE Trans. Inform. Technol. Biomed.*,15(3), 456, May 2011.)

The most important needs in the clinical monitoring of a patient and in healthcare in the case of people who are chronically ill or elderly people require more sophisticated systems based on multiple sensors. Different approaches have been developed such as CodeBlue for electrocardiogram (ECG) monitoring and MobiCare [4], among others. There is much interest in being able to cope with different possible

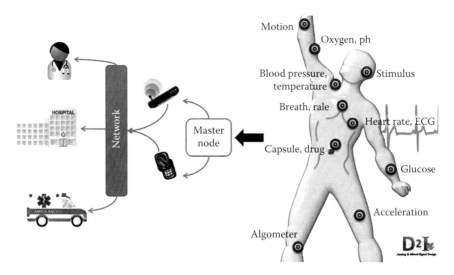

FIGURE 21.3   Typical full WBSNs.

scenarios: fall detection, pulse-oximetry monitoring, ECG monitoring and more. Specific BSNs have been developed for patients with Parkinson's disease [5] based on the SHIMMER platform [6], and there are other examples such as the applications of BSNs in kinesiotherapy activities [7] and mobility support [8].

The traditional approach, where patients are monitored during hospital or surgery visits, would be replaced by continuous and remote monitoring, which could have a great impact on patients' quality of life.

This is the concept of pervasive monitoring [9]. Such an approach should represent a lower cost for health systems, thanks to the reduced number of visits and hospital stays patients will require. Different scenarios can be envisaged in the continuous search to meet technological challenges through miniaturisation, intelligence and autonomy [10]. As has been pointed out, special interest has been taken in developments that address chronic illness [11], where traditional approaches are not capable of ruling out sudden death events and particularly in the case of cardiovascular illness [12]. The chance to detect symptoms early in patients who are at risk, to monitor patients who are following a course of treatment or to track how a disease is progressing offers the possibility of preventing the worst scenarios. In particular, it is of special interest to track the medical parameters of patients who are not following the traditional clinical observation in the hospital. The trick is to monitor patients in their routine daily activities. Another scenario focuses on elderly patients, who constitute a population at risk; as people live longer, the need for medical resources increases. The possibility of monitoring this substantial proportion of the population, remotely, from their homes, will reduce medical cost [13]. However, BSN monitoring is also opening up new options of interest in hospitals. People who undergo surgery may need to be tracked via intensive monitoring immediately before and after the operation, and not just while they are confined to a bed. Furthermore, the possibility of monitoring the patient's specific intervention zone after surgery is increasingly becoming a topic of interest [14]. In the framework of a hospital, there is also the possible development of smart textiles, as considered in [15].

However, the development of such solutions is quite complex. Different papers have been published recently that focus on applications in different fields. These include, for instance, approaches that present multi-sensing platforms for patients who could be monitored in their daily lives at home [16]. A variety of sensors are placed in a monitored home laboratory where different activities are tracked to provide information regarding the motion of the patient, but this is not limited to an option based on wearable sensors. The authors combine such bodily devices with ambient sensors programmed for specific algorithms that have the capability to identify just one activity at a time.

Other approaches have focused on single units that consist of different multi-sensors [17]. In such approaches, a single unit placed on the chest of the patient is able to monitor body motion, activity intensity and other parameters such as heart rate or provide ECG data, EEG data, blood pressure, O_2 saturation, etc.

As has been pointed out, several approaches are based on a wireless sensor network (WSN) and a BSN; it is important to distinguish between such technologies because the challenges they face are quite different. The combination of the two can be defined as a WBSN, composed of a master node – which can be a portable device and acts as a central control – and other smaller slave sensor nodes, which are positioned around or inside the body, as depicted in Figure 21.3.

While a WSN is designed to cover a wide range from a few metres to several kilometres, a BSN is focused on a range of a few millimetres to a few centimetres: strictly around the body. A BSN is more oriented to the use of few sensors, constrained by the available space where the electronics module could be placed. So, such options must be miniaturised as far as possible, a challenge that does not exist for WSNs. The slave sensor nodes are therefore more tightly constrained in terms of size but also in terms of power consumption, and the power supply is of particular interest in such systems. For a WSN, the power supply can be replaced frequently and also relatively easily. If we consider a BSN, and in the particular case of implantable devices, the power supply could be inaccessible and very difficult to replace. That is the main reason for the increasing interest in doing without batteries as a power source and in other solutions based on miniaturised power sources that harvest ambient or human energy.

Another key aspect that must be taken into account is data protection and how the data are treated in order to meet privacy and security requirements. In the case of WSNs, a lower level of data transfer security is required, but in the case of BSNs, a very high level of security must be ensured in order to protect patients' information without compromising their safety. Patient safety could be jeopardised if caregivers cannot access urgent health data because they have no access to the decryption keys, for example. In [18], sensitive patient data are managed in such a way that primary caregivers are able to access data by following a specific privacy protocol. Another approach is envisaged in [9], where a specific medical sensor network (MSN) and a patient area network (PAN) define the different security layers that are connected via a back-end security layer that ensures the security of the system. One key aspect is the quality of service (QoS) [19], that is, what ensures the quality of the data transmitted from the sensor.

BSN solutions are not just conceived as discrete external resources. Specific and highly complex solutions are envisaged in the search for devices that are truly implantable in the body and not just positioned on the body. The power consumption is therefore one of the main issues to consider [20] in a BSN mote. Such systems have been powered but external devices where the approach is different than it is for implantable devices [21], as is presented as follows. Ideally, the change towards removing the use of batteries would be a priority. Concerns about the role of batteries are also presented in [22], where computing platforms based on electronic textiles are considered as one of the fields of future development.

Batteries have a limited lifetime that may be as long as 10 years in some applications, but the ideal approach would be to remove this element and ensure the reliability and operability of the systems without the need to use an element that must be replaced. This is particularly salient when considering implantable devices. The possibility of ensuring a long-term working life of such devices, without the use of the batteries, can improve the quality of life of patients without the need for any surgery. Nowadays, the concept of new e-health systems already imposes very low power consumption restrictions on the electronic instrumentation, processing devices and communication systems in order to extend the operating life of batteries. If the batteries are removed, then the system must be powered by different kinds of energy source present in the immediate environment, that is, the body itself or specific solutions that will depend on the placement of the sensors. This situation produces new challenges for engineers. Systems that rely on just one power source could be a problem. If that energy source is not always available, the power could fail. So, different approaches should be considered. In one case, if a battery is still used, that power source could be combined with other possible scavenging sources in such a way that, in terms of the operating protocol of the system, the battery lifetime is extended. This could be achieved by recharging the battery when enough energy is recovered, or the battery could be in an open-load configuration in which the required operating energy is supplied by the scavenger module (Figure 21.4). If the battery is removed altogether, the system must rely on a combination of different scavenger energy sources.

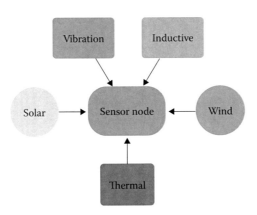

FIGURE 21.4   Combined power sources in a sensor node.

The types of sensors, their location and the amount of data that must be processed and transmitted define the power budget that must be considered in order to define the power module.

The approach depicted in Figure 21.1 shows not only the monitoring objective of such a system; it depicts the concept of a closed loop. In such a system, monitoring the patient is not the only objective; the system also has to be able to generate electrical stimulation. This is a new approach. In some sense, the monitoring sensor nodes must take measurements periodically, while sensors that also have a treatment functionality can operate either periodically or be event driven, which thereby defines different power consumption scenarios. The typical paradigm example of a sensor node with functionality is the artificial pancreas [23]. Another key example is related to nerve stimulation and recording, as in previous work [24] but for which bioamplifier circuits and stimulator output stage circuits are being developed nowadays [25].

Recently there have been different approaches to implantable solutions that vary from digital hearing aids and cochlear implants [26], through cardiac application-specific integrated circuits (ASICs) [27], neurostimulation devices [28] and artificial retina [29] to biomechanical implants [30] and beyond.

A conceptual map of implantable sensors in a body is depicted in Figure 21.5, paying attention to the most relevant disease processes based on [31]. The evolution of semiconductor technology, with low-voltage low-power electronics, allows the integration of several implantable devices for different functions. This new generation will have the capacity to overcome the three main barriers that existed when these systems are conceived: energy and power dissipation, the processing and communication of the measured data and bio-compatibility.

Many examples are currently being developed in the field of implantable devices. Some of them are presented in what follows, paying special attention to the power

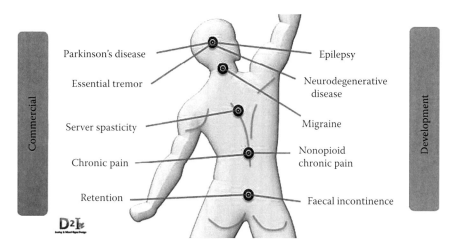

FIGURE 21.5 A conceptual mapping of implantable sensors. (Redrawn from Oesterle, S. et al., New interfaces to body through implantable-system integration, in *Proceedings of the International Solid-State Circuits Conference ISSCC*, San Francisco, CA, February 2011.)

involved, power supply and operating ranges. In the field of pacemakers, [32] presents a programmable implantable micro-stimulator with wireless telemetry for endocardial stimulation in order to detect and correct cardiac arrhythmias. That option lists the global power consumption as 48 μW, but it relies on a rechargeable battery based on radio-frequency (RF) coupling. As follows, an implantable device that is placed deep in the body is presented.

Another interesting case is presented in [33]. There, a transcutaneous implantable device is developed based on an IR link just a few centimetre from the emitter. The device is completely autonomous and battery-free, and it is powered through an alternating current (AC) signal operating at the industrial, scientific and medical (ISM) radio band. It incorporates low-frequency operation at 13.56 MHz, with a global power consumption of 270 μW and an ASIC size of 1.4×1.4 mm^2. An interesting approach is presented in [34]. There, for the particular case of glucose monitoring, a non-invasive solution is presented based on an active contact lens, whose size is limited to 0.5×0.5 mm^2. In that case, the implant is powered by a rectified 2.4 GHz RF power signal source at a distance of 15 cm with a global power consumption of 3 μW. Some specific examples in the case of neuroscience applications are also introduced. In [35], a complementary metal–oxide–semiconductor (CMOS) implementation is presented that operates as a stimulator of the dorsal root ganglion, which has been experimentally probed with rats. A battery-free solution is envisaged with an induced 1 MHz RF signal at a distance of 18 mm, at the standard medical implanted communication system (MICS) frequency of 402 MHz. Another example of an inductively powered neural SoC system is presented in [36]. In a 0.5 μm technology, a 4.9×3.3 mm^2 SoC is designed for a 32-channel wireless integrated neural recording system, with a power dissipation of 5.85 mW.

There are also other implantable devices that are more oriented towards biomechanical applications, in particular, the resources needed for monitoring prosthetic implants that are used in human beings to replace joints and bones (Figure 21.6). Such implants are designed to have a duration of 20 years, but degradation can lead to the necessity to replace them, with all the associated inconveniences and risks to the patients. The possibility of integrating electronics to monitor fatigue into the implants has clear advantages in protecting against premature mechanical failures. The amount of wear that implants suffer depends on many variables but in particular on the level of activity and weight of the patient. The typical locations of such implants are knee joints, carpals, and tarsal, scapula and hip-socket joints. Arthroplasty applications with monitoring electronics must ensure genuine autonomous long-term operation, in terms of capturing measurement data that are to be transmitted and powering the system. Interesting applications have been reported where self-powered systems are based on piezoelectric transducers that use mechanical deformation to sense the strain but are also a harvesting power source for the system [37]. In such cases, human motion is used to power the electronics and the battery is removed, as is presented later. In [38], a knee implant is presented based on the use of piezoceramics that is able to deliver a maximum average electrical power of 12 mW at the tibial base plate. In [39], based on another approach, the estimated power is 1.8 mW. The power levels are low and so the power consumption of the integrated electronics must be very low [40].

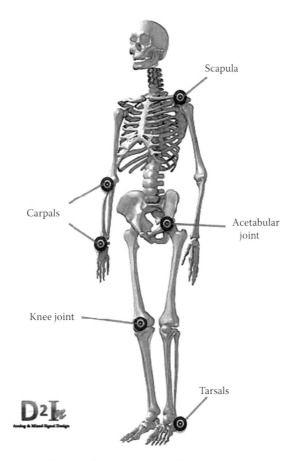

FIGURE 21.6 Implantables for biomechanical applications.

So, different possibilities for overcoming the powering barrier are considered as a function of the type of device and its placement, based on induced RF signals, PZTs, micro-electro-mechanical systems (MEMS) generators or other possibilities, as are presented in this chapter.

21.3   WHAT ARE THE MAIN POWER REQUIREMENTS FOR A BODY SENSOR?

The design of an implantable device must contemplate different modules for their implementation, such as those represented in Figure 21.7. The energy levels that are involved in the development of an integrated solution will vary in terms of the final application and the level of intelligence of the device. A system that must work continuously is not the same as a system that works in bursts. Four modules are usually present in an implantable device (Figure 21.7): (a) the sensor or sensors involved, which fix the type of measurement and the complexity of the front-end instrumentation, which is the signal conditioning; (b) the data processing; (c) the wireless module; and (d) the power management unit. If the system has to close the loop, the

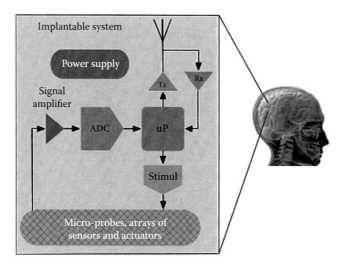

FIGURE 21.7   Electronic modules for an implantable system. (Based on Muller, R. et al., *IEEE J. Solid-State Circuits,* 47(1), 232, January 2012.)

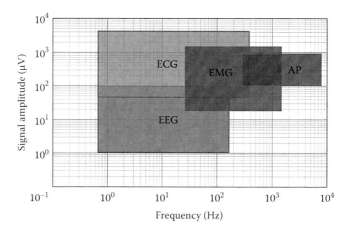

FIGURE 21.8   Typical voltage signals and frequencies for some biopotential signals.

stimulation electronics – based on direct current (DC)–DC converters and drivers – are also involved in the power budget.

The design of the front-end electronics [41] depends on the type of medical signals that must be considered. It can be stated that these signals may vary from a few µV to several mV, with a frequency band that varies from a few Hz to kHz, as depicted in Figure 21.8. Typically, electroencephalogram (EEG) signals range from 1 µV to 50 µV and have low frequencies from 1 to 100 Hz. On the other side, ECG signals present a higher voltage levels, in the range of mV, than the EEG. Electromyogram (EMG) signals and action potentials present the highest frequencies: up to 10 kHz.

The power consumption levels (in terms of energy per operation) depend on the electronics involved. If a microprocessor is needed, then a typical power consumption of 300 μW for commercial solutions such as the Texas Instruments® MSP430 microprocessor is far from the desired power budget. Favourable examples of specific microprocessors have been presented in [42] and [43]. [42] presented the Phoenix Processor, in a 180 nm technology, with a power consumption of 226 nW but with emphasis on the standby consumption, which is as low as 35.4 pW. It has an area of 915 × 915 μm² and operates at 0.5 V. It evolved from [43], where the subthreshold operating region is explored.

Analog-to-digital converter (ADC) conversion in terms of the needs of sample conversion rates and the sub-microwatt transmitters (as a function of the distance between the implanted device and the exterior on the one hand and the data transmission rate on the other) are two of the key elements of these designs. Some pertinent examples show how research aims to find ADCs with lower power consumption. In [44], ultralow-power front-end signal conditioning is implemented for an implantable 16-channel neurosensor array, with a maximum power consumption of 50 μW per channel. The ADC is based on a commercial solution derived by Analog Devices®, a 12-bit AD7495 micro-SO8 standard packaged chip. The complete system has a total consumption of 12 mW. Specific examples of ADCs for implantable applications have been also derived as in [45]. There, a 400 nW SAR ADC converter, with 8 bits of ENOB and 80 kS/s, is presented in 180 nm technology. The implantable blood pressure–sensing micro-system developed in [46] achieves 10-bit resolution with an integrated cyclic ADC converter with 11-bit resolution and a power consumption of 12 μW at 2 V. The pressure sensor has a full-power dissipation of 300 μW. In [47], an ADC converter, based on an integrated ADC, for a blood glucose monitoring implant, presents even better performance at just 10.2 nJ per sample, with 10 ENOB, operating at very low frequencies. Another example, in this case in the field of neural signal acquisition, is presented in [48] where the ADC has a power consumption of 240 nW based on a boxcar sampling ADC [49], operating at 20 kS/s.

Another interesting example, in this case for an intraocular wireless implantable pressure and temperature sensor, is presented in [50], with a total power consumption of 2.3 μW at 1.5 V during continuous monitoring.

Regarding the wireless module, there are different approaches depending on the type of signal captured and the amount of data related to the types of complex signal that define a wide range of data transmission constraints from a few samples by second, as in the case of heart rating (typically 25 Sps), or scenarios with data transmission rates of several Mbps, such as medical imaging [51] or recent developments in endoscopic capsules [52,53].

The review in [54] analyses in detail the parameterisation of the different implantable devices envisaged in Figure 21.5 in terms of the three key variables: the size of the implantable, the power involved and the functionality and performance. Two main approaches to communications are considered depending on the type of placement of the implant in the body: inductive coupling [55], when there is short distance between the implant and the exterior of the body, and far-field electromagnetic communication [56]. The approach for a subcutaneous implant based on inductive links, where coil misalignment and the effects of the geometry have a great impact on performance [57], is not the same as solutions operating at high frequencies following the MICS protocol [58,59]. For inductive link circuits, the operating frequencies are in the range of a few MHz, typically in the 13.56 MHz band, following the ISM protocol, with characteristic data modulation methods for transmission, such as BPSK, PWM-ASK, ASK or OOK, and a data transfer rate that is not very fast: from a few kbps (100) to several Mbps (1.6) [60].

As stated previously, the approaches for a WSN and an implantable BSN are quite different. In [59], a modification of the MICS protocol ensures the correct delivery of the data, thus overcoming issues related to the attenuation of signals across the human body.

In [61], a high-speed OQPSK is presented with a bit rate of 4.16 Mbps, based on an inductive link, with high power levels and multiple coils, operating at 1 MHz for the power link and at the ISM 13.56 MHz frequency for the data. In [62], an implant OOK transmitter is presented that operates at 2.4 GHz in 180 nm technology and that is able to transmit at 136 Mbps with a power consumption of 3 mW. As is pointed out in [40], for the particular case of an orthopaedic implant, the power consumption for the full electronics must be low. The total power consumption of the full electronics is not stated, but the DC–DC that recovers energy from PZTs and the ADC converter that presents a quiescent power consumption of 150 nW and an operating consumption of 12.5 µW at 1.8 V for a sample frequency of 4 kHz and a ENOB > 7 bits merit special attention.

Future implantable BSNs aim to have very low power consumption, in the 1 µW or nW range. Smaller solutions in the mm^3 volume range are trends aimed at future bionic human beings [63], where new interfaces in the brain will allow control of prosthetic devices, as in [64].

21.4   BATTERIES, SUPERCAPACITORS AND FUEL CELLS

The type of energy source required to power a sensor node varies according to the approach. The use of batteries could be a limitation for the envisaged sensor nodes of the future, either on the outside of or implanted within the human body, as in Figure 21.3. The use of large batteries ensures the duration of the system, but the sensor nodes may be too large and heavy. So, smaller solutions with a high enough energy density are needed, combined with ultralow-power electronics solutions that ensure a trade-off between the autonomy of the system and the smart functionality of the sensor, in terms of the sensor, signal processing and communication modules.

Ultralow-power electronics for biomedical applications based on lithium-ion batteries are common for non-implantable solutions, such as BSNs, and also in implantable solutions [65,66]. Lithium-ion batteries are divided into two types: (a) single-use batteries, which are placed, for instance, in drug pumps, cardiac defibrillators and pacemakers, and (b) rechargeable batteries, which are used in artificial hearts. Both types of battery present an adequate energy density, around 1440–3600 J/cm^3, and unlike some other batteries, they do not present the memory effect; that is, they do not need to be discharged completely before a recharge phase. These batteries have a better lifecycle than other types, typically with 20,000 discharges and recharge cycles but with a finite lifetime limitation, which is typically of several years for a battery of 1 cm^3. A key aspect for these batteries is the need for battery-management circuitry to ensure the range of operation, as lithium-ion batteries are extremely sensitive to overvoltage (maximum 4.2 V) and deep discharge (minimum 2 V), and to ensure high energy efficiencies. In [67], application is reported with an average power efficiency of 89.7% and a voltage accuracy of 99.9%, conceived for biomedical applications.

Other approaches are under development. Interest is especially focused on fuel cells, such as the methanol fuel cell [68], but these also have their drawbacks. One is the need to replace the external reactant and the oxidant, which is analogous to the problem of recharging batteries. Although higher levels of energy are expected, based on the use of fuels such as methanol with an energy density of 17,600 J/cm^3, the design issues are highly complex and proving very expensive.

Supercapacitors are another field that is being explored as an option for biomedical sensors instead of batteries, but they have a low energy density, which is a problem for systems that require a constant power source for long periods of time. In [69], a sensor is powered via a supercapacitor that is charged wirelessly, and in [70], an energy management ASIC is implemented and tested to manage supercapacitors for implants.

A comparison of the different energy sources is depicted in Figure 21.9 [71].

A section devoted to how energy in the human body can be used to power the electronic systems is presented as follows, but first we consider another topic that links the human body and fuel cells. From the concept of a fuel cell, the biogenerator for implantable devices has emerged. In some ways, the basic concept is the use of fluids in the body as a fuel source for the fuel cell, which would be an inexhaustible energy source. An interesting approach is the use of glucose as a fuel source and the oxygen dissolved in blood, as in [72] and [73]. Advanced approaches also explore a

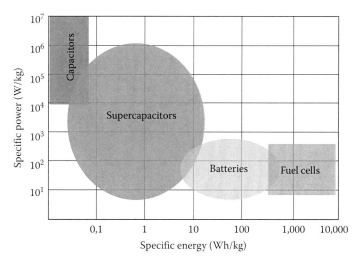

FIGURE 21.9 Power vs. energy for different energy storage elements. (Derived from CAP-XX [Australia] Pty Ltd, Power vs. Energy, [Online] Available: http://www.cap-xx.com.)

shift to the use of white blood cell capacities in biofuel cells [74] or approaches such as that in [75], where the fuel cell is based on the use of a microorganism to convert the chemical energy of glucose into electrical energy, in a PDMS structure.

Whatever the future holds, the development of such sensor nodes that are (quasi) independent of the use of batteries offers the chance to explore alternative power sources.

21.5   HOW TO EXTRACT ENERGY FROM THE BODY

This issue concerns the power sources that are available in and around the human body. For instance, is it possible to recover energy from human bodily motion? Is it possible to recover energy from the body itself? These are two questions addressed in this section. First, we introduce the concept of a harvester system, based on an energy harvester, an energy storage element and power management regulation (Figure 21.10). Four types of energy harvester based on ambient sources are introduced when this system is described, that is, systems that acquire electrical energy from environmental energy sources: mechanical, solar, thermal and RF (the details of which are beyond the scope of this chapter). The power management unit has two main roles: (a) to generate a raw DC voltage from the harvester unit in which it essentially acts as a conversion module and (b) to regulate the output voltage from the energy storage unit to the sensor node electronics. The energy storage unit receives energy from the conversion module and stores it. This module can be a supercapacitor, a thin film battery, etc.

Attention is especially focused on body harvesting. The energy can be generated passively or actively. The natural motion of the body, where the conversion is based on mechanical (vibration) energy to electrical energy, is one of the main research topics [76]. In this context, the power that can be harvested when a human being walks or runs has been studied at different locations, with an average of 0.5 mW/cm^3 for hip, chest, elbow, upper arm and head and a maximum of 10 mW/cm^3 for the ankle and knee. The mechanical energy can be converted to electrical energy based, for instance, on electromagnetic [76], electrostatic [77] or piezoelectric [78] principles.

There are several references to types of design that convert this mechanical energy into electrical energy based on the three methods of transduction just mentioned [79], which are summarised in Table 21.1. The piezoelectric approach is of special interest because high voltages are obtained for low strains. Several examples are presented in the literature. In particular, in [80], piezoelectric micro-generators are envisaged for drug delivery devices or dental applications.

Examples of power generated actively are human power through peddling activity [81] and the Freeplay® or AladdinPower® rechargeable products (see Table 21.2). A complete analysis of how extracting energy when walking alone is done in [82].

FIGURE 21.10   Schematic of an energy harvesting solution.

TABLE 21.1
Energy Generated by Mechanical-to-Electrical Generators

Type	Theoretical Max. (mJ/cm³)	Practical Max. (mJ/cm³)
Electromagnetic	400	4
Electrostatic	44	4
Piezoelectric	335	17.7

Sources: Derived from Roundy, S. et al., Micro-electrostatic vibration-to-electricity converters, *Proceedings of IMECE'02 2002 ASME International Mechanical Engineering Congress and Exposition,* pp. 17–22, New Orleans, LA, 2002.

TABLE 21.2
Examples of Human Power Generated Actively

Activity	Power Generation (W)
Finger (pushing pen)	0.3
Legs (cycling at 25 km/h)	100
Hand and arm (Freeplay)	21
Hand (AladdinPower)	3.6

A prototype has been validated, but with an excessively large size for our purposes. An extended analysis in terms of the placement of the harvester and the type of everyday activity is presented in [83]. Based on a commercial cantilever beam harvester from Midé Technologies [84], the PEH20W Volture, these different scenarios have been analysed. The greatest level of power extracted was from a shank at the instep of the foot while the subject runs fast (28.74 µW), and the lowest level of power (0.02 µW) was also from a shank but while performing knee rehabilitation exercises sitting down. As expected, the maximum levels are obtained for activities and placements where higher amplitudes and impacts are produced.

An example of a wearable wireless sensor based on a kinetic energy harvester is presented in [85]. Electromagnetic transduction is used for an average human motion at 0.5 Hz, with a simple architecture implemented via commercial components and a supercapacitor, but not used as a final storage element of the system. It is used to transfer the charge to a smaller supply capacitor, thus improving system start-up. In [86], a MEMS piezoelectric generator is used to harvest energy from vibrations; it also uses supercapacitors as storage elements. An example of a MEMS designed for implantable devices is given in [87], where it is stated that the micro-generator is able to generate more energy per unit volume than conventional batteries, that is, an RMS power of 390 µW for 1 mm² of footprint area and a thickness of 500 µm, which is smaller than the volume of a typical battery in a pacemaker.

The other main source of power from the human body is heat, which is limited by the Carnot efficiency, which states that the maximum power that can be recovered is in the range of 2.4–4.8 W, where other possible sources are also shown [88]. In the case of body heat, thermoelectric generators are used [89]. Specific designs must be considered when working with thermoelectric generators in these conditions, as the voltages generated are very small and DC–DC converters with a specific step-up are conceived in order to boost the voltage [90]. An example of a thermogenerator is ThermoLife®, which generates 60 μW/cm² for a temperature difference of 5°C [91].

It should also be stated that the recovery depends on the specific placement of the thermoelectric generator. Placing it in the neck is not the same as placing it in the head, both of which are parts of the body that are warmer than other zones. The recovery range is 200–320 mW for the neck and 600–960 mW for the head (three times the surface area of the neck).

21.6   MULTI-HARVESTING ASIC CONCEPT

At this point, a sensor node could be powered by following different approaches and the choice depends on the placement of the sensor, which defines its accessibility, its size and its weight. A battery, a supercapacitor and a fuel cell are possible choices on the one hand. On the other, the use of some kind of energy-harvesting source that recovers energy from the body, mostly based on vibration-to-electrical conversion or thermoelectric generation, is also envisaged. However, there are some other approaches that can also be considered depending on the placement of the sensor; these include the possibility of using environmental resources to harvest energy from for the sensor node, such as light [92,93] or radio waves [94]. The approach that was introduced in the aforementioned is RF coupling, both as a way to supply energy to the sensor node and to use the RF link for communication purposes, typically for short distances between the master and slave modules and typically in the 13.56 MHz ISM band. For instance, at 4 MHz and a distance of 25 mm, for subcutaneous powering of a biomedical device, the energy recovered was 5 mW [95].

A particular example of this is introduced in [96]. There, the rectifier module is designed to work at the 915 MHz ISM bandwidth (only in region 2). The performance of the RF source is quite small, from a typical 4 μW/cm² for GSM to 1 μW/cm² for the WiFi band. For coils, typical values are lower than 1 μW/cm² but with as much as 1 mW for close inductive coils (a few cm). In the 915 MHz ISM bandwidth, at 1.1 m, the energy recovered is around 20 μW [97].

New approaches are being developed. In particular, the use of ultrasonic powering instead of RF powering is of great interest. In [98] and [99], 1 V is generated with a power capability of 21.4 nW.

At this point, the issue that we need to consider is the possibility of placing or using more than just one energy harvester. This approach, combining indoor solar cells, RF coupling at the ISM 13.56 MHz band and mechanical vibration based on piezoelectric transducer, was presented in [100], together with experimental results. The system does not rely on only one harvesting source, and the objective is to use different energy harvesters. In [101], a specific power management unit is designed for a multisource and multi-load energy-harvesting system with the aim of establishing a minimum number of conversion stages and magnetic components and of developing a specific algorithm. In [102], an approach to the envisaged platform architecture for a multi-harvesting WSN is presented. In [103], a 1 cm^3 die-stacked platform is presented where it is stated that the harvested energy from a solar-cell layer (40 nW) is used to recharge a 0.6 μAh battery layer. It is also stated that the system was validated via an external TEG module and a microbial fuel cell, but no more details are reported. The power management unit is presented as a flexible module able to work from various energy sources, but not with all the different sources at the same time. So, potential integration is demonstrated, but full multi-harvesting operation of the device is not.

Another interesting example commented on in [40] has the particularity that it combines the possibility of recovering energy from the motion of a human being, based on PZTs, with recovering energy from the RF signal, depending on the operating mode, but no more details are reported. In [104], the combination of a vibration energy harvester and charging circuitry for a lithium-ion battery is presented.

The architecture proposed in [100,108], depicted in Figure 21.11, is used to demonstrate the feasibility of combining different harvester sources for a 1 cm^3 multi-sensor node. Electromagnetic coupling, with an indoor solar-cell module and a piezoelectric generator, was used. The platform envisaged is defined as a multi-harvesting power chip (MHCP) with a total power consumption of 160 μW. The integrated circuit is designed with 0.13 μm technology, which is a low-voltage technology: up to 3.3 V. The capacity of a small system to recover a few microwatts from the energy present in the environment of the multi-sensor node, combining vibration, light and RF, is a proof of concept and recovers a total power of more than 1 mW. The system is initially analysed with each different source working alone. The system recovers 360 μW from the piezoelectric generator, which is a commercial PZT (QP40W) [105], operating at 7 m/s^2 at 80 Hz. Then, two indoor solar cells [106] under indoor conditions of 1500 lx are used, with a total harvesting power of 2.76 mW. Finally, an RF link generator (TRF7960) emitting at full power (200 mW) with a distance between the base station and the antenna receiver of 25 mm is used and recovers 4.5 mW. The total power expected to be recovered from the combination of the three harvested sources would be 7.8 mW, but as shown in Figure 21.12, the experimentally recovered power was 6.2 mW. The difference of 1.6 mW between the theoretical total and the actual amount of power that was recovered is due to variations in the effective load conditions for the light and RF modules. This suggests research is needed into the trade-off between the implementation of peak power-tracking circuitry for the light and RF modules and the improvement that could be achieved, compared with the cost in silicon area and the power consumption of each module.

The three main elements in the design of the MHCP ASIC are as follows: (a) A bandgap reference circuit, which is based on peak regulation, presented in more detail in [94]. (b) A linear dropout (LDO) regulator, based on the bandgap reference circuit, which uses a PMOS switch and an error amplifier [100]. Both elements are

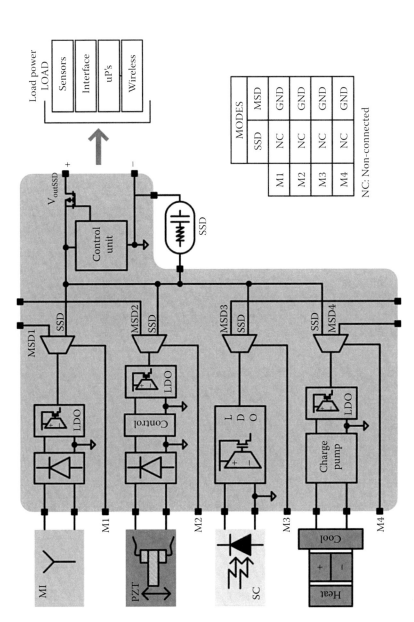

FIGURE 21.11   MHCP concept. (From Colomer-Fararrons, J. et al., *IEEE Trans. Indust. Electron.*, 58(9), 4250, September 2011; Colomer-Fararrons, J. and Miribel-Catala, P., *A CMOS Self-Powered Front-End Architectures for Subcutaneous Event-Detector Devices: Three-Electrodes Amperometric Biosensor Approach*, Springer, 2011.)

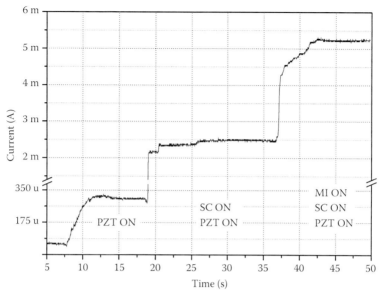

FIGURE 21.12 MHPC power source addition. (From Colomer-Farrarons, J. et al., *IEEE Trans. Indust. Electron.*, 58(9), 4250, September 2011; Colomer-Farrarons, J. and Miribel-Catala, P., *A CMOS Self-Powered Front-End Architectures for Subcutaneous Event-Detector Devices: Three-Electrodes Amperometric Biosensor Approach*, Springer, 2011.)

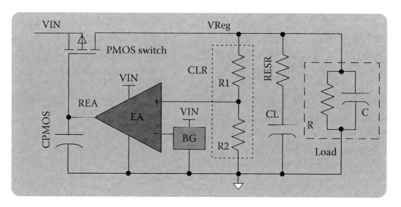

FIGURE 21.13 Full LDO schematic with the adopted bandgap reference circuit. (From Colomer-Farrarons, J. et al., *IEEE Trans. Indust. Electron.*, 58(9), 4250, September 2011; Colomer-Farrarons, J. and Miribel-Catala, P., *A CMOS Self-Powered Front-End Architectures for Subcutaneous Event-Detector Devices: Three-Electrodes Amperometric Biosensor Approach*, Springer, 2011.)

depicted in Figure 21.13. The PMOS switch of the LDO regulator is placed at the output stage of the simple two-stage amplifier, where MN1 and MN2 define the ground current. These elements have a power requirement of 30 μW. Finally, (c) two integrated rectifiers that must be placed in the MHCP chip, one for AC/DC rectification of the piezoelectric generator signal and the other one for the RF-coupled signal [87].

The MHPC has its own control unit that plays the role of the power management unit, based on the concept presented in [107]. The system combines a control switch for each energy harvester channel. In that way, it is possible to operate each channel independently as a single harvesting channel with independent storage elements, multiple storage device (MSD) mode of operation [100], or combining of all them into

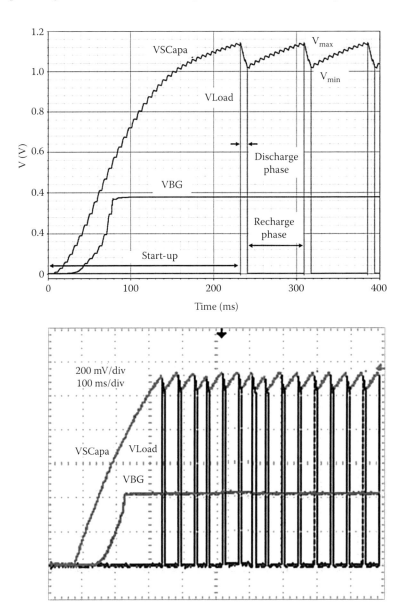

FIGURE 21.14 Operation of the PMU circuitry. (Colomer-Farrarons, J., Miribel-Catala, P., Saiz-Vela, A., Puig-Vidal, M., Samitier, J. *IEEE Transactions on Industrial Electronics*, 55(9), 3249, September 2008.).

a single storage device (SSD) mode of operation, as depicted in Figure 22.11. In SSD mode, the power management unit (PMU) controls the energy stored in the single storage element (a supercapacitor) and transfers energy when an adequate voltage is reached in the capacitor. When the voltage reaches the low value threshold (V_{min}), the system opens the charge transfer to the load until the maximum voltage level (V_{max}) is reached again. The PMU also incorporates power-on reset (POR). This circuit is used to reduce the power consumption of the module during the start-up phase; see Figure 21.14, where the simulated and experimental performance (with only the piezoelectric power generator operating) is depicted, with an external capacitor of 47 μF.

21.7 IMPLANTABLE DEVICE

In this section, generic CMOS architecture for an implantable device [108] is presented. Nowadays, interest in nano-biosensors is increasing in the field of medical diagnosis. The development of such devices and the telemedicine environments that can be derived from them have great market potential, as has been pointed out before. Different approaches are required for discrete, small cm³ devices and for implantable devices, and the performance, communication capabilities, etc., are very different. The size of the implantable device is envisaged as that of a capsule, ideally less than 4.5 cm long and 2.5 cm in diameter, following the same philosophy as some subcutaneous implantable contraceptive devices such as Norplant®, Jadelle® and Implanon®. One proposal is an event-detector implantable system or an event detector that works as an alarm. When the concentration analysed moves out of the range of accepted values and reaches a threshold value, the alarm is activated. The proposed generic implantable architecture is presented in Figure 21.15. It is composed of three biosensor electrodes, an antenna and the electronic modules.

Such a system combines different modules. There is the antenna and the AC/DC module that is used to supply energy to the device (inductive powering) and the communication set-up (backscattering) based on amplitude modulation (AM). Then, a low-voltage and low-power potentiostat is integrated.

The biosensor is the only part of the implantable device interacting with the biological environment. It detects the desired target generating an electrical signal. The design of the biosensor must be carefully selected in function of the type of sample to be detected and the final size of the device where the sensor will be used. Several biosensor configurations formed by simple two, three or four

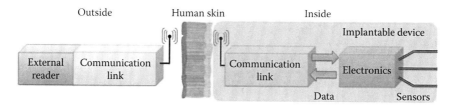

FIGURE 21.15 Generic implantable detector. (Colomer-Farrarons, J. and Miribel-Catala, P., *A CMOS Self-Powered Front-End Architectures for Subcutaneous Event-Detector Devices: Three-Electrodes Amperometric Biosensor Approach*, Springer, 2011.)

electrodes can be used for single target detection, and more complex array structures of microsensors can be introduced for multi-analyte detection.

The antenna and its associated electronics are used for two main aspects in the scenario presented in Figure 21.15. Firstly, it is used to supply energy to the implantable device (inductive powering) working together with an integrated AC/DC module. Secondly, the same antenna transmits the information to the external reader through the electronic communications set-up (backscattering). In this scenario, one antenna is used for both power generation and communication. This reduces the antenna operation frequency to tens of MHz due to the inductive power drop caused by the human skin. Moreover, the amount of transmitted information is limited and the size of the antenna is considerably big. Other scenario considers the use of two antennas in the same implantable device, one for the communication and the other for powering. In this case, the communication link can be established around a hundred of MHz (usually in the 400 MHz ISM band), allowing higher communication rates and reducing the size of the antenna. On the other size, the second antenna is focused to power the electronics through a dedicated inductive link operating at lower frequencies than the communication antenna. In that way, each antenna can be optimised for its functionality.

Then, the integrated electronics is introduced to drive the biosensor and to generate the data to be transmitted. Usually a low-voltage, low-power potentiostat circuit or similar instrumentations are used to control the implantable sensors.

Finally, an analogue lock-in amplifier can be integrated into the modulation/ data processing module. In this case, an FRA approach is followed. In this kind of implementation, there is a multiple trade-off between different factors: complexity and the functionalities of the electronic modules, area and power consumption for the desired measurements. A fully integrated DSP solution, such as a digital lock-in amplifier, would therefore present a major challenge. The design aims at very low power consumption, working with very low power supply. Following this

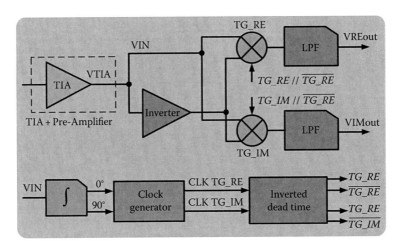

FIGURE 21.16   Lock-in amplifier block diagram. (Colomer-Farrarons, J., Miribel-Català, P., Rodríguez, I., and Samitier, J. *Proceedings of the 35th International Conference of Industrial Electronics, IECON*, 4401, Porto, 2009.)

assumption, an analogue lock-in amplifier is derived (Figure 21.16) that is completely integrated and designed from commercially available technology, such as an electronic interface with implantable biosensors in low-frequency applications. The integrated lock-in is based on two synchronous demodulated channels. Both channels are used simultaneously to find two DC components, which are proportional to the real and imaginary components, VREout and VIMout, after a low-pass filter is placed in each channel after the rectifier stage. The magnitude and phase of the electrochemical cell can then be obtained using (21.1) and (21.2):

$$\text{Magnitude} = \sqrt{\text{VREout}^2 + \text{VIMout}^2} \tag{21.1}$$

$$\text{Phase} = \text{Tan}^{-1}\left(\frac{\text{VIMout}}{\text{VREout}}\right) \tag{21.2}$$

21.8 ADDING VALUE: BEYOND THE SCIENTIFIC APPROACH TO TECHNOLOGY TRANSFER

21.8.1 Introduction

Within a modern, knowledge-driven economy, knowledge transfer is about transferring good ideas, research results and skills between universities, other research organisations, businesses and the wider community to enable innovative new products and services to be developed. The importance of knowledge and technology transfer and of research commercialisation is increasing dramatically as they become a key factor in economic growth. Within such a framework, universities and laboratories are asked to provide technical advances in an increasingly competitive global setting. At the same time, governments and policymakers work to stimulate a significant and effective increase in innovation throughout the economy. The role of research and development (R&D) is vital in the new 'knowledge market', so businesses must access public R&D facilities and innovative projects or develop their own. As a result, companies might create new products and services with which they hope to beat the competition and may be even markets in which to compete.

Nowadays, the commercialisation of university-generated knowledge is closely linked to the emerging, highly scientific component of technology present in key industries. Since universities generate a large share of the scientific production, the interface between academia and industry has had to be brought into sharp focus. To facilitate technology transfer from the university to the commercial sector, the active involvement of university innovators with strong incentives from the technology market is required. At the same time, steps need to be taken to ensure that the disincentives from within the university environment are not excessive [109]. Thus, in order to reach the final link of commercialisation in the university-based research value chain, significant academic freedom to interact with the industry is highly recommendable, including significant involvement in new firms. Moreover, recent findings suggest that the convergence of technologies and the educational background of researchers will ensure successful technology transfer and commercialisation.

Nevertheless, various authors claim that current data do not support the widely held expectations that emerging technologies, such as nanotechnology and biotechnology, will have a revolutionary impact on healthcare and economic development [110]. Notwithstanding, advances in nanotechnology could increase productivity and enhance the speed and coverage of diagnosis and therapy, following the well-established incremental pattern of technological change and 'creative accumulation' that builds upon, rather than disrupts, previous drug or medical device development.

21.8.2   Technology Transfer and Commercialisation of New Medical Devices

Products based on nanotechnology are already in use and analysts expect markets to grow by hundreds of billions of euros during the present decade. Thus, after a long R&D incubation period, several industrial sectors are emerging as early adopters of nanotech-enabled products, and findings suggest that the Bio&Health market will be among the most challenging of markets over the next few years.

In this context, nanobiotechnology* is a rapidly advancing area of scientific and technological opportunities that lead to advances in the food industry, energy production, safety purposes in the environmental field and medicine. This new discipline is located at the interface of physical and biological sciences and has the potential to revolutionise medicine when the tools, ideas and materials of nanoscience and biology are combined.

In the nanomedicine† case, there is a wide range of technologies that can be applied to medical devices, materials, procedures and treatment modes. A closer look at nanomedicine introduces emerging nanomedical techniques such as nanosurgery, tissue engineering, nanoparticle-enabled diagnostics and targeted drug delivery. According to an expert group of the European Medicines Evaluation Agency (EMEA), the majority of current commercial applications of nanotechnology in medicine are devoted to drug delivery. Novel applications of nanotechnology include tissue replacement, transport across biological barriers, remote control of nanoprobes, integrated implantable sensory nanoelectronic systems and multifunctional chemical structures for targeting of disease.

Findings suggest that nanobiotechnology commercialisation places special emphasis on nanomedicine outputs, where research and medical applications are heavily funded by governments and the private sector. The focus on nanomedicine enhances the high-value chain for applications of this kind that are emerging from converging technologies in which nanotechnology is establishing itself as a new industrial revolution and as a global economic model of 'green growth'. Although nanotechnology and biotechnology exhibit similar technological evolutionary patterns [111], nanotechnology has the potential to affect a broader range of industrial sectors than biotechnology [112]. In another study [113], it was found that nanotechnology is linked to a variety of

---

* Nanobiotechnology involves the processing, manufacturing and packaging of organic or biomaterial devices or assemblies, in which the dimension of at least one functional component lies between 1 and 100 nm.
† For the purposes of this chapter, nanomedicine can be considered as the application of nanotechnology to health. It exploits the improved and often novel physical, chemical and biological properties of materials at the nanometric scale. Nanomedicine has a potential impact on the prevention, early and reliable diagnosis and treatment of diseases.

industries and in particular to those with higher-than-average R&D activity. These findings suggest that nanotechnology is connected in a variety of ways to a diverse set of industries. What remains unclear in the literature, however, is whether nanotechnology transfer between different actors takes unique forms. In particular, very little is known about nanotechnology transfer from universities to firms. Any further development of nanotechnology will depend largely on the degree to which existing firms and industries are able to identify commercial applications. This will, in turn, depend on the degree to which the scientific knowledge that is being created can be transferred from public sector research to the private sector [114]. Though nanotechnology remains very much oriented to basic research, it is still important to understand whether there is already a demand from the private sector for this potentially revolutionary technology.

The application of engineering disciplines and technology to the medical field, which is widely known as biomedical engineering (BME), combines engineering expertise with the medical knowledge of the physician to improve patient healthcare through the design of medical devices. Still in its infancy, much of the work in the discipline involves R&D and it is, therefore, crucial that health institutions, research institutes and manufacturers work together efficiently. In particular, multidisciplinary research groups and technology transfer offices play a key role in the development of new nano-enabled implantable biomedical devices through an advanced understanding of the microstructure/property relationship for biocompatible materials and also of their effect on the structure/performance of such devices. To proceed further, a general framework is required that can facilitate an understanding of the technical and medical requirements, so that new tools and methods might be developed. Moreover, in BME there is a pressing need to ensure close university–hospital–industry–administration cooperation, while specific tools and procedures are developed for use by clinicians. Drawing on the authors' experience, in this section we seek to demonstrate the importance of cooperation and collaboration between the four stakeholders and citizens involved in the innovation process that leads to the development of new nano-related medical products that are ready for the market.

According to [115], current research will provide significant breakthroughs in the near future in the realms of bioreactors, biocompatible materials and nano-biosensors; for instance, the interaction between medicine and technology is allowing the development of diagnostic devices that detect or monitor pathogens, ions, diseases, etc. Today, the integration of rapid advances in areas such as microelectronics, microfluidics, microsensors and biocompatible materials is allowing the development of implantable biodevices such as lab-on-chip and point-of-care devices [116]. As a result, continuous monitoring systems are available to develop faster and cheaper clinical tasks – especially when compared to standard methods.

In particular, forecasts suggest a growing demand for implantable and injectable delivery systems. The implantable/injectable drug delivery market could be segmented into two main categories: needle-free drug delivery and other injectable/implantable devices. There could be significant increases in revenues in areas such as vaccines and insulin. It is in this context that we present a case study of innovation management and technology transfer and commercialisation based on integrated front-end architecture for in vivo detection. The system is designed to be implanted under the human skin. Its powering and the communication between this device and an external primary transmitter are based on an inductive link. The architecture presented is designed for two different approaches to defining a true/false alarm system: based either on ampero-metric or on impedance nano-biosensors. For the purposes of this chapter, we use this biomedical device to analyse the nanomedicine sector and, more specifically, in vivo targeted disease monitoring. Among the diseases that might be monitored, we focus on diabetes since its incidence and prevalence are increasing worldwide, which reflects lifestyle changes and aging populations, *and especially because* the World Health Organisation estimates that the number of diabetics will exceed 350 million by 2030.

For in vivo implantable biomedical devices, we are able to examine an ambitious approach covering the entire value chain (from basic research, through engineering and technology to industry), the infrastructure required and the implications to society of these similar current market challenges. In this instance, the entire value chain is hosted by the university system, which highlights the social turnover of public research investment. We also consider the extent to which recent technological innovations in the biomedical industry have been based on academic research and the time lags between investment in such academic research projects and the industrial application of their findings. Results of academic research are so widely disseminated and their effects so fundamental, subtle and widespread that it is often difficult to identify and measure the links between academic research and industrial innovation. Nevertheless, there is convincing evidence, particularly from industries such as those of drugs, instruments and information processing, that the contribution of academic research to industrial innovation has been considerable [117].

21.8.3   NEW ROLE FOR THE UNIVERSITY IN RESEARCH COMMERCIALISATION

It is widely accepted that the university's role has evolved from one of performing conventional research and education functions to serving as an innovation-promoting knowledge hub [118]. In this role, universities have become deeply embedded in the regional innovation system and are key actors in promoting technological innovation and economic development in their regions of influence [119]. Today, universities actively seek to foster interactions and spillovers so as to link research with application and commercialisation. As a result, the processes of the creation, acquisition, diffusion and deployment of knowledge are at the core of the university's functions.

Medical innovations depend heavily on the breaking down of barriers that have long prevailed in the academic world in the form of the boundaries that have seen the discipline coalesce into separate departments. To be specific, some of the most significant breakthroughs in the life sciences have come from the realm of the physical sciences [120]. Here again, to strengthen medical and clinical infrastructures geographically and organisationally, schools with university departments have generated new opportunities for the transfer of instrumentation and techniques across disciplinary boundaries.

One way to ensure success in such attempts at cross-disciplinary interaction is to examine the way scientific knowledge flows between engineers, researchers and physicians while involved in efforts to develop or improve diagnostic devices. Clearly, one of the main characteristics of a biomedical project is its multidisciplinary context and the need to foster the integration of various dimensions of knowledge. Figure 21.17 shows the R&D value chain leading to a marketable nano-enabled implantable device for in vivo biomedical analysis.

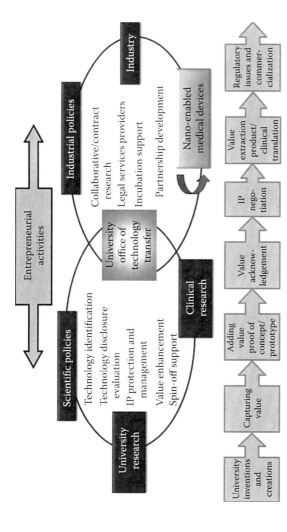

FIGURE 21.17 Overview of the technology transfer process and value chain from initial research to the commercialisation of a new medical device. (Juanola-Feliu, E., Colomer-Farrarons, J., Miribel-Català, P., Samitier, J., Valls-Pasola, J. Market challenges facing academic research in commercializing nano-enabled implantable devices for in-vivo biomedical analysis. *Technovation, Special Issue: The future of Nanotechnologies* 32(3–4), 157–256, 2012.)

Basic and applied research is hosted within university departments and research institutes, where the convergence of science and engineering disciplines at the nanoscale is the current trend for solving new scientific and technological challenges. The university services that promote technology transfer and innovation management play a key role in the subsequent industrialisation of research. In this regard, the sharing of facilities at science and technology parks combined with incubation support for technology enhancement adds value to the process. Clinical research is developed at university hospitals in a multidisciplinary-based, technology feedback process that leads to the development of safe, useful biomedical products. Market penetration is supported by business agencies that provide funding and intellectual property (IP) protection and help with successful commercialisation, as well as by ethical observatories. As a result, the product's technological attributes include a continuous flow of product ideas and technologies embodied in the final product in which the organisational environment and interdisciplinary research define their social and economic assets. In this way, firms obtain ideas that later lead to the creation of new products and processes, and the university identifies new areas of research.

When employing a method for identifying creative scientific research accomplishments in two fields (nanotechnology and human genetics), it was found that creative accomplishments are associated with small group size, organisational contexts that enjoy sufficient access to a complementary variety of technical skills, stable research sponsorship, timely access to extramural skills and resources and the facilitation of leadership [121].

21.8.4 NEW ECOSYSTEM FOR NANO-ENABLED BIOMEDICAL INNOVATIONS

Over the last four decades, research in BME has led to the manufacture of cutting-edge medical instruments. For example, the introduction of endoscopes into surgical practice is considered one of the biggest success stories in the history of medicine. However, a key issue for successful biomedical research in order to develop appropriate medical instruments or procedures is the ability to understand effectively the requirements of the medical practitioner [122]. Furthermore, the two main actors involved in the development of new technologies, namely, universities and industry, need to collaborate and cooperate to a much greater extent. Hence, the process of the flow of knowledge between the various stakeholders involved in the design of medical instruments is of utmost importance. Figure 21.18 shows the framework for a balanced innovation ecosystem in BME in which nanotechnology is gaining increasing relevance, as manipulation at the nanoscale provides new innovative products and processes, thanks to cross-fertilised knowledge [123]. Thus, a general framework should be delineated to facilitate the understanding of both technical and medical requirements.

As a result, new tools and procedures can be developed for the public good by ensuring close cooperation between the university, hospitals, industry and administration. To this end, it is vital to understand the mechanisms and channels of knowledge sharing and transfer within the university–hospital–industry–administration–citizenship 5-helix engine.

There are close research ties between industry and academic institutions in the biomedical field. This observation might be further amplified by noting that there is

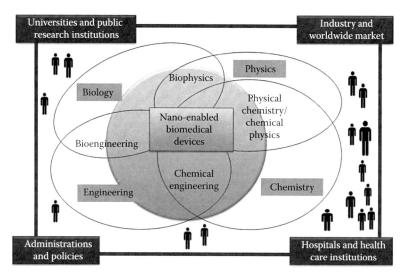

FIGURE 21.18 Innovation ecosystem in biomedical engineering for nano-enabled medical devices: cross-fertilisation and blurred disciplines. (Juanola-Feliu, E., Colomer-Farrarons, J., Miribel-Català, P., Samitier, J., Valls-Pasola, J. Market challenges facing academic research in commercializing nano-enabled implantable devices for in-vivo biomedical analysis. *Technovation, Special Issue: The Future of Nanotechnologies* 32(3–4), 157–256, 2012.)

no significant distinction in their production of medical devices between the more academic institutions, such as hospitals and firms [124]. When alliance activity was examined in a large sample of biotechnology firms [125], it was found that many young biotechnology firms were exploiting this framework by acting as intermediaries in tripartite alliance chains: They enter by establishing partnerships with public upstream research institutions, and later they initiate a commercialisation process by establishing alliances with downstream firms. Although this analysis is undertaken in the biotechnology sector, it is believed that the processes described are relevant to other, science-driven high-technology industries such as the emerging field of nanotechnology. Further evidence for strengthening public–private alliances comes from the new role of the university: Many universities have seen their traditional missions of educating students and advancing understanding evolve to include the patenting and commercialisation of research discoveries, as they become much more proactive in their commercial efforts, particularly over the last few decades [126]. Actually, the modes of interaction between university researchers and firms cover a broad range of formal and informal means of enhancing technology transfer from universities to firms. The common approach in the literature is to focus on tangible outcomes such as patenting and the licensing of research results, but in this work, we argue that a broad perspective of the outcomes of technology transfer should be used including both tangible and intangible outcomes. This means technology transfer also involves the related knowledge whereby firms receive ideas that later lead to new products and processes and universities can identify new research areas [127]. Whatever the case, the rate of commercialisation of university inventions is very low, particularly in those instances where there is a lack of adequately trained staff and insufficient capacity to process inventions in the University Office of Technology Transfer (UOTT). High-tech inventions originating in university labs may well need market space/niche identification, new market creation and the translation of the lab result into an "investor-friendly' business plan [128]. Although technology transfer to the new biomedical firm involves certain familiar components, including the background of the entrepreneur and the 'spin-off' effect, previous studies suggest that the technology transfer process is quite specific in the field of biomedicine [129].

21.9   CONCLUSIONS

This chapter introduces the idea of using the energy generated by the human body to drive ultralow-power electronics to develop new portable biodevices. In that way, it would be possible to envisage new autonomous smart healthcare portable instruments that are able to easily adapt to the daily human activity. The importance of harvesting energy from the normal human activities introduces the enormous advantage of reducing the bulky batteries from the small handheld devices, allowing the miniaturisation and weight reduction of the portable instruments.

Several body-centred power sources are introduced as well as the amount of possible generated energy. The concept of WBSN is described that is focused on the most interesting measuring points for health-care applications.

Moreover, a detailed study of energy storage elements is carried out in order to map several options when designing low-size ultralow-power systems.

A new multi-harvesting architecture concept is described in detail to show the idea of using more than one different power generator to drive a load. In that way, the addition of different energies coming from different sources can warrant a better long-term operation.

Effective management of emergent technology sources will be important for the growth of regional economies and workforces, in relation to new technology commercialisation. In parallel with governments increasing investment in science, there needs to be an effective two-way link between research and the market, to ensure that good research produces good business. This chapter has sought to contribute to the analysis of the interaction between the various factors involved in the process of technology diffusion by extending the discussion to the commercialisation of cutting-edge technologies. Lessons learnt from similar ecosystems and public-funded innovation processes may strengthen the network links between R&D agents – science and technology parks, research institutes and centres, hospitals, technology platforms and incubators – as they explore and confront the new scientific and market challenges presented by the nanotech life sciences.

REFERENCES

1. Yang, G.-Z. *Body Sensor Networks*, Springer, New York, 2006.
2. (Sherman) Shen, X., Misic, J., Kato, N., Langenorfer, P., Lin, X. Emerging technologies and applications of wireless communication in healthcare. *Journal of Communications and Networks* 13(2), 81–85, 2011, doi:10.1109/JCN.2011.6157406.

3. Dey, A.K., Estrin, D. Perspectives on pervasive health from some of the field's leading researchers. *Pervasive Computing, IEEE* 10(2), 4–7, 2011, doi:10.1109/MPRV.2011.29.
4. Available on: http://www.wix.com/mcheung00/mobicare
5. Taub, D.M., Leeb, S.B., Lupton, E.C., Hinman, R.T., Zeisel, J., Blackler, S. The escort system: A safety monitor for people living with Alzheimer's disease. *Pervasive Computing, IEEE* 10(2), 68–77, 2011, doi:10.1109/MPRV.2010.44.
6. Available on: http://www.shimmer-research.com/
7. Guraliuc, A.R., Barsocchi, P., Potortì, F., Nepa, P. Limb movements classification using wearable wireless transceivers. *IEEE Transactions on Information Technology in Biomedicine* 15(3), 474–480, 2011, doi:10.1109/TITB.2011.2118763.
8. González-Valenzuela, S., Chen, M., Leung, V.C.M. Mobility support for health monitoring at home using wearable sensors. *IEEE Transactions on Information Technology in Biomedicine* 15(4), 539–549, 2011, doi:10.1109/TITB.2010.2104326.
9. Garcia-Morchon, O., Falck, T., Heer, T., Wehrle, K. Security for pervasive medical sensor networks. *6th Annual International Conference on Mobile and Ubiquitous Systems: Networking and Services, MobiQuitous, 2009. MobiQuitous'09*, pp. 1–10, 2009, doi:10.4108/ICST.MOBIQUITOUS2009.6832.
10. Penders, J., Gyselinckx, B., Vullers, R., De Nil, M., Nimmala, V., van de Molengraft, J., Yazicioglu, F., Torfs, T., Leonov, V., Merken, P., Van Hoof, C. Human++: From technology to emerging health monitoring concepts. *Proceedings of the International Summer School and Symposium on Medical Devices and Biosensors*, pp. 94–98, Hong Kong, China, 2008.
11. Koutkias, V.G., Chouvarda, I., Triantafyllidis, A., Malousi, A., Giaglis, G.D., Maglaveras, N. A personalized frame work for medication treatment management in chronic care. *IEEE Transactions on Information Technology in Biomedicine* 14(2), 464–472, 2010, doi:10.1109/TITB.2009.2036367.
12. Hai-ying, Z., Kun-mean, H. Pervasive cardiac monitoring system for remote continuous heart care. *2010 4th International Conference on Bioinformatics and Biomedical Engineering (iCBBE)*, pp. 1–4, Chengdu, 2010, doi:10.1109/ICBBE.2010.5514669.
13. Zwifel, P., Felder, S., Meiers, M. Ageing of population and health care expenditure: a red herring? *Health Economics* 8, 485–496, 1999.
14. Allen, M.G. Micromachined endovascularly-implantable wireless aneurysm pressure sensors: From concept to clinic, *Proceedings in the 13th International Conference on Solid-State Sensors, Actuators and Microsystems, Transducers*, pp. 275–278, Seoul, Korea, 2005.
15. López, G., Custodio, V., Moreno, J.I. LOBIN:E-textile and wireless-sensor-network-based platform for healthcare monitoring in future hospital environments. *IEEE Transactions on Information Technology in Biomedicine* 14(6), 1446–1458, 2010, doi:10.1109/TITB.2010.2058812.
16. ElHelw, M., Pansiot, J., McIlwraith, D., Ali, R., Lo, B., Atallah, L. An integrated multi-sensing framework for pervasive healthcare monitoring. *3rd International Conference on Pervasive Computing Technologies for Healthcare, 2009. Pervasive Health 2009.* Department of Computing, Imperial College, London, U.K., pp. 1–7, April 1–3, 2009.
17. Chuo, Y., Marzencki, M., Hung, B., Jaggernauth, C., Tavakolian, K., Lin, P., Kaminska, B. Mechanically flexible wireless multisensor platform for human physical activity and vitals monitoring. *IEEE Transactions on Biomedical Circuits and Systems* 4(5), 281–294, 2010, doi:10.1109/TBCAS.2010.2052616.
18. Salih, R.M., Othmane, L.B., Lilien, L. Privacy protection in pervasive healthcare monitoring systems with active bundles. *2011 Ninth IEEE International Symposium on Parallel and Distributed Processing with Applications Workshops (ISPAW)*, pp. 311–315, Busan, 2011, doi:10.1109/ISPAW.2011.60.
19. Yifeng, H., Wenwu Z., Ling, G. Optimal resource allocation for pervasive health monitoring systems with body sensor networks. *IEEE Transactions on Mobile Computing* 10(11), 1558–1575, 2011, doi:10.1109/TMC.2011.83.

20. Mandal, S., Turicchia, L., Sarpeshkar, R. A low-power, battery-free tag for body sensor networks. *Pervasive Computing, IEEE* 9(1), 71–77, 2010, doi:10.1109/MPRV.2010.1.
21. Qiang, F., Shuenn-Yuh, L., Permana, H., Ghorbani, K., Cosic, I. Developing a wireless implantable body sensor network in MICS band. *IEEE Transactions on Information Technology in Biomedicine* 15(4), 567–576, doi:10.1109/TITB.2011.2153865.
22. Nenggan, Z., Zhaohui, W., Man, L., Yang, L.T. Enhancing battery efficiency for pervasive health-monitoring systems based on electronic textiles. *IEEE Transactions on Information Technology in Biomedicine* 14(2), 350–359, 2010, doi:10.1109/TITB.2009.2034972.
23. Ricotti, L., Assaf, T., Menciasii, A., Dario, P. A novel strategy for long-term implantable pancreas, in *Proceedings of the International IEEE Conference Engineering in Medicine and Biology Society*, pp. 2849–2853, Boston, MA, 2011.
24. Kovacs, G.T.A., Storment, C.W., Rosen, J.M. Regeneration microelectrode array for peripheral nerve recording and stimulation. *IEEE Transactions on Biomedical Engineering* 39(9), 893–902, September 1992.
25. FitzGerald, J. et al. A regenerative microchannel neural interface for recording from and stimulating peripheral axons in vivo. *Journal of Neural Engineering* 9(1), X–X, January 2012.
26. Kim, S., Lee, J.-Y., Song, S.-J., Cho, N., Yoo, H.-J. An energy-efficient analog front-end circuit for a Sub-1-V digital hearing aid chip. *IEEE Solid-State Circuits*, 41(4), 876–882, April 2006.
27. Wong, L.S.Y. et al. A very low-power CMOS mixed-signal IC for implantable pacemaker applications. *IEEE Journal of Solid-State Circuits* 39(12), 2446–2456, December 2004.
28. Rothermel, A. et al. A CMOS chip with active pixel array and specific test feature for subretinal implantation. *IEEE Journal of Solid-State Circuits* 44(1), XX–XX, January 2009.
29. Otha, J. *Smart CMOS Image Sensors and Applications*. CRC Press, Boca Raton, FL, 2007.
30. Liu, Y., Chakrabartty, S., Gkinosatis, D.S., Mohanty, A.K., Lajnef, N. Multi-walled carbon nanotubes/poly(L-lactide) nanocomposite strain sensor for biomechanical implants, in *Proceedings on the IEEE Biomedical Circuits and Systems, BioCAS*, Montreal, Québec Canadá, 2007.
31. Oesterle, S., Gerrish, P., Cong, P. New interfaces to body through implantable-system integration, in *Proceedings of the International Solid-State Circuits Conference ISSCC*, San Francisco, CA, February 2011.
32. Lee, S.-Y., Su, Y.-C., Liang, M.-C., Hong, J.-H., Hsieh, C.-H., Yang, C.-M., Chen, Y.-Y., Lai, H.-Y., Lin, J.-W., Fang, Q. A programmable implantable micro-stimulator Soc with wireless Telemetry: application in closed-loop endocardial stimulation for cardiac pacemaker, in *Proceedings of the International Solid-State Circuits Conference ISSCC*, San Francisco, CA, pp. 44–45, February 2011.
33. Lange, S., Xu, H., Lang, C., Pless, H., Becker, J., Tiedkte, H.-J., Hennig, E., Ortmanns, M. An ac-powered optical receiver consuming 270 µw for transcutaneous 2Mb/s data transfer, in *Proceedings of the International Solid-State Circuits Conference ISSCC*, San Francisco, CA, pp. 304–305, February 2011.
34. Liao, Y.-T., Yao, H., Parviz, B., Otis, B. A 3 µW wirelessly powered CMOS glucose sensor for an active contact lens, in *Proceedings of the International Solid-State Circuits Conference ISSCC*, San Francisco, CA, pp. 38–39, February 2011.
35. Chiu, H.W., Lin, M.L., Lin, C.W., Ho, I.-H., Lin, W.T., Fang, P.H., Lee, Y.C., Wen, Y.R., Lu, S.S. Pain control on demand based on pulsed radio-frequency stimulation of the dorsal root ganglion using a batteryless implantable CMOS SoC. *IEEE Transactions on Biomedical Circuits and Systems* 4(6), 350–359, December 2010.

36. Lee, S.B., Lee, H.-M., Kiani, M., Jow, U.M., Ghovanloo, M. An inductively powered scalable 32-channel wireless neural recording system-on-a-chip for neuroscience applications, *IEEE Transactions on Biomedical Circuits and Systems* 4(6), 360–371, December 2010.

37. Chen, H., Liu, M., Jia, C., Wang, Z. Power harvesting using PZT ceramics embedded in orthopedic implants. *IEEE Transactions on Ultrasonics Ferroelectrics and Frequency Control*, 56(9), 2010–2014, September 2009.

38. Almouahed, S., Gouriou, M., Hamitouche, C., Stindel, E., Roux, C. The use of Piezo ceramics as electrical energy harvesters within instrumented knee implant during walking. *IEEE/ASME Transactions on Mechatronics* 16(5), 799–807, October 2011.

39. Lahuec, C., Almouahed, S., Arzel, M., Gupta, D., Hamitouche, C., Jézéquel, M., Stindel, E., Roux, C. A self-powered telemetry system to estimate the postoperative instability of a knee implant. *IEEE Transactions on Biomedical Engineering*, 58(3), part:2, 822–825, March 2011.

40. Chen, H., Liu, M., Hao, W., Chen, Y., Jia, C., Zhang, C., Wang, Z. Low-power circuits for the bidirectional wireless monitoring system of the orthopedic implants. *IEEE Transactions on Biomedical Circuits and Systems* 3(6), 437–443, October 2009.

41. Yazicioglu, R.F., Vah-Hoof, C., Puers, R. *Biopotential Readout Circuits for Portable Acquisition Systems*. Analog Circuits and Signal Processing Series, Springer, New York, 2009.

42. Hanson, S., Seok, M., Lin, Y.-S., Foo, Z.Y., Kim, D., Lee, Y., Liu, N., Sylvester, D., Blaauw, D. A low-voltage processor for sensing applications with Picowatt standby mode. *IEEE Journal of Solid-State Circuits* 44(4), 1145–1155, April 2009.

43. Zhai, B., Pant, S., Nazhandali, L., Hanson, S., Olson, J., Reeves, A., Minuth, M., Helfand, R., Austin, T., Sylvester, D., Blaauw, D. Energy-efficient subthreshold processor design. *IEEE Transactions on Very Large Scale Integration (VLSI) Systems* 17(8), 1127–1137, August 2009.

44. Song, Y.-K., Borton, D.A., Park, S., Patterson, W.R., Bull, C.W., Laiwalla, F., Mislow, J., Simeral, J.D., Donoghue, J.P., Nurmikko, A.V. Active microelectronic neurosensor arrays for implantable brain communication interfaces. *IEEE Transactions on Neural Systems and Rehabilitation Engineering* 17(4), 339–345, August 2009.

45. Cheong, J.H., Chan, K.L., Khannur, P.B., Tiew, K.T., Je, M. A400-nW19.5-fJ/ Conversion-Step8-ENOB80-kS/sSARADCin0.18-µmCMOS. *IEEE Transactions on Circuits and Systems II: Express Briefs* 58(7), 407–411, July 2011.

46. Cong, P., Chaimanonart, N., Ko, W.H., Young, D.J. A wireless and batteryless 10-Bit implantable blood pressure sensing microsystem with adaptive RF powering for real-time laboratory mice monitoring. *IEEE Transactions of Solid-State Circuits* 44(12), 3631–3644, December 2009.

47. Trung, N.T., Häfliger, P. Time domain ADC for blood glucose implant. *Electronics Letters* 47(26), S18–S20, December 2011.

48. Muller, R., Gambini, S., Rabaey, J.M. A 0.013 mm², 5 µW, DC-coupled neural signal acquisition IC with 0.5 V supply. *IEEE Journal of Solid-State Circuits* 47(1) 232–243, January 2012.

49. Ezekwe, C.D., Boser, B.E. A mode-matching ΣΔ closed-loop vibratory gyroscope readout interface with a 0.004 deg/s/rtHz noise floor over a 50 Hz band. *IEEE Journal of Solid-State Circuits* 43(12), 3039–3048, December 2008.

50. Shih, Y.C., Shen, T., Otis, B.P. A 2.3 µW wireless intraocular pressure/temperature monitor. *IEEE Journal of Solid-State Circuits* 46(11), 2592–2601, November 2011.

51. Liu, Y.-H. A 3.5-mW 15-Mbps O-QPSK transmitter for real-time wireless medical imaging applications, in *Proceedings of the IEEE Custom Integrated Circuits Conference*, pp. 599–602, San Jose, CA, September 2008.

52. Diao, S., Gao, Y., Toh, W., Alper, C., Zheng, Y., Je, M., Heng, C.H. A low-power, high data-rate CMOS ASK transmitter for wireless capsule endoscopy. in *Proceedings of the Defense Science Research Conference and Expo, DSR*, pp.1–4, Singapore, August 2011.

53. Wang, Q., Wolf, K., Plettermeier, D. An UWB capsule endoscope antenna design for biomedical communications. in *Proceedings of the International Symposium on Applied Sciences in Biomedical and Communication Technologies, ISABEL*, pp.1–6, Rome, Italy, November 2010.

54. R. Bashirullah. Wireless implants. *IEEE Microwave Magazine*, 11(7), S14–S23, December 2010 Supplement.

55. Lenaerts, B., Puers, R. *Omnidirectional Inductive Powering for Biomedical Implants*. Analog Circuits and Signal Processing Series, Springer, Berlin, Germany, 2009.

56. Alomainy, A., Hao, Y. Modeling and characterization of biotelemetric radio channel from ingested implants considering organ contents. *IEEE Transactions on Antennas and Propagation*, 57(4), 999–1005, April 2009.

57. Fotopoulou, K., Flynn, B.W. Wireless power transfer in loosely coupled links: Coil misalignment model. *IEEE Transactions on Magnetics*, 47(2), 416–430, February 2011.

58. Fang, Q., Lee, S.Y., Permana, H., Ghorbani, K., Cosic, I. Developing a wireless implantable body sensor network in MICS band. *IEEE Transactions on Information Technology in Biomedicine* 14(4), 567–576, July 2011.

59. Abouei, J., Brown, J.D., Plataniotis, K.N., Pasupathy, S. Energy efficiency and reliability in wireless biomedical implant systems. *IEEE Transactions on Information Technology in Biomedicine* 15(3), 456–466, May 2011.

60. Iniewski, K. (Ed.). *VLSI Circuits for Biomedical Applications*. Artecht House, Inc. Norwood, MA, Chapter 3, 2008.

61. Simard, G., Sawan, M., Massicotte, D. High-speed OQPSK and efficient power transfer through inductive link for biomedical implants. *IEEE Transactions on Biomedical Circuits and Systems* 4(3), 192–200, June 2010.

62. Jung, J., Zhu, S., Liu, P., Chen, Y.E. Deukhyoun Heo "22-pJ/bit energy-efficient 2.4-GHz implantable OOK transmitter for wireless biotelemetry systems: in vitro experiments using rat skin-mimic. *IEEE Transactions on Microwave Theory and Techniques*, 58(12), 4102–4111, December 2010.

63. Carmena, J.M. Becoming bionic. *IEEE Spectrum* 49(3), 24–29, March 2012.

64. Fifer, M.S., Acharya, S., Benz, H.L., Mollazadeh, M., Crone, N.E., Thakor, N.V. Toward electrocorticographic control of a dexterous upper limb prosthesis: building brain-machine interfaces. *IEEE Pulse* 3(1), 38–42, 2012.

65. Spillman, D.M., Takeuhi, E.S. Lithium ion batteries for medical devices. *Proceedings of the Fourteenth Annual Battery Conference on Applications and Advances*, pp. 203–208, Long Beach, CA, 1999.

66. Rubino, R.S., Gan, H., Takeuchi, E.S. Implantable medical applications of lithium-ion technology. *Proceedings of the Seventeenth Annual Battery Conference on Applications and Advances*, pp. 123–127, Long Beach, CA, 2002.

67. Do Valle, D., Wentz, C.T., Sarpeshkar, R. An area and power-efficient analog Li-ion battery charger circuit. *IEEE Transactions on Biomedical Circuits and Systems* 5(2), 131–137, April 2011.

68. Yuming, Y., Liang, Y.C., Kui, Y., Ong, C.K. Low-power fuel delivery with concentration regulation for micro direct methanol fuel cell. *IEEE Transactions on Industry Applications*, 47(3), 1470–1479, March 2011.

69. Pandey, A., Allos, F., Hu, A.P., Budgett, D. Integration of supercapacitors into wirelessly charged biomedical sensors. *Proceedings of the 6th IEEE Conference on Industrial Electronics and Applications*, pp. 56–61, Beijing, 2011.

70. Shanchez, W., Sodini, C., Dawson, J.L. An energy management IC for bio-implants using ultracapacitors for energy storage. *Proceedings of the IEEE Symposium on VLSI Circuits*, pp. 63–64, Honolulu, HI, 2010.

71. CAP-XX (Australia) Pty Ltd, Power vs. Energy, [Online] Available: http://www.cap-xx.com

72. Ravariu, C., Ionescu-Tirgoviste, C., Ravariu, F. Glucose biofuels properties in the bloodstream in conjunction with the beta cell electro-physiology. *Proceedings on the International Conference on Clean Electrical Power*, pp. 124–127, Capri, Italy, 2009.

73. Stetten, F.V., Kerzenmacher, S., Lorenz, A., Chokkalingam, V., Miyakawa, N., Zengerle, R., Ducree, J. A one-compartment, direct glucose fuel cell for powering long-term medical implants. *Proceedings of the 19th International Conference on Micro Electro Mechanical Systems, MEMS*, pp. 934–937, Istanbul, Turkey, 2006.

74. Justin, G.A., Zhang, Y., Sun, M., Sclabassi, R. An investigation of the ability of white blood cells to generate electricity in biofuel cells. *Proceedings of the IEEE 31st Annual Northeast Bioengineering Conference*, pp. 277–278, Hoboen, NJ, 2005.

75. Siu, C.-P.-B., Chiao, M. A microfabricated PDMS microbial fuel cell. *Journal of Microelectromechanical Systems* 17(6), 1329–1341, December 2008.

76. Romero, E., Warrington, R.O., Neuman, M.R. Powering biomedical devices with body motion. *Proceedings of the International Conference of the IEEE Engineering in Medicine and Biology Society (EMBC)*, pp. 3747–4750, Buenos Aires, AG, September 2010.

77. Kiziroglou, M.E., He, C., Yeatman, E.M. Rolling rod electrostatic microgenerator. *IEEE Transactions on Industrial Electronics* 56(4), 1101–1108, April 2009.

78. Le, T.T., Han, J., von Jouanne, A., Mayaram, K., Fiez, T.S. Piezoelectric micro-power generation interface circuits. *IEEE Journal of Solid-State Circuits* 41(6), 1411–1420, June 2006.

79. Roundy, S., Wright, P.K., Pister, K.S.J. Micro-electrostatic vibration-to-electricity converters. *Proceedings of IMECE'02 2002 ASME International Mechanical Engineering Congress and Exposition*, pp. 17–22, New Orleans, LA, 2002.

80. Mhetre, M.R., Nagdeo, N.S., Abhyankar, H.K. Micro energy harvesting for biomedical applications: A review. *Proceedings of the International Conference on Electronics Computer Technology*, pp. 1–5, Kanyakumari, India, 2011.

81. Kazazian, T., Jansen, A.J. Eco-design and human-powered products. *Proceedings of the Electronics Goes Green 2004*, September 6–10, Berlin, 2004.

82. Donelan, J.M., Naing, V., Qingguo, L. Biomechanical energy harvesting. *Proceedings of the IEEE Radio and Wireless Symposium*, pp.1–4, San Diego, CA, 2009.

83. Olivares, A., Olivares, G., Gloesekoetter, P., Górriz, J.M., Ramírez, J. A study of vibration-based energy harvesting in activities of daily living. *Proceedings of the Pervasive Computing Technologies for Healthcare*, Munchen, Germany, March 2010.

84. Available on Internet: htlp://www.mide.comlproducts/volture/peh20w.php

85. Olivo, J., Brunelli, D., Benini, L. A kinetic energy harvester with fast start-up for wearable body-monitoring sensors. *Proceedings of the 4th International Conference on Pervasive Computing Technologies for Healthcare*, pp. 1–7, Munich, Germany, 2007.

86. Zainal Abidin, H.E., Hamzah, A.A., Yeop Majlis, B. Design of interdigitated structures supercapacitor for powering biomedical devices. *Proceedings of the IEEE Regional Symposium on Micro and nanoelectronics*, pp. 88–91, Kota Kinabalu, Malaysia, 2011.

87. Martíne-Quijada, J., Chowdhury, S. Body-motion driven MEMS generator for implantable biomedical devices. *Proceedings of the Canadian Conference on Electrical and Computer Engineering*, pp.164–167, Vancouver, BC, 2007.

88. Paradiso, J.A., Starner, T. Energy scavenging for mobile and wireless electronics. *IEEE Pervasive Computing* 4(1), 18–27, January–March 2005.

89. Lay-Ekuakille, A., Vendramin, G., Trotta, A., Mazzotta, G. Thermoelectric generator design based on power from body heat for biomedical autonomous devices. *Proceedings of the IEEE International workshop on Medical Measurements and Applications*, pp. 1–4, Cetraro, Italy, 2009.

90. Ramadass, Y.K., Chandrakasan, A.P. A battery-less thermoelectric energy harvesting interface circuit with 35 mV startup voltage. *IEEE Journal of Solid-State Circuits* 46(1), 333–341, January 2011.

91. Roundy, S., Steingart, D., Frechette, L., Wright, P., Rabaey, J. Power sources for wireless sensors networks. in *Proceedings of the 1st European Workshop on Wireless Sensors Networks*, pp. 1–17, Berlin, Germany, January 2004.

92. Nasiri, A., Zabalawi, S.A., Mandic, G. Indoor power harvesting using photovoltaic cells for low-power applications. *IEEE Transactions on Industrial Electronics* 56(11), 4502–4509, November 2009.

93. Bertacchini, A., Dondi, D., Larcher, L., Pavan, P. Performance analysis of solar energy harvesting circuits for autonomous sensors. in *Proceedings of the 34th Annual Conference of IEEE Industrial Electronics*, pp. 2655–2660, Orlando, FL, November 2008.

94. Sogorb, T., Llario, J.V., Pelegri, J., Lajara, R., Alberola, J. Studying the feasibility of energy harvesting from broadcast RF station for WSN. in *Proceedings of the IEEE Instrumentation and Measurement Technology Conference*, pp. 1360–1363, Victoria, BC, May 2008.

95. Sauer, C., Stanacevic, M., Cauwenberghs, G., Thakor, N. Power harvesting and telemetry in CMOS for implant devices. *IEEE Transactions on Circuits and Systems* 52(12), 2605–2613, December 2005.

96. Zhang, X., Jiang, H., Zhang, L., Zhang, C., Wang, Z., Chen, X. An energy-efficient ASIC for wireless body sensor networks in medical applications. *IEEE Transactions on Biomedical Circuits and Systems*, 4(1), 11–18, April 2010.

97. Yeatman, E.M. Energy scavenging for wireless sensor nodes. in *Proceedings of the 2nd International Workshop on Advances in Sensors and Interface*, pp. 1–4, Bari, Italy, June 2007.

98. Zhu, Y., Moheimani, S.O.R., Yuce, M.R. Ultrasonic energy transmission and conversion using a 2-D MEMS resonator. *IEEE Device Letters* 31(4), 374–376, April 2010.

99. Zhu, Y., Moheimani, S.O.R., Yuce, M.R. A 2-DOF MEMS ultrasonic energy harvester. *IEEE Sensors Journal* 11(1), 155–161, January 2011.

100. Colomer-Farrarons, J., Miribel-Catala, P., Saiz-Vela, A., Samitier, J. A multiharvested self-powered system in a low-voltage low-power technology. *IEEE Transactions on Industrial Electronics* 58(9), 4250–4263, September 2011.

101. Saggini, S., Mattavelli, P. Power management in multi-source multi-load energy harvesting systems, *Proceedings of the European Conference on Power Electronics and Applications*, pp. 1–10, Barcelona, Spain, 2009.

102. Christmann, J.F., Beigné, E., Condemine, C., Willemin, J. An innovative and efficient energy harvesting platform architecture for autonomous microsystems. *Proceedings of the International NEWCAS Conference*, pp.173–176, Montreal, Québec, Canada, 2010.

103. Kim, G., Bang, S., Kim, Y., Lee, I., Dutta, P., Sylvester, D., Blaauw, D. A modular 1 mm3 die-stacked sensing platform with optical communication and multi-modal energy harvesting. *Proceedings of the IEEE International Solid-State Circuits Conference, ISSCC*, pp. 402–404, San Francisco, CA, February 2012.

104. Torres, E.O., Rincon-Mora, G.A. Electrostatic energy-harvesting and battery-charging CMOS system prototype. *IEEE Transactions on Circuits and Systems I: Regular Papers* 56(9), 1938–1948, September 2009.

105. Midé Engineering Smart Technologies. Mide Technology Corp., Medford, MA., USA., [Online]. Available: http://www.mide.com/

106. IXYS Efficiency through Technology. [Online]. Available: http://www.ixys.com

107. Colomer-Farrarons, J., Miribel-Catala, P., Saiz-Vela, A., Puig-Vidal, M., Samitier, J. Power-conditioning circuitry for a self powered system based on micro PZT generators in a 0.13 µm low-voltage low-power technology. *IEEE Transactions on Industrial Electronics*, 55(9), 3249–3257, September 2008.

108. Colomer-Farrarons, J., Miribel-Catala, P. *A CMOS Self-Powered Front-End Architectures for Subcutaneous Event-Detector Devices: Three-Electrodes Amperometric Biosensor Approach*, Springer, 2011.

109. Goldfarb, B., Henrekson, M. Bottom-up versus top-down policies towards the commercialization of university intellectual property. *Research Policy* 32, 639–658, 2002.

110. Nightingale, P., Hopkins, M., Mahdi, S., Martin, P.A., Craft, A. The myth of the biotech revolution: an assessment of technological, clinical and organisational change. *Research Policy* 36, 566–589, 2007.

111. Rothaermel, F.T., Thursby, M. The nanotech versus the biotech revolution: sources of productivity in incumbent firm research. *Research Policy* 36(6), 832–849, 2007.

112. Youtie, J., Iacopetta, M., Graham, S. Assessing the nature of nanotechnology: can we uncover an emerging general purpose technology? *Journal of Technology Transfer* 33(3), 315–329, 2008.

113. Nikulainen, T. Identifying nanotechnological linkages in the Finnish economy—An explorative study. *Technology Analysis and Strategic Management* 22(5), 513–531, 2010.

114. Nikulainen, T., Palmberg, C. Transferring science-based technologies to industry—Does nanotechnology make a difference? *Technovation* 30, 3–11, 2010.

115. Gennesys. *Whyte paper: Grand European Initiative on Nanoscience and Nanotechnology Using Neutron and Synchrotron Radiation Sources*. Max-Planck-Institut für Metallforschung, Stuttgart, Germany, 2009. http://www.mf.mpg. de/mpg/websiteMetallforschung/pdf/02_Veroeffentlichungen/GENNESYS/ GENNESYS_2009.pdf (accessed on 21, March 13).

116. Barretino, D. Design considerations and recent advances in CMOS-based microsystems for point-of-care clinical diagnostics. *Proceedings of the IEEE International Symposium on Circuits and Systems*, pp. 4362–4365, Island of Kos, Greece, 2006.

117. Mansfield, E. Academic research and industrial innovation. *Research Policy* 20, 1–12, 1991.

118. Youtie, J., Shapira, P. Building an innovation hub: a case study of the transformation of university roles in regional technological and economic development. *Research Policy* 37, 1188–1204, 2008.

119. Juanola-Feliu, E., Samitier, J. Barcelona: Regions in transformation; science, technology parks and urban areas as catalysts. *Proceedings of the XXVI IASP World Conference on Science and Technology Parks*. The Research Triangle Park, Raleigh, North Caroline, 2009.

120. Rosenberg, N. Some critical episodes in the progress of medical innovation: An Anglo-American perspective. *Research Policy* 38, 234–242, 2009.

121. Heinze, T., Shapira, P., Rogers, J.D., Senker, J.M. Organizational and institutional influences on creativity in scientific research. *Research Policy* 38, 610–623, 2009.

122. Arntzen-Bechina, A.A., Leguy C.A.D. An insight into knowledge flow in biomedical engineering science. *The Electronic Journal of Knowledge Management* 5(2), 153–160, 2007.

123. Juanola-Feliu, E., Colomer-Farrarons, J., Miribel-Català, P., Samitier, J., Valls-Pasola, J. Market challenges facing academic research in commercializing nano-enabled implantable devices for in-vivo biomedical analysis. *Technovation, Special Issue: The future of Nanotechnologies* 32(3–4), 157–256, 2012.

124. Mackenzie, M., Cambrosio, A., Keating, P. The commercial application of a scientific discovery: the case of the hybridoma technique. *Research Policy* 17, 155–170, 2008.

125. Stuart, T.E., Ozdemir, S.Z., Ding, W.W. Vertical alliance networks: The case of university–biotechnology–pharmaceutical alliance chains. *Research Policy* 36, 477–498, 2007.

126. Di Gregorio, D., Shane, S. Why do some universities generate more start-ups than others? *Research Policy* 32, 209–227, 2003.

127. Landry, R., Amara, N., Ouimet, M. Determinants of knowledge transfer: evidence from Canadian university researchers in natural sciences and engineering. *The Journal of Technology Transfer* 32(6), 561–592, 2007.

128. Swamidass, P.M., Vulasa, V. Why university inventions rarely produce income? Bottlenecks in university technology transfer, *Journal of Technology Transfer* 34, 343–363, 2009, doi:10.1007/s10961-008-9097-8.

129. Roberts, E.B., Hauftman O. The process of technology transfer to the new biomedical and pharmaceutical firm. *Research Policy* 15, 107–119, 1986.

130. Colomer-Farrarons, J., Miribel-Català, P., Rodríguez, I., and Samitier, J. CMOS Front-End Architecture for In-Vivo Biomedical Implantable Devices. *Proceedings of the 35th International Conference of Industrial Electronics, IECON*, 4401–4408, Porto, 2009.

22 ADC Basics and Design Techniques for a Low-Power ADC for Biomedical Applications

Seung-Tak Ryu

CONTENTS

22.1   INTRODUCTION

Advanced semiconductor technologies and sophisticated low-power circuit design techniques are allowing the realization of portable, wearable or implantable bio-medical devices in response to increasing interest in healthy living. The physiological signals collected from sensors in biomedical systems are usually very small, on the order of tens of μV to tens of mV, and have considerable noise and offset (Zou et al. 2009). After noise/offset filtering and amplification with the analogue front end (Liew et al. 2009; Yazicioglu et al. 2011), the signal is converted into digital form via an analogue-to-digital converter (ADC) for back-end digital signal processing. Considering the bandwidth of physiological signals (typically less than a few kHz) and the dynamic range (DR) (around 60 dB), ADCs for biomedical applications are usually designed for a resolution of about 8b ~ 12b and a conversion rate of 1 k ~ 100 kHz (Yang and Sarpeshkar 2006; Zou et al. 2009). These specifications of resolution and speed can be achieved relatively easily with modern design techniques and processes. However, one of the major concerns in circuit design for portable/implantable biomedical systems is realizing extremely low power consumption. Given that wearable health-care systems sometimes need to monitor signals from the body for more than several days and battery replacement for implanted devices can be very troublesome, the increasing demand for low power consumption is well justified. In addition, considering the possibility of damage to tissues (such as brain cells) by the heat generated from excessive power consumption, the importance of low power consumption has been recognized. Thus, all the circuits for biomedical applications, including those for the ADC, should consume as little power as possible.

The purpose of this chapter is to share essential knowledge on ADCs and to introduce ADC structures and low-power design techniques that are suitable for biomedical applications. In Section 22.2, basic sampling theory and the operational principles of quantizers will be reviewed. In Section 22.3, major sources of nonideality (noise and nonlinearity) in ADC and their effects on performance will be investigated with various performance metrics. Section 22.4 discusses suitable ADC structures for low-power biomedical applications and explains the operational principle of the successive approximation register (SAR) ADC in detail, as it is considered to be the most suitable structure. Section 22.5 introduces recent advances in low-power design techniques in SAR ADCs. Section 22.6 concludes this chapter with notes on performance trends of low-power SAR ADCs.

22.2 OVERVIEW OF ADC BASICS

22.2.1 ADC BLOCK DIAGRAM AND SIGNALS

Operation of an ADC can be classified into two functions: sampling and quantization, as shown in Figure 22.1. Sampling is a process that periodically captures the input signal, as depicted on the right side of the figure with small circles on V_{in}. An ideally sampled signal implies a signal with zero time duration with the same amplitude as the input. However, zero time duration is not possible in the real world; moreover, we actually do not want such a property, because the sampled signal should be held for a certain time duration for reliable operation of the following circuits such as quantizers. Thus, real sampling is done by a sample-and-hold (S/H or track-and-hold [T/H]) circuit that conducts sampling at the clock edge and holds the signal until the next sampling. An example of an ideal S/H output is shown in the figure with a dashed line. The sampled and held signal is then converted into a symbolic digital code (ADC out in the figure) by the following quantizer.

22.2.2 SAMPLING

Figure 22.2 illustrates the ideal sampling operation in both the time domain (upper) and frequency domain (lower). Sampling operation in the time domain can

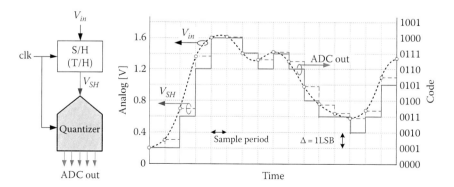

FIGURE 22.1 Functional block diagram and signals of ADC.

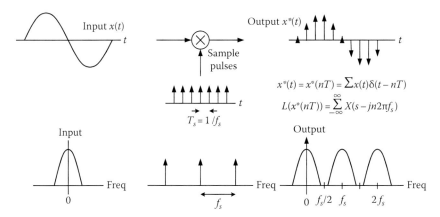

FIGURE 22.2   Understanding of sampling in time and frequency domain.

be interpreted as multiplication of the input signal, $x(t)$, with a periodic sampling impulse train (sampling period in the figure is T_s). The ideal resultant output has its value at every T_s with zero pulse width: $x*(t) = \Sigma x(t)\delta(t-nT_s)$. Since multiplication in the time domain is equivalent to convolution in the frequency domain, the signal spectrum repeats every sampling frequency, f_s: $X(s) = \Sigma X(s-jn2\pi f_s)$. As can be inferred from the output spectrum illustrated in Figure 22.2, if the input signal spectrum is wider than half of the sampling frequency, $f_s/2$, the spectrum centred at f_s in the second Nyquist zone ($f_s/2 \sim 3/2\,f_s$) invades the first Nyquist zone and ruins the original signal information; this mechanism is known as aliasing. Therefore, in order to preserve the original signal information after sampling, the sampling frequency must be higher than twofold the signal bandwidth (Nyquist sampling).

   As mentioned earlier, sampling is done by the S/H or T/H circuit. In principle, T/H can be implemented with one switch and one capacitor, as shown in Figure 22.3. While the clock signal, CLK, is high, the switch is closed (on) and the output, V_{out}, tracks the input signal. When CLK falls to a low level (at falling edge), the input signal is sampled on the capacitor C_H. When CLK = 0, the switch is off and the sampled signal is held on the capacitor. By virtue of this simple circuit structure, T/H is a typical choice for high-speed applications. However, various nonidealities from the real switch, the major drawback of this structure, limit its use in high-resolution ADCs. A more detailed discussion will be provided in Section 22.3.2.

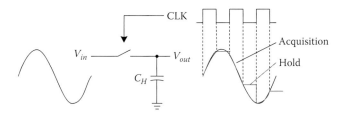

FIGURE 22.3   T/H and the signals.

22.2.3 QUANTIZATION

The sampled and held signal by a T/H is then converted into digital code by a quantizer. The quantizer determines the amplitude range to which the signal belongs and generates a digital code of the corresponding range. In general, the signal full scale is evenly divided into 2^N ranges in an N-bit resolution quantizer, and each divided range is assigned a binary-weighted code. Figure 22.4 shows transfer curves of 1b, 2b and 4b quantizers. The 1b quantizer divides the full scale, V_{ref}, into two ranges with a decision threshold (reference) level of $V_{ref}/2$; that is, if the input is lower than $V_{ref}/2$, the output is 0, and otherwise the output is 1. Similarly, the 4b quantizer divides the full scale by 16 and assigns 4-bit binary codes. The ideal step size of N-bit ADC is thus $V_{ref}/2^N$, and as N increases, the ADC can better represent the original information with less error (note that 4b quantization shows a rather linear transfer curve, while the 1b quantizer hardly tracks a straight line). The error between the ideal transfer curve (straight line) and the quantizer's stair-shaped line is called the quantization error.

It is known that the quantization error can be considered as random noise with negligible correlation with the input signal when the ADC resolution is above 4b (Razavi 1994). Therefore, quantization error is often called quantization noise, and it is the fundamental limit source on the ADC signal-to-noise ratio (SNR). As would be expected, as N increases, the quantization noise reduces and SNR improves. This will be discussed further in Section 22.3.

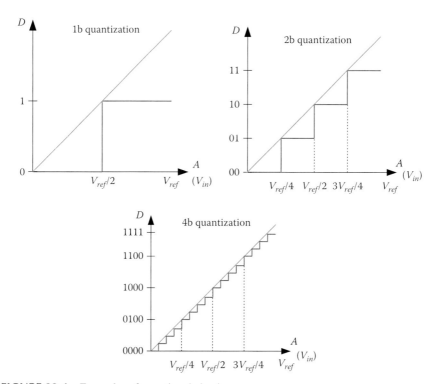

FIGURE 22.4 Examples of quantizer behaviour.

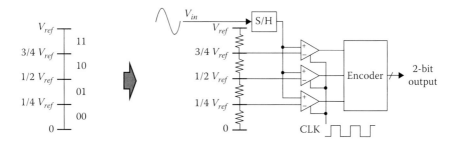

FIGURE 22.5   Circuit realization of a 2b quantizer with flash ADC configuration.

The most straightforward circuit realization of the quantization concept is a flash ADC, shown in Figure 22.5 for a 2-bit resolution example. In order to detect the position of the input signal, all the required reference levels are generated by using a resistor string, and the input signal is compared with them by using a comparator array. If the sampled signal is placed between $1/2\ V_{ref}$ and $3/4\ V_{ref}$, for instance, the comparators will generate 0, 1, 1 from top to bottom. The following encoder converts this code into binary code of 10 as the final output.

A flash ADC is a straightforward realization of the quantizer concept; however, the structure is not suitable for all specifications due to its hardware complexity in high-resolution design. The number of comparators increases exponentially (2^N) as the resolution increases. Depending on the target specifications such as resolution, speed and power consumption, various types of ADCs can be chosen. Figure 22.6 shows a brief spectrum of the resolution and speed of various types of modern ADCs. From the figure, for example, one can see that a sigma-delta (SD) ADC

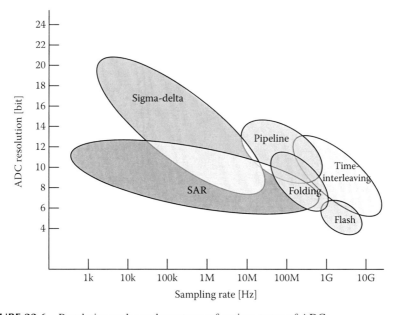

FIGURE 22.6   Resolution and speed spectrum of various types of ADCs.

is suitable for high-resolution low-speed applications and the time-interleaving technique is appropriate for very high-speed applications with relatively moderate resolution. For low-frequency applications, such as biomedical systems, SD ADC and SAR ADC structures might be considered with respect to resolution and power consumption.

22.3 NONIDEALITIES

In general, the ADCs are known to be one of the most difficult circuit blocks to design. In order to successfully implement an ADC, designers must understand the performance limitations from the circuit nonidealities. This section discusses the non-ideality issues. Analogue circuits contain various noise and nonlinearity (distortion) sources. Figure 22.7 illustrates the effects of noise and distortion in analogue signal processing circuits, including the ADC, with transfer-function and frequency-domain characteristics. If an ADC is ideal, its transfer curve will be a perfectly straight line and the frequency spectrum for a pure sinusoidal signal will show a single tone only. However, in reality, noise cannot be avoided, and thus, the transfer curve becomes blurry (seen in the time-domain ramp) and the frequency spectrum has a noise floor, that is, drop in the SNR and DR. If the transfer curve becomes nonlinear, the output signal's linearity is deteriorated, and thus, the output will show harmonic distortions in the frequency domain. This degrades the total harmonic distortion (THD) and spur-free DR (SFDR), which will be explained in Section 22.3.3. In a real circuit, noise and distortion cannot be avoided, and thus, the generalized transfer function and frequency response are expected to be similar to those depicted in the last figure in Figure 22.7. Performance metrics of the SNR + distortion ratio (SNDR) and the effective number of bits (ENOBs) take into account both noise and distortion.

22.3.1 NOISE

Fundamental noises in the ADC are quantization noise, kT/C noise and sampling jitter. As mentioned earlier, quantization noise arises from the nonzero quantization step size. kT/C noise originates from the sampling circuit, and the sampling jitter is mostly due to random timing error in the clock source.

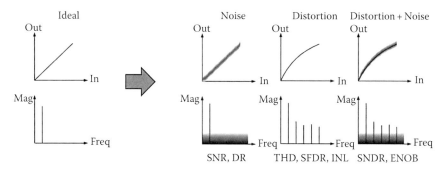

FIGURE 22.7 Effects of nonidealities (noise and distortion).

22.3.1.1   Noise by Quantization

From Figure 22.4, we see that the amount of quantization error ranges from 0 to the quantization step size Δ ($=V_{ref}/2^N$). However, if we recall that each digital code represents a certain signal range rather than the lowest value of the range, each code can be interpreted as representing the centre value of each segment. The quantization error then ranges between $-\Delta/2$ and $+\Delta/2$. This modification can be justified because the error ranging from $0 \sim \Delta$ has a DC term (offset) of $\Delta/2$, which is not random. Interestingly, this quantization error can reasonably be considered as additive white noise with a uniform probability distribution function (PDF) between $-\Delta/2$ and $+\Delta/2$. Thus, the quantization error is often referred to as quantization noise. From this assumption, the noise power can be calculated as (22.1):

$$V_q^2 = \int_{-\Delta/2}^{\Delta/2} x^2 \frac{1}{\Delta} dx = \frac{\Delta^2}{12} = \frac{V_{ref}^2}{12 \cdot 2^{2N}} \tag{22.1}$$

For a single-tone sinusoidal input with a full scale of V_{ref}, the quantization-noise-limited SNR (SQNR) is calculated to be $3/2 \cdot 2^{2N}$. In dB scale, SQNR is described as given in (22.2):

$$10 \log \text{SQNR} = 6.02N + 1.76 \, [\text{dB}] \tag{22.2}$$

This equation states that the ADC's SNR improves by 6.02 dB as the resolution increases by 1b when no other noise exists. For example, even when other noise sources are absent, a 10b ADC cannot achieve an SNR higher than 62 dB.

22.3.1.2   Noise from Sampling (*kT/C* Noise)

All electronic devices that have resistance within them generate thermal noise. In a sampling circuit, the on-resistance of the sampling switch becomes the source of thermal noise, and the noise from it is sampled on the capacitor together with the input signal. The sampled noise, of course, degrades the SNR of the sampling circuit.

An equivalent T/H circuit shown in Figure 22.3 is redrawn in Figure 22.8 including the thermal noise source from the on-resistance, R, of the switch. In order to exclude the input signal effect, $V_{IN} = 0$ is assumed.

The power spectral density of the thermal noise voltage from a resistor R is $4\,kTR$ [V^2/Hz], where k is the Boltzmann constant and T is the absolute temperature. In a room temperature, kT is about 4.1×10^{-21} [FV^2]. While the switch is on, the noise voltage from R is transferred to the output via the RC low-pass network with a transfer function $H(s) = 1/(sRC + 1)$. The total noise power at the output of the RC low-pass filter is calculated by integrating the output power up to infinity frequency, as in (22.3). The result, kT/C, states that the total output noise power is independent of the switch resistance and depends only on the sampling capacitance. The reason for this is that the power spectral density of the noise voltage is proportional to R while the bandwidth is inversely proportional to RC. Thus, the effect of R is cancelled out and only the capacitor's effect remains:

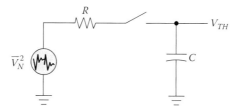

FIGURE 22.8 Equivalent circuit of noise sampling in T/H.

$$V_{n,out}^2 = \int_0^\infty |H(s)|^2 \, 4kTR\,df$$

$$= \int_0^\infty \frac{4kTR}{4\pi^2 R^2 C^2 f^2 + 1}\,df$$

$$= kT/C \tag{22.3}$$

Now we need to calculate the 'sampled' noise power. Recalling that the signal components beyond the $fs/2$ are folded back to the first Nyquist zone in the frequency domain (aliasing), we can state that the sampled signal power is the same as that of the signal before aliasing (i.e. original signal). This means that, regardless of the sampling frequency, the sampled noise power in the signal band (first Nyquist zone) is still kT/C.

By using this result, the required sampling capacitor value for a given SNR target can be calculated. For instance, if we design an N-bit ADC such that the kT/C-limited SNR is the same as the SQNR, we can write Equation 22.4 by approximating SQNR as $6N$. Equation 22.4 states the total SNR drops by 3 dB as the sampling capacitor reduces by 2 because of the doubled noise power. From (22.4), the required minimum sampling capacitance can be calculated as delineated in (22.5):

$$SNR_{kT/C} = 10\log\left(\frac{V_{ref}^2/8}{kT/C}\right) > 6N \tag{22.4}$$

$$C > \frac{10^{0.6N} \cdot 8kT}{V_{ref}^2} \tag{22.5}$$

For example, for a case of $V_{ref} = 1$ V, the sampling capacitor C should be larger than $8\,kT \cdot 10^{0.6N}$ for an N-bit ADC with a 3 dB SNR drop from the ideal SNR value. For a 10-bit ADC, the sampling capacitor should be larger than 33 fF, and for a 14b ADC, 8.3 pF is required. As 1-bit resolution increases, the required sampling capacitor increases fourfold. However, considering other noises and matching, sampling capacitors are usually designed with greater margins.

22.3.1.3 Noise by Sampling Jitter

Another important noise in a data converter is the sample timing error. Sampled amplitude error caused by the sample time shift is illustrated in Figure 22.9. As the amplitude errors, e_1 and e_2, indicate, as the input signal changes more rapidly, the error amplitude increases more for the same timing error. Thus, the sampling jitter effect becomes more serious in high-speed applications. The SNR equation determined by the sampling jitter is shown in (22.6). The SNR by jitter is a function of the signal frequency, f_{in}, and the rms jitter, Δt_{rms}:

$$SNR_{jitter} = -20\log\left(2\pi f_{in} \times \Delta t_{rms}\right) \text{ [dB]} \tag{22.6}$$

By using (22.6), it is possible to calculate the required rms clock jitter. For a 100 kHz input signal, which is of much higher speed than the biomedical signals, and a target DR (SNR) of 60 dB, the required rms jitter value is about 1.6 ns. Most commercial clock sources guarantee much better jitter performance than that required for biomedical applications (Kester 2005). However, if the clock signal needs to be generated on chip by using relaxation oscillators or ring oscillators, the jitter performance should be carefully verified to guarantee the ADC performance.

22.3.2 Nonlinearity

Nonlinearity (deviation of transfer curve from the ideal straight line, either globally or locally) sources in the data converter can be categorized into two parts: device nonlinearity and device mismatch. Examples of device nonlinearity can be found from the signal-dependent on-resistance of switches and transconductance variation of differential pairs, and they result in global nonlinearity of the transfer curve. Resistance mismatch in the resistor ladder and random offsets in the comparator array in a flash ADC are good examples of device mismatch. These mismatches shift the decision thresholds (refer to $1/4\ V_{ref} \sim 3/4\ V_{ref}$ in Figure 22.5) and make the code width non-uniform for a ramp input. This code-by-code error can be categorized

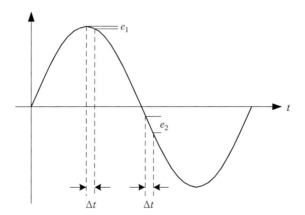

FIGURE 22.9 Sample jitter–induced amplitude error.

as local error as compared with the global transfer curve distortion. However, this code-by-code error also contributes to the overall transfer curve nonlinearity by the integrated local errors along the full signal range.

22.3.2.1 Circuit Nonlinearity

Various examples of the circuit nonlinearity can be found from a T/H. Figure 22.10 shows a T/H circuit with identification of various error sources: on-resistance nonlinearity, clock feedthrough, charge injection, nonlinear parasitic capacitance and signal-dependent sample time (Song 2012). Most of these errors arise from the nonconstant threshold voltage and the parasitic junction capacitance of the switch transistor on the input signal. Nonlinearity by these errors can be ignored if the signal change is small (note that linear approximation is acceptable in a small signal circuit analysis). However, if the signal range increases, global change appears in the transfer curve, and the T/H structure in Figure 22.10 becomes not suitable for high-resolution applications. Similar nonlinear behaviour can also be found when an operational amplifier does not have sufficient gain or when an open-loop amplifier is used for large signal amplification. The nonlinearity problem in T/H can be reduced by using a clock boosting scheme (Abo and Gray 1999). Recent research shows that T/H based sampling can achieve 10b linearity at a 100 MS/s conversion rate by using a switch linearization technique (Liu et al. 2010b). For higher linearity than 10b, a parasitic-insensitive bottom-plate sampling technique is usually recommended (Li et al. 1984; Song 2012).

22.3.2.2 Device Mismatch: Decision Threshold Shift (Global and Local)

In the flash ADC shown in Figure 22.5, the decision thresholds are generated by dividing the full-scale reference, V_{ref}, using a series of resistors (resistor string), and the sampled input signal is compared with those references using comparators. Now, let us consider an example of a nonideal condition where the resistor values become larger as they approach V_{ref}, as shown Figure 22.11. The reference step sizes increase as the reference voltage increases, and the output code widths are not uniform. Code 000 has the narrowest width and code 111 has the widest width. This leads to distortion of the ADC transfer curve, similar to a log function. Note that the comparator offsets also contribute to this decision error. Unlike the resistor string error, which still guarantees reference monotonicity, comparator offset may destroy the reference

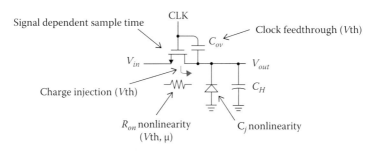

FIGURE 22.10   Nonlinearity sources in T/H.

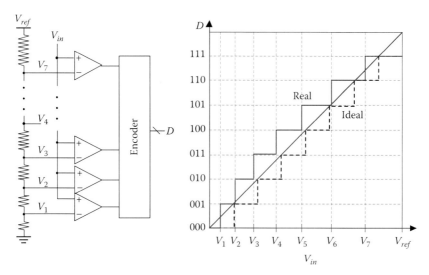

FIGURE 22.11 Nonlinearity from reference shifts.

voltage monotonicity. To sum up, any errors in devices that contribute to the decision threshold affect the transfer curve distortion.

22.3.2.3 Settling Error

Besides the device nonlinearity and mismatch, limited settling speed of the signal and references also deteriorates the ADC linearity. Figure 22.12 shows typical settling behaviour of a one-pole system, which can be assumed to be a T/H output. If A/D conversion is conducted before the output signal V_O settles to its final value, V_{target}, the ADC will suffer a certain amount of overall gain error. If the same phenomenon occurs in a two-step ADC where LSBs are achieved from the residue signal (=remaining signal after MSBs decision), LSB segments will suffer from repeated gain error in the transfer curve whenever the MSB segment changes, as Figure 22.12 demonstrates. With respect to the settling speed, all the circuits in the ADC should

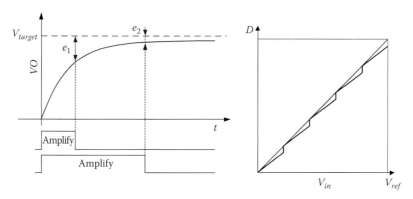

FIGURE 22.12 Inaccurate settling error and the effect on the transfer function in a multistage ADC.

be fast enough so that the unresolved signal segment can be converted into digital code with negligible decision error. Thus, in a high-speed ADC, settling accuracy is a very important issue and becomes a major bottleneck in low-power realizations.

22.3.3 PERFORMANCE METRICS

22.3.3.1 Static Linearity: DNL and INL

In Section 22.3.2, we discussed nonlinearity from the non-uniform code widths and the overall transfer curve change. The performance metrics that represent these static nonlinearities are differential nonlinearity (DNL) and integral nonlinearity (INL). DNL represents code width nonidealities for each code by normalizing the measured code width to the ideal step size. Equation 22.7 depicts the DNL for the k_{th} code, where Δ_{ideal} is the ideal step size (=1 LSB) and Δ_{real_k} is the measured width of the k_{th} code. The DNL represents local error, and its unit is LSB. Positive DNL means the code width is wider than the ideal value and negative DNL indicates narrow width. −1 DNL means code is missing:

$$DNL(k) = \frac{\Delta_{real\_k} - \Delta_{ideal}}{\Delta_{ideal}}[LSB] \tag{22.7}$$

INL is the running sum of the DNL, that is, the INL for code k is calculated by summing the DNL values from code 0 to code k, as (22.8) shows. Since the INL is the accumulated error through the whole input range, the maximum INL value represents how much the actual transfer curve deviates from the ideal straight line. Note that the INL is strongly related to SFDR, because spurs (harmonic distortions) are generated by transfer curve distortion:

$$INL(k) = \sum_{i=0}^{i=k} DNL(i)[LSB] \tag{22.8}$$

Figure 22.13 shows an example of a nonideal transfer curve and the resultant DNL and INL profiles.

22.3.3.2 Dynamic Performance: SNR, SFDR, SNDR, ENOB and DR

Dynamic performances of an ADC (SNR, SFDR, SNDR, THD, ENOB and DR) are explained with the FFT results shown in Figure 22.14.

As discussed earlier, the SNR is the power ratio of the input signal (S) and the total noise (N). In general, harmonic distortions are considered as a part of the signal, rather than noise, in the SNR calculation, because distortion is strongly correlated with the input signal. SFDR is the power ratio of the fundamental input and the largest spur when the input signal is the maximum. This often represents the system linearity. SNDR is the power ratio between the input signal and the power sum of the unwanted components such as distortion and noise. If other conditions are ideal, an N-bit ADC's SNR will be determined as 6.02N + 1.76 dB by the quantization noise.

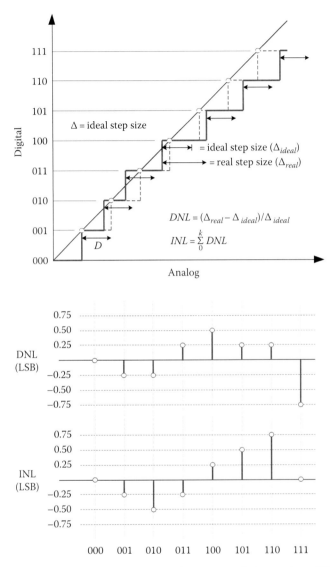

FIGURE 22.13 Example of a nonideal transfer curve and the resultant DNL and INL profiles.

However, as discussed earlier, a real ADC's SNR is degraded by other noises such as kT/C noise and nonideal quantization step sizes. ENOB calculates the effective resolution based on the SQNR equation (Equation 22.2) with the measured SNR. Thus, the effective resolution is found from the following equation: ENOB = (SNR (in dB) – 1.76)/6.02. Recently, the ENOB has been calculated from the SNDR instead of the SNR in order to reflect all nonidealities. For example, if an ADC shows a measured SNDR of 53 dB, the ENOB of the ADC is 8.5. DR is the power ratio between

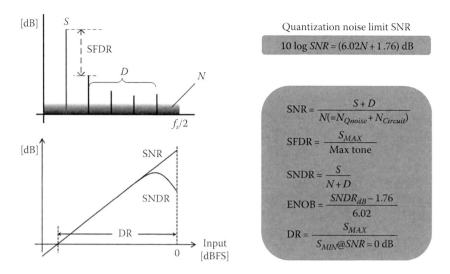

FIGURE 22.14 Definition of dynamic performances.

the largest signal and the smallest detectable signal. At the smallest detectable signal, the signal power is the same as the total noise power (SNR = 0).

22.3.3.3 Figure of Merit

In order to represent the performances of various ADCs, a universal measure is often necessary. As Figure 22.6 shows, there is a clear relationship between resolution and conversion speed. Walden (1999) observed that the ADCs with state-of-the-art class performance show similar values of $2^{ENOB} \times f_{sample}$. Thus, this value can be used as the universal measure of the performance. Including the power consumption, the ADC figure of merit (FOM) is defined as (22.9) and is commonly used at present:

$$FOM = \frac{Power}{2^{ENOB} \times f_{sample}} \quad [Joule/conversion\text{-}step] \quad (22.9)$$

However, note that applying this to all types of ADCs with equal weight might not be fair because of the different mechanisms and performance limitations in ADCs with different specifications (Murmann 2008). Several recently reported ADCs for biomedical applications record less than 10 fJ/conversion step. This will be discussed in Section 22.4.

22.4 LOW-POWER ADCS FOR BIOMEDICAL APPLICATIONS

22.4.1 CANDIDATES

Since ADCs for biomedical applications require 8b ~ 12b resolution with less than a 100 kS/s conversion rate, high-speed-oriented ADCs such as flash and pipelined ADCs are not suitable (refer to Figure 22.6). In this section, we survey suitable ADC structures for low-power biomedical applications.

22.4.1.1 SAR ADC

As shown in Figure 22.15, the SAR ADC has a simple structure with S/H (T/H), DAC, comparator and some digital logic (SAR logic). The lower figure shows the code decision procedure (sampling is not shown). After the input is sampled by S/H, the sampled signal, V_{SH}, is successively compared with the updated DAC output, V_{DAC}, at every decision phase. At the MSB decision phase, P_1, the DAC generates $V_{ref}/2$. Since $V_{SH} < V_{ref}/2$, MSB $(D_1) = 0$, and this indicates that the input signal is between 0 and $V_{ref}/2$ as the grey bar at the end of P_1 column shows. Thus, in P_2, DAC generates 1/4 V_{ref}, the centre of the determined input range in P_1, for MSB-1 bit decision. Since $V_{SH} > V_{DAC}$, D_2 (MSB-1 bit) is 1, and the determined input range is now from 1/4 V_{ref} to 1/2 V_{ref}. The remaining decisions in P_3 and P_4 follow the same binary search algorithm by continuously dividing the remaining range by half. Thus, an N-bit SAR ADC completes the full conversion after N decisions, and SAR logic controls the decision procedure. Compared with flash ADCs that finish one conversion in one clock, the SAR ADC is much slower. However, in biomedical applications, high-speed ADCs over 100 kS/s are not likely to be required. Furthermore, 100 kS/s operation with around 10b resolution can be designed easily

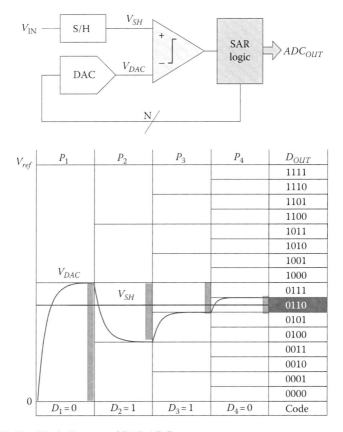

FIGURE 22.15 Block diagram of SAR ADC.

with recent CMOS technologies. A rather important design point is low-power consumption. The SAR ADC's circuit structure is much simpler than that of other high-performance ADCs such as flash and pipelined ADCs. Moreover, all the circuit components in Figure 22.15 can be implemented with no static power consumption, which is very attractive for low-power design. In these regards, the SAR ADC is considered to be the most suitable structure for biomedical applications, and the majority of reported ADCs for biomedical application in recent publications are SAR ADCs (Wu and Xu 2006; Zou et al. 2009; Yazicioglu et al. 2011).

22.4.1.2 Integrating ADC

Figure 22.16 shows the simplest hardware implementation of an integrating ADC. It is composed of an integrator (capacitor + current source), a comparator and a counter. As soon as the input sampling is completed by the switch, SW_{IN}, the sampling capacitor starts to discharge the stored charge ($Q = V_{in} \times C$) with a constant rate by the current source I, and the counter starts to count. The counter stops when the comparator input, Vx, reaches the reference level (ground in the figure) while discharging. As shown in the lower figure, since the discharging time is proportional to the input magnitude, the counter output directly represents the input magnitude. This structure is very compact and thus suitable for small-size applications such as column ADCs in image sensors. Owing to the monotonic characteristic of the integrator, DNL is usually excellent. However, due to the linear counting property, the counting clock must be at least 2^N times faster than the sample rate for an N-bit ADC. Because of the

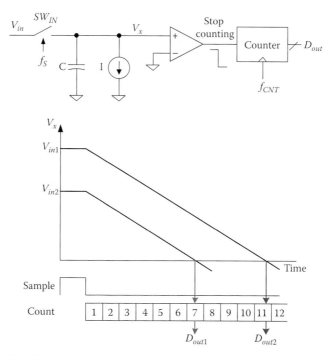

FIGURE 22.16   Circuit diagram of integrating ADC and its conversion principle.

high counting rate, it is difficult to use the structure for high resolution such as 10b or higher. In addition, even though the structure is very compact, a high-frequency clock in the high-resolution ADC can consume considerable power. If an application requires small size with relatively low power consumption, such as an application that has massive channel readout, a single-slope ADC may be suitable.

22.4.1.3   Sigma-Delta (SD) ADC

An SD ADC has two important characteristics that enhance the SNR over Nyquist ADCs: oversampling and noise shaping. In II-2 and III-1, we discussed that all the signal and noise components that exceed half the sampling frequency (say, sample frequency = f_{Nyq}) are folded back below $f_{Nyq}/2$ by aliasing with no total power change. This implies that if the sampling frequency is increased by M times, the noise-power spectral density reduces proportionally: that is, M-times oversampling over the Nyquist rate enhances the SNR by $10 \log M$ (dB) in a fixed signal band. This oversampling effect is illustrated in the left figure in Figure 22.17. Noise shaping in an SD ADC is conducted by high-pass filtering of the quantization noise, as Figure 22.17 (a) shows. Owing to this, the noise density in the signal band greatly attenuates, and therefore, the SNR is further improved. Note that, as can be inferred from the figure, in order to take full advantage of the oversampling and noise shaping, a digital filter must be implemented to cut off the out-of-band noise.

Noise shaping is conducted by the closed-loop SD modulator shown in Figure 22.17 (b). If the feedforward block has a low-pass characteristic, the quantization noise, e_q, added to the quantizer undergoes high-pass filtering. A complete SD ADC is completed with a back-end digital filter. The SNR of an SD ADC enhances as the oversampling ratio (OSR) M increases, as the order of the low-pass filter increases and as the quantization resolution increases. Quantitative analyses of the SNR as a function of OSR, filter order and quantizer resolution can be found in many textbooks such as Schreier and Temes (2004). Owing to these properties, the SD ADC has become the major structure for audio applications where over 100 dB SNR is often required and the signal frequency is relatively low, for example, 20 kHz (good conditions for oversampling).

In biomedical applications, considering the signal bandwidth, the SD ADC is applicable to achieve a high SNR (Wang and Theogarajan 2011), especially when an analogue front end is not applicable. However, the static power consumption by the analogue loop filter and the excessive overhead by the oversampling and digital filter are major drawbacks for low-power implementations. In the system point of view, however, if a given system has to deal with very small signals but a high SNR is required without low-noise amplifier and/or antialiasing filter, the SD ADC may be a good choice (Soundarapandian and Berarducci 2010).

22.4.2   SAR ADC: Schematic and Operation

Considering the requirements for biomedical applications with respect to resolution, speed and power consumption, SAR ADC seems the most suitable structure and recent paper trends reflect it (Wu and Xu 2006; Pang et al. 2009; Zou et al. 2009; Chang et al. 2011; Lu et al. 2011; Yazicioglu et al. 2011). In this section, detailed operations of SAR ADC are explained.

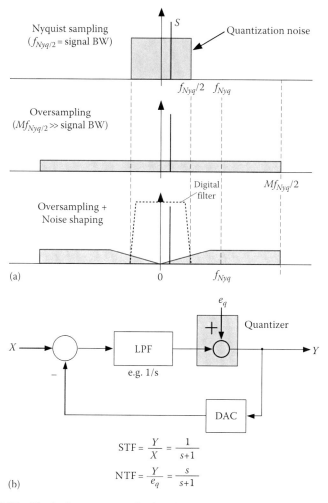

FIGURE 22.17 Block diagram, transfer function (a) and frequency response (b) of SD modulator.

22.4.2.1 CDAC

As shown in Figure 22.15, the SAR ADC is composed of S/H, DAC, comparator and SAR logic. The DAC is the most important building block in a SAR ADC because it determines the ADC linearity and contributes to the kT/C noise. A capacitor array DAC (CDAC) is the common choice for SAR ADCs since it does not consume static power and shows good matching. The CDAC generates a reference voltage for every decision step, as shown in Figure 22.15. The operational principle of the CDAC can be explained with a simple capacitive voltage divider, as shown in Figure 22.18 (Maloberti 2007). If k unit capacitors (C_u's) out of a total 2^N C_u's are switched to V_{ref} from 0, the output voltage V_{DAC} changes by $k/2^N$ V_{ref}. Here, k can range from 0 to $2^N - 1$. Thus, V_{DAC} does exactly the DAC operation by dividing the reference

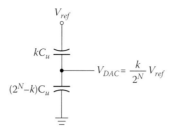

FIGURE 22.18   Operational principle of a capacitor DAC.

V_{ref} by 2^N steps. If C_u's are not exactly matched to their ideal values due to random errors, the step sizes of V_{DAC} will deviate from the desired values, and therefore, the linearity (DNL, INL) will be degraded. Thus, ensuring capacitor matching is a very importance issue in SAR ADC design.

22.4.2.2   S/H + CDAC Merger

Figure 22.15 shows a dedicated S/H in a SAR ADC. However, if we recall that switched-capacitor circuits have an inherent sampling capability, as T/H showed, the S/H function can be embedded into the CDAC. Figure 22.19 shows the merging principle of S/H in a CDAC. In the sample phase, all the 2^N C_u's of the CDAC are connected to the input signal via their bottom plates while their top plates are connected to GND. As a virtual intermediate step, if the bottom plates are connected to GND while the top plates are open, the output voltage (top-plate voltage) V_{out} becomes $-V_{in}$. After that, the DAC operation is performed by connecting k C_u's to V_{ref}, as in Figure 22.18. As a result, the output voltage shifts from $-V_{in}$ by $k/2^N$ V_{ref}, that is, $V_{out} = V_{DAC} - V_{in}$. By virtue of the input sampling with the CDAC, a subtraction operation is also conducted. Therefore, the comparator always compares V_{out} with 0. This merger makes the SAR ADC more compact. Since the comparator's input references a constant voltage, signal-dependent offset variation can be negligible.

22.4.2.3   Comparator

In order to achieve a low-power characteristic, most SAR ADCs use dynamic latches for the comparator that has no static power consumption. Figure 22.20 shows the most widely used dynamic comparator structure (Wicht et al. 2004). It consists of a differential input pair (M_1, M_2) and cross-coupled inverters (M_3–M_4 and M_5–M_6). When CLK = 0, the tail switch M_0 is off (no static current) and the reset switches,

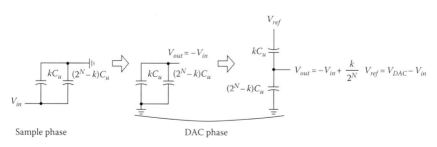

Sample phase   DAC phase

FIGURE 22.19   Merger of S/H and DAC.

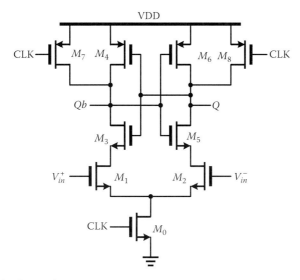

FIGURE 22.20   Dynamic comparator.

M_7 and M_8, are on to reset the outputs, Q and Qb, to VDD. When CLK transitions to high, the differential pair turns on, because the source coupled node of the input pair is connected to GND and generates signal-dependent drain currents. The differential signal currents are transferred to the output node via M_3 and M_5 and discharge the output nodes Q and Qb. By the positive feedback formed by the two inverters $(M_3 \sim M_6)$, the output difference is rapidly amplified, and eventually Q and Qb reach complementary logic levels. Once this latching operation is complete, no static current exists, because one of the two transistors in both inverters is off after latching. Other low-power comparator structures can be found in recent studies (Wicht et al. 2004; Agnes et al. 2008; Goll and Zimmermann 2009; Lee et al. 2009).

22.4.2.4   Overall Architecture and Operation

A 4b SAR ADC with a capacitor DAC is shown in Figure 22.21. S/H is merged in the CDAC, and thus, the comparator compares $V_{DAC} - V_{SH}$ with 0. Control of the DAC

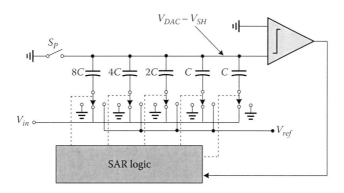

FIGURE 22.21   Circuit diagram of SAR ADC.

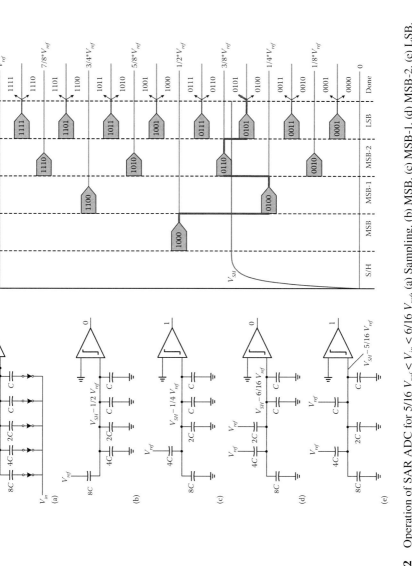

FIGURE 22.22 Operation of SAR ADC for $5/16\ V_{ref} < V_{in} < 6/16\ V_{ref}$. (a) Sampling. (b) MSB. (c) MSB-1. (d) MSB-2. (e) LSB.

and comparator is conducted by SAR logic. Detailed operation of the SAR ADC is shown in Figure 22.22 for the case of $5/16\ V_{ref} < V_{in} < 6/16\ V_{ref}$. Note that waveforms of V_{DAC} and $V_{in}\ (=V_{SH})$ are separated for convenient explanation.

22.5   DESIGN TECHNIQUES OF LOW-POWER SAR ADC FOR BIOMEDICAL APPLICATIONS

Since SAR ADCs can be designed without static power consumption, power consumption P can be described with a switching power equation (22.10) (Rabaey et al. 2003), where C is the total capacitance, V is the switching voltage (usually supply voltage), f is the operating frequency and α is the activity probability:

$$P = \alpha C V^2 f \qquad (22.10)$$

In order to reduce power consumption, we can reduce the total capacitance, supply voltage or conversion frequency. In the following subsections, various low-power design techniques for SAR ADCs will be discussed.

22.5.1   TOTAL CAPACITANCE REDUCTION

As shown in Figure 22.21, an N-bit full binary CDAC needs 2^N unit capacitors. In common CMOS processes, the allowable minimum capacitance for a dedicated capacitor structure is limited to a certain value by the design rule. For example, in a modern CMOS process, metal–insulator–metal (MiM) capacitors are required to have about 20 fF. Thus, if a 10b CDAC is designed in a full binary fashion, the total capacitance becomes about 20 pF, which is much larger than the kT/C noise requirement. This results in high power consumption and large chip area. In order to alleviate the large capacitance problem of the full binary CDAC, a segmented architecture can be used. Figure 22.23 shows a 6-bit DAC that is composed of a 4b coarse CDAC

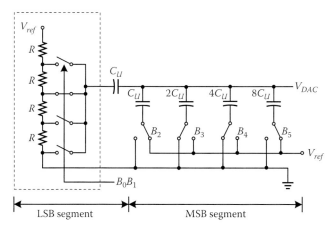

FIGURE 22.23   CDAC + RDAC structure for total capacitance reduction.

and 2b fine RDAC. From a 4b CDAC, additional (finer) levels are generated by controlling the bottom-plate voltage of the termination capacitor, which was formerly fixed to GND. Since the termination capacitor C_U can change V_{DAC} by 1/16 V_{ref} when it is connected to V_{ref}, finer voltages can be generated by dividing V_{ref} with the RDAC without increasing the total number of Cu. However, one drawback of this structure is that the R-string DAC consumes static power, which is unfavourable in battery-based low-power applications.

A segmented DAC can also be implemented by using two CDACs, as shown in Figure 22.24. Note that the bridge capacitor C_A must not be C_U because the equivalent capacitance seen from node V_{DAC} to the LSB segment including C_A must be C_U for the binary-weighted MSB segment. For example, C_A in Figure 22.24 must satisfy condition (22.11), and thus, the ideal value is given by (22.12):

$$C_A // 4C_U = C_U \qquad (22.11)$$

$$C_A = \frac{4}{3} C_U \qquad (22.12)$$

However, because the value (22.12) is not an integer multiple of C_U, it is difficult to guarantee the required ratio to C_U. In order to solve this C_A problem, the termination capacitor in the LSB segment can be eliminated, as in Agnes et al. (2008). Another issue with the CDAC + CDAC structure is the parasitic capacitance at the top node of the LSB segment. If parasitic capacitance is added to the top node of the LSB segment, C_{eq} becomes larger than C_U, and this results in linearity error, similar to the transfer curve in Figure 22.12. In order to solve this problem, self-calibration may be required (Yoshioka et al. 2010).

In contrast to old technologies where dedicated MiM capacitors are preferred, in advanced CMOS processes, by virtue of the reduced spacing rule between metal patterns, even a single metal layer can be used to implement a high-density capacitor

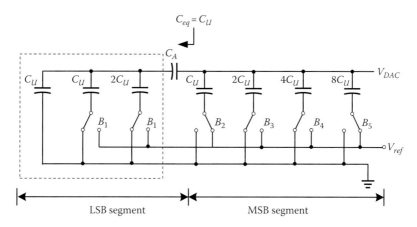

FIGURE 22.24 CDAC + CDAC structure for total capacitance reduction.

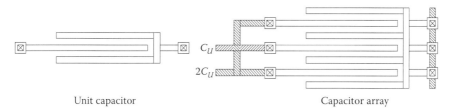

Unit capacitor   Capacitor array

FIGURE 22.25 Metal finger capacitor.

by using the vertical parasitic capacitance, as shown in Figure 22.25. The structure is referred to by various names, such as a metal–oxide–metal (MOM) capacitor, vertical native (VN) capacitor, metal-comb capacitor or metal finger capacitor. By using this structure, a very small unit capacitor with less than 1 fF can be implemented with a reliable matching characteristic. A binary CDAC can be implemented by using metal fingers in a common centroid fashion. In Harpe et al. (2010), an 8b ADC was implemented with less than 1 fF unit capacitance. As the CMOS process improves, the total capacitance for the SAR ADC approaches the kT/C noise limit.

22.5.2 ENERGY-EFFICIENT DAC SWITCHING SCHEMES

Referring to the operation of a conventional SAR ADC shown in Figure 22.22, the largest capacitor, $8C$, is charged to V_{ref} for MSB decision and then switched back (discharged) to 0 in the MSB-1 bit decision phase if the MSB was 0. (Note that the equivalent capacitance seen from V_{ref} is not $8C$ but $4C$ because of $8C//8C$. However, for convenience of explanation here, we just consider the switching capacitor.) Since $4C$ has to be charged to V_{ref} for the MSB-1 bit decision, discharging of $8C$ is an excessive charge loss. In efforts to save this loss of the conventional switching scheme, various modified switching schemes have been reported. These methods are summarized in Table 22.1 for 3-bit differential CDACs, and their operations are explained as follows. The differential input signal is assumed to be $-3/8\ V_{ref} < V_{in\_diff} < -2/8\ V_{ref}$. Note that 0, $1/2V_{ref}$ and V_{ref} in a single-ended signal correspond to $-V_{ref}$, 0 and $+V_{ref}$, respectively, in differential signalling. Thus, the single-ended input condition in Figure 22.22 corresponds to a positive-side signal in differential signalling here. For simplicity, the input sampling is not shown and only the CDAC operation is shown. Each column in the table represents the CDAC operation in the MSB, MSB-1 and MSB-2 bit decision phase, respectively.

22.5.2.1 Conventional Switching

Before discussing power-saving switching schemes, the conventional switching operation is reviewed for a fully differential CDAC with two references, V_{ref} and 0 (refer to Table 22.1). Assume that the upper CDAC samples the positive input, while the lower one samples the negative input. The operation of the upper CDAC is then exactly the same as shown in Figure 22.22 and the lower CDAC is the opposite. In the MSB decision phase, the upper CDAC connects $4C$ to V_{ref} and other capacitors to 0 to generate $V_{ref}/2$. The lower DAC performs exactly the opposite operation and generates the same level, $V_{ref}/2$. By doing this, a differential 0 is generated (which is the centre of $-V_{ref}$ and $+V_{ref}$, the differential full range) for the MSB decision. Unless otherwise stated, the following explanations describe the upper-side DAC operation, because that of the lower CDAC is exactly the opposite. Since the differential input is lower than the centre, differential $-1/2\ V_{ref}$ is generated by returning the $4C$ in the upper CDAC to 0 (switchback), while $2C$ is charged to V_{ref}. Thus, switchback to 0 of $4C$ creates unwanted charge loss.

22.5.2.2   Split-Capacitor Switching

In order to prevent charge loss from $4C$, $4C$ can be split into two $2Cs$ (Ginsburg and Chandrakasan 2005). Compared with the conventional CDAC operation in the MSB-1 bit decision phase, only $2C$ out of $4C$ is discharged, while the remaining capacitors retain the current state. This split-capacitor scheme reduces the charge loss in the switchback operation. However, the overhead due to the additional switching logic must be carefully considered, especially when the unit capacitors are very small as in recent designs, such as that in Harpe et al. (2010).

22.5.2.3   Energy-Saving Switching

This technique splits every capacitor into a pair so that each capacitor pair can be controlled to be either charged or discharged (Chang et al. 2007; Pang et al. 2009). For MSB decision, all the capacitor pairs are split into two directions, to V_{ref} and to 0, to generate $1/2\ V_{ref}$ at both inputs of the comparator (differential 0). After decision of MSB with this state, the MSB-1 bit decision follows by generating $1/4\ V_{ref}$ from the upper CDAC. This is performed by connecting the $2C$ that has been connected to V_{ref} to 0. Since each capacitor pair can be controlled to either side, the switchback operation is not necessary. $3/8\ V_{ref}$ generation at the upper CDAC for the MSB-2 bit decision is conducted by connecting $1C$ from 0 to V_{ref}. Eliminating the switchback operation achieves power savings. Note that even though the total capacitance is the same as that in the conventional schemes, this energy-saving switching has just 3 capacitor pairs, whereas the conventional structure has 4 capacitors. Total power consumption by the energy-saving switching technique can be further improved by partially utilizing Vcm-based tri-level switching or set-and-down schemes (Pang et al. 2009), which are described in the following.

22.5.2.4 V_{CM}-Based Tri-Level Switching

In the aforementioned energy-saving switching scheme, $1/2\ V_{ref}$ is generated by splitting capacitors into two from the conventional CDAC. However, the total capacitance for CDAC was not reduced from the conventional design. In V_{CM}-based tri-level switching, an additional reference of V_{CM} $(=1/2\ V_{ref})$ is used instead of splitting each capacitor (Chen et al. 2009; Hariprasath et al. 2010; Zhu et al. 2010; Cho et al. 2011). By connecting all capacitors to V_{CM}, the centre level of the reference is generated for the MSB decision (without connecting capacitors to either V_{ref} or 0). After the MSB is determined, the $2C$ in the upper CDAC is switched from V_{CM} to 0. Note that the DAC step size is reduced by half $(1/2\ V_{ref})$. Therefore, the total capacitance required for the CDAC could be reduced by half compared with the conventional and the energy-saving switching schemes, as

TABLE 22.1

Summary of Energy-Efficient DAC Switching Techniques

	MSB Decision ⟹	MSB-1 Decision ⟹	MSB-2 Decision
Conventional	V_{ref} 0 0 0 4 — 2 — 1 — 1 — 4 — 2 — 1 — 1 — V_{ref} V_{ref} V_{ref}	0 V_{ref} 0 0 4 — 2 — 1 — 1 — 4 — 2 — 1 — 1 — V_{ref} 0 V_{ref} V_{ref}	0 V_{ref} V_{ref} 0 4 — 2 — 1 — 1 — 4 — 2 — 1 — 1 — V_{ref} 0 0 V_{ref}
Split capacitor	V_{ref} V_{ref} 0 0 0 2 — 2 — 2 — 1 — 1 — 2 — 2 — 2 — 1 — 1 — 0 0 V_{ref} V_{ref} V_{ref}	0 V_{ref} 0 0 0 2 — 2 — 2 — 1 — 1 — 2 — 2 — 2 — 1 — 1 — V_{ref} 0 V_{ref} V_{ref} V_{ref}	0 V_{ref} 0 V_{ref} 0 2 — 2 — 2 — 1 — 1 — 2 — 2 — 2 — 1 — 1 — V_{ref} 0 V_{ref} 0 V_{ref}
Energy saving	V_{ref} V_{ref} V_{ref} 2 — 1 — 1 — 0 0 0 V_{ref} V_{ref} V_{ref} 2 — 1 — 1 — 0 0 0	0 V_{ref} V_{ref} 2 — 1 — 1 — 0 0 0 V_{ref} V_{ref} V_{ref} 2 — 1 — 1 — V_{ref} 0 0	0 V_{ref} V_{ref} 2 — 1 — 1 — 0 V_{ref} 0 V_{ref} 0 V_{ref} 2 — 1 — 1 — V_{ref} 0 0
V_{CM}-base tri-level	V_{CM} V_{CM} V_{CM} 2 — 1 — 1 — 2 — 1 — 1 — V_{CM} V_{CM} V_{CM}	0 V_{CM} V_{CM} 2 — 1 — 1 — 2 — 1 — 1 — V_{ref} V_{CM} V_{CM}	0 V_{ref} V_{CM} 2 — 1 — 1 — 2 — 1 — 1 — V_{ref} 0 V_{CM}
Set-and-down	V_{ref} V_{ref} V_{ref} 2 — 1 — 1 — 2 — 1 — 1 — V_{ref} V_{ref} V_{ref}	0 V_{ref} V_{ref} 2 — 1 — 1 — 2 — 1 — 1 — V_{ref} V_{ref} V_{ref}	0 V_{ref} V_{ref} 2 — 1 — 1 — 2 — 1 — 1 — V_{ref} 0 V_{ref}

shown in Table 22.1. Reduced voltage swing reduces the total power consumption as well. However, in practical use, this method needs generation of an additional reference, V_{CM}, which might impose considerable overhead. Another caution is that there is a possibility that V_{CM} is not enough to turn on the NMOS or PMOS switch(es). This can increase the complexity of the switch control circuit or reduce the settling speed.

22.5.2.5 Set-and-Down Switching

The set-and-down switching (or monotonic switching) technique reduces the total capacitance as much as V_{CM}-based tri-level switching does while utilizing just two references, 0 and V_{ref} (Liu et al. 2010a). These advantages are achieved by utilizing unbalanced switching: this technique controls just one CDAC out of the differential pair at a time. By doing this, the DAC step size can be reduced by half compared with other two-level referencing CDACs, even with a reduced number of capacitors. For the MSB decision, the reference of differential 0 is generated by connecting all capacitors (both upper and lower DACs) to V_{ref}. In the MSB-1 bit decision, only the $2C$ in the upper CDAC is switched from V_{ref} to 0, while its negative counterpart stays the same. As a result, the differential voltage change is the same as that in the MSB-1 bit decision in the V_{CM}-based CDAC control scheme. One problem with this scheme is that the input common level of the comparator continuously drops as the decision proceeds and creates signal-dependent comparator offset variation. This drawback may limit application of this technique for high-resolution design.

In Hariprasath et al. (2010), energy efficiencies of the aforementioned switching schemes are compared and the results are as follows: conventional (highest power) < split-capacitor < energy saving < set-and-down < Vcm-based tri-level switching (lowest power). Note that this comparison was made without considering overheads from additional reference generation and/or modified switching logic. Thus, comparison results using real designs may be different from the estimation.

22.5.2.6 Stepwise Charging

Various switching schemes that reduce the CDAC switching power have been discussed earlier. In the following, stepwise charging, a scheme that reduces the switching power consumption in charging/discharging a single capacitor, is introduced. Figure 22.26 shows a stepwise charging system with 3 uniform charging steps. Stepwise charging of load capacitor C_L in N equivalent steps reduces the energy dissipation by a factor of N, $E_{diss} = C_L V_{ref}^2 / N$, and thus, the power consumption in a SAR ADC can be reduced by applying this method (van Elzakker et al. 2008). Its operation is as follows: when C_L is discharged from V_{ref} to 0, charge tank capacitors, C_{T1} and C_{T2}, receive one charge packet from C_L, respectively. Thus, charge loss occurs only when C_L discharges from $V_{ref}/3$ to 0. When charging from 0 to V_{ref}, C_{T1} and C_{T2} return the received charge to C_L and there is only energy loss from V_{ref} by one charge packet. The voltage levels of C_{T1} and C_{T2} are automatically stabilized to their ideal values as long as the leakage can be ignored. However, overhead from the digital part for the switch control should be taken into account. One design utilizing this method showed an FOM of 4.4 fJ/conversion step in ISSCC 2008, (van Elzakker et al. 2008), which is still one of the best performances achieved globally to date.

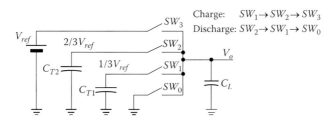

FIGURE 22.26   Stepwise charging.

22.5.3   Other Design Techniques for Low Power Consumption

22.5.3.1   Low-Voltage Design

The simplest technique to reduce the switching power consumption might be reducing the switching voltage (supply voltage). If the supply voltage is reduced by half, the power consumption drops by 1/4 from (22.11).

One problem with a reduced supply, however, is a switch turn-on problem. For example, the V_{GS} of the switch in the T/H shown in Figure 22.10 will be reduced as the clock voltage reduces. In order to solve this problem, a clock boosting scheme (Hong and Lee 2007; Chang et al. 2011; Lu et al. 2011) is popularly applied. As shown from the waveform in Figure 22.27, the switch cannot be turned on when $V_{IN} + V_{TH}$ is higher than V_{DD} unless boosting is used. A conceptual T/H circuit that utilizes a clock boosting scheme is shown in the same figure. During ϕ_2, T/H is in a hold phase and the boosting capacitor, C_{BST}, is charged to V_{BST}. In the tracking phase, ϕ_1, V_{BST} on C_{BST} is added on V_{IN} and applied to the switch gate. The switch's V_{GS} is then always V_{BST} regardless of V_{IN} and this makes the turn-on resistance constant. V_{BST} is usually V_{DD}.

When we reduce the supply voltage, we should consider the SNR change. In old designs, lowering the supply voltage resulted in power savings. However, in advanced processes, a reduction of supply voltage may not always guarantee power reduction. This is because the capacitor matching has been greatly enhanced. If the total capacitance of the CDAC can be designed for a kT/C noise limit under a normal supply voltage while guaranteeing the required matching, supply reduction by 1/2 will require four times greater capacitance for the same SNR (Equation 22.5). This results in identical power consumption as before. Thus, when we plan to shrink the supply voltage, we need to check if the capacitance is already low enough relative to the kT/C noise limit.

Furthermore, in low-speed applications, signal leakage from the sampling capacitor should be considered. In order to prevent leakage current, very high off-resistance is desirable. Off-resistance can be increased by using negative-side boosting, which makes V_{GS} lower than 0 (Lu et al. 2011), or by increasing the threshold voltage using the body effect (Tai et al. 2012).

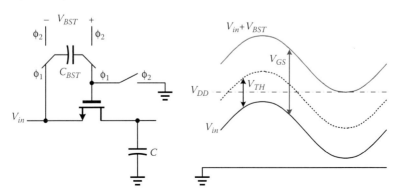

FIGURE 22.27   Concept of clock boosting.

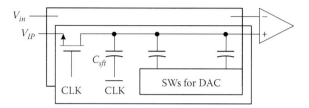

FIGURE 22.28 Input level shifting with C_{sft} for higher input common-mode level of the comparator.

Another problem with a low-voltage design is guaranteeing the comparator's operation. For a lower-level input signal, the comparator's input can be in an off state or in a very weak inversion region, which results in a serious speed drop. In order to avoid this problem, an input level shifting technique can be applied, as shown in Figure 22.28 (Harpe et al. 2010). Other capacitors are for CDAC and C_{sft} is for level shifting. Once the input sampling is done by CLK = 0, C_{sft}'s bottom plate is connected to the supply level. The comparator's inputs are consequently shifted up by VDD × $C_{sft}/(C_{sft} + C_{DAC\_total})$, where C_{DAC_total} is the total capacitance of the CDAC. One drawback of this technique is signal attenuation at the comparator input due to the DAC gain reduction by C_{sft}.

Recently, many research groups have achieved impressive low-power results by using low supply voltage (Chang et al. 2011; Lu et al. 2011; Yip and Chandrakasan 2011; Shikata et al. 2012; Tai et al. 2012).

However, from the perspective of system design, we have to consider how to provide such low supply voltages. If the system supplies the standardized supply level only, such as 1.2 V, the ADC cannot work at such a low-power mode. In order to bring such a low-voltage ADC to real applications, a dedicated power-efficient internal DC/DC converter may be needed.

22.5.3.2 Dual Comparators

Another important noise source in a SAR ADC is the comparator. In order to reduce the noise from the comparator, a large-sized input pair is required. However, a large comparator consumes high power. In order to alleviate this noise-power trade-off problem, a dual comparator technique has been proposed (Giannini et al. 2008). A low-power (but high noise) comparator is used for most decisions from MSB, because the noise effect is not as critical for large step sizes. Considering the risk that the comparator noise can affect the decision accuracy in the LSBs decision, a low-noise (but high power) comparator is used for LSBs decision. Since MSB might include decision error due to the noise, the LSB result can also be wrong. In order to correct the decision error generated in the MSBs decision phases, a redundant decision step is added at the end of the LSB decision for error correction (Figure 22.29). One practical problem with this structure is comparator offset mismatch. Comparator offset must thus be calibrated.

22.5.3.3 Clock Generation for Signal Leakage Avoidance

In low-frequency applications, the sampled signal on a capacitor can leak, while bit decision steps proceed. As a simple example, for a single sampled input, the stored signal at the LSB decision phase can be smaller than that at the MSB decision phase.

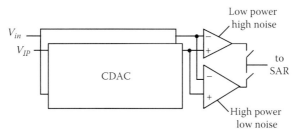

FIGURE 22.29   Two-comparator solution for low-power design.

FIGURE 22.30   Clocking scheme for low-speed ADC.

This results in an SNR drop. In order to prevent this problem, code conversion can be completed in short time despite the slow sample rate, as Figure 22.30 illustrates (Wu and Xu 2006). When conversion is required for the n-th sample, the oscillator enabled signal, OSC_EN, becomes 1 to generate the ADC clock, ADC_CLK. Once the conversion output $D(n)$ is generated, OSC_EN returns to 0 to disable the clock. This disables other logics until the next conversion is required.

Such a clock generator can be embedded on chip in some applications. For low frequency in particular, the relaxation oscillator shown in Figure 22.31 is a popular choice. When OSC_EN = 0, the capacitor C is reset to 0 and the current source is disabled so that the clock generator is off. When OSC_EN = 1, the PMOS transistor connected to the current source turns on and the NMOS switch for capacitor reset turns off. The capacitor C is then charged by the current source I and the voltage V_c increases linearly. If V_c hits the inverter logic threshold, V_{TH_INV}, CLKP turns on after two inverter delays, and the following D-FF is triggered. After two additional inverter delays, the capacitor is discharged and V_c returns to 0 in short time. After two more inverter delays, CLKP returns to 0. If the discharging time of V_c is negligibly short, the operation period of V_c, T_{ramp}, is $C \cdot V_{TH\_INV}/I$, and the clock frequency becomes $2/T_{ramp}$.

22.5.3.4   Clock Frequency Reduction

Once the signal bandwidth is known, usually the minimum sampling frequency is set by the Nyquist sampling theorem. Thus, it is not easy to reduce f in Equation 22.10. However, in some applications where the signal characteristic is well known, adaptive selection of the sampling frequency may be possible. One example can be found in an electrocardiogram (ECG) signal acquisition system (Yazicioglu et al. 2011). An ECG signal is quite periodical, with specific peaks P, Q, R, S and T, as shown in the upper figure in Figure 22.32. The signal bandwidth is below 500 Hz. However, we can observe that the signal component between $Q \sim R \sim S$ contains relatively high

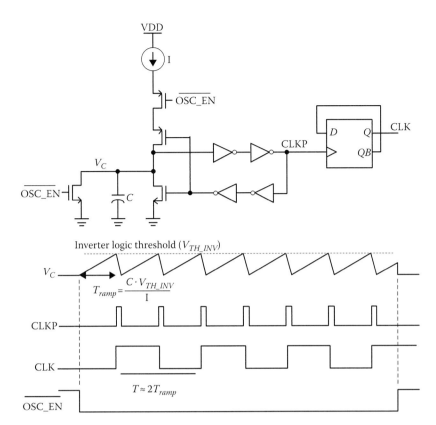

FIGURE 22.31   Relaxation oscillator and its waveforms.

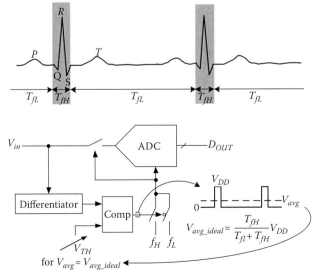

FIGURE 22.32   EEG signal and concept of adaptive sampling.

frequency and the remaining part shows much lower-frequency behaviour. Motivated by this characteristic, the adaptive sampling scheme uses two sampling frequencies. For the time period $Q \sim S$ where the signal change is fast, a high-frequency sample clock, f_H, is used, while the rest part is sampled at much lower frequency, f_L. In Yazicioglu et al. (2011), f_H and f_L are 1024 and 64 Hz, respectively.

An important technical issue in this method is how to detect the f_H period from the signal. In this work, a differentiator senses the signal rate and the result is compared with the threshold (V_{TH}). The group delay of the differentiator must be designed to be sufficiently short in order not to miss the signal information. The comparison threshold is adaptively controlled by using the nature of the ECG signal: since the ideal ratio of the time period of f_H (T_{fH}) and f_L (T_{fL}) is known, the low-pass filtered comparator output (V_{avg}) is used to control the comparison threshold (V_{TH}) such that V_{avg} is close to its ideal value (V_{avg_ideal}). One thing we should consider is that, since recent SAR ADCs claim leakage-level power consumption at 1 KS/s operation, such a scheme would not reduce the ADC core power in biomedical applications. However, due to the reduced clock rate in most operations, following digital block can save power.

22.6 PERFORMANCE TRENDS AND CONCLUSIONS

In this chapter, ADC basic theories have been reviewed and strengths and weaknesses of several ADC structures have been compared for biomedical applications. The SAR ADC has been introduced as the most suitable structure for low-power biomedical applications as it does not consume static power. The operational principle of the SAR ADC was discussed in detail and various conventional and recently reported low-power design techniques have been introduced.

Low-power design techniques for SAR ADCs are quite mature these days owing to advanced CMOS processes and competitive research. Table 22.2 lists recently

TABLE 22.2
Performance Trends of Low-Power SAR ADCs

	Resol [bit]	Speed [kS/s]	Supply [V]	Power [uW]	ENOB [bit]	FOM [fJ/cs]	Process [m]
Hong and Lee (2007)	8	200	0.9	2.47	7.44	65	180n
Chang et al. (2007)	8	500	1	7.75	7.5	86	180n
van Elzakker et al. (2008)	10	1000	1	1.9	8.75	4.4	65n
Lee et al. (2009)	10	100	0.6	1.3	8.7	31	180n
Pang et al. (2009)	10	500	1	42	9.4	124	180n
Yip and Chandrakasan (2011)	10	20	0.55	0.206	8.84	22.4	65n
Cheong et al. (2011)	10	80	1	0.4	8	19.5	180n
Shikata et al. (2012)	8	1100	0.5	1.2	7.5	6.3	40n
Lu et al. (2011)	10	1	0.5	0.0025	8.52	6.8	180n
Chang et al. (2011)	8	31	0.5	0.087	7.2	20	250n
Tai et al. (2012)	10	100	0.35	0.17	9.06	3.2	90n

reported low-power SAR ADCs for biomedical applications or SAR ADCs with similar specifications. All the ADCs that are stated to be for biomedical applications have 8b or 10b resolution. Although several examples with higher resolution, such as 11 ~ 12b, can be found in biomedical systems (Zou et al. 2009; Yoshioka et al. 2010), since it was difficult to find their ADC-only performances, they are not included in the table. All the ADCs in Table 22.2 have supply voltage less than or equal to 1 V. In particular, most recently reported ADCs have 0.5 V or less supply and their FOMs are in a range of several to several tens of fJ/conversion step. Harpe's 1 V 8b SAR ADC designed for 10 MS/s in a 90 nm CMOS (Harpe et al. 2011) shows that the power consumption of a SAR ADC can approach the level of leakage power consumption when the sampling frequency is lower than 500 Hz. Lu et al. (Lu et al. 2011) demonstrated that a 10b ADC in 180 nm CMOS can work at 1 KS/s while only consuming 2.5 nW power, which is even lower than the leakage power in Harpe et al. (2011). This seems to be made possible by the low-speed dedicated design, as compared with Harpe et al. (2011). A slow ADC does not require power-consuming high-speed circuit design. In addition, Lu's design used a process with longer gate length under a lower supply voltage of 0.5 V, which gives less leakage power. The major reason why the SAR ADC can consume this leakage-level power is that it has no static power, and thus, this justifies the application of SAR ADCs in ultra-low-power biomedical systems. However, once again, if the design can be optimized including analogue front ends such as antialias filter and low-noise amplifier, other ADCs such as SD ADC can be another solution.

Finally, different approaches that do not follow the traditional sampling theory such as adaptive sampling (Yazicioglu et al. 2011) and an asynchronous ADC (Trakimas and Sonkusale 2008) can be attractive research directions.

ACKNOWLEDGEMENT

The author gratefully acknowledges the support from the National Research Foundation of Korea funded by the Korean government (MEST) under grant NRF-2012–0001965.

REFERENCES

Abo, A. M. and P. R. Gray (1999). A 1.5-V, 10-bit, 14.3-MS/s CMOS pipeline analog-to-digital converter. *IEEE Journal of Solid-State Circuits* **34**(5): 599–606.

Agnes, A., E. Bonizzoni et al. (2008). A 9.4-ENOB 1V 3.8mW 100kS/s SAR ADC with Time-Domain Comparator. In *Solid-State Circuits Conference, 2008 (ISSCC 2008)*, Digest of Technical Papers, IEEE International.

Chang, S.-I., K. Al-Ashmouny et al. (2011). A 0.5V 20fJ/conversion-step rail-to-rail SAR ADC with programmable time-delayed control units for low-power biomedical application. In *Proceedings of the ESSCIRC (ESSCIRC), 2011*, Helsinki, Finland.

Chang, Y.-K., C.-S. Wang et al. (2007). A 8-bit 500-KS/s low power SAR ADC for bio-medical applications. In *Solid-State Circuits Conference, 2007*. ASSCC '07. IEEE Asian.

Chen, Y., S. Tsukamoto et al. (2009). A 9b 100MS/s 1.46mW SAR ADC in 65nm CMOS. In *Solid-State Circuits Conference, A-SSCC 2009*, San Francisco, CA, IEEE Asian.

Cheong, J. H., K. L. Chan et al. (2011). A 400-nW 19.5-fJ/Conversion-Step 8-ENOB 80-kS/s SAR ADC in 0.18-mm CMOS. *IEEE Transactions on Circuits and Systems II: Express Briefs* **58**(7): 407–411.

Cho, S.-H., C.-K. Lee et al. (2011). A 550-mW 10-b 40-MS/s SAR ADC with multistep addition-only digital error correction. *IEEE Journal of Solid-State Circuits* **46**(8): 1881–1892.

Giannini, V., P. Nuzzo et al. (2008). An 820mW 9b 40MS/s Noise-Tolerant Dynamic-SAR ADC in 90nm Digital CMOS. In *Solid-State Circuits Conference, 2008 (ISSCC 2008). Digest of Technical Papers, IEEE International*.

Ginsburg, B. P. and A. P. Chandrakasan (2005). An energy-efficient charge recycling approach for a SAR converter with capacitive DAC. In *IEEE International Symposium on Circuits and Systems, 2005. ISCAS 2005*, Kobe, Japan.

Goll, B. and H. Zimmermann (2009). A 65nm CMOS comparator with modified latch to achieve 7GHz/1.3mW at 1.2V and 700MHz/47mW at 0.6V. In *IEEE International Solid-State Circuits Conference – Digest of Technical Papers, 2009 (ISSCC 2009)*, San Francisco, CA.

Hariprasath, V., J. Guerber et al. (2010). Merged capacitor switching based SAR ADC with highest switching energy-efficiency. *Electronics Letters* **46**(9): 620–621.

Harpe, P., Z. Cui et al. (2010). A 30fJ/conversion-step 8b 0-to-10MS/s asynchronous SAR ADC in 90nm CMOS. In *Solid-State Circuits Conference Digest of Technical Papers (ISSCC), 2010 IEEE International*, San Francisco, CA.

Harpe, P. J. A., C. Zhou et al. (2011). A 26 mW 8 bit 10 MS/s asynchronous SAR ADC for low energy radios. *IEEE Journal of Solid-State Circuits* **46**(7): 1585–1595.

Hong, H.-C. and G.-M. Lee (2007). A 65-fJ/conversion-step 0.9-V 200-kS/s rail-to-rail 8-bit successive approximation ADC. *IEEE Journal of Solid-State Circuits* **42**(10): 2161–2168.

Kester, W. A. (2005). *Data Conversion Handbook*, Elsevier, Burlington, MA.

Lee, S.-K., S.-J. Park et al. (2009). A 1.3mW 0.6V 8.7-ENOB successive approximation ADC in a 0.18mm CMOS. In *VLSI Circuits, 2009 Symposium on*.

Li, P. W., M. J. Chin et al. (1984). A ratio-independent algorithmic analog-to-digital conversion technique. *IEEE Journal of Solid-State Circuits* **19**(6): 828–836.

Liew, W.-S., X. Zou et al. (2009). A 1-V 60-mW 16-channel interface chip for implantable neural recording. *Custom Integrated Circuits Conference, 2009. CICC '09. IEEE*.

Liu, C.-C., S.-J. Chang et al. (2010a). A 10-bit 50-MS/s SAR ADC with a monotonic capacitor switching procedure. *IEEE Journal of Solid-State Circuits* **45**(4): 731–740.

Liu, C.-C., S.-J. Chang et al. (2010b). A 10b 100MS/s 1.13mW SAR ADC with binary-scaled error compensation. In *IEEE International Solid-State Circuits Conference Digest of Technical Papers (ISSCC), 2010*, San Francisco, CA.

Lu, T.-C., L.-D. Van et al. (2011). A 0.5V 1KS/s 2.5nW 8.52-ENOB 6.8fJ/conversion-step SAR ADC for biomedical applications. *Custom Integrated Circuits Conference (CICC), 2011 IEEE*, San Jose, CA.

Maloberti, F. (2007). *Data Converters*, Springer, Dordrecht, the Netherlands.

Murmann, B. (2008). A/D converter trends: Power dissipation, scaling and digitally assisted architectures. *IEEE Custom Integrated Circuits Conference (CICC 2008)*, San Jose, CA.

Pang, W.-Y., C.-S. Wang et al. (2009). A 10-bit 500-KS/s low power SAR ADC with splitting comparator for bio-medical applications. In *Solid-State Circuits Conference, 2009. A-SSCC 2009. IEEE Asian*.

Rabaey, J. M., A. P. Chandrakasan et al. (2003). *Digital Integrated Circuits: A Design Perspective*, Pearson Education, Upper Saddle River, NJ.

Razavi, B. (1994). *Principles of Data Conversion System Design*, Wiley-IEEE Press, New York.

Schreier, R. and G. C. Temes (2004). *Understanding Delta-Sigma Data Converters*, Wiley-IEEE Press, New York.

Shikata, A., R. Sekimoto et al. (2012). A 0.5 V 1.1 MS/sec 6.3 fJ/Conversion-Step SAR-ADC with tri-level comparator in 40 nm CMOS. *IEEE Journal of Solid-State Circuits* **47**(4): 1022–1030.

Song, B.-S. (2012). *Micro CMOS Design*, CRC Press, Boca Raton, FL.

Soundarapandian, K. and M. Berarducci (2010). Analog front-end design for ECG systems using delta-sigma ADCs, Texas Instruments.

Tai, H.-Y., H.-W. Chen et al. (2012). A 3.2fJ/c.-s. 0.35V 10b 100KS/s SAR ADC in 90nm CMOS. In *VLSI Circuits, 2012 Symposium on*.

Trakimas, M. and S. Sonkusale (2008). A 0.8 V asynchronous ADC for energy constrained sensing applications. In *IEEE Custom Integrated Circuits Conference (CICC 2008)*, San Jose, CA.

van Elzakker, M., E. van Tuijl et al. (2008). A 1.9mW 4.4fJ/Conversion-step 10b 1MS/s Charge-Redistribution ADC. In *Solid-State Circuits Conference, 2008*. ISSCC 2008. Digest of Technical Papers. IEEE International.

Walden, R. H. (1999). Analog-to-digital converter survey and analysis. *IEEE Journal on Selected Areas in Communications* **17**(4): 539–550.

Wang, L. and L. Theogarajan (2011). An 18mW 79dB-DR 20KHz-BW MASH DS modulator utilizing self-biased amplifiers for biomedical applications. In *Custom Integrated Circuits Conference (CICC), 2011*, San Jose, CA, IEEE.

Wicht, B., T. Nirschl et al. (2004). Yield and speed optimization of a latch-type voltage sense amplifier. *IEEE Journal of Solid-State Circuits* **39**(7): 1148–1158.

Wu, H. and Y. P. Xu (2006). A 1V 2.3mW biomedical signal acquisition IC. In *Solid-State Circuits Conference (ISSCC 2006)*. Digest of Technical Papers. IEEE International.

Yang, H. Y. and R. Sarpeshkar (2006). A bio-inspired ultra-energy-efficient analog-to-digital converter for biomedical applications. *IEEE Transactions on Circuits and Systems I-Regular Papers* **53**(11): 2349–2356.

Yazicioglu, R. F., S. Kim et al. (2011). A 30 mW Analog signal processor ASIC for portable biopotential signal monitoring. *IEEE Journal of Solid-State Circuits* **46**(1): 209–223.

Yip, M. and A. P. Chandrakasan (2011). A resolution-reconfigurable 5-to-10b 0.4-to-1V power scalable SAR ADC. *Solid-State Circuits Conference Digest of Technical Papers (ISSCC), 2011*, San Francisco, CA, IEEE International.

Yoshioka, M., K. Ishikawa et al. (2010). A 10b 50MS/s 820mW SAR ADC with on-chip digital calibration. In *IEEE International Solid-State Circuits Conference Digest of Technical Papers (ISSCC), 2010*, San Francisco, CA.

Zhu, Y., C.-H. Chan et al. (2010). A 10-bit 100-MS/s reference-free SAR ADC in 90 nm CMOS. *IEEE Journal of Solid-State Circuits* **45**(6): 1111–1121.

Zou, X. D., X. Y. Xu et al. (2009). A 1-V 450-nW fully integrated programmable biomedical sensor interface chip. *IEEE Journal of Solid-State Circuits* **44**(4): 1067–1077.

23 Ultra-Low-Power Techniques in Small Autonomous Implants and Sensor Nodes

Benoit Gosselin, Sébastien Roy
and Farhad Sheikh Hosseini

CONTENTS

23.1 INTRODUCTION

There is growing interest in the development of autonomous small-scale devices capable of sensing/actuating and performing communication and processing tasks in personnel health-care technology. Such devices form the basis of many novel diagnostic and prosthetic systems and are often networked to form the so-called sensor networks, which constitute a very active field of research. Similarly, autonomous implants within the human body that serve to capture data or to stimulate cells are essentially specialized sensor/actuator nodes, whether or not they are networked. Such implants are typically powered by a battery or through a transdermal inductive link with a device that is external to the body. It is well known that extremely low power consumption is a critical feature in order to maximize the operational life expectancy of a node, an implant or an entire network of such devices powered by batteries. Alternatives to batteries generally fall under the energy-scavenging category with the exception of the inductive link discussed earlier. In all such cases, the power available is likely to be very low. Multiple power-saving schemes have been proposed and are in use in several power-conscious devices. Possibly one of the most common is the introduction of inactive or sleep states whenever appropriate. Indeed, an entire electronic device can be shut down when there is no work to perform, or certain components can be turned on and off depending on dynamic demand. The power consumption of devices in sleep or standby modes can be orders of magnitude lower than their active consumption.

We present in the following two case studies illustrating the utilization of low-power design techniques, both being relevant to personal healthcare and medical research. The first one is concerned with the implementation of monitoring devices incorporating several parallel sensing channels, such devices being specifically applicable to brain–computer interfacing technology. The second case study centres on the broad and increasingly active area of wireless sensor networks (WSNs) and, more specifically, on the wake-up radio (WUR) concept that allows an autonomous device to remain in deep sleep until needed, when it can then be woken up by a specific RF signal received through a passive radio. Brain–computer interfacing technology aims at collecting bioelectrical signals from hundreds of electrodes to take snapshots of the activity occurring in specific areas of the brain. Indeed, each electrode generates weak analogue signals that need to be amplified, further processed and digitized to be transferable into the world of computing for analysis or storage. Such functions are conducted in dedicated sensory circuits providing low-noise amplifiers (LNAs), filters and data converters, thus dissipating a considerable amount of power. However, the sensitivity of brain tissues restricts excessive heat dissipation in the nearby circuits, necessitating low power consumption. Indeed, data reduction and data compression schemes that can decrease the overall output data rates of sensory circuits are becoming critical building blocks in such monitoring devices. On the other hand, different power-scheduling schemes have been proposed in order to power up the sensory circuits only when needed. Specifically, activity-based schemes exploit the intermittent nature of biopotential recordings along with their low duty cycles by powering up the sensory building blocks only during occurrences of biopotentials. Such a strategy requires the utilization of accurate signal detectors along with a smart power management approach. Typically, most recording building blocks will remain in an idle state, draining practically no power until biopotentials occur.

The first case study covers dedicated electronic recording strategies to address the challenge of operating large numbers of recording channels to gather the neural information from several neurons within very low-power constraints and an appropriately compact form factor. Specifically, we cover system-level-based strategies for smart power management, and we present dedicated sensory circuits topologies for extracting and separating multiple biopotential modalities with high energy efficiency. A practical implementation is presented to illustrate the application of the described strategies. Such application consists of a discrete-time analogue front end that leverages highly energy-efficient low-noise sensory circuits and dedicated system-level power-saving schemes.

The second case study is in the context of sensor networks where communication itself is often a dominant factor in the power budget. For example, Texas Instruments' low-power CC2240 consumes 18.8 mA when in receive mode and 17.4 mA when transmitting a 0 dBm signal. Furthermore, it can be shown that the receive side consumes more power on average than the transmit side, because the receiver is constantly 'listening' while transmissions are typically rare and short. This picture is however heavily influenced by the communication protocol in use, including the medium access mechanism. Typical solutions include various forms of synchronization where it is possible to turn the receive radios on only during certain predetermined intervals. More sophisticated dynamic evolutions of this concept are termed 'rendezvous' protocols.

The holy grail in this realm is the WUR, a receiver that can normally operate in an entirely passive fashion, thereby consuming very little power, until reception of a radio impulse having certain characteristics 'wakes' it. In other words, the energy of the received pulse is used to trigger a power-on mode for subsequent reception of a packet. Only through such a mechanism can it be ensured that the receive radio is only turned on when receiving bits. However, it has been estimated in the literature that the power consumption of the WUR should be below 50 µA for this scheme to be effective and competitive with the best rendezvous protocols. Very few realizations of wake-up devices are reported in the literature. This case study presents the first reported design having power dissipation below 40 µW. It consists in a complete wake-up device, including an RF detector and an address decoder, having an average power consumption of less than 20 µW. One of the unique features of this design is the use of pulse-width modulation (PWM) instead of the more common on–off keying (OOK) schemes. This choice has opened up significant power-saving opportunities. While the sphere of applications of WURs and energy-efficient WSNs in general is very broad, there are many medical contexts where such technology, allowing an autonomous device to consume significant power only when polled, could be advantageously leveraged. This includes in-body devices with an autonomous limited power source and body-area networks designed to monitor vital signs.

23.2 MICROSYSTEMS FOR BIOPOTENTIAL RECORDING

Nowadays, neural interfacing microsystems capable of continuously monitoring large groups of neurons are being actively researched by leveraging the recent advancements in neurosciences, microelectronics, communications and microfabrication. Such monitoring microdevices are pursuing two critical objectives for prosthetic applications and advanced research tools: (1) replacing hardwired connections with a wireless link to eliminate cable tethering and infections and (2) enabling the local processing of neural signals to improve signal integrity. A suitable interface to the cortex must enable chronic utilization and high resolution through the simultaneous sampling of the activity of hundreds of neurons. In prosthetic applications, there are severe limitations on size, weight and power consumption of monitoring implants in order to limit invasiveness and heat dissipation in surrounding tissues.

A major effort has recently been directed towards designing neural recording circuitry consuming very few microwatts per channel by leveraging low-power circuit techniques and smart data/power management. Indeed, low-power sensory circuits, energy-efficient system-level architectures and dedicated on-chip management strategies are necessary means for achieving high resolution while addressing stringent power requirements.

23.2.1 MULTICHANNEL SYSTEM ARCHITECTURES

Gathering the sampled neural activity from hundreds of channels, digitizing it and sending it wirelessly to a base station in an efficient fashion is very challenging and thus requires a dedicated system architecture. A straightforward approach consists in sharing a fast digitizer between several sensory channels. In such a scheme, the sensory channels are directed towards an analogue-to-digital converter (ADC) by employing time-division multiplexing (TDM) in the analogue domain. The challenge with such an approach consists in minimizing power consumption from the high-speed unity gain buffers, sample-and-hold amplifier (SHA) circuit and ADC. Indeed, unity gain buffers presenting wide bandwidth, much higher than the maximum frequency of the incoming signal (f_{max}), are needed to drive the SHA circuit or the ADC within small TDM time intervals, a requirement that results in high power consumption. According to [1], the buffers and the SHA circuit must feature a low-pass cut-off frequency (f_{-3dB}) that is at least five times higher than f_{max} in order to achieve a tracking error smaller than 1/2 LSB, for a 10-bit representation. Moreover, the analogue multiplexers must be designed carefully in order to avoid excessive crosstalk and distortion. Examples of such system-level configurations are presented in [2,3].

Another approach consists in providing one low-power, low-rate ADC for each sensory channel and performing TDM in the digital domain. In contrast with the first scheme described earlier, such an approach has the advantage of avoiding the need for several power-consuming unity gain buffers and eliminating interchannel crosstalk. However, great care must be taken in the design of a suitable ADC in order to minimize the chip area. An example of this type of implementation is reported in [4]. A third approach consists in performing digitization off chip to save power and silicon area. Digitization is performed in two phases: The first phase consists in converting the multiplexed analogue samples into time delays, an operation known as analogue-to-time conversion (ATC). Then, after transmitting the ATC signal outside the body where power and size are not highly constrained, the second phase consists in performing time-to-digital conversion (TDC). In addition to saving power, this approach does not require synchronization with a clock signal. However, TDM must be performed in the analogue domain, thus leading to crosstalk. Such an approach is presented in [5].

23.2.2 LOW-POWER SYSTEM-LEVEL STRATEGIES

The need for a parallel arrangement comprising several power-hungry modules, like low-noise sensory circuits, high-speed data converters and wireless transmitters, motivates the application of dedicated system-level approaches for addressing

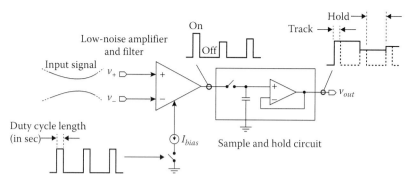

FIGURE 23.1 Simplified representation of a power-scheduling strategy.

excessive power consumption in the targeted highly constrained application. Power scheduling consists in powering up the recording circuits only when necessary (Figure 23.1).

Specifically, current-supply modulation [5,6] and duty cycling [5,7,8] are power-scheduling techniques consisting in switching specific building blocks between an active and an idle state with low duty cycle. The shorter the duty cycling period, the lower the power dissipation in the circuits. The difference between both techniques is that the former uses a low-level bias current in the idle state (Figure 23.2), whereas the latter draws no current in the idle state, thus leading to more power savings (Figure 23.3). However, a duty cycle period that is too short can potentially degrade the reading accuracy of a neural recording channel, because fast intermittent monitoring requires circuits with larger bandwidth, thus allowing more noise to enter the system. The battery-powered sensor interface presented in [8] addresses this

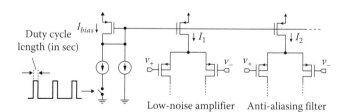

FIGURE 23.2 Simplified schematic of a current-supply modulation scheme.

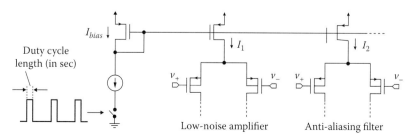

FIGURE 23.3 Simplified schematic of a duty cycling scheme.

trade-off by providing different levels of accuracy and lifetime through the utiliza-tion of a programmable low-pass filter. Such a filter allows selecting between differ-ent input-referred noise levels and duty cycle lengths, which determines the overall accuracy and power consumption. The power-scheduling mechanism employed in [5] consists in putting most of the neural amplifiers that are not being sampled in sleep mode where they draw a fraction of their active current consumption (0.5 μA). Indeed, not turning the neural amplifiers completely off affords time for the neural amplifiers to reach their active state faster, ahead of each sampling instant. This leads to a power reduction of 18% in the analogue multichannel front-end block.

Another technique, the multiplexing of several electrodes towards one LNA, has been demonstrated to save power and silicon area. TDM of four channels towards a sin-gle neural amplifier is achieved in [9]. In the proposed topology, a single LNA is shared between four independent input electrodes in order to decrease the power consump-tion per channel by four. A frequency-division multiplexing scheme is implemented in [10] where the amplitude of the neural activity seen at several individual electrodes is modulated and directed towards a single-wideband neural amplifier. It is shown in [10] that the maximum number of electrodes that can be multiplexed towards a single amplifier is limited by the sum of the thermal noise from each electrode at the input node of the wideband neural amplifier, and it is in the range of 5–10 for typical cases.

On the other hand, activity-based schemes exploit the transient characteristics of neural signals to maximize efficiency. Adaptive sampling [11] is an activity-based technique that affords a significant decrease of the data rate by dynamically varying the sampling rate of an ADC, based on the input signal activity. Reduction factors of seven are reported with this scheme. The algorithm requires the implementation of the second derivative of the signal within dedicated circuits to measure the rate of change of the input. Other activity-based schemes exploit the intermittent nature of neural recordings along with their low duty cycles by powering up the recording building blocks only when neural events occur (Figure 23.4). Such strategies require

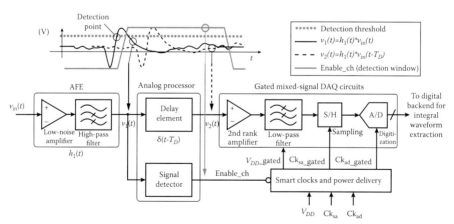

FIGURE 23.4 Activity-based scheme that powers up the recording building blocks only when necessary. To drastically decrease power consumption, this strategy employs an accu-rate biopotential detector along with power scheduling. (From Gosselin, B. et al., *IEEE Trans. Biomed. Circ. Syst.*, 4, 171, 2010.)

the utilization of accurate biopotential detectors along with power scheduling. In such schemes, most recording building blocks remain in an idle state, draining practically no current until neural events occur. In [12], a low-overhead analogue detector is employed to detect neural events and trigger the recording circuitry. Indeed, the approach proposed in [13] uses low-overhead analogue delay elements that are implemented within ultralow-power linear-delay filters to wake up the recording circuits 'ahead' of biopotential occurrences, a critical behaviour to avoid truncated waveforms.

Adaptive mechanisms have also been proposed for achieving high power efficiency. In such approaches, the DC operating points of a circuit are optimally adjusted by employing feedback. A neural amplifier using adaptive biasing is demonstrated in [14]. This closed-loop scheme consists in directly adjusting the signal-to-noise ratio (SNR) of an LNA by changing its bias current in order to set the input-referred noise of the amplifier right above the noise floor of the input electrode, thus avoiding any waste of energy.

23.2.3 Energy-Efficient Sensory Circuit Topologies

The LNA is the main building block of an analogue front end. The LNA must amplify and filter the neural waveforms in order to remove any input DC offset seen across a pair of differential electrodes, thereby maximizing the dynamic range in the recording channel. Indeed, it must provide sufficient gain, appropriate bandwidth, high SNR, excellent linearity, high common-mode rejection ratios (CMRR) and high power supply rejection ratios (PSRR) to provide the expected signal quality. In the case of a multichannel interface, one such sensory circuit per electrode is needed. Therefore, the LNA must consume little power, be of small size, as well as should be scalable to multiple parallel channels. Furthermore, it is essential to optimize the design of the LNA for very low power through a dedicated circuit design methodology, as in [15]. The noise efficiency factor (NEF) has been widely adopted as a main figure of merit to assess the performance of LNAs and compare the several existing topologies together.

Designing sensory circuits that can capture multimodal neural information is critical to gather as much neural information as possible. Table 23.1 shows typical values of amplitude and bandwidth for different neural signals. Appropriate amplifiers can discriminate among multiple types of biopotentials by accommodating different frequency ranges [5,16,17]. They can provide different bandwidth settings by (1) tuning the resistive values in a filter, (2) selecting different capacitors values from an array or (3) changing the operating points of a circuit. In several designs, the high-pass cut-off is changed by varying the gate voltage of a pseudo-resistor [17,18] or that of a weakly inverted MOSFET [19]. In contrast, the low-pass cut-off is changed by varying the operating point of the LNA directly [15] or by changing its capacitive load through the selection of different output capacitors [7].

However, tuneable circuits can present significant distortion and exhibit excessive process-dependent variations. Indeed, the resistance of the MOS devices is highly dependent on the voltage level of the output signal [18]. Linearization circuits techniques have been proposed for performing appropriate biasing of the gate of a MOS

TABLE 23.1

Characteristics of Various Bioelectric Signals

Signal	Bandwidth (Hz)	Signal Range (mVpp)
Electrocardiogram (ECG)	0.05 ~ 256	0.1 ~ 10
Electroencephalogram (EEG)	0.001 ~ 100	0.01 ~ 0.4
Electrocorticograms (ECoG)	0.1 ~ 64	0.02 ~ 0.1
Electromyogram (EMG)	1 ~ 1 K	0.02 ~ 1
LFP	0.001 ~ 200	0.1 ~ 5
Extracellular AP (EAP)	100 ~ 10 K	0.04 ~ 0.5

device and linearizing the MOS resistor [1,18]. Such approaches have been extended further in [8] by replacing any inaccurate current mirrors and source followers by precise closed-loop op-amps, thus achieving high linearity above 74 dB within ±200 mV and enabling process-independent frequency cut-off values. This proposed strategy consists in linearizing a pseudo-resistor by reporting any variations in the input voltage right at the gates of its constituting pMOS devices with unity gain. This maintains the gate to source voltage of the MOSFETs at constant value, thus cancelling any non-linearity while providing adequate DC biasing to set the desired cut-off frequency.

A switched-capacitor (SC) neural amplifier with tuneable characteristics was recently demonstrated in [20] as an alternative to conventional continuous-time circuits. In addition to providing low noise and satisfactory gain, this amplifier can accommodate local field potentials (LFP) and action potentials (AP) through tuning of its clocking frequency, thus implementing different low- and high-pass cut-off frequencies in a straightforward fashion.

23.2.4 Low-Power Discrete-Time Neural Interfacing Front End

An analogue front-end topology employing a power-scheduling strategy is presented. This approach aims at decreasing the average power consumption by leveraging duty cycling. The schematic of the front end is presented in Figure 23.5. In this scheme, an LNA providing a gain of 100 is powered up only for brief time intervals. The output of the LNA is sampled with a low-power SC SHA circuit featuring an embedded low-pass cut-off frequency for limiting the Johnson noise in the channel. This SHA circuit efficiently combines sampling action with low-pass filtering in a compact building block. One challenge in the implementation of the said duty cycling strategy is designing an LNA whose output can settle within a short time frame in order to provide high signal quality as well as low power. Indeed, wide bandwidth must be achieved in the LNA to allow fast settling, but this must be done without adding too much power overhead. The cut-off frequency of the LNA is set to 40 kHz, given a typical recording bandwidth of 8 kHz for extracellular AP. Indeed, the SHA circuit provides an embedded low-pass cut-off frequency to limit the noise power below 8 kHz. Its simulated NEF was found to be as low as 1.3.

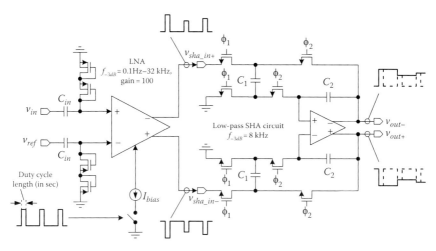

FIGURE 23.5 Analogue front end employing duty cycling.

TABLE 23.2

Performance of the Analogue Front End

Parameter	Value
Supply voltage	1.8 V
Power consumption	<3 μW
Gain	100 V/V
Input-referred noise	31 nV/√Hz
Bandwidth	1 mHz–8 kHz
Process	CMOS 0.35 μm

Source: Gosselin, B., Approaches for the efficient extraction and processing of neural signals in implantable neural interfacing microsystems, in *The 33rd Annual International Conference of the IEEE Engineering in Medicine and Biology Society (EMBC'11)*, Boston MA, 2011, pp. 5855–5859.

The amplifier circuit is simulated using Cadence design tools and CMOS 0.35 μm MOSFET models provided by the foundry. Table 23.2 summarises the performance of the front end.

Before deriving the cut-off frequency of the SHA circuit, its z-transfer function will be obtained. Then, the continuous transfer function of the circuit will be derived from the z-transfer function, and the cut-off frequency of the SC-SHA will be calculated. We will use the single-ended representation of the circuit for simplicity, the schematic of which is shown in Figure 23.6.

First, in sampling mode $(nT–T)$, when ϕ_1 is high and ϕ_2 is low, switches S_1 and S_2 are closed and C_1 is charged to v_{in} (Figure 23.7).

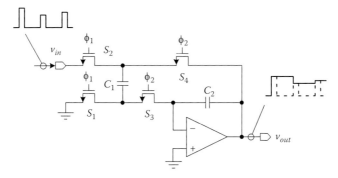

FIGURE 23.6 Single-ended form of the SHA circuit.

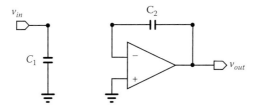

FIGURE 23.7 Equivalent representation of the circuit in sampling mode.

After this step, the expressions of the charges on C_1 and C_2 are as follows:

$$Q_1 = C_1 \cdot v_{in}(nT - T),$$ (23.1)

$$Q_2 = C_2 \cdot v_{out}(nT - T).$$ (23.2)

Then, in the holding mode $(nT - T/2)$, when ϕ_1 is low and ϕ_2 is high, switches S_3 and S_4 are closed (Figure 23.8) and the charge on C_1 is combined with the charge that is already present on C_2. The resulting charge can be written as follows:

$$\left(C_1 + C_2\right) \cdot v_{out}\left(nT - \frac{T}{2}\right) = C_2 \cdot v_{out}(nT - T) + C_1 \cdot v_{in}(nT - T).$$ (23.3)

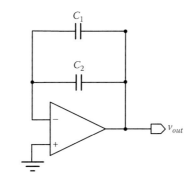

FIGURE 23.8 Equivalent representation of the circuit in holding mode.

We note that once ϕ_1 turns off, the charge on C_2 will remain the same during the next ϕ_1, until ϕ_2 turns on again in the next cycle. Therefore, the charge on C_2 at time (nT), at the end of the next ϕ_1, is equal to that at time $(nT - T/2)$, or mathematically,

$$(C_1 + C_2) \cdot v_{out}\left(nT - \frac{T}{2}\right) = (C_1 + C_2) \cdot v_{out}(nT). \tag{23.4}$$

Then, v_{out} can be obtained by combining (23.4) and (23.5), i.e.

$$(C_1 + C_2) \cdot v_{out}(nT) = C_2 \cdot v_{out}(nT - T) + C_1 \cdot v_{in}(nT - T). \tag{23.5}$$

Thus, v_{out} of the SC-SHA will be

$$v_{out}(nT) = \frac{C_2 \cdot v_{out}(nT - T) + C_1 \cdot v_{in}(nT - T)}{(C_1 + C_2)}. \tag{23.6}$$

The z-transfer function of the circuit is then derived as follows:

$$V_{out}(z) = \frac{C_2 \cdot V_{out}(z) \cdot z^{-1} + C_1 \cdot V_{in}(z) \cdot z^{-1}}{C_1 + C_2} \rightarrow H(z) = \frac{V_{out}(z)}{V_{in}(z)} = \frac{C_1 \cdot z^{-1}}{C_1 + C_2 - C_2 \cdot z^{-1}}$$

$$H(z) = \frac{C_1 \cdot z^{-1}}{C_1 + C_2 - C_2 \cdot z^{-1}} = \frac{z^{-1}}{1 + \dfrac{C_2}{C_1} - \dfrac{C_2}{C_1} \cdot z^{-1}} \rightarrow$$

$$H(z) = \frac{z^{-1}}{1 + \dfrac{C_2}{C_1} \cdot (1 - z^{-1})}. \tag{23.7}$$

To find the frequency response, we use $z = e^{j\omega T}$ in (23.7) that gives

$$H(e^{j\omega T}) = \frac{e^{-j\omega T}}{1 + \dfrac{C_2}{C_1} \cdot (1 - e^{-j\omega T})}, \tag{23.8}$$

where the period T is the inverse of the clock frequency of the SC circuit f_{clk}. The cut-off frequency of the SC circuit can be obtained from the denominator as follows:

$$1 + \frac{C_2}{C_1} \cdot (1 - e^{-j\omega T}) = 0 \rightarrow e^{-j\omega T} = 1 + \frac{C_1}{C_2}. \tag{23.9}$$

If the clock frequency is much higher than the highest frequency component of the signal ($\omega T \ll 1$), we can approximate $e^{-j\omega T}$ with the two first terms of its Taylor expansion. Therefore, the cut-off frequency at -3 dB will approximately be

$$1 - j\omega T = 1 + \frac{C_1}{C_2} \rightarrow j\omega T = -\frac{C_1}{C_2} \rightarrow \omega = -\frac{C_1}{C_2} \cdot \frac{1}{T} \rightarrow$$

$$f_{-3dB} = -\frac{1}{2\pi}\frac{C_1}{C_2} \cdot f_{clk}. \tag{23.10}$$

The schematic of the wideband low-power LNA along with its common-mode feedback (CMFB) circuit is presented in Figure 23.9. This design naturally features wide bandwidth since it employs only seven transistors, which implies minimum parasitic in the signal path, thus making the overall power consumption of the LNA solely determined by the required input-referred noise level. A transconductor (made of M_1–M_2, M_5–M_6 and R_1) converts $v_{in+} - v_{in-}$ into a current, which is deflected, in part, into a transimpedance amplifier (made of M_3–M_4 and R_2) and then translated into an amplified output voltage $v_{out+} - v_{out-}$.

Then, an accurate and a simplified expression are obtained for the gain of the LNA. After that, the cut-off frequency of the LNA will be derived from this simplified gain expression. The half-circuit model of the LNA, the schematic of which is shown in Figure 23.10a, is used to derive an expression for the gain.

From Figure 23.10b, we can obtain the voltages at nodes v_x and v_y by writing the KCL equations at the different nodes. We first obtain i_3 and i_1 as shown in Figure 23.10b:

$$v_{out} = i_3 \cdot R_2 \rightarrow i_3 = \frac{v_{out}}{R_2}, \tag{23.11}$$

and

$$i_1 = \frac{(v_{in} - v_x)}{r_{s1}}. \tag{23.12}$$

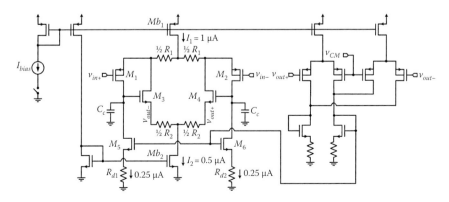

FIGURE 23.9   Schematic of an LNA employing duty cycling.

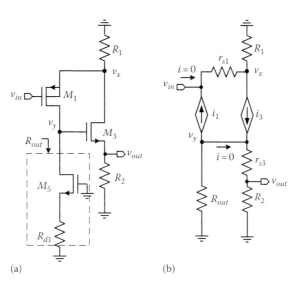

(a)   (b)

FIGURE 23.10 (a) Half-circuit model of the LNA. (b) Small signal model of the half circuit.

where r_{s1} corresponds to the inverse of g_{m1} in the T model of transistor M_1. Then, voltage v_x is obtained as follows:

$$v_x = r_{s1} \cdot \left(\frac{v_{in}}{r_{s1}} + \frac{r_{s3}}{R_{out}} \cdot \left(\frac{v_{out}}{R_2} + \frac{v_{out}}{r_{s3}} \right) \right) \tag{23.13}$$

where r_{s3} corresponds to the inverse of g_{m3} in the T model of transistor M_3 and R_{out} is the equivalent resistance from the drain of M_5. The latter is much larger than R_1 and R_2. Voltage v_y is obtained as follows:

$$v_y = -R_{out} \cdot i_1. \tag{23.14}$$

Then, we obtain the accurate gain expression of the LNA by writing

$$\frac{v_x}{R_1} + i_3 - i_1 = 0 \rightarrow \frac{r_{s1} \cdot \left(\frac{v_{in}}{r_{s1}} + \frac{r_{s3}}{R_{out}} \cdot \left(\frac{v_{out}}{R_2} + \frac{v_{out}}{r_{s3}} \right) \right)}{R_1} + \frac{v_{out}}{R_2} + \frac{r_{s3} \cdot \left(\frac{v_{out}}{R_2} + \frac{v_{out}}{r_{s3}} \right)}{R_{out}} =$$

$$0 \rightarrow A_v = \frac{v_{out}}{v_{in}} = -\frac{1}{R_1} \cdot \left(\frac{1}{\frac{1}{R_2} + \frac{1}{R_{out}} \cdot \left(\frac{r_{s1} \cdot r_{s3}}{R_1 R_2} + \frac{r_{s1}}{R_1} + \frac{r_{s3}}{R_2} + 1 \right)} \right). \tag{23.15}$$

Finally, we replace r_{s1} and r_{s3} with $1/g_{m1}$ and $1/g_{m3}$, respectively, and the gain expression of the LNA becomes

$$
A_v = \frac{v_{out}}{v_{in}} = -\frac{1}{R_1} \cdot \left(\frac{1}{\dfrac{1}{R_2} + \dfrac{1}{R_{out}} \left(\dfrac{1}{g_{m1} \cdot g_{m3} \cdot R_1 R_2} + \dfrac{1}{g_{m1} R_1} + \dfrac{1}{g_{m3} R_2} + 1 \right)} \right). \tag{23.16}
$$

We use the identity $x + y + xy + 1 = (1 + x) \cdot (1 + y)$ to simplify (23.16). Assuming that $x = \dfrac{1}{g_{m1} R_1}$ and $y = \dfrac{1}{g_{m3} R_2}$, the gain expression then becomes

$$
A_v = -\frac{1}{R_1} \cdot \left(\frac{1}{\dfrac{1}{R_2} + \dfrac{1}{R_{out}} \cdot \left(\left(1 + \dfrac{1}{g_{m1} R_1} \right) \cdot \left(1 + \dfrac{1}{g_{m3} R_2} \right) \right)} \right) \tag{23.17}
$$

Defining $\alpha = \left(1 + \dfrac{1}{g_{m1} R_1} \right) \cdot \left(1 + \dfrac{1}{g_{m3} R_2} \right)$, the gain expression can be rewritten as

$$
A_v = -\frac{1}{R_1} \cdot \left(\frac{1}{\dfrac{1}{R_2} + \dfrac{\alpha}{R_{out}}} \right) \rightarrow A_v = -\frac{R_2}{R_1} \cdot \left(\frac{1}{1 + \dfrac{\alpha R_2}{R_{out}}} \right) \tag{23.18}
$$

If $\alpha \cdot R_2 \ll R_{out}$, then the gain is approximately

$$
A_v \approx -\frac{R_2}{R_1} \tag{23.19}
$$

Denoting C_p the capacitor at node v_y and knowing that $\alpha R_2 \ll R_{out}$, we can write $\alpha R_2 \cdot C_p \ll R_{out} \cdot C_p \rightarrow \dfrac{1}{\alpha R_2 \cdot C_p} \gg \dfrac{1}{R_{out} \cdot C_p}$, thus yielding the −3 dB cut-off frequency of the LNA according to

$$
f_{-3dB} = \frac{1}{2\pi} \cdot \frac{1}{\alpha R_2 \cdot C_p} \tag{23.20}
$$

This design is using source degeneration resistors R_{d1}–R_{d2} in the active load M_5–M_6 to decrease the contributed noise from these MOS devices, as in [17]. Moreover, split resistors R_1–R_2 are used along with a single current source in both differential pairs to efficiently cancel the second harmonic terms and achieve high linearity. The total current consumption of this optimized design can be very low since it features only three main branches and employs only seven transistors (not counting the CMFB and bias circuits). Furthermore, the utilization of source degeneration resistors in the MOS current mirror load is significantly lowering the overall input-referred noise. Furthermore, the SHA circuit can be designed for very low power since the sampled signal is much less sensitive to noise after having been amplified by the LNA. For power consumption, the current drained from the power supply scales with the duty cycle length. Indeed, using a duty cycle of 25% in this proposed design reduces the power consumption of the LNA by a factor of 4. Besides, similarly to a duty cycling arrangement, the same discrete-time front end can be incorporated into a TDM strategy where one common wideband LNA would be shared between several neural recording electrodes to save power. In such a configuration, the LNA must roughly provide n times the signal bandwidth ($n \times f_{max}$) in order to process n TDM electrodes. For instance, the proposed LNA could be mutliplexed toward a single LNA of four neural recording electrodes with $f_{max} = 8$ kHz could be multiplexed toward a single LNA since its cut-off frequency is higher than four times the signal bandwidth ($4 \times f_{max} \approx 32$ kHz). Moreover, the proposed SC-SHA circuit can also be shared between four LNAs for reducing the power consumption by a factor of four into the whole front-end.

23.2.5 SUMMARY

In this section, we have reviewed energy-efficient system-level architectures, smart power management strategies and low-power sensory circuit topologies that improve efficiency in power-constrained multichannel recording arrangements. Moreover, the described strategies were illustrated through the design of a practical implementation, namely, a discrete-time neural recording front end employing duty cycling as a means to efficiently decrease its power consumption by 75%.

23.3 WUR FOR WIRELESS SENSOR NETWORKS

A WSN is a collection of low-cost, low-energy computing devices, designated *nodes* or *motes*, each being equipped with one or more sensor, with limited computational and memory resources that operate as a whole to accomplish a specific task. Typical applications include monitoring and data aggregation in fields including military, agricultural, medical and environmental.

Two outstanding characteristics of WSNs are (1) the ad hoc, self-organized and self-healing nature of the network and (2) the extremely low energy budget each sensor node is allocated. The latter constraint is due in part to the small size and cost of the nodes that leave little room for bulky batteries and also to the nature of the applications that require the nodes to remain operational for months, perhaps years after initial deployment without human intervention. Energy sources include small batteries and energy-scavenging mechanisms (solar, mechanical vibrations, heat). In the case of batteries, which is by far the more common solution, the useful life of the network is limited by the first node failure, so network protocols are typically designed to ensure that all nodes are equally active, so that their batteries are drained at an equal rate.

The power consumption of the individual nodes is therefore a topic of the highest importance, which tends to be dominated by the RF transmit and receive functions in the vast majority of applications. Any subsystem of a node that is not needed momentarily is typically put in sleep mode, and the RF subsystem is no exception.

Various techniques have been devised to allow a node to 'wake up' only when it needs to transmit and/or receive a data packet and then go back to sleep once its task is complete. In typical operation, nodes can acquire data from their sensor at periodic intervals, process said data locally (feature extraction, data compression, etc.), forward the data, relay packets that are meant for another node or for the base/aggregation station or receive query/command packets. Multi-hop packet relaying is favoured because it is more energy efficient than transmitting in a single hop over a larger distance.

To allow nodes to be in deep sleep (including the RF section) most of the time, consuming minimal power typically requires protocols that operate in a synchronous or pseudo-synchronous manner to ensure that nodes that need to communicate are awake at the same time [22]. In synchronous operation, the whole network wakes up at periodic intervals; performs whatever acquisition, processing and communication tasks are required; and then goes back to sleep. In pseudo-synchronous or *rendezvous* protocols, nodes attempt to predict at what moment they will be needed and should wake up based on past history. In effect, the network as a whole attempts to establish a dynamic optimum schedule in a totally distributed manner. This obviously implies additional processing (and possibly signalling) overhead to determine the continuously changing sleep interval, as well as missed rendezvous and false alarm wake-ups. In the end, no matter how sophisticated the rendezvous protocol, the receive sections of nodes have to stay awake significantly longer than the packet duration in order to have a fair chance of making the rendezvous. For this reason, the RF receive function typically dominates the power consumption budget in WSN, even though its instantaneous consumption is less than the transmit function.

23.3.1   WUR Concept

The aforementioned discussion highlights how desirable it is to minimize the on-time of the RF receive functions and, if possible, to avoid the complications of rendezvous protocols. The only remaining option, totally asynchronous operation, requires a means to 'wake up' a node remotely on demand, using only an RF signal. Foregoing the need to have the network maintain a communication schedule, each node is equipped with an ultralow-power WUR that continuously monitors the channel. This is in addition to the main RF transceiver unit that can therefore be turned off most of the time. When a node wishes to communicate with a neighbour, it sends a wake-up signal, containing the wake-up code or address of the target, thus waking only the desired node (see Figure 23.11). After successful reception and address decoding of the wake-up signal, the WUR brings the entire node out of sleep mode if the wake-up code matches one of the node's wake-up codes. An acknowledgement packet is then sent to signal the first node that transmission can proceed. After the

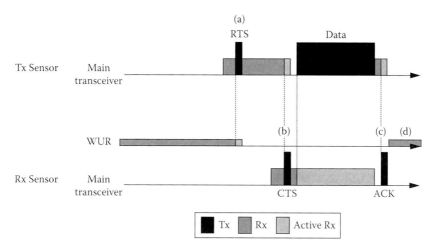

FIGURE 23.11 Example of asynchronous operation presenting a wake-up, acknowledge and receive wake-up behaviour. (a) Transmission of the wake-up code. (b) Reception and processing of the wake-up code and possible wake-up of the node. (c) Acknowledgement of wakeup. (d) Data transmission and return to sleep mode.

data exchange is complete, and after any additional tasks are performed, both nodes can go back to sleep, activating their WUR before doing so.

For asynchronous operation to be effective and competitive with rendezvous protocols, the power consumption of the wake-up device must be extremely low. It has been estimated that the WUR must have a power consumption below 50 μW in order to constitute an attractive alternative [23] and should be below 10 μW to extend the range of WSN applications [24]. In a 2010 survey [25], the feasibility requirements of WURs are listed: low cost, low power usage, low latency, low interference, low missed wake-up rate and appropriate range (with a typical goal being 10 m). As of 2010, no reported implementation meets all these requirements and very few complete, stand-alone and working implementations have been reported to date.

Restricting our attention to complete fully functional sensor nodes, the first reported WUR (in 2007) is described in [26]. Since it is built from off-the-shelf components (including a microcontroller) and consists of a WUR piggyback board in conjunction with Delft University's T-node platform, the power consumption in practice is quite high, above 600 μW, although the design could in principle have operated at 170 μW after resolution of certain practical issues. Furthermore, the range was only 2–3 m. It follows that while this is a conceptually compelling first effort, its interest is mostly academic.

Examples of the current state of the art include the design reported by J. Ansari et al. in 2008 [27] and that of C. Hambeck et al. in 2011 [28]. Based on the Telos node, the design in [27] uses a secondary radio in the 868 MHz band to transmit the wake-up signal, thus shifting some of the complexity and power burden to the transmitter. The receiver employs impedance matching and a five-stage voltage multiplier, resulting in a wake-up circuit that consumes only 876 nA. The entire node in sleep node consumes 12.5 μW that is significantly below the 50 μW barrier. However, the observed range is of the order of 2.5 m, thus falling short of the 10 m objective. The authors of [28] report a WUR chip implementation in 130 nm CMOS, leveraging various low-power techniques to achieve a consumption of 2.4 µW (wake-up unit only) and a sensitivity of −71 dBm at 868 MHz. However, it seems that this extraordinary sensitivity is obtained in part by correlating over very long sequences, thus trading off wake-up time for sensitivity and possibly shifting part of the power burden to the transmitter. In fact, the paper cites a wake-up time of 40–110 ms, which is enormous, and the extraordinary receiver sensitivity is based on correlation over a 7 ms period (thus accumulating energy) with an external surface acoustic wave (SAW) filter.

23.3.2   LOW-POWER WUR DESIGN BASED ON PWM

It has been established earlier that in assessing the quality of a WUR design, multiple variables must be taken into consideration in addition to receiver sensitivity and power consumption, such as wake-up time, bit rate, robustness to interference, range, band of operation and the completeness of the design. As a case study, we examine a design dating from 2008 [29,30] that presents some unusual and innovative characteristics.

The vast majority of reported WUR designs operate at a frequency of 868 MHz, including the latest state-of-the-art ones [27,28]. This allows better range and/or lower power consumption compared with higher frequency bands such as the popular 2.4 GHz unlicensed band. However, it should be noted that 868 MHz does not constitute a globally available band. It is an unlicensed band in Europe, but not in North America (where the ISM band is at 908 MHz instead). Furthermore, while path loss is lower at 868 MHz, antenna gain is also lower and/or antenna size is larger than at 2.4 GHz. It follows that for compact nodes with decent antenna gains AND for designs that can operate globally, 2.4 GHz is a better option.

23.3.3   PWM MODULATION

Aside from operating in the 2.4 GHz band, the design under study also features a highly unusual modulation format. Indeed, nearly all surveyed WUR implementations leverage OOK for simplicity of operation, while the design under study is based on PWM. This is arguably the central innovation of this design, as this choice leads to many benefits. To the best of our knowledge, this modulation has not been used before in single-chip WUR implementations, although an early effort from off-the-shelf components was reported [31], where PWM was used specifically to alleviate complex synchronization requirements.

First, we observe that both OOK and PWM are forms of amplitude modulations that are amenable to mostly passive, extremely simple receiver structures based on, for example, an envelope detector and an integrator. Secondly, unlike OOK, the PWM is self-synchronizing (since there is a pulse for every transmitted bit) and does not require sophisticated clock recovery circuitry. Third, PWM allows arbitrary control of the duty cycle. For example, choosing to transmit zeros as 3 µs pulses and ones as 12 µs pulses, and assuming a symbol period of 20 µs, the mean duty cycle (with equiprobable bits) is 37.5%. This leads to transmit energy savings with respect to constant envelope (frequency and/or phase modulation, duty cycle of 100%) and OOK (duty cycle of 50%).

The fourth and possibly most important benefit of PWM is its natural robustness to both noise and interference. Indeed, because the information is encoded in the duration of the pulses, the effect of additive white Gaussian noise is different, thus leading to higher resistance [32,33]. Additionally, because this modulation is radically different from that of other modulation types used by interfering signals, and its detection is based on pulse duration, it is also robust to interferers.

It is noteworthy that PWM has been forgotten for a long time as a wireless form of modulation because it is inherently not bandwidth efficient. However, bandwidth efficiency is not a primary concern in the unusual context of WSNs.

23.3.4 ARCHITECTURE

The WUR circuit architecture is depicted in Figure 23.12. The RF detector is a form of envelope detector, more specifically a zero-bias Schottky diode voltage doubler. Impedance matching is integrated into this component with lumped reactive elements in order to minimize the form factor. The said matching is optimized for a subset of the 2.4 GHz ISM band, with a worst-case return loss of −10 dB for the first 6 802.11 channels (out of 14).

The amplifier is by far the most power-hungry block of the WUR reception chain. It consists of a two-stage structure (see Figure 23.13) providing an approximate gain

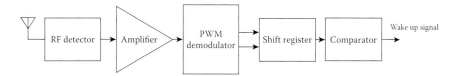

FIGURE 23.12 Architecture of PWM-based WUR. The antenna is not necessarily dedicated to the wake-up device; it can be shared with the main transceiver. The RF detector provides a baseband signal to the demodulator. The amplitude modulation used is PWM since it provides clocking information. The received information is fed to a shift register and its content is compared to the node's address, generating a wake-up signal if there's a match.

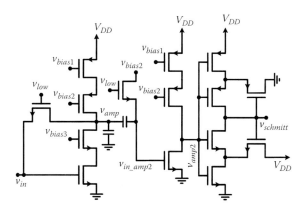

FIGURE 23.13 Two-stage amplifier structure with Schmitt-trigger output.

of 86 dB for a bandwidth of 180 kHz. A Schmitt-trigger inverter at the output provides digitization prior to entering the PWM demodulator.

PWM modulation is performed by an asymmetrical inverter followed by a capacitor, the pair acting as a resettable integrator. Transistors are sized to determine integration time constants such that pulses shorter than 10 µs yield a logical zero and longer pulses yield a logical one. A symbol-rate clock is generated along with the main output of the integrator, and both signals are passed to the shift register.

The shift register is eight-bit long and, together with the comparator, allows matching with a local unique eight-bit address. Thus, only WUR signals targeted at this specific node will be recognized and activate the wake-up line to the core of the node.

23.3.5 PERFORMANCE

The proposed WUR circuit, consisting of a layout in 130 nm CMOS using an IBM process, was validated through SPICE simulation given a 50 kHz PWM signal having a 15% duty cycle for zeros and a 60% duty cycle for ones. It was seen that the device functions adequately with a sensitivity of −55 dBm. This is deemed to correspond to a practical range of between 4 and 5 m for propagation exponents between two and three.

The breakdown of the power consumption per component is given in Table 23.3. It can be seen that all components except the amplifier consume negligible power, thus providing a clear path for further improvement in this respect. With a 1 V power supply, the entire circuit (in simulation) consumes 10 µA.

It is noteworthy that longer addresses than 8 bits could be used with minimal impact on the power budget. Only the current drawn by the comparator stage would change, in a manner that is linear with the address length.

The range could also be substantially increased by making use of directional antennas. One possible scenario involves equipping a node with four directional antennas of moderate gain pointing 90° away from each other with patterns that are wide enough to collectively adequately cover 360° in azimuth. This could provide a gain of approximately 6 dB or more, depending on how directive the antennas

TABLE 23.3

Per Component Breakdown of the Average Power Consumption of the WUR Circuit with 15%/60% Duty Cycle 50 kHz PWM Signal

Building Block	Average Consumption with 50 kHz Signal and Noise (µW)	Average Consumption with Noise Only (µW)
RF detector	0	0
Amplifier	17.80	18.29
PWM demodulator	0.8	0.06
Address decoder	0.02	<0.01
Total	**18.61**	**18.35**

are in elevation. A similar gain could be obtained at the transmitter (for a total gain of 12–15 dB) if the said transmitter 'knows' in what direction the target node is (a likely hypothesis provided the network is capable of geographic self-discovery and self-organization).

Two possible schemes can be devised to integrate such a four-antenna layout with a WUR. In the first scheme, a single WUR circuit is shared by all four antennas. A switching mechanism connects it to a single selected antenna out of the set of four. A continuously running timer would trigger a switching event at regular intervals so that all four antennas are listened to in turn. This has the benefit of requiring only one complete WUR circuit, albeit at the cost of slightly increased complexity and power consumption (the timer and switching mechanism) and a higher chance of missed wake-ups.

In the second scheme, a full WUR circuit is implemented for each antenna so that wake-up signals from any direction can be intercepted at any time. Obviously, the power consumption of this scheme would be much higher. However, it has been observed that most of the power consumption (60%) is in the amplifier bias circuitry. Since this circuitry can be shared among all four WUR components, it can be assumed that the overall power consumption would be $(60\% + 4 \times 40\%) \times 19\ \mu W = 2.2 \times 19\ \mu W = 41.8\ \mu W$, that is still below the 50 µW threshold.

This demonstrates that the multiple-antenna context is feasible, especially in light of potential improvements in the amplification section to make it more power efficient.

23.3.6 SUMMARY

In this section, we have surveyed the state of the art in WUR circuit design as a means to achieve fully asynchronous ultralow-power WSNs. We have also presented a complete stand-alone WUR architecture that, in the form of a layout in 130 nm CMOS, provides an overall power consumption of less than 19 µW in the 2.4 GHz band and with a sensitivity on the order of −55 dBm. This demonstrates that the WUR concept is both feasible and useful, since the power consumption is far below the 50 µW usefulness threshold agreed upon in the literature. Furthermore, ideas could be borrowed from recently reported designs with even lower power consumptions (although they operate at 868 MHz instead of 2.4 GHz) to improve the reviewed architecture. The notion of using multiple directive antennas in order to achieve directive gain and thus greater range and/or RF energy savings was also discussed.

REFERENCES

1. Z. Xiaodan, X. Xiaoyuan, Y. Libin, and L. Yong, A 1-V 450-nW fully integrated programmable biomedical sensor interface chip, *IEEE J. Solid-State Circuits*, 44, 1067–1077, 2009.
2. A. Bonfanti, et al., A multi-channel low-power system-on-chip for single-unit recording and narrowband wireless transmission of neural signal, in *Annual Int. Conf. IEEE Eng. Med. Biol. Soc. (EMBC'10)*, Buenos Aires, Argentina, 2010, pp. 1555–1560.
3. A. M. Sodagar, K. D. Wise, and K. Najafi, A wireless implantable microsystem for multichannel neural recording, *IEEE Trans. Microw. Theory Tech.*, 57, 2565–2573, 2009.
4. B. Gosselin, et al., A mixed-signal multichip neural recording interface with bandwidth reduction, *IEEE Trans. Biomed. Circ. Syst.*, 3, 129–141, 2009.

5. L. Seung Bae, L. Hyung-Min, M. Kiani, J. Uei-Ming, and M. Ghovanloo, An inductively powered scalable 32-channel wireless neural recording system-on-a-chip for neuroscience applications, *IEEE Trans. Biomed. Circ. Syst.*, 4, 360–371, 2010.

6. X. Zhiming, T. Chun-Ming, C. M. Dougherty, and R. Bashirullah, A 20μW neural recording tag with supply-current-modulated AFE in 0.13μm CMOS, in *IEEE Int. Solid-State Circuits Conf. (ISSCC'10)*, San Francisco, CA, 2010, pp. 122–123.

7. M. S. Chae, Z. Yang, M. R. Yuce, L. Hoang, and W. Liu, A 128-channel 6 mW wireless neural recording IC with spike feature extraction and μWB transmitter, *IEEE Trans. Neural Syst. Rehabil. Eng.*, 17, 312–321, 2009.

8. B. Gosselin and M. Ghovanloo, A high-performance analog front-end for an intraoral tongue-operated assistive technology, *IEEE Int. Symp. Circ. Syst.*, Rio de Janeiro, Brazil, 2011, pp. 2613–2616.

9. P. Chung-Ching, X. Zhiming, and R. Bashirullah, Toward energy efficient neural interfaces, *IEEE Trans. Biomed. Eng.*, 56, 2697–2700, 2009.

10. N. Joye, A. Schmid, and Y. Leblebici, Extracellular recording system based on amplitude modulation for CMOS microelectrode arrays, in *IEEE Biomed. Circuits Syst. Conf.*, Paphos, Cyprus, 2010, pp. 102–105.

11. R. F. Yazicioglu, K. Sunyoung, T. Torfs, K. Hyejung, and C. Van Hoof, A 30 μW analog signal processor ASIC for portable biopotential signal monitoring, *IEEE J. Solid-State Circuits*, 46, 209–223, 2011.

12. B. Gosselin and M. Sawan, Circuits techniques and microsystems assembly for intracortical multichannel ENG recording, in *IEEE Custom Integrated Circuits Conf. (CICC'09)*, San Jose, CA, 2009, pp. 97–104.

13. B. Gosselin, M. Sawan, and E. Kerherve, Linear-phase delay filters for ultra-low-power signal processing in neural recording implants, *IEEE Trans. Biomed. Circ. Syst.*, 4, 171–180, 2010.

14. R. Sarpeshkar, et al., Low-power circuits for brain–machine interfaces, *IEEE Trans. Biomed. Circ. Syst.*, 2, 173–183, 2008.

15. R. R. Harrison and C. Charles, A low-power low-noise CMOS amplifier for neural recording applications, *IEEE J. Solid-State Circuits*, 38, 958–965, 2003.

16. M. Mollazadeh, K. Murari, G. CauWenberghs, and N. Thakor, Micropower CMOS integrated low-noise amplification, filtering, and digitization of multimodal neuropotentials, *IEEE Trans. Biomed. Circ. Syst.*, 3, 1–10, 2009.

17. W. Wattanapanitch, M. Fee, and R. Sarpeshkar, An energy-efficient micropower neural recording amplifier, *IEEE Trans. Biomed. Circ. Syst.*, 1, 136–147, 2007.

18. Y. Ming and M. Ghovanloo, A low-noise preamplifier with adjustable gain and bandwidth for biopotential recording applications, in *IEEE Int. Symp. Circ. Syst.*, New Orleans, LA, 2007, pp. 321–324.

19. P. Mohseni and K. Najafi, A fully integrated neural recording amplifier with DC input stabilization, *IEEE Trans. Biomed. Eng.*, 51, 832–837, 2004.

20. L. Jongwoo, M. D. Johnson, and D. R. Kipke, A tunable biquad switched-capacitor amplifier-filter for neural recording, *IEEE Trans. Biomed. Circ. Syst.*, 4, 295–300, 2010.

21. B. Gosselin, Approaches for the efficient extraction and processing of neural signals in implantable neural interfacing microsystems, in *The 33rd Annual International Conference of the IEEE Engineering in Medicine and Biology Society (EMBC'11)*, Boston, MA, 2011, pp. 5855–5859.

22. H. Karl and A. Willig, *Protocols and Architectures for Wireless Sensor Networks*, Chichester, U.K.: Wiley, 2007.

23. E.-Y. Lin, J. Rabaey and A. Wolisz, Power-efficient rendezvous schemes for dense wireless sensor networks, in *Proc. ICC (IEEE Internat. Conf. Commun.)*, Paris, France, 2004, pp. 3769–3776.

24. M. Spinola Durante, Wakeup receiver for wireless sensor networks, Ph.D. dissertation, Institute of Computer Technology, Vienna, University of Technology, Vienna, Austria, 2009.
25. B. Kersten and C.G.U. Okwudire, MAC layer energy concerns in WSNs: Solved?, Unpublished research paper, SAN (System Architecture and Networking group) Seminar, TU Eindhoven, the Netherlands, Feb. 2010.
26. B. Van der Doorn, W. Kavelaars, and K. Langendoen, A prototype low cost wakeup radio for the 868 MHz band, *Int. J. Sen. Netw.*, 5(1), 22–32, 2009.
27. J. Ansari, D. Pankin, and P. Mähönen, Radio-triggered wake-ups with addressing capabilities for extremely low power sensor network applications, in *IEEE Int. Symp. Pers. Indoor, Mobile Radio Commun.*, 2008, pp. 1–5.
28. C. Hambeck, S. Mahlknecht and T. Herndl, A 2.4µW wake-up receiver for wireless sensor nodes with −71dBm sensitivity, in *IEEE Int. Symp. Circuits Syst.*, Rio de Janeiro, Brazil, 2011, pp. 534–537.
29. P. Le-Huy and S. Roy, Low-power 2.4 GHz wake-up radio for wireless sensors, in *Proc. WiMob (Wireless Mobility Conf.)*, Avignon, France, 2008, pp. 13–18.
30. P. Le-Huy and S. Roy, Low-power wake-up radio for wireless sensor networks, *Mobile Netw. Appl.*, 15(2), 226–236, April 2010.
31. S. von der Mark et al., Three stage wakeup scheme for sensor networks, *IEEE MTT-S Int'l. Conf. Microwave and Optoelectronics*, July 2005, pp. 205–208
32. H. S. Black, *Modulation Theory*, New York: Van Nostrand, 1953.
33. A. B. Carlson, *Communication Systems*, 4th edn., New York: McGraw-Hill, 2002.

Index